DYNAMIC LIGHT SCATTERING

WITH APPLICATIONS TO CHEMISTRY, BIOLOGY, AND PHYSICS

BRUCE J. BERNE
Department of Chemistry
Columbia University
New York, New York

and

ROBERT PECORA
Department of Chemistry
Stanford University
Stanford, California

A WILEY-INTERSCIENCE PUBLICATION

JOHN WILEY & SONS, INC. New York · London · Sydney · Toronto

To Naomi, David, and Michael Berne

B. J. B.

Library of Congress Cataloging in Publication Data:

Berne, Bruce J 1940–

Dynamic light scattering.

"A Wiley-Interscience publication."

Includes bibliographies and index.

1. Light–Scattering. 2. Molecules. I. Pecora, Robert, 1938– joint author. II. Title.

QC427.4.B47 535'.4 75-19140

ISBN 0-471-07100-5

Printed in the United States of America

10 9 8 7 6 5 4 3 2 1

PREFACE

This book presents a comprehensive introduction to the principles governing the application of light scattering to the study of problems in chemistry, biology, and physics. With the advent of the laser and its associated detection techniques, light scattering has become an important tool for the study of many problems in these fields. In fact, one recent review cites 513 references to articles on light scattering in the last decade. However, there has to date been no book treating the principles behind the recent developments in this field. Many students and workers have expressed a need for such a book. It is the authors' intention to fulfil this need.

The book stresses the use of light scattering to study the time dependence of thermal fluctuations in fluid systems. Since time-correlation functions are usually used to describe these fluctuations, the book should also be useful as an introduction to the theory of time-correlation functions.

Because of the wide range of applications of light scattering, many topics have by necessity been treated in a perfunctory fashion or have not been treated at all. This was done in order to confine the book to its preordained length. Thus nonlinear light scattering, multiple scattering, and scattering from solids, liquid crystals and turbulent fluids are not treated at all, while scattering from fluids in the critical region, collision-induced scattering and scattering from nonperfect gases are treated in a cursory manner. In addition, details of the experimental methods used in light-scattering experiments are not given, although a summary of the basic ideas behind some of the experimental schemes is presented. Our choice of topics was determined by the knowledge and interests of the authors and is by no means indicative of the relative importance of these topics.

The book has been arranged so that the more elementary material is presented first, while the later chapters contain the more formal and mathematically complex material. Chapters 1 through 8 contain most of the material needed for an elementary understanding of the applications to the study of macromolecules, or comparably sized particles in fluids, and to the motility of microorganisms. Chapter 9 through 14 are concerned with the study of collective (or many particle) effects and include more sophisticated treatments of macromolecules in solution and most of the applications of light scattering to the study of fluids containing small molecules. Chapter 11 presents a self-contained treatment of the formal properties of time-correlation functions and a modern discussion of projection operator techniques which have proven very useful in the treatment of fluctuations. Chapter 15 contains a brief discussion comparing light scattering with other methods for measuring the time dependence of thermal fluctuations in fluids.

It is a pleasure to acknowledge our indebtedness to a considerable number of our colleagues who have read portions of the manuscript and offered criticism and advice. In particular, we are very grateful to Professors Victor Bloomfield, W. A. Flygare, W. H. Gelbart, and W. A. Steele, and Drs. Raymond D. Mountain, Ralph Nossal, and

Dale Schaefer. We are also indebted to several of our present and former students at Stanford and Columbia Universities who have read portions of the manuscript. At Stanford we would like to thank Sergio Aragon, David R. Bauer, Dane R. Jones, and Lewis Miller, and at Columbia, Alan Ganz, John Gethner, Joseph Kushick, and Lawrence Friedhoff.

In the preparation of this book, the authors have benefited from lively and stimulating discussions on light scattering with Professors H. C. Andersen, George Flynn, and Philip Pechukas. The authors owe a special debt of gratitude to Professor Richard Bersohn, who first stimulated their interest in light scattering.

We should like to point out, however, that we have not always followed the advice we have been given and that all errors and shortcomings in the book are our own responsibility.

We wish to acknowledge the skill and patience of Mrs. Jill Castriota in typing the manuscript and in tolerating the large number of additions, deletions, and changes.

We are indebted to the National Science Foundation and the Alfred P. Sloan Foundation for supporting our research programs. In addition, one of us (B. J. B.) has benefited from grants from the Petroleum Research Foundation of the American Chemical Society, NATO, and the award of a John Simon Guggenheim Foundation Fellowship during the tenure of which, as a visiting professor at the Chemistry Department of the University of Tel Aviv, he wrote a segment of this book. In this connection it is a pleasure to thank Professor Joshua Jortner for his gracious hospitality during this visit. In addition, he would like to thank Professors Leon Lederman and Allan Sachs for their hospitality at Nevis Laboratories in Irvington, N.Y., where many pleasant days were spent writing some of the later chapters. One of us (R. P.) wishes to thank Professor S. F. Edward, then of the University of Manchester, for his kind hospitality in Manchester, where a portion of this book was written.

Finally, one of us (B. J. B.) would like to thank his wife, Naomi, for her patience, encouragement, and compassion during this rather trying period. She heroically endured with only an occasional complaint the disruption to family life that the preparation of this book entailed.

BRUCE J. BERNE
ROBERT PECORA

New York, New York
Stanford, California
September 1975

CONTENTS

DYNAMIC LIGHT SCATTERING

CHAPTER 1

INTRODUCTON

1 · 1 HISTORICAL SKETCH

Electromagnetic radiation is one of the most important probes of the structure and dynamics of matter. The absorption of ultraviolet, visible, infrared, and microwave radiation has provided detailed information about electronic, vibrational, and rotational energy levels of molecules and has in some instances enabled the chemist and physicist to determine the structure of complex molecules. Radiofrequency spectroscopy has had an enormous impact on solid-state and molecular physics and physical, inorganic, and organic chemistry. The structure of solids and biological macromolecules has been elucidated by x-ray diffraction experiments. Raman scattering is another spectroscopic technique that provides information similar to that of absorption spectroscopy. When photons impinge on a molecule they can either impart energy to (or gain energy from) the translational, rotational, vibrational, and electronic degrees of freedom of the molecules. They thereby suffer frequency shifts. Thus the frequency spectrum of the scattered light will exhibit resonances at the frequencies corresponding to these transitions. Raman scattering, therefore, provides information about the energy spectra of molecules. This book deals only with the characteristics of the light scattered from translational and rotational degrees of freedom, that is, with what is now commonly called *Rayleigh scattering*.

Recent advances in laser techniques have made possible the measurement of very small frequency shifts in the light scattered from gases, liquids, and solids. Moreover, because of the high intensities of laser sources, it is possible to measure even weakly scattered light. Thus the main difficulties in performing light-scattering experiments encountered in the past are eliminated when lasers are used. This explains the rather remarkable proliferation of laser light-scattering experiments in recent years. The structure and dynamics of such diverse systems as solids, liquid crystals, gels, solutions of biological macromolecules, simple molecular fluids, electrolyte solutions, dispersions of microorganisms, solutions of viruses, membrane vesicles, protoplasm in algae, and colloidal dispersions have now been studied by laser light-scattering techniques.

When light impinges on matter, the electric field of the light induces an oscillating polarization of the electrons in the molecules. The molecules then serve as secondary sources of light and subsequently radiate (scatter) light. The frequency shifts, the angular distribution, the polarization, and the intensity of the scattered light are determined by the size, shape, and molecular interactions in the scattering material. Thus from the light-scattering characteristics of a given system it should be possible, with the aid of electrodynamics and the theory of time-dependent statistical mechanics, to obtain information about the structure and molecular dynamics of the scattering medium.

The basic theory of *Rayleigh scattering* was developed more than a half century ago by Rayleigh, Mie, Smoluchowski, Einstein, and Debye. It is well worth summarizing some of the high points in the history of the field.

Since the experimental studies by Tyndall (1869) on light scattering from aerosols and the initial theoretical work of Rayleigh (1871, 1881), light scattering has been used to study a variety of physical phenomena. These studies concerned scattering from assemblies of noninteracting particles sufficiently small compared to the wavelength of the light to be regarded as point–dipole oscillators. In his 1881 article Rayleigh also presented an approximate theory for particles of any shape and size having a relative refractive index approximately equal to one.[1] Rayleigh (1899) explained the blue color of the sky and the red sunsets as due to the preferential scattering of blue light by the molecules in the atmosphere. In subsequent papers (Rayleigh 1910, 1914, 1918) Rayleigh derived the full formula for spheres of arbitrary size. For these larger particles there are fixed phase relations between the waves scattered from different points of the same particle, but each scattering element of the particle is regarded as an independent dipole oscillator. Debye (1915) made further contributions to the theory of these large particles and extended the calculations to particles of nonspherical shape.

Gans (1925) also contributed to the theory of large particles of relative refractive index[1] approximately equal to one. The theory of such particles is often referred to as the *Rayleigh–Gans theory*. According to Kerker (1969), "Gans' contribution to this method was hardly significant and it seems more appropriate to call it *Rayleigh–Debye scattering.*"

In large particles of relative refractive index much different from one there are not only fixed spatial relations between the scattering elements, there is also a strong dependence of the electric field amplitude on the position in the particle. There are formidable theoretical problems associated with the treatment of these large particles. Only for the case of spheres does there exist a complete solution. Mie (1908), and independently Debye (1909) solved this problem. This type of scattering is now referred to as *Mie scattering*. These problems are discussed at great length in the monographs by Van de Hulst (1957) and by Kerker (1969), and are consequently not considered in this book. Studies of the angular dependence and polarization of the scattered light are now routinely used to study the shapes and sizes of large particles.

Although Rayleigh had developed a theory of light scattering from gases with some success, it was soon found that the intensity of scattering by condensed phases (molecule per molecule) was less than that predicted by his formula by more than one order of magnitude. This effect was correctly attributed to the destructive interference between the wavelets scattered from different molecules, but unfortunately the means of calculating the extent of this interference were not known at that time. Smoluchowski (1908) and Einstein (1910) elegantly circumvented this difficulty by considering the liquid to be a continuous medium in which thermal fluctuations give rise to local inhomogeneities and thereby to density and concentration fluctuations. These authors developed a *fluctuation theory of light scattering*.

According to this theory, the intensity of the scattered light can be calculated from the mean-square fluctuations in density and concentration which in turn can be determined from macroscopic data such as the isothermal compressibility and the concentration-dependence of the osmotic pressure. The intensity of the light is thus obtained without considering the detailed molecular structure of the medium. This phenomenological approach to light scattering has continued to play a very important role in the theory of light scattering, although profound questions regarding the validity of this approach have been raised [see, for example, Fixman (1955), and more recently Felderhof (1974) and references cited therein].

The scattering from a system of particles whose positions are correlated (governed by a pair-correlation function) was investigated by Zernike and Prins (1927) in connection with the theory of x-ray diffraction of liquids. The same theory applies to light scattering from liquids. This theory was developed by Ornstein and Zernike (1914, 1916, and 1926), who extensively applied it to the study of the intense scattering of light that occurs in the fluid critical region (critical opalescence). The marked increase in the turbidity of the fluids near the gas–liquid critical point is a consequence of the fact that the pair-correlation function in a system near its critical point becomes infinitely long-ranged.

In the foregoing phenomenological theory no attempt was made to describe the effects of molecular optical anisotropy on the intensity, angular dependence, and polarization characteristics, of the scattered light. Subsequent work dealt with a molecular theory of independent optically anisotropic scatterers (Cabannes, 1929; Gans, 1921, 1923). Debye and Zimm and co-workers synthesized the *Rayleigh–Debye* and the phenomenological points of view in the 1940s and developed light scattering as a method for studying molecular weights, sizes, shapes, and interactions of macromolecules in solution. The classic papers on the subject are reprinted in McIntyre and Gornick (1964).

All these studies treated only the intensities of the scattered light. There was, however, a parallel development in light scattering which started with the work of Leon Brillouin (1914, 1922), who predicted a doublet in the frequency distribution of the scattered light due to scattering from thermal sound waves in a solid. This doublet is now known as the *Brillouin doublet.*

In the early 1930s Gross conducted a series of light-scattering experiments on liquids observing the Brillouin doublet and a central or *Rayleigh line* whose peak maximum was unshifted. Landau and Placzek (1934) gave a theoretical explanation of the Rayleigh line using a quasi-thermodynamic approach. They showed that the ratio of the integrated intensity of the central line to that of the doublet is given by the heat-capacity ratio (now known as the *Landau–Placzek ratio*):

$$\frac{I_c}{I_d} = \frac{c_P - c_V}{c_V}$$

This field was carried on by only a few workers, mainly in the Soviet Union and India (see, for example, Fabelinskii, 1968 and references cited therein), but it was not until the development of the laser in the early 1960s that these measurements of frequency changes became a major tool for the study of liquids. The modern hydrodynamic theory of light scattering from liquids is described in Chapters 10, 11, 12, and 13.

With the advent of the laser, another type of experiment became possible. In 1964, Pecora showed that the frequency distribution of light scattered from macromolecular solutions would yield values of the macromolecular diffusion coefficient and under certain conditions might be used to study rotational motion and flexing of macromolecules. These frequency changes are so small that conventional monochromators (or "filters") could not be used to resolve the frequency distribution of the scattered light. In 1964, Cummins, Knable, and Yeh used an *optical-mixing* technique to spectrally resolve the light scattered from dilute suspensions of polystyrene spheres. Since this pioneering work applications have proliferated, and *optical-mixing spectroscopy* has become a major field of research for workers in chemistry, physics, and biology.

It is the purpose of this book to describe the theory of light-scattering spectroscopy experiments and its applications to major topics of interest to chemists, physicists, and biologists. The older theories concerned with integrated intensities are described in detail only where they are of importance in understanding spectral distribution experiments. The emphasis throughout is on the use of light scattering to study the dynamics of fluctuations in fluids and not on the electrodynamical theory of the interaction of radiation with matter.

1 · 2 SYNOPSIS

In a light-scattering experiment, light from a laser passes through a polarizer to define the polarization of the incident beam and then impinges on the scattering medium. The scattered light then passes through an analyzer which selects a given polarization and finally enters a detector. The position of the detector defines the *scattering angle* θ. In addition, the intersection of the incident beam and the beam interecepted by the detector defines a scattering region of volume V. This is illustrated in Fig. 1.2.1. Prelaser light-scattering experiments usually used mercury sources. The detector used in

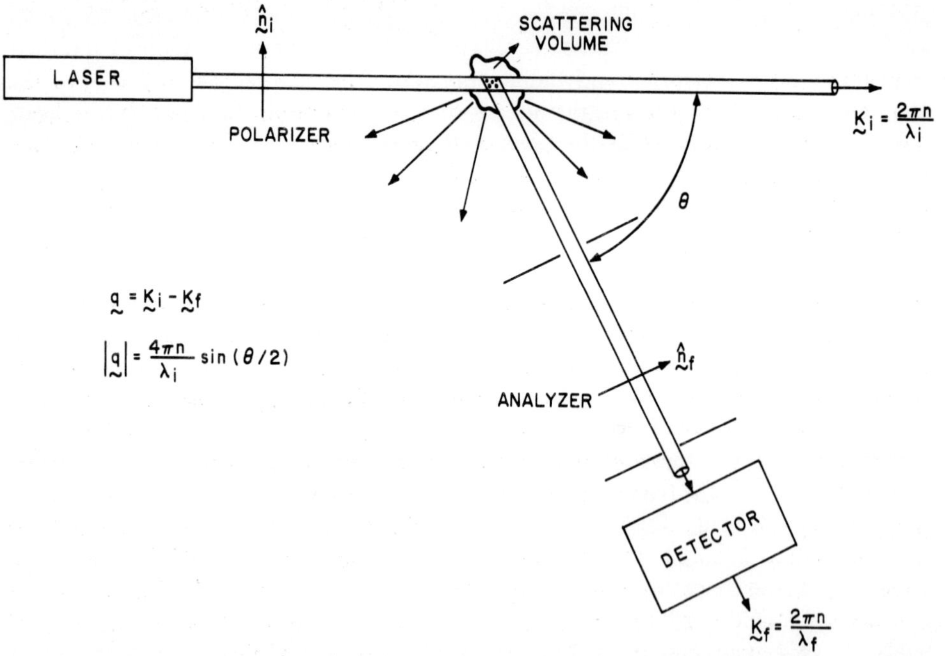

FIG. 1.2.1. A schematic representation of the light-scattering experiment.

these experiments was normally a phototube whose dc output was proportional to the intensity of the scattered light beam. In modern light-scattering experiments the scattered light spectral distribution (or the equivalent) is also measured. In these experiments a photomultiplier is the main detector, but the pre- and postphotomultiplier systems differ depending on the frequency change of the scattered light. The three different methods used, called *filter, homodyne,* and *heterodyne methods,* are schematically illustrated in Fig. 1.2.2. Note that homodyne and heterodyne methods use no monochromator or "filter" between the scattering cell and the photomultiplier. These methods are discussed in Chapter 4.

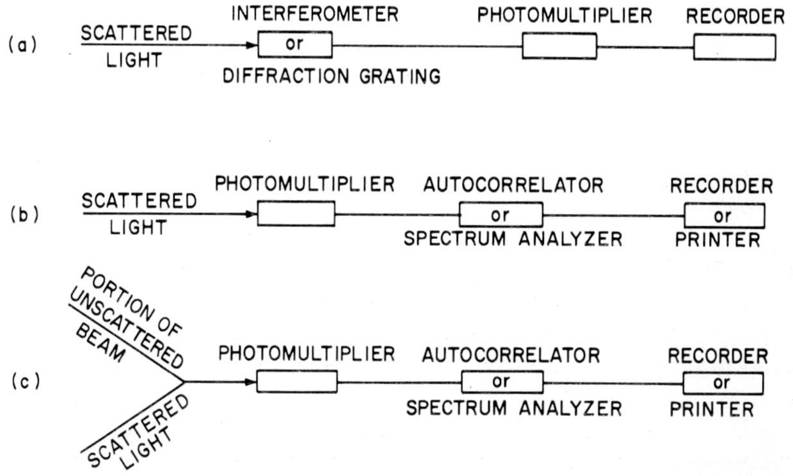

FIG. 1.2.2. Schematic illustration of the various techniques used in light-scattering experiments: (*a*) filter methods; (*b*) homodyne; (*c*) heterodyne.

The spectral characteristics of the scattered light depend on the time scales characterizing the motions of the scatterers. These relationships are discussed in Chapter 3. The quantities measured in light-scattering experiments are the time-correlation function of either the scattered field or the scattered intensity (or their spectral densities). Consequently, time-correlation functions and their spectral densities are central to an understanding of light scattering. They are, therefore, discussed at the outset in in Chapter 2.

The theory of light scattering from the simplest systems—dilute solutions or gases composed of spherical molecules—is presented in Chapter 5. This chapter includes discussions of the applications of light scattering to the study of macromolecular diffusion, electrophoretic motions, and the motility of microorganisms. In Chapter 6, a theory of light scattering from a simple model system in chemical equilibrium is presented. Conditions are given under which it might be possible to measure rate constants for chemical reactions by this method, although there have as yet been no unequivocal experimental results that report measurements of rate constants.[2] An important new technique, fluorescence fluctuation spectroscopy (FFS), is also discussed in this chapter. This technique has been successfully used to measure rate constants for binding of small molecules to macromolecules as well as the diffusion of molecules in membranes. It was thus felt that a treatment of chemical kinetics would be of value to workers in these related areas.

Light scattering can be used to measure rotational time constants for nonspherical molecules in gases and solutions. The theory of scattering from these systems is somewhat more complicated than that from spherical molecules, so that in Chapter 7 several alternative procedures for arriving at some of the results are presented. The mathematical techniques presented in this chapter are useful also for treating related problems such as fluorescence depolarization (Appendix 7.B), electron-spin resonance (ESR), nuclear magnetic relaxation (NMR), and neutron scattering.

When molecules are no longer small compared to the wavelength of light, intramolecular interference becomes important in light-scattering experiments. Since this interference depends on the mass distribution in the molecule, this phenomenon forms the basis for measurements of radii of gyration of macromolecules from integrated intensity measurements. Chapter 8 reviews the theory of light scattering from polymer solutions and also shows how intramolecular interference affects the scattered light frequency dependence and the integrated intensity.

Chapters 9–14 treat systems composed of interacting molecules and the collective modes in these systems. Chapter 9 shows how the long-range Coulomb forces affect light-scattering spectra from solutions. Chapter 10 gives a short treatment of the phenomenological basis of hydrodynamics and then applies it to the calculation of light-scattering spectra. The Brillouin doublet and central line described in Sec. 1.1 as well as the Landau–Placzek ratio are all predicted by this theory.

Chapter 11 reviews the statistical mechanical basis of hydrodynamics and discusses theories that may be used to extend hydrodynamics beyond the "classical" equations discussed in Chapter 10. Chapter 12 applies the statistical mechanical theory to the calculation of depolarized light-scattering spectra from dense liquids where interactions between anisotropic molecules are important.

Chapter 13 includes a short introduction to the theory of nonequilibrium thermodynamics. A discussion of frames of reference in the definition of transport coefficients is given and a systematic theory of diffusion is presented. Fluctuations in electrolyte solutions are analyzed, and the parameters measured in electrophoretic light-scattering experiments are related to conductance and to the transference numbers—quantities usually measured in conventional electrochemistry.

Chapter 14 is devoted to a brief description of collision-induced phenomena and the kinetic theory of gases.

The book concludes with a brief summary of other methods for determining time-correlation functions in Chapter 15.

NOTES

1. By "relative refractive index" is meant the ratio of the refractive index inside a particle to that outside the particle.
2. However, measurement of intramolecular conformational relaxation rates which are discussed in Chapter 8 may in a sense be regarded as a "chemical reaction."

REFERENCES

Brillouin, L., *Comptes Rendus* **158,** 1331 (1914).

Brillouin, L., *Ann. Phys.* **17,** 88 (1922).

Cabannes, J., *La Diffusion Moleculaire de la Lumiere,* Les Presses Universitaires de France, Paris (1929).

Cummins, H. Z., Knable, N., and Yeh, Y., *Phys. Rev. Letts.* **12,** 150 (1964).

Debye, P., *Ann. Phys.* **30,** 755 (1909); Der Lichtdruck auf Kugeln, Ph. D. thesis, Munich (1909).

Debye, P., *Ann. Phys.* **46,** 809 (1915).

Einstein, A., *Ann. Phys.* **33,** 1275 (1910); English translation in *Colloid Chemistry,* J. Alexander, Ed., Vol. 1, p. 323, Reinhold, New York (1926).

Fabelinskii, I. L., *Molecular Scattering of Light,* Plenum, New York (1968).

Felderhof, B. U., preprint (1974) (*Physica,* in press).

Fixman, M., *J. Chem. Phys.* **23,** 2074 (1955).

Gans, R., *Ann. Phys.* **65,** 97 (1921); **17,** 353 (1923); **76,** 29 (1925).

Gross, E., *Nature* **126,** 201 (1930); **129,** 722 (1932).

Kerker, M., *The Scattering of Light and Other Electromagnetic Radiation,* Academic Press, New York (1969), 414.

Landau, L., and Placzek, G., *Phys. Zeit. Sow.* **5,** 172 (1934).

McIntyre, D., and Gornick, F., *Light Scattering from Dilute Polymer Solutions,* Gordon and Breach, New York (1964).

Mie, G., *Ann. Phys.* **25,** 377 (1908).

Ornstein, L. S., and Zernike, F., *Proc. Acad. Sci. Amst.* **17,** 793 (1914).

Ornstein, L. S., and Zernike, F., *Proc. Acad. Sci. Amst.* **19,** 1312, 1321 (1916).

Ornstein, L. S., and Zernike, F., *Phys. Z.* **27,** 761 (1926).

Pecora, R., *J. Chem. Phys.* **40,** 1604 (1964).

Rayleigh, Lord, *Phil. Mag.* **41,** 107, 274, 447 (1871).

Rayleigh, Lord, *Phil. Mag.* **12,** 81 (1881).

Rayleigh, Lord, *Phil. Mag.* **47,** 375 (1899).

Rayleigh, Lord, *Proc. Roy. Soc.* **A 84,** 25 (1910).

Rayleigh, Lord, *Proc. Roy. Soc.* **A 90,** 219 (1914).

Rayleigh, Lord, *Proc. Roy. Soc.* **A 94,** 296 (1918).

Smoluchowski, M., *Ann. Phys.* **25,** 205 (1908).

Tyndall, J., *Phil. Mag.* **37,** 384; **38,** 156 (1869).

Van de Hulst, H. C., *Light Scattering by Small Particles,* Wiley, New York (1957).

Zernike, F., and Prins, T. H., *Z. Physik.* **41,** 184 (1927).

CHAPTER 2

LIGHT SCATTERING AND FLUCTUATIONS

2 · 1 INTRODUCTION

In a light-scattering experiment a monochromatic beam of laser light impinges on a sample and is scattered into a detector placed at an angle θ with respect to the transmitted beam (cf. Fig. 1.2.1). The intersection between the incident beam and the scattered beam defines a volume V, called the *scattering volume* or the *illuminated volume.*

In an idealized light-scattering experiment the incident light is a plane electromagnetic wave

$$\mathbf{E}_i(r, t) = \mathbf{n}_i E_0 \exp i[\mathbf{k}_i \cdot \mathbf{r} - \omega_i t] \tag{2.1.1}$$

of wavelength λ, frequency ω_i, polarization $\mathbf{n}_i$, amplitude E_0, and wave vector $\mathbf{k}_i$, where $\mathbf{k}_i$ is

$$\mathbf{k}_i = \left(\frac{\omega_i}{c}\right)\hat{\mathbf{k}}_i$$

and $\hat{\mathbf{k}}_i$ is a unit vector specifying the direction of propagation of the incident wave. $\mathbf{E}_i(\mathbf{r}, t)$ is the electric field at the point in space $\mathbf{r}$ at time t. When the molecules in the illuminated volume are subjected to this incident electric field their constituent charges experience a force and are thereby accelerated. According to classical electromagnetic theory, an accelerating charge radiates light. The radiated (or scattered) light field at the detector at a given time is the sum (superposition) of the electric fields radiated from all of the charges in the illuminated volume and consequently depends on the exact positions of the charges.

The molecules in the illuminated region are perpetually translating, rotating, and vibrating by virtue of thermal interactions. Because of this motion the positions of the charges are constantly changing so that the total scattered electric field at the detector will fluctuate in time. Implicit in these fluctuations is important structural and dynamical information about the positions and orientations of the molecules. It is the purpose of this book to show how this structural and dynamical information can be obtained from the fluctuations of the scattered field at the detector.

Thermal molecular motion is erratic, so that the total scattered field varies randomly at the detector. A recording of this field will look very much like a noise pattern. Hence it is no wonder that the theory of noise and fluctuations is relevant to the study of light-scattering spectroscopy. Before deriving the fundamental formulas of light scattering we present some of the basic ideas in the theory of noise and stochastic processes.

2 · 2 FLUCTUATIONS AND TIME-CORRELATION FUNCTIONS

In light-scattering experiments, the incident light field is sufficiently weak that the system can be assumed to respond linearly to it. The basic theoretical problem is to describe the response of an equilibrium system to this weak incident field, or more precisely, the changes of the light field (frequency shifts, polarization changes, etc.) due to its interaction with the system. This problem has been solved in general for weak probes. The major result of this theory, which is called *linear response theory,* can be simply stated (Zwanzig, 1965). Whenever two systems are weakly coupled to one another (such as radiation weakly coupled to matter), it is only necessary to know how both systems behave in the absence of the coupling in order to describe the way in which one system responds to the other. Furthermore, the response of one system to the other is completely describable in terms of time-correlation functions of dynamical variables.

Time-dependent *correlation functions* have been familiar for a long time in the theory of noise and stochastic processes (Wax, 1954). In recent years they have become very useful in many areas of statistical physics and spectroscopy. Correlation functions provide a concise method for expressing the degree to which two dynamical properties are correlated over a period of time. In this chapter we discuss some of the basic properties of these functions that are relevant to our understanding of light-scattering spectroscopy.

Let us consider a property A that depends on the positions and momenta of all the particles in the system. By virtue of their thermal motions the particles are constantly jostling around so that their positions and momenta are changing in time, and so too is the property A. Although the constituent particles are moving according to Newton's equations (or Schrödinger's equation), their very number makes their motion appear to be somewhat random. The time-dependence of the property $A(t)$ will generally resemble a noise pattern (cf. Fig. 2.2.1).

As an example, consider the pressure on the wall of a cylinder containing a gas in equilibrium. The pressure on the wall at a given time is proportional to the total force on the wall, which in turn is a function of the distances of all the particles from the wall. As the particles move about, the total force fluctuates in time in a very difinite manner. The pressure is therefore a fluctuating property. Suppose now that we could couple some kind of gauge to the wall that could respond rapidly to the pressure changes. The needle on this gauge would execute an erratic behavior—it would fluctuate. Since molecular motion is very rapid, the needle would jump around very rapidly. What should be reported as the pressure of the gas? The answer is obvious. The gauge should be read at a large number of time intervals and the results should be averaged. An average over a sufficiently long time (a time long compared with the period of the fluctuation) would yield a fairly reliable pressure. By this we mean that if the same average were performed at a different time, essentially the same average value would be obtained.

It is clear from this discussion that the measured bulk property of an equilibrium system is simply a time average

$$\bar{A}(t_0, T) = \frac{1}{T}\int_{t_0}^{t_0+T} dt A(t)$$

where t_0 is the time at which the measurement is initiated and T is the time over which it is averaged. The average becomes meaningful only if T is large compared to the period of fluctuation. The ideal experiment would be one in which A is averaged over an infinite time,

$$\bar{A}(t_0) = \lim_{T\to\infty} \frac{1}{T}\int_{t_0}^{t_0+T} dt A(t)$$

It can be shown that under certain general conditions[1] this infinite time average is independent of t_0. In statistical mechanics it is usually assumed that this is valid. In general a property whose average is independent of t_0 is called a *stationary property* (cf. Section 11.C). In Fig. 2.2.1 we see that the property A fluctuates about this time average, which because of its independence of t_0 can be expressed as

$$\langle A \rangle = \lim_{T\to\infty} \frac{1}{T}\int_{0}^{T} dt A(t) \tag{2.2.1}$$

The noise signal $A(t)$ in Fig. 2.2.1 displays the following features: the property A at

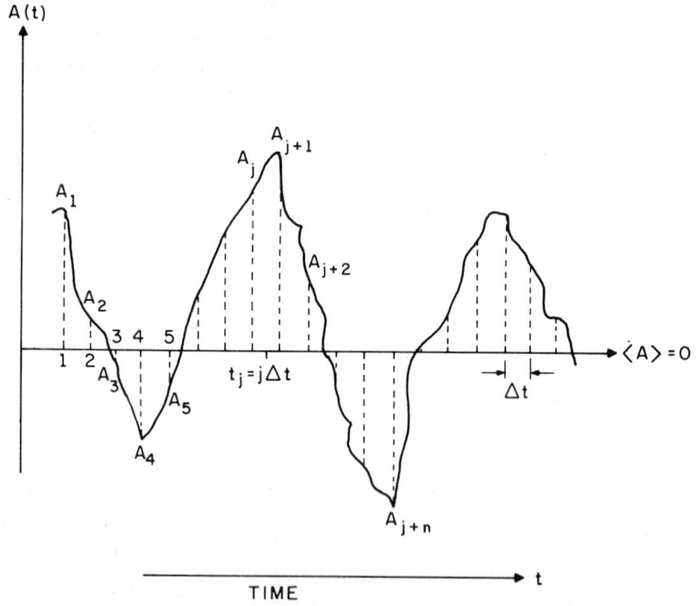

FIG. 2.2.1. The property $A(t)$ fluctuates in time as the molecules move around in the fluid. The time axis is divided into discrete intervals, Δt, and the time average $\langle A \rangle$ is assumed to be zero for convenience.

the two times t, and $t + \tau$ can in general have different values so that $A(t + \tau) \neq A(t)$. Nevertheless when τ is very small compared to times typifying the fluctuations in A, $A(t + \tau)$ will be very close to $A(t)$. As τ increases the deviation of $A(t + \tau)$ from $A(t)$ is more likely to be nonzero. Thus in some sense we can say that the value $A(t + \tau)$ is correlated with $A(t)$ when τ is small but that this correlation is lost as τ becomes large compared with the period of the fluctuations. A measure of this correlation is the autocorrelation function of the property A which is defined by[2]

$$\langle A(0)A(\tau)\rangle = \lim_{T\to\infty} \frac{1}{T}\int_0^T dt\, A(t)A(t + \tau) \tag{2.2.2}$$

Suppose that the time axis is divided into discrete intervals Δt, such that $t = j\Delta t$; $\tau = n\Delta t$; $T = N\Delta t$ and $t + \tau = (j + n)\Delta t$; and suppose further that the property A varies very little over the time interval Δt. From the definition of the integral it then follows that Eqs. (2.2.1) and (2.2.2) can be approximated by

$$\begin{aligned} \langle A\rangle &\cong \lim_{N\to\infty} \frac{1}{N}\sum_{j=1}^{N} A_j \\ \langle A(0)A(\tau)\rangle &\cong \lim_{N\to\infty} \frac{1}{N}\sum_{j=1}^{N} A_jA_{j+n} \end{aligned} \tag{2.2.3}$$

where A_j is the value of the property at the beginning of the jth interval. These sums become better approximations to the infinite time averages as $\Delta t \to 0$.

In optical mixing experiments, a correlator computes time-correlation functions of the scattered field in this discrete manner (see Chapter 4). Of course in any experimental determination the averaging is done over a finite number of steps (finite time).

We introduce the discrete notation in order to clarify the ensuing discussion. What we want to demonstrate is how the time-correlation function varies with time. In Fig. 2.2.1 we present the noise signal $A(t)$. Note that many of the terms in the sum Eq. (2.2.3) are negative. For example, in Fig. 2.2.1 A_jA_{j+n} is negative. Consequently, this sum will involve some cancellation between positive and negative terms. Now consider the case $\langle A(0)A(0)\rangle$. The sum contributing to this is $\sum_j A_jA_j = \sum_j A_j^2$. Since $A_j^2 \geqslant 0$ all the terms in the sum are positive and we expect the total to be large. What this implies is that[3]

$$\sum_{j=1}^{N} A_j^2 \geqslant \sum_{j=1}^{N} A_jA_{j+n}$$

or

$$\langle A(0)^2\rangle \geq \langle A(0)A(\tau)\rangle. \tag{2.2.4}$$

Thus it would appear that the autocorrelation function either remains equal to its initial value for all times τ, in which case A is a constant of the motion (a conserved quantity) or decays from its initial value which is a maximum.

From the foregoing we would expect that the autocorrelation function of a nonconserved, nonperiodic property decays from its initial value $\langle A^2\rangle$. For times τ large compared to the characteristic time for the fluctuation of A, $A(t)$ and $A(t + \tau)$ are expected to become totally uncorrelated; thus

$$\lim_{\tau \to \infty} \langle A(0)A(\tau)\rangle = \langle A(0)\rangle\langle A(\tau)\rangle = \langle A\rangle^2 \tag{2.2.5}$$

so that the time-correlation function of a nonperiodic property decays from $\langle A^2\rangle$ to $\langle A\rangle^2$ in the course of time. This is shown in Fig. 2.2.2.

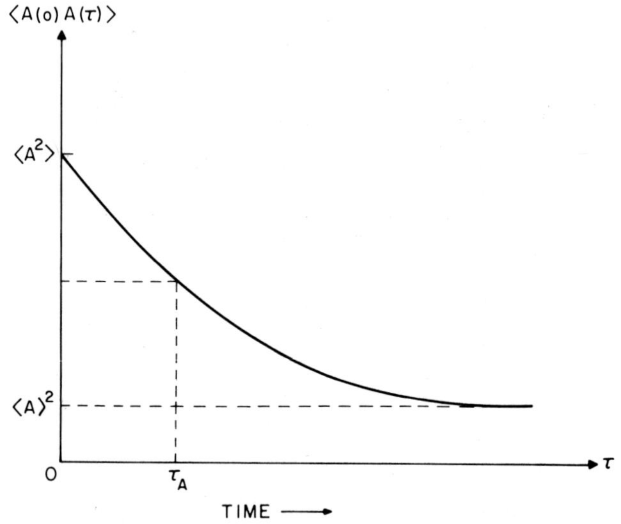

FIG. 2.2.2. The time-correlation function, $\langle A(0)\, A(\tau)\rangle$. Initially this function is $\langle A^2\rangle$. For times very long compared to the correlation time, τ_A, the correlation function decays to $\langle A\rangle^2$. [cf. Eq.(2.2.5)].

An actual experimental noise signal is shown in Fig. 2.2.3a. The particular signal is proportional to the intensity of light scattered from a solution of polystyrene spheres of diameter 1.01 μm. The corresponding (time-averaged) time-correlation function is given in Fig. 2.2.3b. As we show in Chapter 5, the diffusion coefficient of the spheres can be determined from the correlation time of this function. This is discussed in greater detail later.

In many applications the autocorrelation function decays like a single exponential so that

$$\langle A(0)A(\tau)\rangle = \langle A\rangle^2 + \{\langle A^2\rangle - \langle A\rangle^2\} \exp \frac{-\tau}{\tau_r} \tag{2.2.6}$$

where τ_r is called the "relaxation time" or the correlation time of the property. It represents the characteristic decay time of the property. If we define

$$\delta A(t) \equiv A(t) - \langle A\rangle \tag{2.2.7}$$

which is the deviation of the instantaneous value of $A(t)$ from its average value, it is easy to show that[4]

$$\langle \delta A(0)\delta A(\tau)\rangle = \langle A(0)A(\tau)\rangle - \langle A\rangle^2 \tag{2.2.8}$$

and

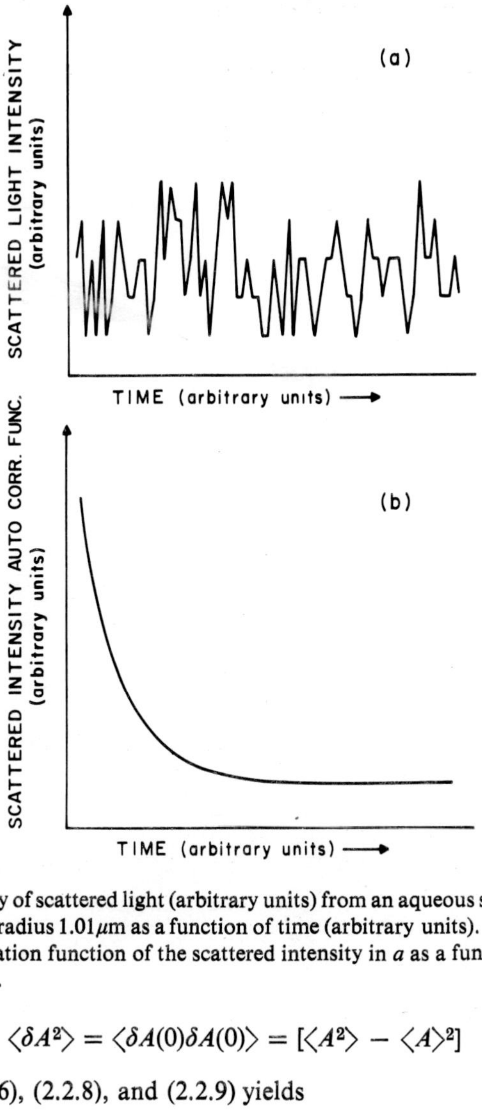

FIG. 2.2.3. (*a*) Intensity of scattered light (arbitrary units) from an aqueous solution of polystyrene spheres of radius 1.01 μm as a function of time (arbitrary units). (*b*) The time-averaged autocorrelation function of the scattered intensity in *a* as a function of time in arbitrary units.

$$\langle \delta A^2 \rangle = \langle \delta A(0)\delta A(0) \rangle = [\langle A^2 \rangle - \langle A \rangle^2] \tag{2.2.9}$$

Combining Eqs. (2.2.6), (2.2.8), and (2.2.9) yields

$$\langle \delta A(0)\delta A(\tau) \rangle = \langle \delta A^2 \rangle \exp\frac{-\tau}{\tau_r} \tag{2.2.10}$$

$\delta A(t)$ is often referred to as a "fluctuation" in that it represents the deviation of the property from its average value. The autocorrelation functions of fluctuations have a simpler structure than the autocorrelation function of the properties themselves because the time invariant part $\langle A \rangle^2$ is removed.

Not all fluctuations decay exponentially. We often want to have some parameter that typifies the time scale for the decay of the correlations. We therefore define the correlation time τ_c to be

$$\tau_c \equiv \int_0^\infty d\tau \frac{\langle \delta A(0)\delta A(\tau) \rangle}{\langle \delta A^2 \rangle} \tag{2.2.11}$$

We note that for exponential decay $\tau_c = \tau_r$. In general, the correlation time will be a some complicated function of all of the relaxation processes contributing to the decay of δA.[5]

Let us summarize; the autocorrelation function is a measure of the similarity between two noise signals $A(t)$ and $A(t + \tau)$. When $\tau = 0$ these two signals are completely in phase with each other and $\langle A(0)A(\tau)\rangle$ is large; as τ increases $A(t)$ and $A(t + \tau)$ get out of phase with each other and the autocorrelation function $\langle A(0)A(\tau)\rangle$ is small.

2 · 3 ENSEMBLE-AVERAGED TIME-CORRELATION FUNCTIONS

The time-correlation functions measured in NMR, neutron, and light-scattering spectroscopy are time averages, whereas in most theoretical calculations what is calculated is the ensemble-averaged time-correlation function. According to Birkhoff's ergodic theorem, these two correlation functions will be identical if the mechanical systems studied are ergodic. Unfortunately, it has not been possible to prove that real systems are ergodic (Uhlenbeck and Ford, 1963). Nevertheless, the predictions of ensemble theory have been so consistent with experiment that throughout the remainder of this book we assume the equivalence between time-averaged and ensemble-averaged time-correlation functions. Hence we devote this section to the definition of the ensemble average both in classical and quantum systems.

The *instantaneous state* of a classical isolated mechanical system of f degrees of freedom is completely specified by f generalized positions $(q_1, \ldots, q_f)$ and f generalized momenta $(p_1, \ldots, p_f)$. Such a state can be represented by a point in a $2f$ dimensional cartesian space with orthogonal coordinate axes labeled by these f positions and f momenta.[6] Given the initial state, $\boldsymbol{\Gamma}_0 \equiv (q_1(0), \ldots, p_f(0))$ of the mechanical system the canonical equations of motion,

$$\left\{\begin{aligned} \dot{q}_i &= \frac{\partial H}{\partial p_i} \\ p_i &= -\frac{\partial H}{\partial q_i} \end{aligned}\right. \qquad i = 1, \ldots f \tag{2.3.1}$$

have unique solutions so that all subsequent states of the system are unambiguously specified for all time. This means that as time goes by the particles move to new positions with new momenta. The specific positions and momenta at any time t, $\boldsymbol{\Gamma}_t \equiv [q_1(t), \ldots, p_f(t)]$ are totally determined by the initial state of the system and the canonical equations (or Newton's equations). Of course it is necessary to specify the Hamiltonian of the system, H. For a conservative system it is possible to choose the generalized coordinates and momenta such that

$$H = T(p_1, \ldots, p_f) + V(q_1, \ldots, q_f) \tag{2.3.2}$$

where T is the kinetic energy and V is the potential energy of the particles. The complete behavior of the system can be represented by a trajectory in phase space, as shown in Fig. 2.3.1*a*.

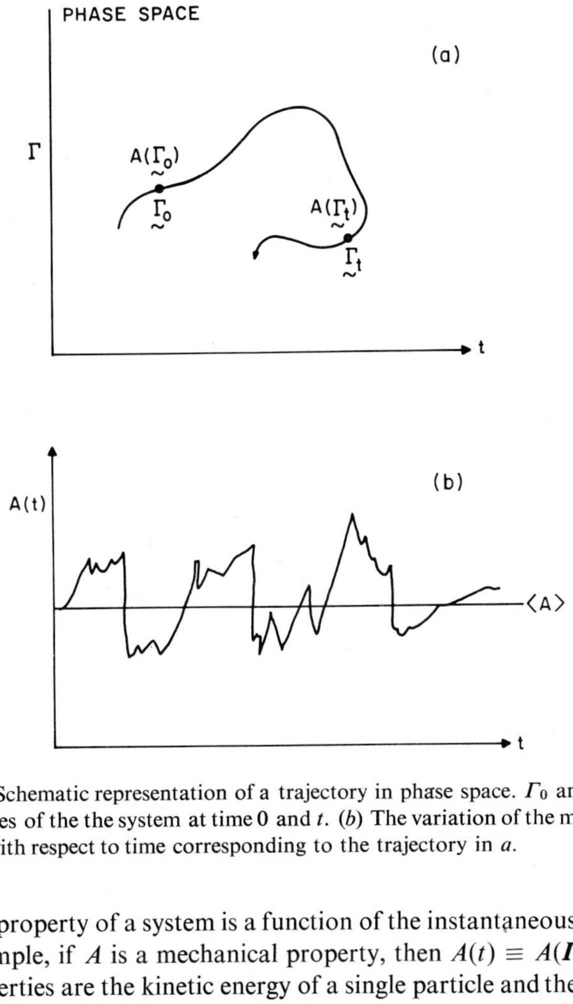

FIG. 2.3.1. (*a*) Schematic representation of a trajectory in phase space. Γ_0 and Γ_t represent the states of the the system at time 0 and t. (*b*) The variation of the mechanical property A with respect to time corresponding to the trajectory in *a*.

A mechanical property of a system is a function of the instantaneous state, $\boldsymbol{\Gamma}_t$, of the system. For example, if A is a mechanical property, then $A(t) \equiv A(\boldsymbol{\Gamma}_t)$. Examples of mechanical properties are the kinetic energy of a single particle and the number density in the neighborhood of a point in the system. As time goes by a mechanical property will change unless it is a "constant" of the motion. The typical behavior of a mechanical property A corresponding to a given trajectory in phase space is illustrated in Fig. 2.3.1*b*.

The classical equilibrium ensemble-averaged time-correlation function of the property A is defined as

$$\langle A(0)A(t)\rangle \equiv \int d\boldsymbol{\Gamma}_0\, \rho(\boldsymbol{\Gamma}_0)A(\boldsymbol{\Gamma}_0)A(\boldsymbol{\Gamma}_t) \tag{2.3.3a}$$

where the product[7] $A(\Gamma_0)A(\Gamma_t)$ depends on the time t and on the initial state Γ_0, and $\rho(\boldsymbol{\Gamma}_0)d\boldsymbol{\Gamma}_0$ is the probability[8] of finding the system in the initial state Γ_0.

In quantum statistical mechanics the time-correlation function of the observable A is defined as

$$\langle A(0)A(t)\rangle = Tr\rho_0\, A \exp\frac{i\hat{H}t}{h}\, A \exp\frac{-i\hat{H}t}{h} = Tr\rho_0\, A(0)A(t) \tag{2.3.3b}$$

where ρ_0 is the equilibrium density matrix, A is the linear Hermitian operator representing the observable A, $\hat{H}$ is the Hamiltonian operator and $A(t) \equiv \exp\frac{i\hat{H}t}{h}\, A \exp\frac{-i\hat{H}t}{h}$ is the linear operator representing[9] the observable A at time t. In quantum mechanics a new feature arises. Because $A(t)$ and $A(0)$ do not, in general, commute, $\langle A(0)A(t)\rangle$ and $\langle A(t)A(0)\rangle$ are different functions of the time (in fact, they are complex conjugate pairs). Because most of the applications in this book involve systems that can be treated classically, we shall not dwell on this point here.

One particular time-correlation function of the many that we have occasion to study in this book is

$$F(\mathbf{q}, t) = \langle \rho_{-\mathbf{q}}(0)\rho_{\mathbf{q}}(t)\rangle \tag{2.3.4}$$

where the property $\rho_{\mathbf{q}}(t)$ is

$$\rho_{\mathbf{q}}(t) = \sum_{j=1}^{N} \exp i\mathbf{q}\cdot\mathbf{r}_j(t) \tag{2.3.5}$$

where $\mathbf{q}$ is an arbitrary vector and $\mathbf{r}_j(t)$ is the center-of-mass position of the jth molecule in the system. We note that $\rho_q(t)$ is the Fourier transform[10] with respect to $\mathbf{r}$ of the property

$$\rho(\mathbf{r}, t) = \sum_{j=1}^{N} \delta(\mathbf{r} - \mathbf{r}_j(t)) \tag{2.3.6}$$

where $\delta(x)$ is the Dirac delta function. This property is the microscopic number density at the point $\mathbf{r}$. This is readily verified since the delta function contributes 1 to any average only if $\mathbf{r}_j(t)$ is in the neighborhood of the point $\mathbf{r}$ and zero otherwise. Thus the sum in Eq. (2.3.6) counts the number of particles in the neighborhood of the point $\mathbf{r}$ at time t, and thereby the number density. $F(\mathbf{q}, t)$ is a measure of the correlation between spatial Fourier components of the number density at two different times. This function can be determined by light-scattering experiments and, as we shall see in subsequent chapters, contains a great deal of information about static and dynamic properties of fluid systems.

2 · 4 THE SPECTRAL DENSITY

The spectral density (or power spectrum) $I_A(\omega)$ of a time-correlation function $\langle A^*(0)$ $A(t)\rangle$ is defined as[11]

$$I_A(\omega) \equiv \frac{1}{2\pi}\int_{-\infty}^{+\infty} dt\; e^{-i\omega t}\,\langle A^*(0)A(t)\rangle \tag{2.4.1}$$

where A^* is the complex conjugate of A.

This quantity plays an important role in much of what follows. In fact, as we shall see, what is sometimes measured in light scattering is the spectral density of the electric field of the scattered light. Let us dwell for a moment on some properties of these functions. Fourier inversion of Eq. (2.4.1) leads to an expression for the time-correlation function in terms of the spectral density.

$$\langle A^*(0)A(t)\rangle = \int_{-\infty}^{+\infty} d\omega\, e^{+i\omega t}\, I_A(\omega) \tag{2.4.2}$$

Thus $\langle A(0)A(t)\rangle$ and $I_A(\omega)$ are Fourier transforms of one another and an experimental determination of one as a function of its arguments is sufficient for the determination of the other. In fact, it is $I_A(\omega)$ rather than the time-correlation function that is often measured in an experiment. Moreover, it should be noted that the equilibrium mean-square value of the property A is found by setting $t = 0$ in the preceding formula so that

$$\langle |A|^2\rangle = \langle |A(0)|^2\rangle = \int_{-\infty}^{+\infty} d\omega I_A(\omega) \tag{2.4.3}$$

Thus $I_A(\omega)d\omega$ can be interpreted as the "amount" of $|A|^2$ in the frequency interval $(\omega, \omega + d\omega)$. This interpretation can be clarified as follows. Figure 2.4.1 schematically

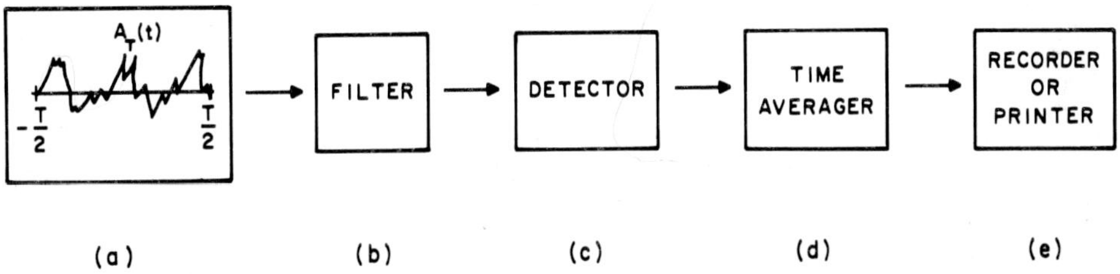

FIG. 2.4.1. Schematic of apparatus for measuring the spectral density of a fluctuating variable.

illustrates the main features of an apparatus for measuring $I_A(\omega)$. The signal $A_T(t)$, measured for a period T, is first passed through a filter which allows through only a very narrow range of frequencies. The signal that gets through the filter, $A_{TO}(t)$, then impinges on a detector whose output is proportional to $|A_{TO}(t)|^2$. The output of the detector is then averaged over the time interval T for which the signal is defined and the result, $\langle |A_{TO}|^2\rangle_T$ is then suitably recorded.

A function $A(t)$ measured over a time interval $(-T/2, T/2)$ can always be expressed in terms of its Fourier components so that

$$A_T(t) = \frac{1}{\sqrt{T}} \sum_n A_n \exp i\omega_n t \tag{2.4.4}$$

where $\omega_n = \dfrac{2\pi}{T} n$ and $\{A_n\}$ are the Fourier coefficients.

On passing through the filter, some of the frequency components of A_T are filtered out. We can describe the action of the filter by a set of numbers F_n which we allow to be zero if the frequency ω_n is not allowed through and to be one otherwise. Thus the signal $A_T(t)$ is transformed by the filter to

$$A_{TO}(t) = \frac{1}{\sqrt{T}} \sum_n F_n A_n \exp i\omega_n t \tag{2.4.5}$$

The output of the detector is proportional to the square of this function

$$|A_{TO}(t)|^2 = \frac{1}{T} \sum_{nn'} F_{n'} F_n A_{n'}{}^* A_n \exp i(\omega_n - \omega_{n'})\, t$$

The averager then performs the time average,

$$\langle |A_{TO}|^2 \rangle = \frac{1}{T} \int_{-T/2}^{T/2} dt\, |A_{TO}(t)|^2$$

$$= \frac{1}{T} \int_{-T/2}^{T/2} \frac{dt}{T} \sum_{n,n'} F_{n'} F_n A_{n'}{}^* A_n \exp i(\omega_n - \omega_{n'})t$$

where Eq. (2.4.5) has been substituted. Performing this integration over t we obtain[12]

$$\langle |A_{TO}|^2 \rangle = \frac{1}{T} \sum_n F_n{}^2 |A_n|^2 \tag{2.4.6}$$

This can be related to the time-correlation function[13]

$$\langle A^*(t) A(t+\tau) \rangle_T = \frac{1}{T} \int_{-T/2}^{T/2} dt\, A_T{}^*(t) A_T(t+\tau) \tag{2.4.7}$$

When the Fourier series expansion, Eq. (2.4.4), is substituted into this, we obtain

$$\langle A^*(t) A(t+\tau) \rangle_T = \sum_{n,n'} \frac{A_n{}'^* A_n}{T} \int_{-T/2}^{T/2} \frac{dt}{T} \exp i(\omega_n - \omega_n{}')\, t \exp i\omega_n \tau$$

But this is simply equal to[14]

$$\langle A^*(t) A(t+\tau) \rangle_T = \sum_n \frac{|A_n|^2}{T} \exp i\omega_n \tau \tag{2.4.8}$$

Multiplication by $\exp -i\omega_m \tau$ and integration over τ from $-T/2$ to $T/2$ then gives

$$|A_m|^2 = \int_{-T/2}^{T/2} d\tau \langle A^*(t) A(t+\tau) \rangle_T \exp -i\omega_m \tau = 2\pi I_A^T(\omega_m) \tag{2.4.9}$$

where $I_A{}^T(\omega_m)$ is by definition the spectral density of the time-correlation function $\langle A^*(t) A(t+\tau) \rangle_T$. Substituting Eq. (2.4.9) into Eq. (2.4.6) we now obtain

$$\langle |A_{TO}|^2 \rangle_T = \frac{2\pi}{T} \sum_n I_A^T(\omega_n)\, F_n{}^2 \tag{2.4.10}$$

This can be written in another form. The separation between adjacent frequencies $\Delta\omega_n = \omega_{n+1} - \omega_n = 2\pi/T$, so that

$$\langle |A_{TO}|^2 \rangle_T = \sum_n \Delta\omega_n I_A^T(\omega_n) F_n{}^2$$

If $T \to \infty$, $\Delta\omega_n \to 0$, the above sum can be written as an integral

$$\lim_{T\to\infty} \langle |A_{TO}|^2 \rangle_T = \int_{-\infty}^{+\infty} d\omega \, I_A(\omega) |F(\omega)|^2 \tag{2.4.11}$$

where $I_A(\omega)$ is the spectral density of A determined by an average over an infinite time [i.e., $I_A(\omega) = \lim_{T\to\infty} I_A{}^T(\omega)$] and $|F(\omega)|^2$ is a filter function. If the filter is a "narrow-band filter

$$|F(\omega)|^2 = \begin{cases} 1 & \omega_0 \le \omega \le \omega_0 + \Delta\omega \\ 0 & \text{otherwise} \end{cases} \tag{2.4.12}$$

then

$$\lim_{T\to\infty} \langle |A_{TO}|^2 \rangle_T = I_A(\omega_0)\, \Delta\omega \tag{2.4.13}$$

Thus from a time-average of $|A_{TO}|^2$ we can obtain the spectral density $I_A(\omega_0)$, and by tuning the filter through different values of ω_0 we can determine the spectrum of the fluctuation A. From its definition we see that $\lim_{T\to\infty} \langle |A_{TO}|^2 \rangle_T$, and correspondingly $I_A(\omega_0)\Delta\omega$ can be interpreted as the "amount" of $|A_T|^2$ passed through the filter, that is, the amount in the frequency interval $(\omega_0, \omega_0 + \Delta\omega)$, thus clarifying our assertion following Eq. (2.4.3).

The foregoing result is quite general. If $E_T(t)$ is the electric field of a scattered light wave, if the filter is an interferometer, grating or prism (all of which are narrow-band filters), and if the detector is a photomultiplier (all photomultipliers are square-law detectors), then according to Eq. (2.4.13) the output is

$$\lim_{T\to\infty} \langle |E_{TO}|^2 \rangle_T = I_E(\omega_0)\Delta\omega \tag{2.4.14}$$

where $I_E(\omega_0)$ is the spectral density of the electric field autocorrelation function and ω_0 is defined by the filter. This experimental situation corresponds to the usual optical spectrometer (see Chapter 4 for further details) which consequently measures the spectral density $I_E(\omega)$. We conclude that if the filter is tuned through all frequencies, the spectral density

$$I_E(\omega) = \frac{1}{2\pi} \int_{-\infty}^{\infty} d\tau \, \langle E^*(t)E(t+\tau) \rangle \exp i\omega\tau \tag{2.4.15}$$

can be determined as a function of frequency, and the time-correlation function can be determined by Fourier inversion. This means that the property $I_E(\omega)$ which is measured in a filter experiment and the correlation function $\langle E^*(0)E(t) \rangle$ measured in heterodyne experiments are Fourier-transform pairs.[15]

The term $I_E(\omega)$ can be measured directly either by filter experiments or by heterodyne methods using a spectrum analyzer instead of a correlator. As we have noted before, the technique of choice depends upon the time scale of the fluctuations.

It is clear from Eq. (2.4.15) and Section 1.2 that the light-scattering spectrum is determined from autocorrelation functions of the electric field at the detector. Thus the goal of any theory of light scattering is to show how important physical properties of the scattering medium can be extracted from the measured time-correlation functions.

NOTES

1. Birkhoff proved this for an ergodic system; that is, a system that is metrically indecomposable (Uhlenbeck and Ford, 1963).
2. The correlation between two different properties A and B is similarly given by the *cross-correlation functions*

$$\langle A(0)B(\tau)\rangle = \lim_{T\to\infty} \frac{1}{T}\int_0^T dt\ A(t)B(t+\tau)$$

and

$$\langle B(0)A(\tau)\rangle = \lim_{T\to\infty} \frac{1}{T}\int^T dt\ B(t)A(t+\tau)$$

3. This inequality can be proved using Schwartz's inequality, according to which

$$|\sum_j A_jB_j|^2 \leqslant |\sum_j A_j^2|\sum B_j^2|$$

If we take $B_j = A_{j+n}$, divide both sides by N^2, take the limit $N \to \infty$ and then recognize that:

$$\text{(a)}\ \lim_{N\to\infty} \frac{1}{N}\sum_{j=1}^N A_j^2 = \langle A^2\rangle$$

$$\text{(b)}\ \lim_{N\to\infty} \frac{1}{N}\sum_{j=1}^N A_{j+n}^2 = \langle A^2\rangle$$

$$\text{(c)}\ \lim_{N\to\infty} \frac{1}{N}\sum_{j=1}^N A_jA_{j+n} = \langle A(0)A(\tau)\rangle$$

it follows that

$$|\langle A(0)A(\tau)\rangle|^2 \leq \langle A^2\rangle^2$$

Since $\langle A(0)A(\tau)\rangle$ is real, the inequality

$$\langle A(0)A(\tau)\rangle \leq \langle A^2\rangle$$

follows. The limit in (b) follows from the presumed independence of the time average from the time at which averaging is initiated.

It should also be noted that the correlation function can under some circumstances be periodic.

4. Substitute $A(t) = \langle A\rangle + \delta A(t)$ into Eq. (2.2.2). Then note that

$$\langle A(0)A(\tau)\rangle = \lim_{T\to\infty} \frac{1}{T}\int_0^T dt\,\{\langle A\rangle^2 + \langle A\rangle[\delta A(t) + \delta A(t+\tau)] + \delta A(t)\delta A(t+\tau)\}$$

Since $\langle A\rangle$ is a constant and $\langle \delta A(t)\rangle = 0 = \langle \delta A(t+\tau)\rangle$;

$$\langle A(0)A(\tau)\rangle = \langle A\rangle^2 + \langle \delta A(0)\delta A(\tau)\rangle$$

5. This definition of $/\tau_c/$ is not always useful. There are cases in which, for instance, the correlation function decays in such a way that $/\tau_c/$ as defined by Eq. (2.2.11) is zero. For example, the correlation function may have positive and negative regions that cancel. In such cases Eq. (2.2.11) does not provide an adequate measure of the decay time.
6. This space is called *phase space*.
7. In many cases the property A may have complex values. In such cases we define the autocorrelation funtion as

$$\langle A^*(0)A(t)\rangle = \int d\Gamma_0\ \rho(\Gamma_0)A^*(\Gamma_0)A(\Gamma t)$$

where A^* is the complex conjugate of A.

8. $\rho(\Gamma_0)$ can be any of the equilibrium ensemble distribution functions. For example, in the canonical ensemble $\rho(\Gamma_0) = Q^{-1}\exp[-\beta H(\Gamma_0)]$ where Q is the canonical partition function, $Q = \int d\Gamma_0\ e^{-\beta H}$.
9. This is found in the Heisenberg representation.
10. The Fourier transform of a function $f(\mathbf{r})$ is denoted $f_{\mathbf{q}}$ where

$$f_{\mathbf{q}} \equiv \int d^3r\, e^{i\mathbf{q}\cdot\mathbf{r}}\, f(\mathbf{r})$$

11. More precisely, the spectral density is defined as

$$I_A(\omega) = \lim_{\varepsilon \to 0+} \frac{1}{2\pi} \int_{-\infty}^{+\infty} dt\, e^{-i\omega t}\, e^{-\varepsilon|t|}\, \langle A^*(0)A(t)\rangle$$

where the factor $e^{-\varepsilon|t|}$ ($\varepsilon > 0$) is introduced to ensure the existence of the integral. $I_A(\omega)$ is the time Fourier transform of $\langle A^*(0)A(t)\rangle$. Equations (2.4.1) and (2.4.2) are Fourier-transform pairs.

12. Where we use the fact that

$$\int_{-T/2}^{T/2} dt \exp i(\omega_n - \omega_{n'})t = T\delta_{n,n'}$$

13. This is the time-correlation function averaged over a finite time T.

14. See Note 12.

15. The average *intensity* I of light passing through the filter is $\frac{c}{8\pi}\langle |E_{TO}|^2\rangle$, where c is the speed of light. It follows that $I(\omega)d\omega$, the intensity of scattered light with frequency between ω and $\omega + d\omega$ is related to $I_E(\omega)$ by

$$I(\omega) = \frac{c}{8\pi} I_E(\omega)$$

that is by a proportionality constant $c/8\pi$. Since we are not interested in absolute intensities, we will always refer to $I_E(\omega)$ as the scattered-intensity spectrum.

REFERENCES

Uhlenbeck, G. E., and Ford G. W., *Lectures in Statistical Mechanics* American Mathematical Society, Providence, Rhode Island (1963). This book contains a lucid account of Birkhoff's theorems and is, moreover, an excellent exposition of the underlying principles of statistical mechanics.

Wax, N. (Ed.), *Selected Papers on Noise and Stochastic Processes,* Dover, New York (1954). This book is a collection of many of the classic papers on noise and stochastic processes.

Zwanzig, R., *Ann. Rev. Phys. Chem.* **16,** 67 (1965). This is an excellent introduction to the subject of linear-response theory and contains references to most of the early papers on the subject.

CHAPTER 3

BASIC LIGHT SCATTERING THEORY

3 · 1 INTRODUCTION

The theory of light scattering can be developed on the basis of quantum field theory. The major results of this theory differ little (and in most cases not at all) from a classical theory of light scattering. Since the major emphasis in this book is on what the light scattering spectrum tells us about physical systems, we do not dwell on the electrodynamical theory. It suffices for our purposes to discuss only the elementary theory, keeping to a minimum the mathematical derivations, which are relegated to Appendix 3.A.

In the classical theory of light scattering an incident electromagnetic field exerts a force on the charges in the scattering volume. These accelerating charges then radiate light. The incident field is said to polarize the medium. When visible light is incident upon the medium the atoms in a subregion of the illuminated volume, small compared to the cube of the incident light wavelength, see essentially the same incident electric field. If many subregions of equal size are considered, the scattered electric field is the superposition of the scattered fields from each of them. If the subregions are optically identical, that is, each has the same dielectric constant, there will be no scattered light in other than the forward direction. This is so because the wavelets scattered from each subregion are identical except for a phase factor that depends on the relative positions of the subregions. If we ignore surface effects it is clear that for a large medium, each subregion can always be paired with another subregion whose scattered field is identical in amplitude but opposite in phase and will thus cancel, leaving no net scattered light in other than the forward direction. If, however, the regions are optically different, that is, have different dielectric constants, then the amplitudes of the light scattered from the different subregions are no longer identical. Complete cancellation will no longer take place, and there will be scattered light in other than the forward direction. Thus in this semimacroscopic view, originally introduced by Einstein, light scattering is a result of local fluctuations in the dielectric constant of the medium (Einstein, 1910). It is easy to understand how such fluctuations may take place. We know from kinetic theory that molecules are constantly translating and rotating so that the instantaneous dielectric constant of a given subregion (which depends on the positions and orientations of the molecules) will fluctuate and thus give rise to light scattering.

3·2 RESULTS FROM ELECTROMAGNETIC THEORY

Consider a nonmagnetic, nonconducting, nonabsorbing medium[1] with average dielectric constant ε_0 (and refractive index $n = \sqrt{\varepsilon_0}$). Let the incident electric field be a plane wave of the form

$$\mathbf{E}_i(r, t) = \mathbf{n}_i E_0 \exp i(\mathbf{k}_i \cdot \mathbf{r} - \omega_i t) \tag{3.2.1}$$

where $\mathbf{n}_i$ is a unit vector in the direction of the incident electric field; E_0 is the field amplitude; $\mathbf{k}_i$ is the propagation vector[2] (or wave vector) and ω_i is the angular frequency.

This plane wave is incident upon a medium that has a *local* dielectric constant

$$\boldsymbol{\varepsilon}(\mathbf{r}, t) = \varepsilon_0 \boldsymbol{I} + \delta\boldsymbol{\varepsilon}(\mathbf{r}, t) \tag{3.2.2}$$

where $\delta\boldsymbol{\varepsilon}(\mathbf{r}, t)$ is the dielectric constant fluctuation tensor at position **r** and time t and **I** is the second-rank unit tensor.

It may be shown by the methods given in Appendix 3.A that the component of the scattered electric field at a large distance R from the scattering volume with polarization $\mathbf{n}_f$, propagation vector $\mathbf{k}_f$, and frequency ω_f is[3]

$$E_s(R, t) = \frac{E_0}{4\pi R \varepsilon_0} \exp ik_f R \int_V d^3r \exp i(\mathbf{q} \cdot \mathbf{r} - i\omega_i t)\,[\mathbf{n}_f \cdot [\mathbf{k}_f \times (\mathbf{k}_f \times (\delta\boldsymbol{\varepsilon}(\mathbf{r}, t) \cdot \mathbf{n}_i)]] \tag{3.2.3}$$

where the subscript V indicates that the integral is over the scattering volume (see Fig. 3.2.1). The vector **q** is defined in terms of the scattering geometry as

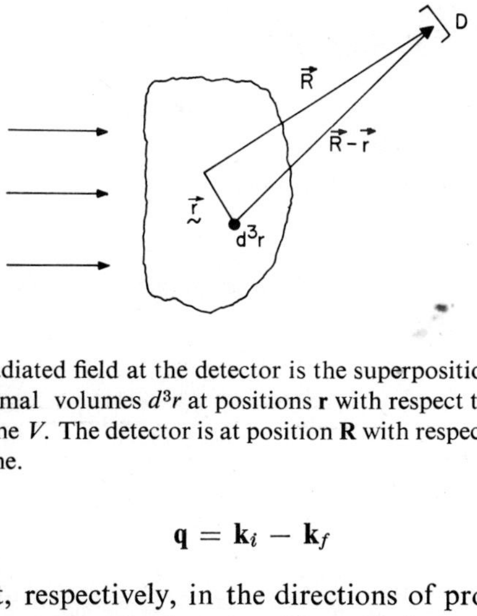

FIG. 3.2.1. The total radiated field at the detector is the superposition of the fields radiated from all infinitesimal volumes d^3r at positions **r** with respect to the center of the illuminated volume V. The detector is at position **R** with respect to the center of the illuminated volume.

$$\mathbf{q} = \mathbf{k}_i - \mathbf{k}_f \tag{3.2.4}$$

where $\mathbf{k}_i$ and $\mathbf{k}_f$ point, respectively, in the directions of propagation of the incident wave and the wave that reaches the detector. The angle between $\mathbf{k}_i$ and $\mathbf{k}_f$ is called the *scattering angle* θ. This is illustrated in Fig. 3.2.2. The magnitudes of $\mathbf{k}_i$ and $\mathbf{k}_f$ are respectively $2\pi n/\lambda_i$ and $2\pi n/\lambda_f$, where λ_i and λ_f are the wave lengths in vacuo of the incident and scattered radiation and n is the refractive index of the scattering medium. It

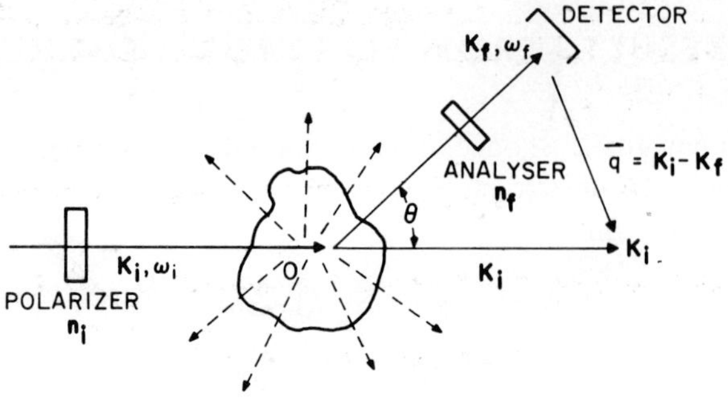

FIG. 3.2.2. Light of polarization $\mathbf{n}_i$ and wave vector $\mathbf{k}_i$ is scattered in all directions. Only scattered light of wave vector $\mathbf{k}_f$ and polarization $\mathbf{n}_f$ arrives at the detector. The scattering vector $\mathbf{q} = \mathbf{k}_i - \mathbf{k}_f$ is defined by the geometry. Since the scattered wave has essentially the same wavelength as the incident wave, $k_f \simeq (2\pi\ n)/\lambda_i = k_i$, it follows from the law of cosines that $q = 2k_i \sin \theta/2$.

is usually the case that the wavelength of the incident light is changed very little in the scattering process so that

$$|\mathbf{k}_i| \cong |\mathbf{k}_f|$$

Thus the triangle in Fig. 3.2.1 is an isosceles triangle and the magnitude of $\mathbf{q}$ can be found from the law of cosines,

$$q^2 = |\mathbf{k}_f - \mathbf{k}_i|^2 = k_f^2 + k_i^2 - 2\mathbf{k}_i \cdot \mathbf{k}_f = 2k_i^2 - 2k_i^2 \cos\theta = 4k_i^2 \sin^2\frac{\theta}{2}$$

$$q = 2k_i \sin\frac{\theta}{2} = \frac{4\pi n}{\lambda_i} \sin\frac{\theta}{2} \tag{3.2.5}$$

This is the Bragg condition. It specifies the wave vector component of the dielectric constant fluctuation that will give rise to scattering at an angle θ.

Equation (3.2.3) can be expressed in terms of the spatial Fourier transform of the dielectric fluctuation

$$\delta\boldsymbol{\varepsilon}(\mathbf{q}, t) = \int_V d^3r \exp i\mathbf{q}\cdot\mathbf{r}\ \delta\boldsymbol{\varepsilon}(\mathbf{r}, t) \tag{3.2.6}$$

as

$$E_s(R, t) = \frac{E_0}{4\pi R\varepsilon_0} \exp i(k_f R - \omega_i t)\ \{\mathbf{n}_f \cdot [\mathbf{k}_f \times \mathbf{k}_f \times (\delta\boldsymbol{\varepsilon}(\mathbf{q}, t) \cdot \mathbf{n}_i)]\} \tag{3.2.7}$$

Equation (3.2.7) can be simplified by working out the vector cross products.[4] Then

$$E_s(R, t) = \frac{-k_f^2 E_0}{4\pi R\varepsilon_0} \exp i(k_f R - \omega_i t)\ \delta\varepsilon_{if}(\mathbf{q}, t) \tag{3.2.8}$$

where

$$\delta\varepsilon_{if}(\mathbf{q}, t) \equiv \mathbf{n}_f \cdot \delta\boldsymbol{\varepsilon}(\mathbf{q}, t) \cdot \mathbf{n}_i \tag{3.2.9}$$

is the component of the dielectric constant fluctuation tensor along the initial and final polarization directions. The time-correlation function (cf. Sections 2.2 and 2.3) of E_s can be evaluated from Eq. (3.2.8)

$$\langle E_s^*(R, 0)E_s(R, t)\rangle = \frac{k_f^4|E_0|^2}{16\pi^2R^2\varepsilon_0^2}\langle\delta\varepsilon_{if}(\mathbf{q}, 0)\,\delta\varepsilon_{if}(\mathbf{q}, t)\rangle \exp -i\omega_i t \tag{3.2.10}$$

The spectral density of light scattered into the detector such that

$$(\mathbf{n}_i, \mathbf{k}_i, \omega_i) \to (\mathbf{n}_f, \mathbf{k}_f, \omega_f)$$

can be determined by substitution of Eq. (3.2.10) into Eq. (2.4.15)

$$I_{if}(\mathbf{q}, \omega_f, R) = \left[\frac{I_0k_f^4}{16\pi^2R^2\varepsilon_0^2}\right]\frac{1}{2\pi}\int_{-\infty}^{+\infty} dt\,\langle\delta\varepsilon_{if}(\mathbf{q}, 0)\delta\varepsilon_{if}(\mathbf{q}, t)\rangle \exp i(\omega_f - \omega_i)t \tag{3.2.11}$$

where $I_0 \equiv |E_0|^2$

Note the following in Eq. (3.2.11): (a) $I_{if} \propto \lambda^{-4}$ (b) $I_{if} \propto R^{-2}$, and (c) I_{if} depends on ω_i and ω_f only through their difference

$$\omega \equiv \omega_i - \omega_f \tag{3.2.12}$$

which is the frequency change in the scattering process. The inverse λ^4 dependence in (a) indicates, for instance, that blue light is scattered more than red light. This results in the blue colors of the sky and oceans. It also indicates that, other things being equal, radio waves would not be scattered as much as visible light. As a consequence of the larger scattering intensities, it is much easier to do scattering experiments with visible light than with the longer wavelength infrared or radio waves. The R^{-2} dependence is just the attenuation expected for a spherical wave. The frequency change occurs only if $\delta\boldsymbol{\varepsilon}(\mathbf{q}, t)$ varies with time,[5] that is, scattering could occur from "frozen" fluctuations but the frequency of the scattered wave would be identical to that of the incident wave.

In Eq. (3.2.11) $I_{if}(\mathbf{q}, \omega_f, R)$ is thus seen to be proportional to the spectral density of the dielectric constant fluctuations, $I_{if}^{\varepsilon}(q, \omega)$

$$I_{if}^{\varepsilon}(\mathbf{q}, \omega) = \frac{A}{2\pi}\int_{-\infty}^{+\infty} dt\, e^{-i\omega t}\,\langle\delta\varepsilon_{if}^*(\mathbf{q}, 0)\,\delta\varepsilon_{if}(\mathbf{q}, t)\rangle \tag{3.2.13}$$

where the proportionality constant is

$$A = \frac{k_f^4 I_0}{16\pi^2R^2\varepsilon_0^2} \tag{3.2.14}$$

In the following chapters the autocorrelation function of the dielectric constant fluctuations is denoted

$$I_{if}^{\varepsilon}(\mathbf{q}, t) = \langle\delta\varepsilon_{if}^*(\mathbf{q}, 0)\,\delta\varepsilon_{if}(\mathbf{q}, t)\rangle \tag{3.2.15}$$

Thus we see that the scattering event that produces the wave vector change $\mathbf{q}$ and frequency shift ω is due entirely to dielectric constant fluctuations of wave vector $\mathbf{q}$ and frequency ω

The scattering event may be viewed in terms of momentum and energy conservation. A "photon" suffers an energy change from $\hbar\omega_i$ to $\hbar\omega_f$ and a momentum change from $\hbar\mathbf{k}_i$ to $\hbar\mathbf{k}_f$, thereby creating (or annihilating) an excitation in the scattering medium of energy $\hbar\omega$ and momentum $\hbar\mathbf{q}$, where from energy and momentum conservation[6]

$$\hbar\omega = \hbar\omega_i - \hbar\omega_f$$
$$\hbar\mathbf{q} = \hbar\mathbf{k}_i - \hbar\mathbf{k}_f$$

In order to calculate a light-scattering spectrum we must have a model for the mechanism by which dielectric fluctuations decay. The remainder of this book is devoted primarily to the study of these fluctuations.

What is sometimes[7] measured in light-scattering spectroscopy is $I_{if}^{\varepsilon}(\mathbf{q}, \omega)$ at a given scattering angle (that is for fixed $\mathbf{q}$) as a function of ω. The spectrum can be quite complicated, and in fact contains a great deal of interesting information. Only with highly monochromatic sources can the spectrum be sufficiently resolved for these experiments to be useful. Prior to the advent of the laser, only the total intensity $I_{if}^{\varepsilon}(\mathbf{q})$ was measured routinely. This is equivalent to integrating the spectrum $I_{if}^{\varepsilon}(\mathbf{q}, \omega)$ over all frequencies, that is, to finding the area under the spectrum. Even this integrated intensity provides important information about the system, as we see later. Note that the integral of Eq. (3.2.13) over frequency is[8]

$$I_{if}^{\varepsilon}(\mathbf{q}) = \int_{-\infty}^{+\infty} d\omega I_{if}^{\varepsilon}(\mathbf{q}, \omega) = \int_{-\infty}^{+\infty} dt\ \delta(t) \langle \delta\varepsilon_{if}^{*}(\mathbf{q}, 0)\, \delta\varepsilon_{if}(\mathbf{q}, t)\rangle$$

so that

$$I_{if}^{\varepsilon}(\mathbf{q}) = \langle |\delta\varepsilon_{if}(\mathbf{q})|^2\rangle \tag{3.2.16}$$

Integrated intensities therefore provide information about the mean-square fluctuations of $\boldsymbol{\varepsilon}$ for given wave vectors.

3 · 3 MOLECULAR APPROACH TO LIGHT SCATTERING

Equation (3.2.13) is an expression for the scattered light spectral density in terms of dielectric constant fluctuations. Nowhere in this treatment was it necessary to determine the explicit dependence of these fluctuations on molecular properties. In fact, this theoretical expression is purely phenomenological. Any attempt to write this formula in molecular terms will necessarily involve some degree of approximation. Nevertheless, a molecular formulation will contribute much to our intuitive understanding of light scattering and will also be useful for practical application.

Let us first consider what happens when an incident monochromatic beam [c.f. Eq. (3.2.1)] impinges on a single molecule which has an anisotropic polarizability specified by a polarizability tensor $\boldsymbol{\alpha}$. The incident light induces a dipole moment

$$\boldsymbol{\mu}(t) = \boldsymbol{\alpha} \cdot \mathbf{E}(t) \tag{3.3.1}$$

which varies with time. According to classical radiation theory a time varying dipole emits electromagnetic radiation. The electric field of the radiated or scattered light (cf. Jackson, 1965) at the detector is proportional to $\hat{\mathbf{k}}_f \times (\hat{\mathbf{k}}_f \times \ddot{\boldsymbol{\mu}}(t'))$ where t' is the retarded time (cf. Appendix 3. A) and $\ddot{\boldsymbol{\mu}}$ is the second time derivative of $\boldsymbol{\mu}$. Following the same kind of argument given in Appendix 3.A and evaluating the vector cross products, it is simple to show that the scattered field at the detector with polarization $\mathbf{n}_f$ is proportional to $\alpha_{if}(t)\, e^{i\mathbf{q}\cdot\mathbf{r}(t)}$ where

$$\alpha_{if}(t) = \mathbf{n}_f \cdot \boldsymbol{\alpha}(t) \cdot \mathbf{n}_i \tag{3.3.2}$$

is the component of the molecular polarizability tensor along $\mathbf{n}_i$ and $\mathbf{n}_f$; $\mathbf{r}(t)$ is the position of the center of mass of the molecule at time t and $\mathbf{q}$ is the usual scattering vector[9] given in Eq. (3.2.4). $\alpha_{if}(t)$ varies in time because the molecule rotates and vibrates, while the phase factor, $e^{i\mathbf{q}\cdot\mathbf{r}(t)}$, varies in time because the molecule translates.

In a fluid, if the molecules are electronically weakly coupled, that is, if the electronic states of the molecules are not perturbed very much by their neighbors, it is reasonable to assume that the light scattered from the assembly of molecules in the illuminated volume will be a superposition of amplitudes scattered from each of the molecules; that is, the scattered field will be proportional to a sum of terms

$$\sum_j{}' \alpha_{if}^j(t) \exp i\mathbf{q} \cdot \mathbf{r}_j(t)$$

where the index stands for the j th molecule in the assembly (and the prime on the sum indicates that the sum is only over molecules in the illuminated volume). This leads to an interference pattern that is modulated by molecular motions. The spectral density of the scattered field will thus be proportional to[10]

$$I_{if}^{\alpha}(\mathbf{q}, \omega) = \frac{1}{2\pi} \int_{-\infty}^{+\infty} dt\, e^{-i\omega t}\, I_{if}^{\alpha}(\mathbf{q}, t) \tag{3.3.3a}$$

where

$$I_{if}^{\alpha}(\mathbf{q}, t) = \langle \delta\alpha_{if}^{*}(\mathbf{q}, 0)\delta\alpha_{if}(\mathbf{q}, t)\rangle \tag{3.3.3b}$$

and

$$\delta\alpha_{if}(\mathbf{q}, t) = \sum_{j=1}^{N}{}' \alpha_{if}^j(t) \exp i\mathbf{q} \cdot \mathbf{r}(t) \tag{3.3.4}$$

is obviously the spatial Fourier component of the polarizability density (cf. Section 2.3).

$$\delta\alpha_{if}(\mathbf{r}, t) = \sum_{j=1}^{N}{}' \alpha_{if}^j(t)\, \delta(\mathbf{r} - \mathbf{r}_j(t)) \tag{3.3.5}$$

This "molecular' theory is clearly an approximation. In general, when two molecules collide they suffer distortions in their electronic charge distributions. These distortions persist for the time it takes a molecule to cross the effective range of the intermolecular potential. This is of order 10^{-13} sec. Thus we expect on the basis of physical intuition that Eq. (3.3.3) will not account for short-time phenomena—what is now called *collision-induced light scattering* (see Chapter 14). These short-time effects lead to very broad bands with exponential wings in the spectrum (Gelbart, 1974), on top of which sit relatively narrow bands relating to the slower motions. These latter bands can be predicted on the basis of Eq. (3.3.3) and therefore lend credence to this approximation. In general, α_{if}^{j} is not the isolated molecule polarizability, but is some renormalized or effective polarizability.

The polarizability component $\alpha_{if}^{j}(t)$ can be regarded as the sum of two terms: (a) the polarizability of the molecule frozen in its equilibrium nuclear configuration and (b) the term linear in the vibrational displacements. Equation (3.3.3) will then consist of four terms (a) a term only involving the rigid molecule polarizability, (b) two cross-terms that are linear in the vibrational displacements, and (c) a term quadratic in the vibrational displacements. The cross terms (b) usually average to zero whereas the term (c) gives rise to the vibration–rotation Raman spectrum. This can be seen as follows. The term (c) contains the vibrational displacements Q in products such as $Q'(0)\,Q(t)$, where Q and Q' belong to different molecules but refer to the same vibrational mode. First, we note that the displacements on different molecules are usually weakly coupled so that the only significant terms derive from autocorrelations of Q on the same molecule. Second, we note that in a normal mode $Q(t) \sim \cos \Omega t$ where Ω is the frequency of the mode. Thus these Raman terms give rise to shifts of frequency of the bands by Ω and $-\Omega$. The former is the anti-Stokes band and the latter, the Stokes band of the Raman spectrum. The dependence of α_{if}^{j} on the vibrational displacements therefore gives rise to several bands corresponding to the different "Raman active" normal modes. A mode is Raman allowed if $\left(\frac{\partial \alpha_{if}}{\partial Q}\right)_0$ is unequal to zero. This is the well-known classical theory of the Raman effect. It incorrectly predicts Stokes and anti-Stokes lines of equal intensity. The quantum theory proceeds in a similar way but correctly accounts for the observed differences in intensity of these bands (cf. Placzek, 1934) The frequency displacements of these vibration–rotation Raman bands are usually in the range 100 to 4000 cm^{-1}. These frequency changes are much greater than those usually observed for terms of type (a) which only involve rotations of the molecules through the dependence of $\alpha_{if}^{j}(t)$ on the instantaneous molecular orientation and translations through the dependence of the phase factors $\exp i\mathbf{q} \cdot \mathbf{r}_j(t)$ on the positions of the molecules.

In this book we are concerned only with that part of the spectrum dependent on pure rotations and translations of molecules. No further discussion of the vibrational Raman scattering is given. Thus Eq. (3.3.3) is used, but always with the rigid-frame polarizabilities. We refer to this type of scattering as *"Rayleigh–Brillouin"* scattering.[11]

3 · 4 SCATTERING GEOMETRIES

The scattering expressions in Sections 3.2 and 3.3 have been written in general tensor notation and hence are independent of any specific laboratory coordinate system used for a scattering experiment. It is, however, convenient for many applications to use specific scattering geometries. In this book we use two basic scattering geometries throughout.[12] We describe below these geometries and write out the components of the dielectric constant fluctuation tensor in each.

The plane defined by the initial and final wave vectors of the light is called the *scattering plane*. It is necessary to define the scattering geometry in relation to the scattering plane. The two geometries used are indicated in Figs. 3.4.1 and 3.4.2. For convenience scattering geometry I will be used in connection with the macroscopic theories of light scattering and scattering geometry II, in connection with the molecular theories.

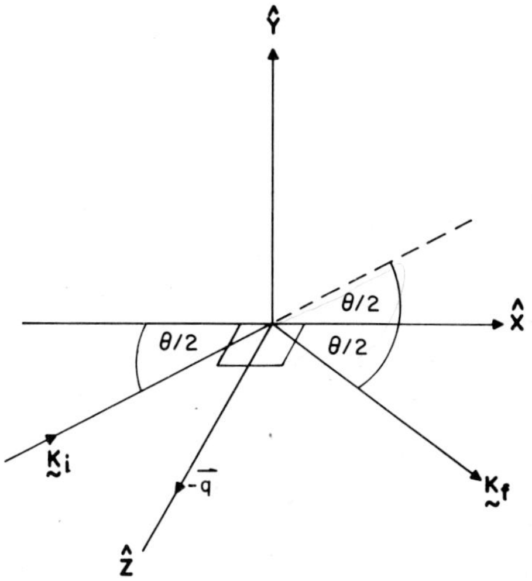

FIG. 3.4.1. *Scattering geometry I.* In this geometry the XZ plane is the scattering plane. The angle ($\mathbf{k}_i$, $\mathbf{k}_f$) is the scattering angle, and the scattering vector $\mathbf{q} = \mathbf{k}_i - \mathbf{k}_f$ is antiparallel to the Z axis.

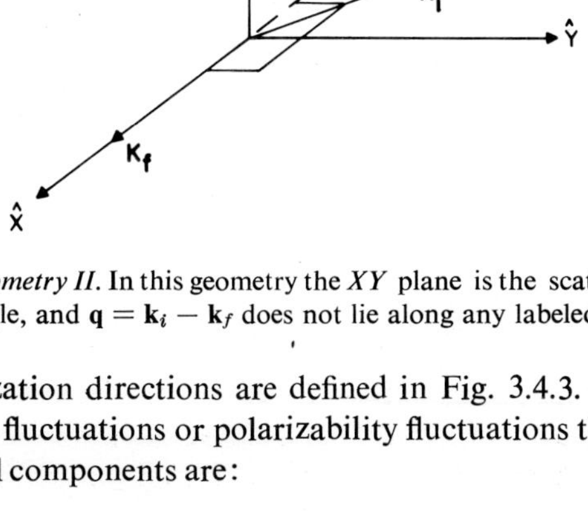

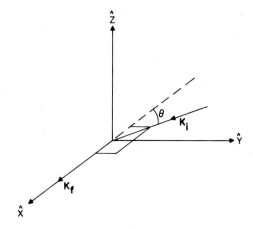

FIG. 3.4.2. *Scattering geometry II.* In this geometry the XY plane is the scattering plane, θ is the scattering angle, and $\mathbf{q} = \mathbf{k}_i - \mathbf{k}_f$ does not lie along any labeled axis.

Four different polarization directions are defined in Fig. 3.4.3. The specific components of the dielectric fluctuations or polarizability fluctuations that are responsible for each of these spectral components are:

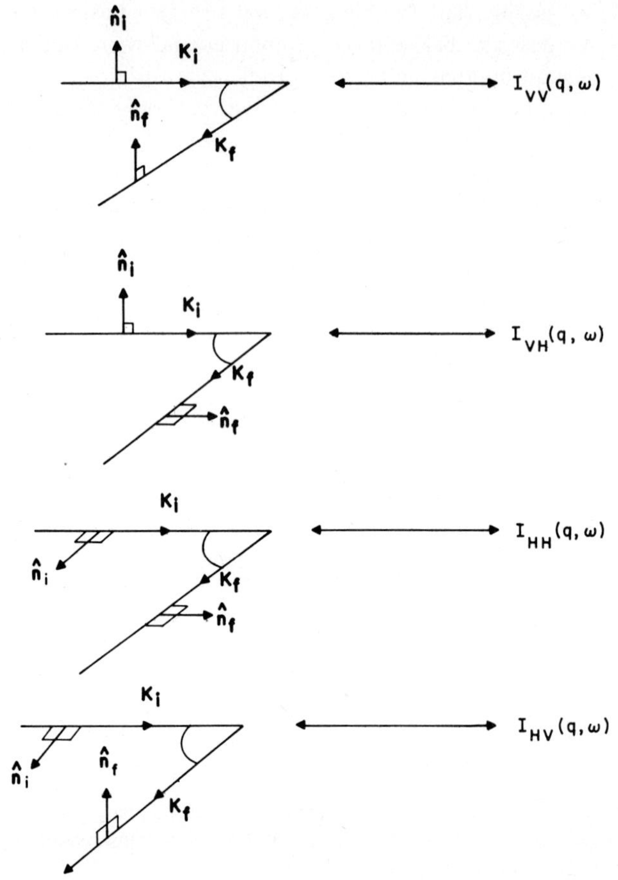

FIG. 3.4.3. Four different pairs of polarization directions commonly used in light-scattering experiments.

Geometry I: $\mathbf{q} = -q\hat{z}$

$$\delta\varepsilon_{VV}(\mathbf{q}, t) = \delta\varepsilon_{YY}(\mathbf{q}, t)$$

$$\delta\varepsilon_{VH}(\mathbf{q}, t) = \delta\varepsilon_{YX}(\mathbf{q}, t) \sin\frac{\theta}{2} - \delta\varepsilon_{YZ}(\mathbf{q}, t)\cos\frac{\theta}{2}$$

$$\delta\varepsilon_{HV}(\mathbf{q}, t) = \delta\varepsilon_{XY}(\mathbf{q}, t) \sin\frac{\theta}{2} + \delta\varepsilon_{ZY}(\mathbf{q}, t)\cos\frac{\theta}{2} \tag{3.4.1}$$

$$\delta\varepsilon_{HH}(\mathbf{q}, t) = \delta\varepsilon_{XX}(\mathbf{q}, t) \sin^2\frac{\theta}{2} - \delta\varepsilon_{ZZ}(\mathbf{q}, t)\cos^2\frac{\theta}{2}$$

$$+ [\delta\varepsilon_{ZX}(\mathbf{q}, t) - \delta\varepsilon_{XZ}(\mathbf{q}, t)] \sin\frac{\theta}{2}\cos\frac{\theta}{2}$$

Geometry II : $\mathbf{q} = q[\hat{x}(1 + \cos\theta) - \hat{y}\sin\theta)]$

$$\delta\varepsilon_{VV}(\mathbf{q}, t) = \delta\varepsilon_{ZZ}(\mathbf{q}, t)$$
$$\delta\varepsilon_{VH}(\mathbf{q}, t) = \delta\varepsilon_{ZY}(\mathbf{q}, t)$$

$$\delta\varepsilon_{HV}(\mathbf{q}, t) = \delta\varepsilon_{XZ}(\mathbf{q}, t) \sin\theta + \delta\varepsilon_{YZ}(\mathbf{q}, t) \cos\theta \tag{3.4.2}$$
$$\delta\varepsilon_{HH}(\mathbf{q}, t) = \delta\varepsilon_{XY}(q, t) \sin\theta + \delta\varepsilon_{YY}(\mathbf{q}, t) \cos\theta$$

The subscripts V and H correspond to directions that are vertical and horizontal with respect to the scattering plane. I_{VV} is sometimes called the *polarized* component and I_{VH} and I_{HV} are usually called the *depolarized* components. Usually $I_{VH} = I_{HV}$. Systems exhibiting optical activity are a common exception to this rule[13]. I_{HH} is often a linear combination of I_{VV} and I_{VH}.

APPENDIX 3.A DERIVATION OF THE SCATTERED FIELD

The Maxwell equations for a nonconducting, nonmagnetic medium may be used to obtain the basic equation for the scattered field Eq. (3.2.4). We follow the treatment of Landau and Lifshitz (1960). For more background on the Maxwell equations the interested reader should consult this text.

For simplicity we consider at the outset a medium with a local dielectric constant tensor

$$\boldsymbol{\varepsilon} = \varepsilon_0 \boldsymbol{I} + \delta\boldsymbol{\varepsilon} \tag{3.A.1}$$

and note that in general the scattered field is much lower in amplitude than the incident field.

If the incident plane wave fields are $\mathbf{E}_i$, $\mathbf{D}_i$, $\mathbf{H}_i$ and scattered fields are $\mathbf{E}_s$, $\mathbf{D}_s$, and $\mathbf{H}_s$.[14] Then the totals of these fields at a point in the scattering medium are

$$\begin{aligned} \mathbf{E} &= \mathbf{E}_i + \mathbf{E}_s \\ \mathbf{D} &= \mathbf{D}_i + \mathbf{D}_s \\ \mathbf{H} &= \mathbf{H}_i + \mathbf{H}_s \end{aligned} \tag{3.A.2}$$

Since $(\mathbf{E}, \mathbf{D}, \mathbf{H})$ and $(\mathbf{E}_i, \mathbf{D}_i, \mathbf{H}_i)$ each satisfy the Maxwell equations it is easy to show that the scattered fields $\mathbf{E}_s$, $\mathbf{D}_s$, and $\mathbf{H}_s$ also obey the Maxwell Equations. Thus

$$\nabla \times \mathbf{E}_s = -\frac{1}{c}\frac{\partial \mathbf{H}_s}{\partial t} \tag{3.A.3}$$

$$\nabla \times \mathbf{H}_s = \frac{1}{c}\frac{\partial \mathbf{D}_s}{\partial t} \tag{3.A.4}$$

$$\nabla \cdot \mathbf{H}_s = 0 \tag{3.A.5}$$

$$\nabla \cdot \mathbf{D}_s = 0 \tag{3.A.6}$$

$\mathbf{H}_s$ may be eliminated from these equations by taking the curl of Eq. (3.A.3) and substituting Eq. (3.A.4) into it;

$$\nabla \times \nabla \times \mathbf{E}_s = \frac{-1}{c^2}\frac{\partial^2 \mathbf{D}_s}{\partial t^2} \tag{3.A.7}$$

The total displacement vector $\mathbf{D}$ and the total electric field vector $\mathbf{E}$ are related through the dielectric constant (Eq. 3.A.1),

$$\begin{aligned}\mathbf{D} &= (\varepsilon_0 \boldsymbol{I} + \delta\boldsymbol{\varepsilon}) \cdot (\mathbf{E}_i + \mathbf{E}_s) \\ &= \varepsilon_0 \mathbf{E}_i + (\delta\boldsymbol{\varepsilon}) \cdot \mathbf{E}_i + \varepsilon_0 \mathbf{E}_s + (\delta\boldsymbol{\varepsilon}) \cdot \mathbf{E}_s \end{aligned} \tag{3.A.8}$$

From Eq. (3.A.2) and the fact that $\mathbf{D}_i = \varepsilon_0 \mathbf{E}_i$, Eq. (3.A.8) becomes

$$\mathbf{D}_s = \varepsilon_0 \mathbf{E}_s + (\delta\boldsymbol{\varepsilon}) \cdot \mathbf{E}_i \tag{3.A.9}$$

In Eq. (3.A.7) we have neglected the second-order term $(\delta\boldsymbol{\varepsilon}) \cdot \mathbf{E}_s$.

Then, solving Eq. (3.A.9) for $\mathbf{E}_s$, substituting the result into Eq. (3.A.7) and using Eq. (3.A.9), we obtain[15] an inhomogeneous wave equation

$$\nabla^2 \mathbf{D}_s - \left(\frac{\varepsilon_0}{c^2}\right) \frac{\partial^2 \mathbf{D}_s}{\partial t^2} = -\nabla \times \nabla \times (\delta\boldsymbol{\varepsilon} \cdot \mathbf{E}_i) \tag{3.A.10}$$

Equation (3.A.10) may be simplified by defining a new vector, $\boldsymbol{\pi}$ (the Hertz vector) by

$$\mathbf{D}_s = \nabla \times \nabla \times \boldsymbol{\pi} \tag{3.A.11}$$

Substituting (3.A.11) into (3.A.10), we see that the Hertz vector satisfies a wave equation, with a simple source term $-(\delta\boldsymbol{\varepsilon}) \cdot \mathbf{E}_i$,

$$\nabla^2 \boldsymbol{\pi} - \left(\frac{\varepsilon_0}{c^2}\right) \frac{\partial^2 \boldsymbol{\pi}}{\partial t^2} = -(\delta\boldsymbol{\varepsilon}) \cdot (\mathbf{E}_i) \tag{3.A.12}$$

The formal solution of Eq. (3.A.12) is

$$\boldsymbol{\pi}(R, t) = \frac{1}{4\pi} \int d^3r \frac{\delta\boldsymbol{\varepsilon}(\mathbf{r}, t')}{|\mathbf{R} - \mathbf{r}|} \cdot \mathbf{E}_i(\mathbf{r}, t') \tag{3.A.13}$$

where $\mathbf{R}$ and $\mathbf{r}$ are defined in Fig. 3.A.1 and t' is the retarded time

$$t' = t - \frac{\sqrt{\varepsilon_0}}{c} |\mathbf{R} - \mathbf{r}| \tag{3.A.14}$$

If Eq. (3.2.1) for $\mathbf{E}_i$ is substituted into Eq. (3.A.13), the operations in Eq. (3.A.11) are performed upon the result to obtain $\mathbf{D}_s$ and, it is noted that at the detector (assumed to be immersed in a medium of dielectric constant ε_0) $\mathbf{D}_s = \varepsilon_0 \mathbf{E}_s$, we obtain

$$\mathbf{E}_s(R, t) = \nabla \times \nabla \times \left[\frac{E_0}{4\pi\varepsilon_0} \int_j d^3r \frac{1}{|\mathbf{R} - \mathbf{r}|} [\delta\boldsymbol{\varepsilon}(\mathbf{r}, t') \cdot \mathbf{n}_i] \exp i(\mathbf{k}_i \cdot \mathbf{r} - \omega_i t')\right. \tag{3.A.15}$$

Since the detector is a large distance from the scattering medium, $|\mathbf{R} - \mathbf{r}|$ may be expanded in a power series

$$|\mathbf{R} - \mathbf{r}| \cong R - \mathbf{r} \cdot \hat{\mathbf{k}}_f + \ldots \tag{3.A.16}$$

where $\hat{\mathbf{k}}_f$ is a unit vector in the direction of $\mathbf{R}$. From Eq. (3.A.14) it follows that

$$t' \cong t - \frac{\sqrt{\varepsilon_0}}{c}(R - \mathbf{r} \cdot \hat{\mathbf{k}}_f) \tag{3.A.17}$$

To proceed, we perform a Fourier analysis of $\delta\boldsymbol{\varepsilon}(\mathbf{r}, t')$ over an interval T which allows us to write

$$\delta\boldsymbol{\varepsilon}(\mathbf{r}, t') = \sum_p \delta\boldsymbol{\varepsilon}_p(\mathbf{r}) \exp i\Omega_p t' \tag{3.A.18}$$

where $\Omega_p = (2\pi/T)p$. The only frequency components Ω_p that contribute to this sum are those typifying the natural rotational and translational motions of the system which are typically smaller than 10^{13}sec^{-1} and which are always small compared with the incident light frequency ω_i; that is, $\omega_i \gg \Omega_p$ for all relevant Ω_p. Substituting Eqs. (3.A.17) and (3.A.18) into Eq. (3.A.15) and defining

$$\begin{cases} \omega_f \equiv \omega_i - \Omega_p \\ \mathbf{k}_p \equiv \dfrac{\sqrt{\varepsilon_0}}{c}\omega_f \hat{\mathbf{k}}_f \\ \mathbf{q}_p \equiv \mathbf{k}_i - \mathbf{k}_p \end{cases} \tag{3.A.19}$$

we find

$$E_s(R, t) = \frac{E_0}{4\pi\varepsilon_0 R} \sum_p \exp i[k_p R - \omega_i t]\, \mathbf{k}_p \times \left[\mathbf{k}_p \times \int_V d^3r \exp i(\mathbf{k}_i - k_p\hat{\mathbf{k}}_f) \cdot r \right.$$
$$\left. \delta\boldsymbol{\varepsilon}_p(\mathbf{r})\, (\exp i\Omega_p t) \cdot \mathbf{n}_i \right] \tag{3.A.20}$$

where we have ignored terms of higher order than $(1/R)$. Now we note that because $\Omega_p \ll \omega_i$, k_P is to a very good approximation

$$k_p \simeq \frac{\sqrt{\varepsilon_0}}{c}\omega_i = k_i \cong k_f \tag{3.A.21}$$

where the second equality follows from $\omega_i = (ck_i/n)$ and where the refractive index is $n = \sqrt{\varepsilon_0}$. It follows that $k_p\hat{\mathbf{k}}_f \simeq k_i\hat{\mathbf{k}}_f$. Thus to this order of approximation $\delta\varepsilon_p$ and $\varepsilon^{i\Omega_p t}$ are the only p-dependent quantities in the sum, so that Eq. (3.A.20) becomes

$$E_s(R, t) = \frac{E_0}{4\pi\varepsilon_0 R} \exp i[k_f R - \omega_i t]\, \mathbf{k}_f \times \left[\mathbf{k}_f \times \int_V d^3r\, (\exp i\mathbf{q} \cdot \mathbf{r})(\delta\boldsymbol{\varepsilon}(\mathbf{r}, t) \cdot \mathbf{n}_i)\right] \tag{3.A.22}$$

where we have substituted Eq. (3.A.18) in the form $\delta\boldsymbol{\varepsilon}(\mathbf{r}, t) = \sum_p \delta\boldsymbol{\varepsilon}_p(\mathbf{r}) \exp i\Omega_p t$, and where we have defined

$$\mathbf{k}_f = k_i\hat{\mathbf{k}}_f \tag{3.A.23}$$
$$\mathbf{q} = \mathbf{k}_i - \mathbf{k}_f$$

It should be noted that $k_f = k_i$ and that $\mathbf{q}$ and $\mathbf{k}_f$ so defined are consistent with the drawing in Fig. (3.2.1) and with Eq. (3.2.5). Note that Eq. (3.A.22) is free of the retarded time and depends only on the real time t.

If we take the component of $\mathbf{E}_s$ in the direction $\mathbf{n}_f$ and set $\varepsilon_0 = 1$, we obtain

$$E_s(R, t) = \frac{E_0}{4\pi R} \exp i[k_f R - \omega_i t] \int_V d^3r\, e^{i\mathbf{q}\cdot\mathbf{r}}\, \mathbf{n}_f \cdot [\mathbf{k}_f \times \mathbf{k}_f \times [\delta\boldsymbol{\varepsilon}(\mathbf{r}, t) \cdot \mathbf{n}_i]] \quad (3.A.24)$$

which is the same as Eq. (3.2.3). Equation (3.2.13) may then be obtained from this by the simple operations indicated in the text.

It should be pointed out that this result and all the other results given in this book refer only to the single scattering of an incident light wave. Multiple scattering is entirely neglected because we have omitted higher order terms in $\delta\boldsymbol{\varepsilon}$ after Eq. (3.A.10).

NOTES

1. Many of the applications of light scattering are to ionic solutions, which are conducting media. However since the ions are massive the free charge density will vary on a much slower time scale than that specified by the laser frequency ($\sim 10^{14}$ Hz). Thus the medium may be considered to be nonconducting as far as this derivation is concerned. Also, since we consider the medium to be nonabsorbing, we are restricted to incident wave lengths which are not resonant with any molecular transitions of the scattering medium. The reader interested in resonant light scattering should consult the appropriate references (Peticolas, 1972, Bauer et al., 1975).
2. The propagation vector is a vector with a direction parallel to the direction of propagation of the wave and magnitude $|\mathbf{k}_i| = 2\pi n/\lambda_i$ where λ_i is the wavelength in vacuo.
3. This formula ignores both multiple scattering and "local field" effects (e.g., see Gelbart, 1974).
4. where we use the identity

$$\mathbf{A} \times (\mathbf{B} \times \mathbf{C}) = \mathbf{B}(\mathbf{A} \cdot \mathbf{C}) - \mathbf{C}(\mathbf{A} \cdot \mathbf{B})$$

and have used the property that $\mathbf{n}_f \cdot \hat{\mathbf{k}}_f = 0$; that is, the final polarization $\mathbf{n}_f$ must be perpendicular to the scattered field vector $\hat{\mathbf{k}}_f$.

5. This may be easily seen from Eq. (3.2.11). If $\langle\delta\varepsilon_{if}(0)\delta\varepsilon_{if}(t)\rangle$ is independent of time, the time integral is reduced to a delta function $\delta(\omega_f - \omega_i)$, and there is no frequency shift. Actually there would be a small shift due to the recoil of the whole system required by momentum conservation. This is a high-order effect that is completely neglected in our treatment of light scattering.
6. This language is appropriate to the quantum theory of light scattering.
7. In the filter technique or heterodyne technique with spectrum analyzer (see Chapter 4).
8. Where we have used the integral representation of the delta function

$$1/2\pi \int_{-\infty}^{+\infty} d\omega\, e^{i\omega t} = \delta(t)$$

9. In the molecular theory it is a subtle problem to show that the correct q which enters the phase factors involves the medium wave vector with the refractive index and not the vacuum wave vectors (e.g., see Felderhoff, 1974).
10. This same result can be derived from the quantum mechanical T matrix formulation of the scattering process (e.g., see Goldberger and Watson, 1964). Also, the integrated intensity is proportional to

$$I_{if}^{\alpha}(\mathbf{q}, 0) = \langle\delta\alpha^*_{if}(\mathbf{q}, 0)\delta\alpha_{if}(\mathbf{q}, 0)\rangle$$

From macroscopic electrodynamics it can be shown that $\varepsilon = 1 + 4\pi\alpha$ where α is the polarizability. It follows that $\delta\varepsilon = 4\pi\delta\alpha$ and consequently that

$$I_{if}^{\varepsilon}(\mathbf{q}, t) = 16\pi^2 I_{if}^{\alpha}(\mathbf{q}, t)$$

11. There is some arbitrariness in what we call Raman and Rayleigh–Brillouin scattering. It should be

noted that terms of type (a) above also include what is sometimes called "pure rotational" Raman scattering. It would, in fact, be reasonable to think of Rayleigh–Brillouin scattering as translation–rotation Raman scattering.

12. For some scattering problems it is useful to consider less restrictive scattering geometries than those used here.
13. For a discussion of the conditions under which this "reciprocity" relation is true see Perrin (1942).
14. **E**, **D**, and **H** are, respectively, the electric, dielectric displacement, and magnetic fields.
15. The vector identity

$$\nabla \times \nabla \times \mathbf{A} = -\nabla^2\mathbf{A} + \nabla(\nabla \cdot \mathbf{A})$$

has also been used.

REFERENCES

Bauer, D. R., Hudson, B., and Pecora, R., *J. Chem. Phys.* **63,** 588 (1975)

Einstein, A., *Ann. Phys.* **33,** 1275 (1910). [English translation in *Colloid Chemistry,* Alexander, J. (ed.), Vol. I, 323–339, Reinhold, New York (1926).]

Felderhoff, B. U. On the propagation and scattering of light in fluids, *Phys.* in press (1974).

Gelbart, W., *Adv. Chem. Phys.* (1974). This is the most recent review that contains a discussion of local field, multiple scattering, and collision-induced effects. References to the important papers are cited in this article. See in particular the references to Komarov and Fisher (1963) and Pecora (1964).

Goldberger, M. L., and Watson, K. M., *Collision Theory,* Wiley, New York (1964).

Jackson, J. D., *Classical Electrodynamics,* Wiley, New York (1965).

Landau L. D., and Lifshitz, E. M., *Electrodynamics of Continuous Media,* Addison Wesley, Reading, Mass. (1960).

Oxtoby, D. W., and Gelbart, W., *J. Chem. Phys.* **60,** 3359 (1974).

Perrin, F., *J. Chem. Phys.* **10,** 415 (1942).

Peticolas, W., *Ann. Rev. Phys. Chem.,* **23,** 93 (1972).

Placzek, G., *Handbuch der Radiologie,* Marx, E. (ed.), *Leipz.* **6**(2), 209–374 (1934). [English Translation in UCRL Trans. No. 526 (L).]

CHAPTER 4

THE LIGHT-SCATTERING EXPERIMENT

4 · 1 INTRODUCTION

It is not the purpose of this chapter to give details or evaluations of the different experimental methods, but only to outline some of the main techniques presently employed so that the different methods of data presentation can be related to some of the theoretical correlation functions described in this book. The reader interested in further details should consult the review articles by Cummins and Swinney (1970), Fleury and Boon (1974), and Cummins and Pike (1974) and the references contained therein.

In a light-scattering experiment a beam of light is focused onto a region of a fluid and is scattered into a detector. Polarizers and analyzers are used to define the polarizations of the incident and scattered light beams, respectively. Physically, the instantaneous scattered field can be regarded as the superposition of waves scattered from the individual scattering centers. This scattered field therefore fluctuates in response to the molecular motions of the scatterers. A variety of detection schemes are used to analyze the time-dependence of these fluctuations. The detection method used in a particular experiment depends on the time scale of these fluctuations. *Filter methods* are used to study relatively rapid molecular dynamic processes, that is, those that occur on a time scale faster than about 10^{-6} sec. *Optical mixing* or *beating methods* are usually used for processes that occur on time scales slower than about 10^{-6} sec.

4 · 2 FILTER TECHNIQUES

The filter method involves the spectral decomposition of the scattered light by a diffraction grating or a Fabry–Perot interferometer. These devices are interposed between the scattering sample and a photomultiplier tube and act as filters, which for a given setting pass only a single frequency component of the scattered light. The filter output is then incident upon a photomultiplier cathode (PM) whose average dc output is proportional to the spectral density of the scattered electric field at the filter frequency. The filter is then swept through a range of frequencies. As discussed in Section 2.4, the PM output is proportional to the spectral density of the scattered electric field E_s

$$I_E(\omega_f) = \frac{1}{2\pi}\int_{-\infty}^{+\infty} dt \exp i\omega_f t \,\langle E_s^*(0)E_s(t)\rangle \tag{4.2.1}$$

where ω_f is the frequency of the scattered light. Now we note from Eq. (3.2.8) that $E_s(t)$ contains an amplitude factor $\delta\varepsilon_{if}(\mathbf{q}, t)$ and a phase factor $\exp i(k_f R - \omega_i t)$. Substitution of Eq. (3.2.8) into Eq. (4.2.1) then shows that the output of the filter is proportional to the spectral density $I_{if}^{\varepsilon}(\mathbf{q}, \omega)$ or $I_{if}^{\alpha}(\mathbf{q}, \omega)$ where $\omega \equiv \omega_i - \omega_f$ is the frequency *shift* of the scattered light [see Eqs. (3.3.3) and (3.2.12)]. Thus the quantity directly measured in filter experiments is proportional to the Fourier transform of the dielectric or polarizability time-correlation functions $I_{if}^{\varepsilon}(\mathbf{q}, t)$ or $I_{if}^{\alpha}(\mathbf{q}, t)$ given, respectively, by Eqs. (3.2.15) and (3.3.3b). In the ensuing chapters we often let $I_{if}(\mathbf{q}, t)$ [or $I_{if}^{(1)}(\mathbf{q}, t)$] represent either $I_{if}^{\alpha}(\mathbf{q}, t)$ or $I_{if}^{\varepsilon}(\mathbf{q}, t)$.

For processes faster than about 10^{-10} sec, diffraction gratings are used as the filter. For the slower processes with relaxation times in the range of about 10^{-6}–10^{-10} sec Fabry–Perot interfermeters are used (See Appendix 4.A). Filter experiments do not have sufficient resolution to study processes slower than about 10^{-6} sec. For fluctuations in the time range 1–10^{-6} sec optical mixing techniques must be used.

4 · 3 OPTICAL MIXING TECHNIQUES

Optical mixing techniques are the optical analogs of the beating techniques developed in radio-frequency spectroscopy (Forrester, 1961). They have made possible the application of light scattering to the study of the dynamics of relatively slow processes such as macromolecular diffusion, the dynamics of fluctuations in the critical region, and the motility of microorganisms.

In optical mixing methods no "filter" is inserted between the scattering medium and the photomultiplier (see Fig. 1.2.2). The scattered light impinges directly on the PM cathode. In the *homodyne* (or self-beat) method only the scattered light impinges on the photocathode, while in the *heterodyne* method a *local oscillator* (usually a small portion of the unscattered laser beam) is mixed with the scattered light on the cathode surface. Since the phototube is a square-law detector, its instantaneous current output is proportional to the square of the incident electric field $i(t) \propto |E(t)|^2$. The square of the electric field is proportional to the intensity of the light (or in quantum language, the number of photons).

The PM output is then usually passed into a hardwire computer called an *autocorrelator,* which calculates its time autocorrelation function

$$\langle i(t)\, i(0)\rangle = B\langle |E(0)|^2 |E(t)|^2\rangle \tag{4.3.1}$$

where B is a proportionality constant.[1] The autocorrelator can be used in either a "digital" or an "analog" mode (see Appendix 4.B). In the digital method one counts and then autocorrelates current pulses (or photons) and in the analog method, one directly autocorrelates the fluctuations in the PM output current. For low light levels the digital method is normally preferred. (Jakeman and Pike, 1968; 1969).

For the purpose of discussing the differences between heterodyne and homodyne scattering we define the two scattered field autocorrelation functions

$$I_1(t) \equiv \langle E_s^*(0)\, E_s(t)\rangle \tag{4.3.2a}$$

$$I_2(t) \equiv \langle |E_s(0)|^2 |E_s(t)|^2 \rangle \tag{4.3.2b}$$

Homodyne Method

Since in the homodyne method only the scattered light impinges on the photocathode, $E(t)$ in Eq. (4.3.1) is equal to the scattered field $E_s(t)$, so that $\langle i(0)i(t) \rangle$ is proportional to $I_2(t)$—which is consequently sometimes called the *homodyne correlation function*. The amplitude of $E_s(t)$, the scattered field, is proportional to the instantaneous dielectric constant fluctuations in the scattering volume and, of course, fluctuates in the same manner. In certain circumstances the homodyne correlation function may be simply expressed in terms of $I_1(t)$ or equivalently $I_{if}^s(\mathbf{q}, t)$ or $I_{if}^\alpha(\mathbf{q}, t)$ of Eqs. (3.2.15) and (3.3.3b), respectively, as we now discuss.

The scattering volume V can be subdivided into subregions of volume small compared to the wavelength of light. Then the scattered field E_s can be regarded as a superposition of fields from each of the subregions, so that

$$E_s = \sum_n E_s^{(n)}$$

where $E_s^{(n)}$ is the scattered field from the nth subregion. As particles move $E_s^{(n)}$ fluctuates. If, as is often the case, the subregions are sufficiently large to permit particle motions in one subregion to be independent of those in another, E_s can be regarded as a sum of independent random variables ($E_s^{(1)}, E_s^{(2)}, \ldots$). In this eventuality, the central limit theorem implies that E_s, which is itself a random variable, must be distributed according to a Gaussian distribution. A Gaussian distribution is completely characterized by its first and second moments. It follows that all higher moments of this distribution are related to the first two moments. From considerations of this kind (see Appendix 4.C) it is easy to show that the quantity in Eq. (4.3.2b), which is a fourth moment of the distribution, is related to $I_1(t)$, which is the second moment, through the equation.

$$I_2(t) = |I_1(0)|^2 + |I_1(t)|^2 \tag{4.3.3}$$

The important assumption in deriving this result is that the scattering volume can be divided into a large number of statistically independent subregions.[2] There are circumstances in which this assumption may be invalid. For example, systems in their critical region have very long correlation lengths; hence care must be exercised in applying Eq. (4.3.3) to critical fluids. In particular, the scattering volume must be chosen to contain a sufficiently large number of *correlation-volumes* to justify the use of the central limit theorem.

Another example involves the scattering from either dilute macromolecular solutions or bacterial dispersions. The scattered field is a superposition of the amplitudes scattered from each of these large independent particles. If the scattering volume V is sufficiently large, there will be on the average a large number $\langle N \rangle$ of particles in V and consequently a large number of terms in the superposition, thereby justifying the use of the central limit theorem to derive Eq. (4.3.3). However if either the concentration c or the volume is small so that $\langle N \rangle = cV$ is small, there might be an insufficient numberof terms in the superposition so that the central limit theorem, and thereby Eq. (4.3.3), is invalid. In Section 5.5 we show that in this eventuality, I_2 contains more information than I_1 and moreover can be used effectively to study additional properties of these solutions. Although these conditions are probably satisfied in the vast majority of applications, the restrictions must be kept in mind in any application of the homodyne technique.

Let us now return to Eq. (4.3.3). The $|I_1(0)|^2$ term is a dc term that determines the baseline for the homodyne correlation function of the scattered light.[3] If, as is often the case, $I_1(t)$ is a sum of exponentials,

$$I_1(t) = \sum_i a_i \exp\frac{-t}{\tau_i} \tag{4.3.4}$$

then in the Gaussian approximation [Eq. (4.3.3)] the corresponding $I_2(t)$ is

$$I_2(t) = \sum_{i,j} a_i a_j \left[1 + \exp -t\left(\frac{1}{\tau_i} + \frac{1}{\tau_j}\right)\right] \tag{4.3.5}$$

If an autocorrelator is used to analyze the PM output the correlation function in Eq. (4.3.5) is directly measured. Sometimes a spectrum analyzer[4] is used instead of an autocorrelator. This device determines the spectral density of the photocurrent from the PM and thus in a homodyne experiment it determines the time Fourier transform of $I_2(t)$, that is,

$$I_2(\omega) = \sum_{i,j} a_i a_j \left[\delta(\omega) + \frac{1}{\pi} \frac{\left(\frac{1}{\tau_i} + \frac{1}{\tau_j}\right)}{\left(\omega^2 + \left(\frac{1}{\tau_i} + \frac{1}{\tau_j}\right)\right)^2}\right] \tag{4.3.6}$$

where the delta function arises from the constant term in Eq. (4.3.5) (which is equivalent to the term $|I_1(0)|^2$ in Eq. (4.3.3)).

Note from Eqs. (4.3.5) that more exponentials may appear in $I_2(t)$ than in $I_1(t)$. Similarly from Eq. (4.3.6), more Lorentzians may appear in $I_2(\omega)$ than in $I_1(\omega)$. For instance, if $I_1(t)$ is composed of two exponentials

$$I_1(t) = a_1 \exp\frac{-t}{\tau_1} + a_2 \exp\frac{-t}{\tau_2} \tag{4.3.7}$$

then $I_2(t)$ is composed of three exponentials

$$I_2(t) = (a_1 + a_2)^2 + a_1^2 \exp -\frac{2t}{\tau_1} + a_2^2 \exp -\frac{2t}{\tau_2} + 2a_1a_2 \exp -t\left(\frac{1}{\tau_1} + \frac{1}{\tau_2}\right) \tag{4.3.8}$$

and $I_2(\omega)$ is composed of three Lorentzians

$$I_2(\omega) = (a_1 + a_2)^2\delta(\omega) + \frac{1}{\pi}\left\{a_1^2 \frac{\frac{2}{\tau_1}}{\omega^2 + \left(\frac{2}{\tau_1}\right)^2} + a_2^2 \frac{\frac{2}{\tau_2}}{\omega^2 + \left(\frac{2}{\tau_2}\right)^2} + 2a_1a_2 \frac{\left(\frac{1}{\tau_1} + \frac{1}{\tau_2}\right)}{\omega^2 + \left(\frac{1}{\tau_1} + \frac{1}{\tau_2}\right)^2}\right\} \tag{4.3.9}$$

For processes with several relaxation times, the analysis of homodyne data is quite difficult. For a one-relaxation time process, however, the homodyne correlation function (or spectral density) is still exponential (Lorentzian), except that the relaxation time that appears is $\tau/2$ rather than τ.

Heterodyne Method

In the heterodyne method, a small portion of the unscattered laser light (local oscillator) is mixed with the scattered light on the photomultiplier cathode (see Fig. 1.2.2*c*). We assume below that the local oscillator varies at the laser frequency.[2] The PM output may then be analyzed with an autocorrelator or a spectrum analyzer.

If $E_{\text{LO}}(t)$ represents the local oscillator electric field, then the electric field at the PM is the superposition of $E_{\text{LO}}(t)$ and $E_s(t)$ and thus the autocorrelation function of the PM output, Eq. (4.3.1), becomes

$$\langle i(0)i(t)\rangle = B\langle |E_{\text{LO}}(t) + E_s(t)|^2 |E_{\text{LO}}(0) + E_s(0)|^2\rangle \tag{4.3.10}$$

By proper choice of the experimental conditions, the amplitude of the local oscillator may be made much greater than the amplitude of the scattered field

$$|E_{\text{LO}}(t)| \gg |E_s(t)|$$

The following approximations are also made: (a) fluctuations of the local oscillator field are negligible and (b) the local oscillator field and the scattered field are statistically independent so that, for example, $\langle I_s I_{\text{LO}}\rangle = \langle I_{\text{LO}}\rangle\langle I_s\rangle$.

With these assumptions the expansion of Eq. (4.3.10) yields 16 terms, 10 of which are zero, three dc, and one negligible (the term $\langle |E_s(0)|^2 |E_s(t)|^2\rangle$). The remaining two terms, the only ones of importance for our consideration, give

$$\langle i(0)i(t)\rangle \cong B[I_{LO}{}^2 + 2I_{LO} Re I_1(t)] \tag{4.3.11}$$

where

$$I_{LO} = \langle |E_{\text{LO}}|^2\rangle$$

is the intensity of the local oscillator signal, and $ReI_1(t)$ is the real part[6] of $I_1(t)$. Thus $ReI_1(t)$ is sometimes called the *heterodyne correlation function*. The background or dc term in the heterdyne experiment is determined by the LO intensity I_{LO}.

Note that the heterodyne technique introduces no extra terms into the scattered field time-correlation function, as contrasted to the homodyne spectrum of a many-exponential process.

Because $E_s(t)$ is proportional to $\delta\varepsilon_{if}(\mathbf{q}, t)$, or correspondingly, $\delta\alpha_{if}(\mathbf{q}, t)$, $I_1(t)$ and $I_2(t)$ are, respectively, proportional to the correlation functions

$$I_{if}^{(1)}(\mathbf{q}, t) = \langle \delta\varepsilon_{if}^*(\mathbf{q}, 0)\, \delta\varepsilon_{if}(\mathbf{q}, t)\rangle \tag{4.3.12a}$$

$$I_{if}^{(2)}(\mathbf{q}, t) = \langle |\delta\varepsilon_{if}(\mathbf{q}, 0)|^2 |\delta\varepsilon_{if}(\mathbf{q}, t)|^2\rangle \tag{4.3.12b}$$

with proportionality constants A and A^2 defined by Eq. (3.2.14).

From Eqs. (4.3.12) it is clear that the homodyne and heterodyne techniques measure different correlation functions of the dielectric constant fluctuations. In the event that the Gaussian approximation applies these two correlation functions are related to each other by Eq. (4.3.3),

$$I^{(2)}_{if}(\mathbf{q}, t) = |I^{(1)}_{if}(\mathbf{q}, 0)|^2 + |I^{(1)}_{if}(\mathbf{q}, t)|^2 \tag{4.3.13}$$

Exceptions to Eq. (4.3.13) are discussed in Chapter 5.

It should be noted that if a heterodyne experiment is performed with a spectrum analyzer rather than an autocorrelator, the resulting spectrum is proportional to[7]

$$I^{(1)}_{if}(\mathbf{q}, \omega) \equiv \frac{1}{2\pi}\int_{-\infty}^{+\infty} dt\, e^{-i\omega t}\, Re I^{(1)}_{if}(\mathbf{q}, t) \tag{4.3.14}$$

We shall call this the *heterodyne spectrum.* Comparing Eq. (4.3.14) with $I^{(1)}_{if}(\mathbf{q}, \omega)$ the spectrum measured by filter methods [see Eq. (4.2.1)], we see that only when $I^{(1)}_{if}(\mathbf{q}, t)$ is a real function of the time do the two experiments give the same spectrum. When currents are present in a system the resulting spectra are somewhat different, as we see in Chapter 5.

APPENDIX 4.A FABRY-PEROT INTERFEROMETER

A Fabry-Perot interferometer consists of two plane dielectric mirrors held parallel to each other (see Fig. 4.A.1). The inner surfaces of the mirrors are highly reflecting ($\approx$ 98%). In light-scattering experiments, light usually enters the interferometer normal to the mirrors and is then reflected back and forth in the interferometer cavity. The condition for survival in the cavity of a wave of wavelength λ is that an integral number of half wavelengths must fit in the cavity, that is,

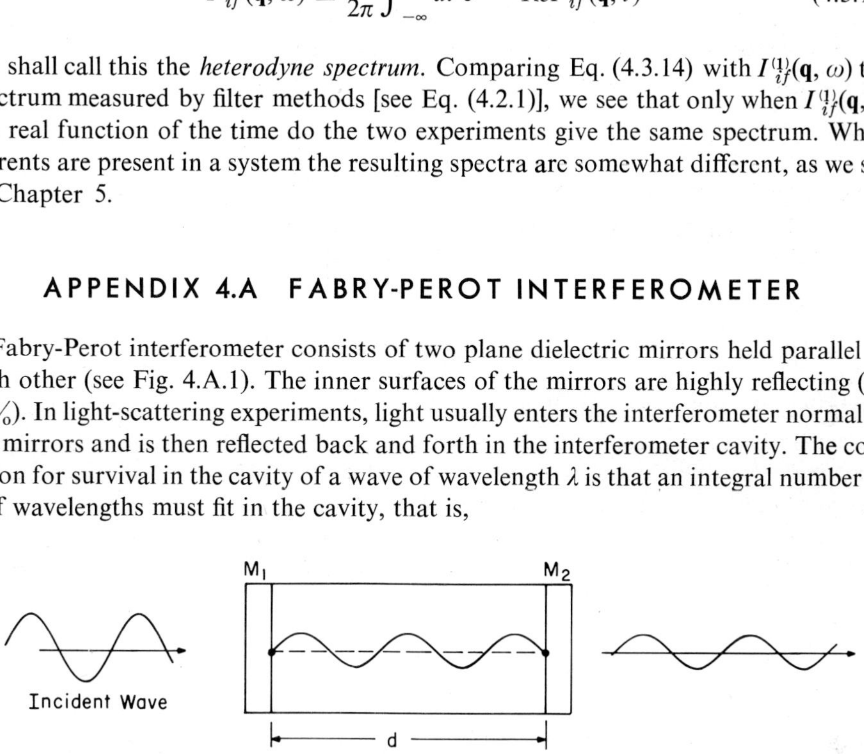

FIG. 4.A.1. Light enters the interferometer cavity normal to the mirror M . The wave is reflected back and forth in the cavity. Only waves for which a halfintegral number of wavelengths fit in the cavity survive. The mirrors are only partially reflecting so that the output wave contains only these allowed wavelengths.

$$m\left(\frac{\lambda}{2}\right) = d \tag{4.A.1}$$

where m is an integer and d the spacing between the mirrors. At this wavelength there will be a maximum in the intensity of the transmitted light (see Section 7.6.1 of Born and Wolf, 1964). If d is 0.1 cm and $\lambda = 5000$ Å, then $m = 2 \times 10^3$. The difference between successive wavelengths, $\Delta\lambda$, that can pass through the interferometer for this value of d is $\Delta\lambda \sim 2.5$Å. The corresponding frequency range is $\Delta\nu \sim 3 \times 10^{11}$ Hz. This is large compared to the width of a typical Rayleigh-Brillouin line. Thus a proper setting of d in the interferometer will select only one frequency component (wavelength) of the light-scattering spectrum. By varying d one can detect the spectral density of the scattered light as a function of frequency (see Section 4.2).

There are two common methods for sweeping the spectrum: (a) physically moving the plates piezoelectrically, thereby changing d and (b) changing λ (in the cavity) by changing the index of refraction n of the medium between the plates by, say, pumping a gas into the cavity.[8] Moving the plates piezoelectrically has the disadvantage that it might upset the alignment of the plates. The index of refraction change method has the disadvantage of being time-consuming.

Some workers who want especially high resolution use spherical mirrors rather than the flat mirrors described above. Spherical mirrors increase the light-gathering power and also allow easier scanning of a spectrum.

There are two important parameters that characterize Fabry–Perot interferometers, the *free spectral range* (FSR) and the *finesse*. The FSR is essentially the frequency spacing between adjacent cavity modes. For a flat-plate interferometer, it is given by

$$\Delta\nu = \frac{c}{2nd}$$

and the finesse F is

$$F = \left(\frac{\Delta\nu}{\delta\nu}\right)$$

where $\delta\nu$ is the full width at half maximum of the instrumental linewidth of the interferometer. That there is any width at all is related to the fact that no mirror is perfectly reflecting or flat. The higher F is, the more clearly the interferometer distinguishes between different cavity modes. In addition, the impinging spectral linewidth should be small compared with the FSR otherwise the interferometer will give overlapping orders. Thus we desire an interferometer with large F and relatively large $\Delta\nu$.

APPENDIX 4.B OPTICAL MIXING EXPERIMENTS

A brief description of some of the important features of optical mixing experiments is given in this appendix. We include discussions of the concepts of "coherence" and "coherence area," their application to the calculation of predetection signal to noise ratios in optical mixing experiments, and a brief discussion of digital (photon counting) methods.

4 · B · 1 COHERENCE AREA

The *degree of coherence* of a light wave is essentially a measure of how close the wave is to a pure (or monochromatic) sine wave. Most waves are of limited duration and hence consist of many-frequency Fourier components. Such waves are not fully coherent. (They may be "partially" coherent.) Noncoherent waves have phase and amplitude fluctuations that are random in space and time (Born and Wolf, 1964; Lipson and Lipson, 1969).

The concept of "coherence area" is central in considerations of the signal-to-noise ratio in optical mixing experiments. When light from an extended source impinges on a screen, a diffraction pattern is produced which depends among other things, on the extent of the source; that is, the intensity maxima and minima depend on the source dimensions. Figure 4.B.1 illustrates this for a one-dimensional case. Assume that each

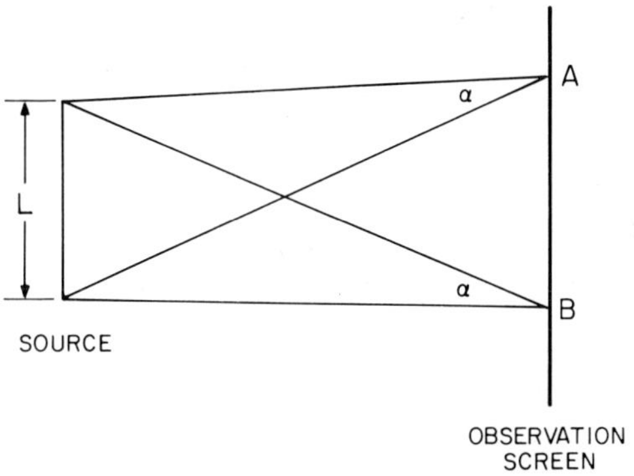

FIG. 4.B.1: A source of dimension L emits light that falls on the observation screen.

point on the source L radiates incoherently, that is, that light emitted from this point has random phase. Thus at point A on the screen the intensity will result from the superposition of waves arising from every point on the source. Consider now point B. If B is very close to A the signal at B would be almost identical to and "coherent" with that at A. In fact, when $A = B$, they are identical. How far apart must A and B be before this coherence disappears? This "coherence distance" is formally defined as the distance over which the spatial correlation function of the electric fields of the signals at A and B is significantly different from zero. In other words, it is the separation distance beyond which this correlation function $\langle E(A)E(B)\rangle$ has appreciably decayed.

A reasonable estimate of the coherence distance for the one-dimensional case shown in Fig. 4.B.1 may be obtained in the following way. First calculate the electric field at a point A on the screen. This field is the sum of those fields emanating from each point of the source. Thus

$$E(A) = \sum_i E(i)$$

where $E(i)$ is the field of the wave arriving at A from point i on the source. The field at B may be calculated in a similar manner,

$$E(B) = \sum_i E'(i)$$

where the field of the signal arriving at B from point i of the source $E'(i)$ differs from that arriving at A from the same point i by a phase factor which depends on the different path lengths traveled by the signal from point i to points A and B. It may then be shown by computing $\langle E(A)E(B)\rangle$ that the distance over which the signals at A and B are correlated (Lipson and Lipson, 1969) is roughly

$$l_c \approx \frac{\lambda}{\alpha} \tag{4.B.1a}$$

where α is the angle of the source subtended at the screen (see Fig. 4.B.1) and λ is the average wavelength of the radiation. If points A and B are separated by more than l_c the signals at A and B will not be appreciably correlated.

If the source were three-dimensional and the observation plane a flat surface such as the cathode of a photomultiplier tube, then one could define an *area* around the point A such that the signals at all points within this area are partially coherent with those at A. This area would be called the *coherence area.*

Estimates of the coherence area for typical light-scattering experiments have been given by several authors (Forrester, et al., 1955; Degiorgio and Lastovka, 1971). A useful estimate is

$$A_{\text{coh}} \approx \frac{\lambda^2}{\Omega} \tag{4.B.1b}$$

where Ω is the solid angle subtended by the source at the detector. Thus the smaller Ω, the larger A_{coh}.

4 · B · 2 PREDETECTION SIGNAL TO NOISE RATIOS IN ANALOG OPTICAL MIXING EXPERIMENTS

In an *"analog"* optical mixing experiment one measures either the time autocorrelation function of the photomultiplier output current or its corresponding spectral density. In the former case, the photomultiplier output is analyzed by an autocorrelator and in the latter, by a spectrum analyzer.

First let us consider what happens when the PM area A consists of

$$N = \frac{A}{A_{\text{coh}}}$$

coherence areas where A_{coh} is the coherence area.

The total PM current can be expressed as

$$i(t) = \sum_{j=1}^{N} i_j(t)$$

where $i_j(t)$ is the current from the jth coherence area due to the scattered light. Furthermore, $i_j(t)$ can be expressed as

$$i_j(t) = \langle i_c \rangle + \delta i_j(t)$$

where $\langle i_c \rangle$ is the average current contribution of a coherence area and $\delta i_j(t)$ is the fluctuation of the current from the jth coherence area. The autocorrelation function of the PM output is

$$\langle i(0)i(t)\rangle = \langle \sum_{j=1}^{N} \sum_{k=1}^{N} [\langle i_c\rangle + \delta i_j(0)][\langle i_c\rangle + \delta i_k(t)]\rangle + \langle i_s(0)i_s(t)\rangle \tag{4.B.2a}$$

where in Eq. (4.B.2.a) we have added a shot-noise term $\langle i_s(0)i_s(t)\rangle$. The shot-noise term can be shown to be (Cummins and Swinney, 1970)

$$\langle i_s(0)i_s(t)\rangle = Ne\,\langle i_c\rangle\,\delta(t) \tag{4.B.2b}$$

where e is the electronic charge. Expanding the term in brackets in Eq. (4.B.2a) and noting that δi_j and δi_k are independent if $j \neq k$ (because by definition different coherence areas are uncorrelated), Eq. (4.B.2a) can be expressed as

$$\langle i(0)i(t)\rangle = N^2\langle i_c\rangle^2 + N\langle \delta i_c(0)\,\delta i_c(t)\rangle + Ne\langle i_c\rangle\,\delta(t) \tag{4.B.2c}$$

where $\delta i_c(t)$ is the current fluctuation due to one typical coherence area.[9]

The term $\langle \delta i_c(0)\delta i_c(t)\rangle$ gives the time-dependence of the fluctuations measured by light scattering. In general, for a homodyne experiment

$$\frac{\langle \delta i_c(0)\delta i_c(t)\rangle}{\langle i_c\rangle^2} = \frac{I_2(t) - |I_1(0)|^2}{|I_1(0)|^2} \tag{4.B.2d}$$

where I_1 and I_2 are defined by Eqs. (4.3.2a) and (4.3.2b). In the Gaussian approximation Eq. (4.B.2d) reduces to

$$\frac{\langle \delta i_c(0)\,\delta i_c(t)\rangle}{\langle i_c\rangle^2} = \frac{|I_1(t)|^2}{|I_1(0)|^2} \tag{4.B.2e}$$

Defining $i_0 = N\langle i_c\rangle$, we can write Eq. (4.B.2c) in the Gaussian approximation as

$$\langle i(0)i(t)\rangle = i_0^2\left[1 + \frac{1}{N}\frac{|I_1(t)|^2}{|I_1(0)|^2}\right] + ei_0\delta(t) \tag{4.B.2f}$$

In a spectrum analyzer, the time Fourier transform $I(\omega)$ of Eq. (4.B.2f) is determined

$$I(\omega) = i_0^2\left[\delta(\omega) + \frac{1}{2\pi N}\int_{-\infty}^{+\infty} dt\,\exp i\omega t\,\frac{|I_1(t)|^2}{|I_1(0)|^2}\right] + \frac{ei_0}{2\pi} \tag{4.B.2g}$$

Thus in spectrum analysis the shot noise contributes a constant (or dc) background $\dfrac{ei_0}{2\pi}$.

The predetection signal-to-noise ratio is the ratio of the coefficients of the integral containing $|I_1(t)|^2/|I_1(0)|^2$ to the shot-noise term, that is,

$$\frac{s}{n} = \frac{i_0}{Ne} = \frac{\langle i_c\rangle}{e} \tag{4.B.2h}$$

Thus for $N > 1$, s/n does not vary with N.

What happpens when the illuminated area of the photocathode is less than one coherence area? In this case the fluctuating part of $\langle i(0)i(t)\rangle$ is proportional to i_0^2 (the square of the total average dc photocurrent). Thus, the signal-to-noise ratio is now [cf. Eq. (4.B.2f)]

$$\frac{s}{n} = \frac{i_0}{e} = \frac{N\langle i_c \rangle}{e} \tag{4.B.2i}$$

Since i_0 increases linearly with the area A (or equivalently with $N = A/A_{coh}$), it may be seen from Eqs. (4.B.2h) and (4.B.2i) that s/n increases with A until $A = A_{coh}$ and then remains constant as A is increased beyond one coherence area. The important quantity in optical mixing experiments is therefore the signal per coherence area. It is advantageous in any given experiment to make the coherence area as large as possible. This may be done by making the scattering volume small, which is accomplished by focusing the laser beam to as small a volume as possible. This makes the solid angle Ω of Eq. (4.B.1b) small, thereby increasing A_{coh}.

Similar considerations apply to heterodyne spectroscopy. The predetection s/n ratio again depends on the signal per coherence area. In this case, however, the s/n also depends on the square of the heterodyning mixing efficiency. The heterodyne mixing efficiency represents the degree to which the scattered field and local oscillator field have matched phase fronts over an area equal to a coherence area (see, e.g., Benedek, 1969). The heterodyne mixing efficiency is unity if the local oscillator and the scattered light have wave fronts which match exactly in phase. If, however, the wave fronts are tilted or distorted relative to one another, the mixing efficiency will fall to a very small value. For strong scattered light signals, small values of the mixing efficiency may be tolerated, but for weak signals accurate matching must be achieved to obtain the high mixing efficiency necessary for signal processing. For the strong signal case, the local oscillator is usually taken to be unshifted laser light scattered from imperfections in the scattering cell.

The above considerations deal only with the predetection signal-to-noise ratio. The measuring device (spectrum analyzer or autocorrelator) will, for instance, introduce further errors. In addition, thermal noise in the electrical circuits has been ignored. In general, multichannel spectrum analyzers and autocorrelators give the best postdetection signal-to-noise ratios.

4 · B · 3 DIGITAL (PHOTOCOUNT) AUTOCORRELATION TECHNIQUES

In analog detection methods (See Section 4.B.2) the photomultiplier output is treated as a continuous variable. At low light-scattering levels, the signal-to-noise ratio in these methods may become small because of various sources of postdetection noise in the system (thermal or Johnson noise, PM dark current, etc.). Under these conditions it becomes advantageous to use the digital or photocount autocorrelation method.

In this technique one counts the number of photomultiplier output pulses in a given time and computes the time-autocorrelation function of these photocounts. Since each pulse, in the ideal case, corresponds to one arriving scattered photon, one measures the time-autocorrelation function of the number of photons arriving at the detector.

Thus if $n(t)$ is the number of photons arriving in the time interval between t and $t + \delta t$, the digital method gives

$$C(t) \equiv \langle n(t)n(0)\rangle \tag{4.B.3a}$$

The photocount rate is related to the instantaneous light intensity $I(R, t)$ which falls upon position R of the PM cathode at time t by the relation

$$n(t) = \sigma \int_A d^2R \int_t^{t+\delta t} dt' \, I(R, t') \tag{4.B.3b}$$

where σ is a constant related to the quantum efficiency of the photocathode and the spatial integration is over the illuminated area of the photocathode A. Although it is not explicitly indicated, $n(t)$ in Eq. (4.B.3b) depends on the sampling interval δt. From Eqs. (4.B.3a, b), the photocount correlation function is given by

$$C(t) = \sigma^2 \int_t^{t+\delta t} dt' \int_t^{t+\delta t} dt'' \int_A d^2R' \int_A d^2R'' \, \langle I(R', t')\, I(R'', t'')\rangle \tag{4.B.3c}$$

The integrations over the photocathode area A in Eq. (4.B.3c) depend on the coherence properties of the scattered light.

In the Gaussian approximation $C(t)$ for a photocount *homodyne*[10] experiment reduces to (Jakeman and Pike, 1969)

$$C(t) = \langle n\rangle^2 \left[1 + f(A)\frac{|I_1(t)|^2}{|I_1(0)|^2}\right] \tag{4.B.3d}$$

where $\langle n\rangle \equiv \sigma(\delta t)\langle I\rangle$ is the average number of photocounts in the time interval δt, $I_1(t)$ is given by Eq. (4.3.2a) and $f(A)$ is a spatial coherence factor which depends on the number of coherence areas viewed and the sampling interval δt.

In practice $f(A)$ is usually determined by a fit to experimental data, whereas $\langle n\rangle$ is usually directly measured. In the case where $I_1(t)$ is a single exponential

$$I_1(t) = a \exp\frac{-t}{\tau} \tag{4.B.3e}$$

the photocount correlation function in the Gaussian approximation (Eq. (4.B.3d)) becomes

$$C(t) = \langle n\rangle^2 \left[1 + f(A) \exp\frac{-2t}{\tau}\right] \tag{4.B.3f}$$

Since $\langle n\rangle$ may be directly measured from the total count rate, only two parameters, $f(A)$ and τ, must be obtained from the experimental time-correlation function points. This is an advantage over the analog homodyne method in which, in the corresponding case, the measured $I_2(t)$ is fit to the form

$$I_2(t) = A + B \exp\frac{-2t}{\tau}$$

where A, B, and τ are parameters. Thus in the analog case the baseline A must in addition be obtained by a fit to the experimental time-correlation function points.

In measurements of the photocount correlation function an amplifier–discriminator system is placed after the PM. This system rejects small signals corresponding to noise in the circuits and amplifies and standardizes the PM pulses. The pulses are then counted and the photocount correlation function computed.

In order to attain fast sampling times and simplify the autocorrelator electronics, some workers do not use the "ideal" or "full" correlation methods described above, but instead use "clipping" and sometimes "scaling" methods.

A clipped correlator measures the correlation function $\langle n_k(0)n(t)\rangle$ where $n_k(t)$ is the clipped signal derived from the original $n(t)$ by

$$n_k(t) = \begin{cases} 1 & \text{when } n(t) > k \\ 0 & \text{when } n(t) \leq k \end{cases}$$

The integer k is called the *clipping level.* Thus instead of performing the multibit multiplication $n(0)n(t)$ and averaging, one has only to "multiply" $n(t)$ by 0 or 1. This "multiplication" may be achieved simply by using a "gate." That is, $n(t)$ is either passed or not passed into a storage register, depending on whether n_k is zero or one. Clipping simplifies the autocorrelator electronics considerably.

In the Gaussian approximation, the clipped autocorrelation function can be written in the form of Eq. (4.B.3d), where $f(A)$ now depends also on the clipping level, and the factor $\langle n\rangle^2$ is replaced by $\langle n_k\rangle\langle n\rangle$ (Jakeman and Pike, 1969). The clipping method has the disadvantage of being restricted to cases in which the Gaussian approximation applies. In cases in which the Gaussian approximation breaks down, (see Section 5.5) the "scaling method is preferable (Koppel and Schaefer, 1973). In the scaling method one measures $\langle n_0^{(s)}(0)n(t)\rangle$, where $n_0^{(s)} = 1$ if the sth count occurs in the given time interval and zero otherwise. The scaling autocorrelation function is simply related to the intensity-autocorrelation function even for non-Gaussian scattered light if there is a sufficiently small count rate.

In conclusion let us note that there is an error due to the finite time average [cf. Eq. (2.4.7)] in the determination of the full autocorrelation functions. If T is the averaging time and τ_c is the correlation tme, then there are effectively $(T/2\tau_c)$ "independent' samplings contributing to the correlation function. The relative error is proportional to the reciprocal of the square root of the number or $\pm(2\tau_c/T)^{1/2}$. A formula often used for an estimate of this error is

$$\Delta_T(t) = \pm\left(\frac{2\tau_c}{T}\right)^{1/2}[1 - C(t)]$$

where $C(t)$ is the normalized correlation function. Thus one must average over many correlation times in order to obtain reasonably small errors. This error is always present in addition to the other noise effects mentioned above.

APPENDIX 4.C THE GAUSSIAN APPROXIMATION

In general, the homodyne and heterodyne experiments yield different information about a system. Yet it is often asserted in the literature that precisely the same information is contained in both of these experiments. This assertion is based on the assumption that the complex property of Eq. (3.3.4) is a Gaussian stochastic variable. What does this mean? Note that Eq. (3.3.4) involves a property of the form

$$\psi(\mathbf{q}, t) = \sum_{j=1}^{N} a_j \exp i\mathbf{q} \cdot \mathbf{r}_j(t),$$

which fluctuates as the particles in the system move around. If we know the joint probability that

$$\psi_0 \leq \psi(\mathbf{q}, 0) \leq \psi_0 + d\psi_0$$
$$\psi \leq \psi(\mathbf{q}, t) \leq \psi + d\psi$$

that is, if we know the joint probability that the fluctuating quantity $\psi(\mathbf{q}, t)$ has a value between ψ_0 and $\psi_0 + d\psi_0$ at time zero and between ψ and $\psi + d\psi$ at time t, then we can determine[11] the correlation functions $I^{(1)}_{if}$ and $I^{(2)}_{if}$ of Eq. (4.3.12),

$$I^{(1)}_{if}(\mathbf{q}, t) = \int d\psi_0 \int d\psi \psi_0 P(\psi_0, \psi_j t)\psi$$

$$I^{(2)}_{if}(\mathbf{q}, t) = \int d\psi_0 \int d\psi |\psi_0|^2 P(\psi_0, \psi_j t)|\psi|^2$$

Here $P(\psi_0, \psi_j t)$ is the joint probability distribution. If $\psi(\mathbf{q}, t)$ is a Gaussian stochastic variable, it can be shown (see, e.g., Wang and Uhlenbeck, 1945) that

$$P(\psi_0, \psi, t) = [2\pi\langle|\psi|^2\rangle]^{-1}[1 - \hat{I}^{(1)}_{if}(\mathbf{q}, t)]^{-1/2} \exp -\psi_0^2/2\langle|\psi|^2\rangle]$$
$$\times \exp\left[\frac{(\psi - \psi_0\hat{I}_1(\mathbf{q}, t))^2}{2\langle|\psi|^2\rangle[1 - \hat{I}_1^2(\mathbf{q}, t)]}\right]$$

where

$$\hat{I}_1(\mathbf{q}, t) \equiv I^{(1)}_{if}(\mathbf{q}, t)/I^{(1)}_{if}(\mathbf{q}, 0)$$

If this expression is substituted into $I^{(2)}_{if}$ and the integration is carried out we obtain Eq. (4.3.2),

$$I^{(2)}_{if}(\mathbf{q}, t) = |I^{(1)}_{if}(\mathbf{q}, 0)|^2 + |I^{(1)}_{if}(\mathbf{q}, t)|^2$$

The assumption that ψ is a Gaussian random variable leads to the prediction that $|I^{(1)}_{if}|$ can be obtained from $I^{(2)}_{if}$. Homodyne detection then gives precisely the same information as heterodyne detection.

This Gaussian assumption is used so frequently in the interpretation of light scattering that it has become a traditional assumption. However it is not generally valid. We show this in the ensuing sections.

NOTES

1. B contains quantities related to the efficiency of the PM tube.
2. Relative deviations from Eq. (4.3.3) will be of order $1/n$ where n is the number of subregions.
3. This is the case even when the Gaussian approximation does not apply. For very long times

$$\langle|E_s(0)|^2|E_s(t)|^2\rangle \rightarrow \langle|E_s|^2\rangle^2$$

4. In some sense a spectrum analyzer is equivalent to the insertion of an electronic filter between the PM and an averager.
5. Sometimes the unscattered light is modulated to change its frequency from the laser frequency.
6. In most applications $I_1(t)$ is a real even function of the time. It is not a real function, for example, for a system of charged particles in an external electric field or for any system in which there is a flow (cf. Section 5.8).
7. The full spectrum follows from Eqs. (4.3.11), (4.3.12a), and (3.2.14) and is

$$I'_{if}(\omega) = B[I_{LO}^2\delta(\omega) + 2AI_{LO}I_{if}^{(1)\prime}(\omega)]$$

where we have omitted the contribution of the PM noise spectrum (see Appendix 4.B).
8. Since $\lambda = c/nv$, a change in n changes λ.
9. It is assumed that all coherence areas are equivalent.
10. The heterodyne case is discussed by Jakeman, 1970.
11. Let $\psi_0 = x_0 + iy_0$ and $\psi = x + iy$ where (x_0, y_0) and (x, y) are, respectively, the real and imaginary parts of ψ_0 and ψ. Then $d\psi = dxdy$ and $d\psi_0 = dx_0dy_0$ are elements of area in the complex ψ and ψ_0 planes.

REFERENCES

Benedek, G. B., in *Polarization; Matiere and Rayonnement, Les* Presses Universitaires de France, Paris (1969).

Born, M. and Wolf, E. *Principles of Optics,* Macmillan, New York (1964). This book contains a detailed discussion of Fabry–Perot interferometers and discussions of the concept of coherence.

Cummins, H. Z. and Pike, E. R. (eds.), *Photon Correlation and Light Beating Spectroscopy,* Plenum, New York (1974). This book contains lecture notes on various aspects of optical mixing techniques, including discussions of precision, noise, the relative merits of digital and analog techniques, and the design of autocorrelators.

Cummins, H. and Swinney, H. L., *Prog. Opt.* **8,** 133 (1970). Optical mixing techniques are reviewed in this article. The reader should particularly note the references to the work of Benedek et al., Pike et al., and Chen.

Degiorgio, V. and Lastovka, J. B., *Phys. Rev.* **A4,** 2033 (1971).

Fleury, P. A. and Boon, J. P., *Adv. Chem. Phys.* **24** (1974). This is a recent review article on light scattering and gives extensive references to work on both filter and optical mixing techniques.

Forrester, A. T., *J. Opt. Soc. Am.* **51,** 253 (1961). See also Forrester, A. T., Gudmandsen, R. A., and Johnson, P. O., *Phys. Rev.* **99,** 1691 (1955).

Jakeman, E. *J. Phys.* **3A,** 201 (1970).

Jakeman, E. and Pike, E. R. *J. Phys.* **A1,** 128 (1968); **2,** 115, 411 (1969).

Koppel, D. E. and Schaefer, D. W., *Appl. Phys. Lett.* **22,** 36 (1973).

Lipson, S. G., and Lipson, H. *Optical Physics,* Cambridge University Press, *London,* Cambridge, 1969.

Wang, M. C. and Uhlenbeck, G. E., *Rev. Mod. Phys.* **17,** 323 (1945).

CHAPTER 5

MODEL SYSTEMS OF SPHERICAL MOLECULES

5 · 1 INTRODUCTION

The light scattered from complicated systems has spectral features that would be difficult to understand if it were not for the fact that certain simple classical model systems exist for which the spectral features can be completely predicted.

How does the light scattered from a system containing anisotropic molecules differ from that scattered from a system containing spherical molecules? What are the effects of collective motion on the spectrum? How can we separate simple single-molecule dynamics from collective motions in condensed media? What are the similarities and differences between homodyne and heterodyne spectroscopy? These are only some of the many questions that we want to answer. The answers to all of these questions can best be given in stages. Only after intuition is built by studying simple models will it be possible for the reader to understand the answers. This chapter is consequently devoted to a study of some simple classical models used frequently in the interpretation of light-scattering spectra.

5 · 2 SPHERICAL MOLECULES

The components of the induced dipole moment in a nonspherical molecule are given by

$$\begin{aligned} \mu_x &= \alpha_{xx}E_x + \alpha_{xy}E_y + \alpha_{xz}E_z \\ \mu_y &= \alpha_{yx}E_x + \alpha_{yy}E_y + \alpha_{yz}E_z \\ \mu_z &= \alpha_{zx}E_x + \alpha_{zy}E_y + \alpha_{zz}E_z \end{aligned} \tag{5.2.1}$$

where $\alpha_{xx}, \ldots, \alpha_{zz}$ are components of the polarizability tensor, $\boldsymbol{\alpha}$, referred to a set of coordinate axes (x, y, z) fixed in the laboratory, and E_x, E_y, E_z are the components of the applied electric field.

The above set of equations can be expressed in matrix form

$$\boldsymbol{\mu} = \boldsymbol{\alpha} \cdot \mathbf{E} \tag{5.2.2}$$

where $\boldsymbol{\mu}$ and $\mathbf{E}$ are the column vectors

$$\boldsymbol{\mu} \equiv \begin{pmatrix} \mu_x \\ \mu_y \\ \mu_z \end{pmatrix} \quad ; \quad \mathbf{E} \equiv \begin{pmatrix} E_x \\ E_y \\ E_z \end{pmatrix} \tag{5.2.3}$$

and $\boldsymbol{\alpha}$ is the matrix

$$\boldsymbol{\alpha} \equiv \begin{pmatrix} \alpha_{xx} & \alpha_{xy} & \alpha_{xz} \\ \alpha_{yx} & \alpha_{yy} & \alpha_{yz} \\ \alpha_{zx} & \alpha_{zy} & \alpha_{zz} \end{pmatrix} \tag{5.2.4}$$

in the laboratory coordinate system. A convenient way to express the above equations is

$$\mu_\alpha = \alpha_{\alpha\beta} E_\beta \tag{5.2.5}$$

where *repeated indices are summed*[1] (β is summed) and α, β run from 1 to 3 where the components x, y, z are labeled 1, 2, 3, respectively.

The simplest case to treat in light scattering is that of spherical molecules. In this case the induced dipole moment is always parallel to the applied electric field so that

$$\boldsymbol{\mu} = \alpha \mathbf{E} \tag{5.2.6}$$

where α is a scalar quantity. Comparison with Eq. (5.2.1) shows that for a spherical molecule $\alpha_{xx} = \alpha_{yy} = \alpha_{zz} = \alpha$, and all off-diagonal elements of the polarizability are zero. In component form this is expressed as

$$\alpha_{\alpha\beta} = \alpha \delta_{\alpha\beta} \tag{5.2.7}$$

where $\delta_{\alpha\beta}$ is the Kronecker delta symbol.[2] In matrix notation this can be expressed as $\boldsymbol{\alpha} = \alpha \boldsymbol{I}$ where $\boldsymbol{I}$ is the unit matrix.

The time-correlation function that occurs in light scattering [cf. Eq. (3.3.3)] involves the molecular polarizability through the quantity [see Eq. (3.3.2)]

$$\alpha_{if} = \mathbf{n}_i \cdot \boldsymbol{\alpha} \cdot \mathbf{n}_f = (\mathbf{n}_i)_\alpha \alpha_{\alpha\beta} (\mathbf{n}_f)_\beta \tag{5.2.8}$$

where repeated indices imply summation. Substitution of Eq. (5.2.7) leads to

$$\alpha_{if} = (\mathbf{n}_i)_\alpha \alpha \delta_{\alpha\beta} (\mathbf{n}_f)_\beta = \alpha (\mathbf{n}_i)_\alpha (\mathbf{n}_f)_\alpha \tag{5.2.9}$$

Because repeated indices are to be summed it follows that

$$(\mathbf{n}_i)_\alpha (\mathbf{n}_f)_\alpha = \sum_{\alpha=1}^{3} (\mathbf{n}_i)_\alpha (\mathbf{n}_f)_\alpha = \mathbf{n}_i \cdot \mathbf{n}_f \tag{5.2.10}$$

so that for spherical molecules

$$\alpha_{if} = (\mathbf{n}_i \cdot \mathbf{n}_f)\, \alpha \tag{5.2.11}$$

It immediately follows from Eqs. (5.2.11) and (3.3.4) that

$$\delta\alpha_{if}(\mathbf{q}, t) = (\mathbf{n}_i \cdot \mathbf{n}_f)\alpha \sum_{j=1}^{N}{}' \exp i[\mathbf{q} \cdot \mathbf{r}_j(t)] \tag{5.2.12}$$

where, as before, the prime denotes that the sum is only over particles in the illuminated volume. Substitution of this into the formulas for homodyne and heterodyne scattering [Eq. (4.3.2)] shows that $I_{if}^{(2)}(t)$ and $I_{if}^{(1)}(t)$ are respectively proportional to[3]

$$F_2(\mathbf{q}, t) = \langle |\psi^*(\mathbf{q}, 0)|^2 |\psi(\mathbf{q}, t)|^2 \rangle \tag{5.2.13}$$

$$F_1(\mathbf{q}, t) = \langle \psi^*(\mathbf{q}, 0)\psi(\mathbf{q}, t) \rangle \tag{5.2.14}$$

where ψ is simply

$$\psi(\mathbf{q}, t) \equiv \sum_{j=1}^{N}{}' \exp i\mathbf{q} \cdot \mathbf{r}_j(t) \tag{5.2.15}$$

It is important to remember that the sum in ψ is only over particles which are in the scattering volume V at time t (cf. Section 3.3). When a particle leaves V it ceases to contribute to the scattering until it re-enters V. Obviously, $\psi(\mathbf{q}, t)$ can also be expressed as

$$\psi(\mathbf{q}, t) = \int_V d^3r \sum_{j=1}^{N} \delta(\mathbf{r} - \mathbf{r}_j(t))\, e^{i\mathbf{q}\cdot\mathbf{r}} \tag{5.2.16}$$

where the sum over particles is now unrestricted; that is, it goes over all particles in the sample cell and not only over those particles in the illuminated volume V of the cell. The volume integral, however, is evaluated over V so that $\psi(\mathbf{q}, t)$ still contains only contribution from particles in V. The sum $\sum_j \delta(\mathbf{r} - \mathbf{r}_j(t))$ is the instantaneous number density $\rho(\mathbf{r}, t)$ at point $\mathbf{r}$ at time t (cf. Section 2.3). This can be written as $\rho(\mathbf{r}, t) = \rho_0 + \delta\rho(\mathbf{r}, t)$, where ρ_0 is the average number density and $\delta\rho(\mathbf{r}, t)$ is the number density fluctuation at $(\mathbf{r}, t)$. Substitution of this into Eq. (5.2.16) then gives

$$\psi(\mathbf{q}, t) = \int_V d^3r\, \delta\rho(\mathbf{r}, t)\, e^{i\mathbf{q}\cdot\mathbf{r}} = \delta\rho(\mathbf{q}, t) \tag{5.2.17}$$

where the term due to ρ_0 is zero[4] for $q \neq 0$. Thus $\psi(\mathbf{q}, t)$ is equal to a Fourier transform of the number density fluctuation; that is, $\delta\rho(\mathbf{q}, t)$.

The light-scattering spectrum which is related to $I_{if}^{\alpha}(\mathbf{q}, t)$ by Eq. (3.3.3) consequently probes how a density fluctuation $\delta\rho(\mathbf{q})$ spontaneously arises and decays due to the thermal motion of the molecules. Density disturbances in macroscopic systems can propagate in the form of sound waves. It follows that light scattering in pure fluids and mixtures will eventually require the use of thermodynamic and hydrodynamic models. In this chapter we do not deal with these complicated theories (see Chapters 9–13); but rather with the simplest possible systems that do not require these theories. Examples of such systems are dilute macromolecular solutions, ideal gases, and bacterial dispersions.

In Eq. (5.2.15) it is very important to bear in mind that the sum is only over those molecules which are in the illuminated volume V at times zero and t. In order to make this fact explicit we define a quantity having the following properties

$$b_j(t) = \begin{cases} 1 & j \in V \\ 0 & j \notin V \end{cases} \tag{5.2.18}$$

where $b_j(t)$ tells us whether or not particle j is in V at time t. The sum of $b_j(t)$ over all particles in the total system N then gives

$$N(t) = \sum_{j=1}^{N} b_j(t) \tag{5.2.17}$$

where N(t) is the number of particles in V at time t. Now Eq. (5.2.15) can be written

$$\psi(\mathbf{q}, t) = \sum_{j=1}^{N} b_j(t) \exp i\mathbf{q} \cdot \mathbf{r}_j(t) \tag{5.2.20}$$

where the sum goes over not only the particles in the illuminated region but over all the molecules in the cell.

Corresponding to the three different polarization directions in the geometries described in Chapter 3, we find for spherical molecules that

$$\begin{aligned} I_{VV}(\mathbf{q}, t) &= \alpha^2 F_1(\mathbf{q}, t) \\ I_{VH}(\mathbf{q}, t) &= I_{HV}(\mathbf{q}, t) = 0 \\ I_{HH}(\mathbf{q}, t) &= \cos^2\theta\, I_{VV}(\mathbf{q}, t) \end{aligned} \tag{5.2.21}$$

Consequently, the light scattered from spherical molecules is not expected to be depolarized.[5] Nevertheless as we see in Chapter 14, even inert gas atoms depolarize the light (i.e., $I_{VH} \neq 0$); this arises from the anisotropy induced by collisions, an effect which is discussed in Section 10.1 and Chapter 14.

5 · 3 DILUTE SOLUTIONS AND PARTICLE INDEPENDENCE

It is typical of solutions of macromolecules that: (a) the polarizability of a macromolecule is enormous by comparison to the polarizability of a solvent molecule and (b) macromolecules move much more slowly than solvent molecules. From (a) it can be concluded that the macromolecules will be far more efficient scatterers of light than individual solvent molecules.[6] Furthermore, according to (b) the macromolecules will contribute a slowly fluctuating field at the detector compared to the solvent so that the macromolecular motion should be temporally separable from the solvent motion.

Because of these considerations, the macromolecules dominate the long-time behavior of F_1 and F_2, and it is only necessary[7] to sum over these molecules in Eq. (5.2.20) so that $\psi(\mathbf{q}, t)$ becomes

$$\psi(\mathbf{q}, t) = \sum_{j=1}^{N} b_j(t) \exp i\mathbf{q} \cdot \mathbf{r}_j(t)$$

where the sum only goes over the N macromolecules in the system and $\mathbf{r}_j(t)$ is the center-of-mass position of macromolecule j at time t.

In sufficiently dilute solutions, the macromolecules so rarely encounter each other that we can assume their positions to be statistically independent.[8] In this eventuality Eq. (5.2.14) simplifies to

$$F_1(\mathbf{q}, t) = \langle \sum_{j=1}^{N} b_j(0)\, b_j(t) \exp i\mathbf{q} \cdot [\mathbf{r}_j(t) - \mathbf{r}_j(0)] \rangle \tag{5.3.1}$$

This is an example of a *self-correlation* function where only properties of the same molecule are correlated. These considerations also apply to rarefied gases since their collisions are so infrequent that particles behave independently.

5 · 4 HETERODYNE CORRELATION FUNCTION FOR PARTICLE DIFFUSION

The heterodyne correlation function [cf. Eq. (4.3.12)] for dilute *macromolecular solutions* is proportional to Eq. (5.3.1). To understand the time scales characterizing the decay of this function it is necessary to consider each of the factors in $F_1(\mathbf{q}, t)$.

The only particles that contribute to $F_1(\mathbf{q}, t)$ are those which are initially in V, that is, those for which $b_j(0) = 1$. The product $b_j(0)b_j(t)$ is initially 1, and jumps from 1 to 0 when particle j leaves V. Thus the time scale for the variation of $b_j(0)b_j(t)$ is simply given by the time it takes macromolecule j to move the distance L where L is a characteristic dimension of the scattering volume (typically smaller than 1 mm). A particle diffuses a distance L on the time scale

$$\tau = L^2/D \tag{5.4.1a}$$

where D is the diffusion coefficient of the particle,[9] so that τ is the characteristic time of $b_j(0)b_j(t)$. The quantity $\exp i\mathbf{q} \cdot [\mathbf{r}_j(t) - \mathbf{r}_j(0)]$ deviates considerably from 1 only for times such that the displacement $\mathbf{r}_j(t) - \mathbf{r}_j(0)$ becomes comparable to the length q^{-1}. For a diffusing particle this time scale[9] is

$$\tau_q = (q^2 D)^{-1} \tag{5.4.1b}$$

Comparing these two time scales we see that

$$\frac{\tau}{\tau_q} = (qL)^2 \tag{5.4.1c}$$

For typical light-scattering experiments $q \sim 10^5\ \text{cm}^{-1}$ and $L \sim .01$ cm so that $\tau/\tau_q \sim 10^6$. This means that $b_j(0)b_j(t)$ varies on a much slower time scale than $\exp i\mathbf{q} \cdot [\mathbf{r}_j(\mathbf{t}) - \mathbf{r}_j(0)]$. It is therefore permissible to set $b_j(0)b_j(t)$ equal to its initial value $b_j(0)b_j(0)$. Since $b_j(0)$ can have only one of the two values 0, or 1, $b_j^2(0) = b_j(0)$, and

$$F_1(\mathbf{q}, t) = \langle \sum_{j} b_j(0) \exp i\mathbf{q} \cdot [\mathbf{r}_j(t) - \mathbf{r}_j(0)] \rangle$$

The quantity $\exp i\mathbf{q} \cdot [\mathbf{r}_j(t) - \mathbf{r}_j(0)]$ depends on the displacement $[\mathbf{r}_j(t) - \mathbf{r}_j(0)]$ and is expected to have the same statistical properties regardless of whether particle j is in the scattering volume at $t = 0$. Consequently, we expect this quantity to be statistically independent of $b_j(0)$, so that

$$F_1(\mathbf{q}, t) = \sum_j \langle b_j(0)\rangle \langle \exp i\mathbf{q} \cdot [\mathbf{r}_j(t) - \mathbf{r}_j(0)]\rangle$$

The quantity

$$F_s(\mathbf{q}, t) \equiv \langle \exp i\mathbf{q} \cdot [\mathbf{r}_j(t) - \mathbf{r}_j(0)]\rangle \tag{5.4.2}$$

should be identical for each particle j because it represents an ensemble average. $F_s(\mathbf{q}, t)$ can therefore be factored out of the above sum. Moreover $\langle \sum_j b_j(0)\rangle$ is simply $\langle N\rangle$, the average number of particles in V[cf. Eq. (5.2.17)] so that $F_1(\mathbf{q}, t)$ becomes

$$F_1(\mathbf{q}, t) = \langle N\rangle F_s(\mathbf{q}, t) \tag{5.4.3}$$

The quantity $F_s(\mathbf{q}, t)$ defined in Eq. (5.4.2) appears so frequently that it has been given the name *self-intermediate scattering function.* It is related to the probability distribution $G_s(\mathbf{R}, t)$ for a particle to suffer a displacement $\mathbf{R}$ in the time t

$$G_s(\mathbf{R}, t) = \langle \delta(\mathbf{R} - [\mathbf{r}_j(t) - \mathbf{r}_j(0)])\rangle \tag{5.4.4}$$

That this quantity is the distribution function for particle displacements follows from the properties of the delta function. The delta function has the property that those members of the ensemble for which the displacement

$$\Delta\mathbf{r}_j(t) \equiv \mathbf{r}_j(t) - \mathbf{r}_j(0)$$

is in the neighborhood of the point $\mathbf{R}$ are assigned a value of 1, whereas those for which $\Delta\mathbf{r}_j(t)$ is not in this neighborhood of R contribute zero. Therefore $G_s(\mathbf{R}, t)d^3R$ can be regarded as the probability that particle j will suffer a displacement in the neighborhood d^3R of the point R in time t. Particle j is not unique. Any macromolecule could have been chosen, and the same $G_s(R, t)$ would result.

$G_s(\mathbf{R}, t)$ appears so frequently that it has been given the name *Van-Hove self space-time correlation function* after Leon Van-Hove, who first demonstrated its relationship to neutron scattering (Van Hove, 1954). It was asserted above that $G_s(\mathbf{R}, t)$ is related to $F_s(\mathbf{q}, t)$. What is this relationship?

The spatial Fourier transform of $G_s(\mathbf{R}, t)$ is

$$\int d^3R e^{i\mathbf{q}\cdot\mathbf{R}} \langle \delta(\mathbf{R} - [\mathbf{r}_j(t) - \mathbf{r}_j(0)])\rangle$$

Commuting the integral with the ensemble average and exploiting the property of the delta function gives

$$\langle \exp i\mathbf{q} \cdot [\mathbf{r}_j(t) - \mathbf{r}_j(0)]\rangle$$

which is simply the intermediate scattering function. We conclude that

$$F_s(\mathbf{q}, t) = \int d^3R e^{i\mathbf{q}\cdot\mathbf{R}} G_s(\mathbf{R}, t) \tag{5.4.5}$$

In probability theory the Fourier transform of a probability distribution function is called the *characteristic function* of the distribution. Thus $F_s(\mathbf{q}, t)$ is the characteristic function of $G_s(\mathbf{R}, t)$. $F_s(\mathbf{q}, t)$ can be determined from light scattering since, $G_s(\mathbf{R}, t)$ can be determined by an inverse Fourier transform,

$$G_s(\mathbf{R}, t) = (2\pi)^{-3} \int d^3q\, e^{-i\mathbf{q}\cdot\mathbf{R}}\, F_s(\mathbf{q}, t) \tag{5.4.6}$$

For this purpose it would be necessary to determine $F_s(\mathbf{q}, t)$ as a function of $\mathbf{q}$ for all $\mathbf{q}$. However, because only a limited range of q can be probed, great caution must be exercised in the interpretation of such results.

In order to proceed it is necessary to determine either $G_s(\mathbf{R}, t)$ or equivalently $F_s(\mathbf{q}, t)$ for the dilute macromolecular solution. This requires a model. The first model we discuss is based on simple diffusion theory.

From its definition and the fact that $\Delta r_j(0) = 0$,

$$G_s(\mathbf{R},)0 = \langle\delta(\mathbf{R})\rangle = \delta(\mathbf{R}) \tag{5.4.7}$$

and

$$F_s(\mathbf{q}, 0) = \langle\exp i\mathbf{q} \cdot [\mathbf{r}_j(0) - \mathbf{r}_j(0)]\rangle = 1 \tag{5.4.8}$$

Consequently $G_s(\mathbf{R}, t)d^3R$ can be regarded as the probability of finding a particle in the neighborhood d^3R of the point $\mathbf{R}$ at time t given that initially ($t = 0$) the particle was in the neighborhood of the origin. Suppose we prepare a solution such that at time $t = 0$ a macromolecule is in the neighborhood of the origin. As time progresses it is expected that this macromolecule will execute small excursions—perform a random walk—so that after a time t this particle will "diffuse" into the neighborhood of the point $\mathbf{R}$ with probability $G_s(\mathbf{R}, t)d^3R$. It is well known from the theory of the random walk that the diffusion equation [for times long compared to the velocity correlation time, see Section (5.9)] describes this probability. Then $G_s(\mathbf{R}, t)$ can, to a very good approximation, be regarded as the solution to the diffusion equation[10]

$$\frac{\partial}{\partial t} G_s(\mathbf{R}, t) = D\nabla^2 G_s(\mathbf{R}, t) \tag{5.4.9}$$

subject to the initial condition [Eq. (5.4.7)] where D is the coefficient of self-diffusion.

The spatial Fourier transform of Eq. (5.4.9) is

$$\frac{\partial}{\partial t} F_s(\mathbf{q}, t) = -q^2 D F_s(\mathbf{q}, t) \tag{5.4.10}$$

The solution of this equation subject to the boundary condition $F_s(\mathbf{q}, 0) = 1$ of Eq. (5.4.8) is

$$F_s(\mathbf{q}, t) = \exp(-q^2 D t) = \exp\frac{-t}{\tau_q} \tag{5.4.11}$$

where $\tau_q = (q^2 D)^{-1}$ turns out to be the relaxation time defined in Eq. (5.4.1b). Thus we conclude that the heterodyne correlation function is an exponentially decaying function of time with time constant τ_q.

According to the Einstein relation the self-diffusion coefficient is

$$D = \frac{k_B T}{\zeta} \tag{5.4.12}$$

where the friction constant ζ is in the Stokes approximation (for stick-boundary conditions)

$$\zeta = 6\pi\eta a \tag{5.4.13}$$

where η is the viscosity of the solvent and a is the radius of the spherical macromolecule (e.g., see Landau and Lifshitz, 1959). Since water has a viscosity $\eta \sim 0.01$ poise, the coefficient of self-diffusion in aqueous solution at room temperature is $D \sim 2 \times 10^{-13} a^{-1}$. The wave number q in light-scattering experiments is given by Eq. (3.2.5). For visible light and a scattering angle $\theta \sim 90°$, $q \sim 10^5 \text{cm}^{-1}$. In typical light-scattering experiments it follows that $q^2 D \sim 10^{-3} a^{-1}$ and that the correlation time τ_q of $F_s(\mathbf{q}, t)$ is of order $\tau_q \sim a \times 10^3$ so that $F_s(\mathbf{q}, t)$ decays more slowly the larger the macromolecule. For spheres of radius .01 μm, $\tau_q \sim 10^{-3}$sec.

The heterodyne correlation function and its corresponding spectrum[11] follows from Eq. (5.4.3)

$$\begin{aligned} F_1(\mathbf{q}, t) &= \langle N \rangle \exp -q^2 D|t| \\ F_1(\mathbf{q}, \omega) &= \pi^{-1} \langle N \rangle \left[\frac{q^2 D}{\omega^2 + [q^2 D]^2}\right] \end{aligned} \tag{5.4.14}$$

Heterodyne spectroscopy can be used to measure the diffusion coefficient of a macromolecule and thereby the radius of thc particle. This is done by either measuring the time constant τ_q in $F_1(\mathbf{q}, t)$ or the half-width at half maximum ω_q of the spectrum where

$$\begin{aligned} \tau_q &= (q^2 D)^{-1} \\ \omega_q &= q^2 D \end{aligned} \tag{5.4.15}$$

and plotting τ_q^{-1} or ω_q against $q^2 = 4k_i^2 \sin^2 \theta/2$. The slope of this plot gives the coefficient of self-diffusion.

On the basis of Eq. (4.3.3) we can predict the homodyne correlation function for diffusion in the Gaussian approximation,

$$F_2(\mathbf{q}, t) = \langle N \rangle^2 + \langle N \rangle^2 \exp - 2q^2 D t \tag{5.4.16}$$

and its associated spectrum[12]

$$F_2(\mathbf{q}, \omega) = \langle N \rangle^2 \delta(\omega) + \pi^{-1} \langle N \rangle^2 \left[\frac{2q^2 D}{\omega^2 + [2q^2 D]^2}\right] \tag{5.4.17}$$

Newman et al. (1974) applied this method to the study of a highly monodisperse sample of single-stranded circular DNA from the *fd* bacteriophage. By performing light scattering and sedimentation experiments, they determined the molecular weight of this molecule. An example of their homodyne correlation function is given in Fig. 5.4.1. These data can be fit to an exponential function.

Prior to the development of light scattering techniques, diffusion coefficients were determined from optical studies of concentration gradients. The most widely used

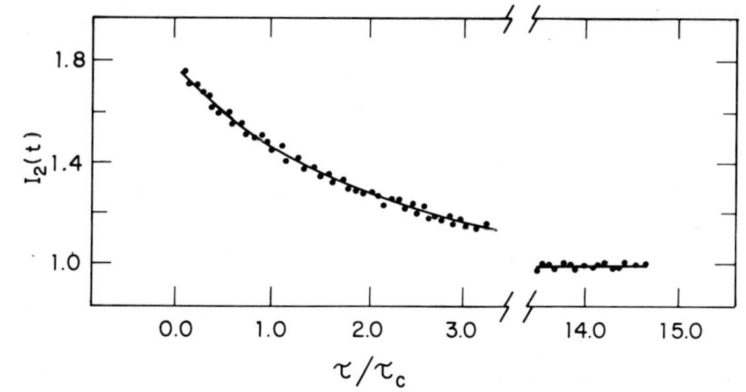

FIG. 5.4.1. Homodyne correlation function $I_2(t)$ obtained at a scattering angle of 60° for a solution containing 0.17 mg/cm³ of *fd* DNA in SCC (0.15 m NaCl, 0.015M Nacitrate, pH = 8) as a function of τ/τ_c where $\tau_c = (q^2 D)^{-1}$ is the correlation time. (From Newman and Swinney, 1974.)

method employed Schlieren optical systems (Van Holde, 1971). Optical mixing techniques give diffusion coefficients more rapidly (in a matter of minutes) and easily than these other methods. When combined with sedimentation data, the diffusion data provide accurate molecular weights.

In sedimentation experiments the sedimentation velocity u divided by the centrifugal field strength $(\omega^2 r)$, is called the *sedimentation coefficient s*. Here ω is the angular speed of the centrifuge rotor and r is the distance of the sample from the axis of rotation. The velocity of sedimentation is found as follows. There are three forces on a sedimenting particle of mass m: (a) a centrifugal force $m\omega^2 r$, (b) a buoyancy force $-m_0\omega^2 r$, where m_0 is the mass of the displaced fluid, and (c) a frictional force $-\zeta u$. Balancing these three forces gives

$$m\omega^2 r - m_0\omega^2 r - \zeta u = 0$$

The mass of the displaced fluid m_0 is the product of the polymer mass, its *specific* volume $\bar{v}$ and the fluid density ρ, so that $m_0 = m\bar{v}\rho$. Thus the sedimentation coefficient s, which is measured in sedimentation velocity experiments, depends on m through

$$s = \frac{m(1 - \bar{v}\rho)}{\zeta} \tag{5.4.18}$$

Combining this with Eq. (5.4.12) for the diffusion coefficient gives

$$\frac{s}{D} = \beta m(1 - \bar{v}\rho) \tag{5.4.19}$$

where $\beta = (k_B T)^{-1}$. The polymer mass and thereby its molecular weight can be determined if the specific volume of the polymer is known. This last property, while not easy to determine precisely, is often known with adequate accuracy for many polymers. Nevertheless this combination of light scattering and sedimentation data is very useful in determining polymer molecular weights (Dubin et al., 1970; Foord et al., 1972).

Measurements of macromolecular diffusion coefficients are now common. Many groups are engaged in using these measurements to study problems of chemical and biological interest (e.g., see the review of Berne and Pecora, 1974). We mention here some of the more novel recent studies.

Koppel (1973) has combined the light-scattering method with zonal sedimentation in a sucrose gradient to study the ribosomes of *Escherichia coli*. During centrifugation in a sucrose gradient, the macromolecules separate according to their sedimentation properties into physically distinct bands. The light-scattering analysis is then carried out immediately after centrifugation directly on the different bands and thus on different macromolecular components. By using the sedimentation coefficient, the tabulated partial specific volumes and the diffusion coefficient in Eq. (5.4.19), Koppel obtained the molecular weight of the *E. coli* 70*s* ribosomes and the 50*s* and 30*s* ribosome subunits. Thus in a single experiment Koppel was able to determine the diffusion coefficients, sedimentation coefficients, molecular weights, and relative concentrations of each of the components of a complex system.

Other authors have used the light-scattering method to study aggregation (see Berne and Pecora, 1974 for references). Since the diffusion coefficient of a molecular aggregate is smaller than that of the monomer and the aggregate should in fact contribute more to the total time correlation function because of its higher molecular weight, we expect light scattering to be very sensitive to small amounts of aggregate in a system. Wilson et al. (1974) used this fact to detect the onset of aggregation of hemoglobin *S* molecules. The erythrocyte sickling phenomenon of sickle cell anemia is caused by this aggregation.

Most macromolecules when dissolved in salt solutions acquire charges that are shielded by an atmosphere of counterions. This ion atmosphere affects the diffusion coefficient of the macromolecule and hence the light-scattering time-correlation function. Electrolyte solutions are discussed in Chapters 9 and 13. Recent measurements of diffusion coefficients have been made by several groups. Lee and Schurr (1974) have studied poly-*L*-lysine-HBr. Schleich and Yeh (1973) have performed similar studies on poly-*L*-proline. Raj and Flygare (1974) have studied bovine serum albumin (BSA) and find that at high ionic strength and low pH the diffusion constant decreases. This they attribute to the expansion of the molecule.

The treatment of diffusion given in this section is valid only for the analysis of solutions in the limit of inifinite dilution. We return to the question of diffusion in several sections of this book. In Section 9.2 a simple theory of diffusion in electrolyte solutions is discussed. In Section 10.6 the coupling between diffusion and heat conduction is treated in some detail. In section 11.6 a microscopic description of diffusion is given. Finally in Sections 13.5 and 13.6 a detailed treatment of diffusion in binary and ternary solutions of nonelectrolytes and electrolytes is presented. The concentration-dependence of the diffusion coefficient is considered in Section 13.5. These sections are based on the theory of nonequilibrium thermodynamics and are thus relegated to the chapter on this subject. Particular attention should be given to these sections by any reader interested in the analysis of diffusion.

5 · 5 HOMODYNE SPECTRUM FOR VERY DILUTE SOLUTIONS

The homodyne correlation function for a dilute solution of diffusing macromolecules follows directly from Eqs. (5.2.20) and (5.2.13),

$$F_2(\mathbf{q}, t) = \langle \sum_{j,k,l,m=1}^{N} b_j(0)b_k(0)b_l(t)b_m(t) \exp i\mathbf{q} \cdot [\mathbf{r}_k(0) - \mathbf{r}_j(0) + \mathbf{r}_l(t) - \mathbf{r}_m(t)]\rangle \tag{5.5.1}$$

These terms simplify considerably for dilute solutions when the positions of different molecules are statistically independent. Then the term

$$\langle b_j(0)b_k(0)b_l(t)b_m(t) \exp i\mathbf{q} \cdot [\mathbf{r}_l(t) - \mathbf{r}_j(0)] \exp -i\mathbf{q} \cdot [\mathbf{r}_m(t) - r_k(0)]\rangle$$

will be zero if any of the four particle indices are distinct, as we now show. Suppose particle l is distinct. The term factorizes due to the assumed particle independence, into

$$\langle b_l(t) \exp i\mathbf{q} \cdot \mathbf{r}_l(t)\rangle\langle b_j(0)b_k(0)\mathrm{b}_m(t) \exp -i\mathbf{q} \cdot \mathbf{r}_j(0) \exp -i\mathbf{q} \cdot [\mathbf{r}_m(t) - \mathbf{r}_k(0)]\rangle$$

The first factor is simply the ensemble average of the quantity $\exp(i\mathbf{q} \cdot \mathbf{r}_l)$. If the system is homogeneous the particles are distributed randomly so that the probability of finding particle l in the neighborhood d^3r is d^3rV^{-1} where V is the illuminated volume of the sample (not the volume of the scattering cell). This is

$$\langle \exp i\mathbf{q} \cdot \mathbf{r}_l(t)\rangle = V^{-1} \int_V d^3r e^{i\mathbf{q}\cdot\mathbf{r}} \rightarrow \delta(\mathbf{q}) \tag{5.5.2}$$

For scattering in other than the forward direction ($\mathrm{q} \neq 0$), this quantity is zero. Consequently only two kinds of terms survive; those for which (a) $j = k,\ l = m$ and (b) $j = l,\ m = k,\ j \neq m$. In the former case we obtain the term

$$\langle b_j^2(0)\ b_l^2(t)\rangle$$

whereas in the latter case we obtain

$$\langle [b_j(0)b_j(t) \exp i\mathbf{q} \cdot \Delta\mathbf{r}_j(t)]\ [b_k(0)b_k(t) \exp -i\mathbf{q} \cdot \Delta r_k(t)]\rangle$$

where $j \neq k$ and $\Delta\mathbf{r}_j(t) \equiv \mathbf{r}_j(t) - \mathbf{r}_j(0)$. Two points are in order. Because particles j and k are statistically independent, the above product factorizes. Furthermore because $\exp i\mathbf{q} \cdot \Delta\mathbf{r}_j(t)$ fluctuates much faster than any correlation function involving the b_j's, it is permissible to replace the correlation function of the b_j's by its initial value whenever it is a factor in a product involving $\exp i\mathbf{q} \cdot \Delta\mathbf{r}_j(t)$. It follows that the above term reduces to (cf. Section 5.4)

$$\langle b_j^2(0) \exp i\mathbf{q} \cdot \Delta\mathbf{r}_j(t)\rangle\ \langle b_k^2(0) \exp -i\mathbf{q} \cdot \Delta\mathbf{r}_k(t)\rangle$$

As before $b_j^2(0) = b_j(0)$ and $b_k^2(0) = b_k(0)$ and moreover $\exp i\mathbf{q} \cdot \Delta r_j$ is statistically independent of $b_j(0)$ so that the above term simplifies further to

$$\langle b_j(0)\ b_k(0)\rangle |F_s(\mathbf{q}, t)|^2$$

It follows from these considerations that for dilute polymer solutions

$$F_2(\mathbf{q}, t) = \langle \sum_{j,l=1}^{N} b_j^2(0)b_l^2(t)\rangle + \langle \sum_{j\neq k=1}^{N} b_j(0)b_k(0)\rangle |F_s(\mathbf{q}, t)|^2 \tag{5.5.3}$$

Because $b_l^2(t) = b_l(t)$, and because $\sum_l b_l(t) = N(t)$,

$$\left\langle \sum_{j,l=1}^{N} b_j^2(0) b_l^2(t) \right\rangle = \left\langle \sum_{j,l=1}^{N} b_j(0) b_l(t) \right\rangle = \langle N(0)\, N(t) \rangle \tag{5.5.4a}$$

Similarly,

$$\sum_{j \neq k=1}^{N} \langle b_j(0)\, b_k(0) \rangle = \langle N(N-1) \rangle \tag{5.5.4b}$$

Combining these results, we obtain

$$F_2(\mathbf{q}, t) = \langle N(0) N(t) \rangle + \langle N(N-1) \rangle |F_s(\mathbf{q}, t)|^2 \tag{5.5.5}$$

The number of particles in the illuminated region can be expressed as

$$N(t) = \langle N \rangle + \delta N(t)$$

where $\delta N(t)$ is the deviation of the number of particles from the average number. Because $\langle \delta N(t) \rangle = 0$,

$$\langle N(0)\, N(t) \rangle = \langle N \rangle^2 + \langle \delta N(0)\, \delta N(t) \rangle \tag{5.5.6}$$

The probability P_N that N noninteracting particles will be in the illuminated region V at any instant is the Poisson distribution[13]

$$P_N = \frac{\langle N \rangle^N}{N!} \exp - \langle N \rangle$$

This distribution function has the following moments:

$$\begin{aligned} &\text{(a)} \quad \langle N^2 \rangle = \langle N \rangle^2 + \langle N \rangle \\ &\text{(b)} \quad \langle \delta N^2 \rangle = \langle [N - \langle N \rangle]^2 \rangle = \langle N^2 \rangle - \langle N \rangle^2 = \langle N \rangle \\ &\text{(c)} \quad \langle N(N-1) \rangle = \langle N^2 \rangle - \langle N \rangle = \langle N \rangle^2 \end{aligned} \tag{5.5.7}$$

Using these relations Eq. (5.5.5) becomes (Schaefer and Berne, 1972)

$$F_2(\mathbf{q}, t) = \underbrace{\langle N \rangle^2 \{1 + |F_s(\mathbf{q}, t)|^2\}}_{\text{Gaussian approximation}} + \underbrace{\langle \delta N(0)\, \delta N(t) \rangle}_{\text{extra term}} \tag{5.5.8}$$

That the Gaussian approximation gives the indicated term follows directly from a substitution of Eq. (5.4.2) into Eq. (4.3.3). It should be observed that Eq. (5.5.8) contains an extra term $\langle \delta N(0)\, \delta N(t) \rangle$. This term can be regarded as a deviation from the Gaussian approximation. It explicitly depends on fluctuations, $\delta N(t)$, in the number of particles in the scattering volume.[14] These fluctuations occur on a time scale τ [cf. Eq. (5.4.1a)] which characterizes the time required for a particle to traverse the scattering volume. The Gaussian term decays on a time scale τ_q, which characterizes the time required for the particle to traverse the distance q^{-1}. As we have seen, $\tau_q \ll \tau$ so that $F_2(\mathbf{q}, t)$ decays in two stages. First the Gaussian term decays from its initial value of 2 $\langle N\rangle^2$ to its relaxed value $\langle N\rangle^2$ during which time $\langle\delta N(0)\ \delta N(t)\rangle$ remains equal to its initial value $\langle\delta N^2\rangle = \langle N\rangle$.

Then for times of order τ, $\langle\delta N(0)\ \delta N(t)\rangle$ decays from its initial value $\langle N\rangle$ to zero. This is summarized below

$$F_2(\mathbf{q}, t) = \begin{cases} 2\langle N\rangle^2 + \langle N\rangle & t = 0 \\ \langle N\rangle^2 + \langle N\rangle & \tau_q \ll t \ll \tau \\ \langle N\rangle^2 & t \gg \tau \end{cases} \tag{5.5.9}$$

The correction to the Gaussian approximation is clearly of order $\langle \mathrm{N}\rangle^{-1}$. Thus for sufficiently concentrated solutions this correction can be ignored.

In the diffusion approximation

$$F_2(\mathbf{q}, t) = \langle N\rangle^2\,[1 + \exp(-2q^2Dt)] + \langle\delta N(0)\ \delta N(t)\rangle \tag{5.5.10}$$

where $\langle\delta N(0)\ \delta N(t)\rangle$ is related to the *probability after-effect function* defined by Chandrasekar (1943). The first term in Eq. (5.5.10) has already been derived in Section. 5.4. [cf. Eq. (5.4.16)]. The evaluation of $\langle\delta N(0)\ \delta N(t)\rangle$ requires integrations over the scattering volume. This function is discussed in Section 5.7.

We observe from Eqs. (5.5.8) and (5.5.10) that light-scattering experiments on dilute solutions can yield, in principle, important information concerning $\langle\delta N(0)\ \delta N(t)\rangle$ in addition to $F_s(\mathbf{q}, t)$. Moreover we conclude from Eq. (5.5.9) the experimentally relevant fact that the apparent background for time $t \ll \tau$ in homodyne experiments is not the hitherto expected background $\langle N\rangle^2$ but contains an additional factor $\langle N\rangle$. This apparent background is of major importance in studies of solutions sufficiently dilute that $\langle N\rangle \sim \langle N\rangle^2$. It decays to the true background $\langle N\rangle^2$ in a time t long compared to τ_q.

In many biological applications of light scattering the solutions are sufficiently dilute to satisfy the assumption of particle independence but still sufficiently concentrated that $\langle\delta N(0)\ \delta N(t)\rangle$ can be neglected compared with $\langle N\rangle^2$. The Gaussian approximation is then valid. In most of this book we deal with this type of solution, and consequently automatically assume the validity of the Gaussian approximation unless otherwise stated. It is then necessary to present only F_1 since F_2 can be computed directly from F_1.

5 · 6 DILUTE GASES

A dilute gas contains molecules which only seldomly collide. Thus at sufficiently low pressures the gas molecules may be regarded as independent. A typical gas molecule moves on a linear trajectory with velocity $\mathbf{V}$ between collisions with its neighbors. The average length of a linear trajectory is called the *mean free path*, Λ. All that is required to predict the light-scattering spectrum is the function $F_s(\mathbf{q}, t)$ [cf. Eq. (5.4.2)].

The particle j must move a distance $\Delta\mathbf{r}_j$ of the order of q^{-1} to cause $F_s(\mathbf{q}, t)$ to decay. If the distance q^{-1} is small compared to the mean free path Λ ($q\Lambda \gg 1$), then particle j will move the required distance q^{-1} along a linear trajectory with velocity $\mathbf{V}_j$. The displacement of the particle j is then

$$\mathbf{r}_j(t) - \mathbf{r}_j(0) = \mathbf{V}_j t \tag{5.6.1}$$

and $F_s(\mathbf{q}, t)$ becomes

$$F_s(q, t) = \langle \exp i\mathbf{q} \cdot [\mathbf{r}_j(t) - \mathbf{r}_j(0)] \rangle = \langle \exp i\mathbf{q} \cdot \mathbf{V}_j t \rangle \tag{5.6.2}$$

The molecular velocities are distributed according to the Maxwell distribution function

$$P(\mathbf{V}) = \left[\frac{M}{2\pi k_B T}\right]^{3/2} \exp - \frac{MV^2}{2k_B T} \tag{5.6.3}$$

It follows from Eq. (5.6.2), where the bracket indicates an average over the Maxwell distribution, that

$$F_s(\mathbf{q}, t) = \left[\frac{M}{2\pi k_B T}\right]^{3/2} \int d^3V \exp - \frac{MV^2}{2k_B T} \exp i\mathbf{q} \cdot \mathbf{V}t \tag{5.6.4}$$

This integral simplifies considerably if we choose coordinate axes such that the z axis is along the direction of $\mathbf{q}$. Then $\mathbf{q} \cdot \mathbf{V} = qV_z$. The integral over V_x and V_y can be performed so that

$$F_s(\mathbf{q}, t) = [M/2\pi k_B T]^{1/2} \int_{-\infty}^{+\infty} dV_z \exp \frac{-MV_z^2}{2k_B T} \exp iqV_z t \tag{5.6.5}$$

This may be regarded as the Fourier transform of the Maxwell distribution with Fourier variable (qt). The integral can be evaluated by completing the squares, giving

$$F_s(\mathbf{q}, t) = \exp - \frac{1}{2} q^2 \langle V_z^2 \rangle t^2 \tag{5.6.6}$$

where

$$\langle V_z^2 \rangle = \frac{1}{3} \langle V_x^2 + V_y^2 + V_z^2 \rangle = \frac{k_B T}{M} \tag{5.6.7}$$

is the mean square value of the z component of the velocity (which is of order $10^8 \text{cm}^2/\text{sec}^2$ at room temperature). Note that $F_s(\mathbf{q}, t)$ decays on a time scale

$$\tau_q = [q \langle V_z^2 \rangle^{1/2}]^{-1} \tag{5.6.8}$$

Since $q^2 \sim 10^{10} \text{cm}^{-2}$, $\tau_q \sim 10^{-9}$ sec. The heterodyne correlation function can be expressed as[15]

$$F_1(\mathbf{q}, t) = \langle N \rangle \exp - \frac{1}{6} q^2 \langle V^2 \rangle t^2 = \langle N \rangle \exp \frac{-t^2}{\tau_q^2} \tag{5.6.9}$$

If F_1 is plotted aqainst qt it will be independent of scattering angle. This provides a simple test of free particle motion.

In most gas studies, the densities are sufficiently high that $\langle N \rangle$ is very large and

$\langle \delta N(0)\, \delta N(t) \rangle$ can be disregarded, so that the Gaussian approximation should suffice for $F_2(\mathbf{q}, t)$.[16]

In the opposite extreme the gas particle j will suffer a very large number of collisions in moving the distance q^{-1}; that is, $q\Lambda \ll 1$, and $F_s(\mathbf{q}, t)$ cannot be computed on the basis of linear trajectories. Because it suffers so many collisions, the particle may be regarded as a random walker, and $F_s(\mathbf{q}, t)$ can be calculated in the diffusion approximation; which gives Eq. (5.4.11). Then if $F_s(\mathbf{q}, t)$ is plotted against $q^2 t$ it will be independent of scattering angle. These observations provide a simple test of whether a particle is freely moving or diffusing. It is important to note, however, that in the limit $(q\Lambda \ll 1)$ where collisions are important the gas cannot be regarded as a system of independent particles. This means that the full formula for the scattering involving the density fluctuations $\langle \delta\rho_{-\mathbf{q}}(0)\, \delta\rho_{\mathbf{q}}(t) \rangle$ must be used. In order to compute the function and thereby the spectrum it is necessary to use kinetic equations (such as the Boltzmann equation) from the kinetic theory of gases (see Chapter 14).

So far we have mentioned the two limits $(q\Lambda \ll 1)$ and $(q\Lambda \gg 1)$. Actually these limits can be explored by varying the pressure and thereby Λ, at constant scattering angle, or by varying the scattering angle θ and thereby q, at constant pressure. Low-angle scattering gives small values of q $(\sim \sin \theta/2)$ and therefore can give the $(q\Lambda \ll 1)$ limit whereas high angle scattering gives large values of q and can therefore give the $(q\Lambda \gg 1)$ limit. Light scattering therefore provides us with a tool for the study of the kinetic theory of gases. At large angles light scatering explores the almost collisionless regime, at intermediate angles it explores the kinetic theory regime, whereas at small angles it explores the hydrodynamic regime (Nelkin and Yip, 1966; Greytak and Benedek, 1966). A short discussion of this is given in Chapter 14.

In dense gases or liquids, the values of q accessible to light scattering $(0 < q < 10^5\ \text{cm}^{-1})$ are always such that $q\Lambda \ll 1$ so that in dense systems light scattering probes only the hydrodynamic behavior.

5 · 7 MOTILE MICROORGANISMS

Berge et al. (1967) have demonstrated the feasibility of using light scattering to study the motions of motile microorganisms. They found that the spectrum of light scattered from live spermatozoa was much broader than from dead spermatozoa. Recently Nossal et al. (1971) have studied the motility of *E. coli* which swim by beating flagella. They have gone one step further than Berge et al. and have reported an explicit determination of the speed distribution of swimming bacteria.

Motile microorganisms are, needless to say, far more complex than molecules. They move in a very complicated way. It seems to be true that once a microorganism starts moving in a given direction it persists with constant velocity in that direction for a distance long compared with typical ralues of q^{-1}. That is to say, the "mean free path" of a microorganism is in general long compared with q^{-1}. This then is the analog of the ideal gas where to a very good approximation "collisions ' (changes in swimming direction) can be ignored. The difference is that the velocity distribution of motile microorganisms is not Maxwellian.

Equation (5.4.2) is applicable to this problem.[17] If $P(\mathbf{V})$ is the velocity-distribution function of the microorganisms, then

$$F_s(\mathbf{q}, t) = \int d^3V P(\mathbf{V}) \exp i\mathbf{q} \cdot \mathbf{V}t \tag{5.7.1}$$

If the coordinate axes are chosen such that $\mathbf{q}$ lies along the z axis and the velocity is expressed in spherical polar coordinates.

$$F_s(\mathbf{q}, t) = \int_0^{2\pi} d\phi \int_0^{\pi} d\theta \sin\theta \int_0^{\infty} dV V^2 P(\mathbf{V}) \exp iq\, Vt \cos\theta \tag{5.7.2}$$

Because there is no symmetry-breaking field like a concentration gradient, it is equally likely for a bacterium to move off in one direction as another. Therefore $P(\mathbf{V})$ should not depend on the direction of $\mathbf{V}$ but only on its magnitude $|\mathbf{V}|$. Thus the angular integrations can be carried through, yielding

$$F_s(\mathbf{q}, t) = 4\pi \int_0^{\infty} dV V^2 P(V) \left[\frac{\sin qVt}{qVt}\right] \tag{5.7.3}$$

The quantity

$$W(V)dV \equiv 4\pi V^2 P(V) dV \tag{5.7.4}$$

is simply the probability that an organism will be found with a speed between V and $V + dV$. $W(V)$ is an even function of V; that is, $W(-V) = W(V)$.

It should be noted that $F_s(\mathbf{q}, t)$ depends on q and t only through the combination $x \equiv qt$

$$F_s(\mathbf{q}, t) = F_s(x) \tag{5.7.5}$$

$$F_s(x) = \int_0^{\infty} dV W(V) \frac{\sin xV}{xV} \tag{5.7.6}$$

The Fourier sine transform of this function yields the swiming speed distribution

$$W(V) = \frac{2V}{\pi} \int_0^{\infty} dx x F_s(x) \sin xV \tag{5.7.7}$$

Thus in principle the speed distribution function of motile microorganisms can be determined from light-scattering experiments.

This application is illustrated in Fig. 5.7.1. Equation (5.7.5) implies that $F_1(\mathbf{q}, t)$ should be a function of the variable $x = qt$ so that if the $F_1(\mathbf{q}, t)$ data taken at different scattering angles are plotted as a function of x or equivalently $t \sin\theta/2$ they should superimpose. Figure 5.7.1a shows scattering data from a motile sample, taken at three different angles. The experiment was done in the homodyne mode and $F_1(\mathbf{q}, t)$ was found from the Gaussian approximation [see Eq. (4.3.2)]. In Fig. 5.7.1b the data are plotted as a function of $x = qt$, and Eq. (5.7.5) is corroborated. The speed distribution of the microoganisms is found by sine transforming the data in Fig. 5.7.1b. The resulting distribution is shown in Fig. 5.7.1c.

Nossal and Chen (1971) report that the microorganisms in the sample to which Fig.

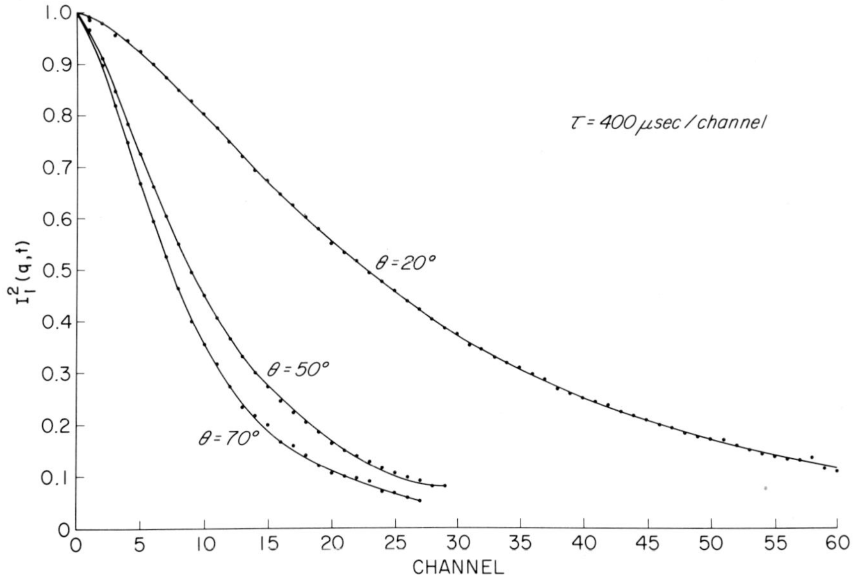

FIG. 5.7.1a. The square of heterodyne time correlation function $F_1^2(\mathbf{q}, t)$ extracted from a homodyne experiment (using the Gaussian approximation [Eq. (5.5.8)] for motile *E. coli* k_{12} in *L*-broth, $T = 25°C$ taken at different scattering angles. Concentration of bacteria ≈ 10 /cm³. Number of bacteria in scattering volume ≈ 10 . (From Nossal et al., 1971.)

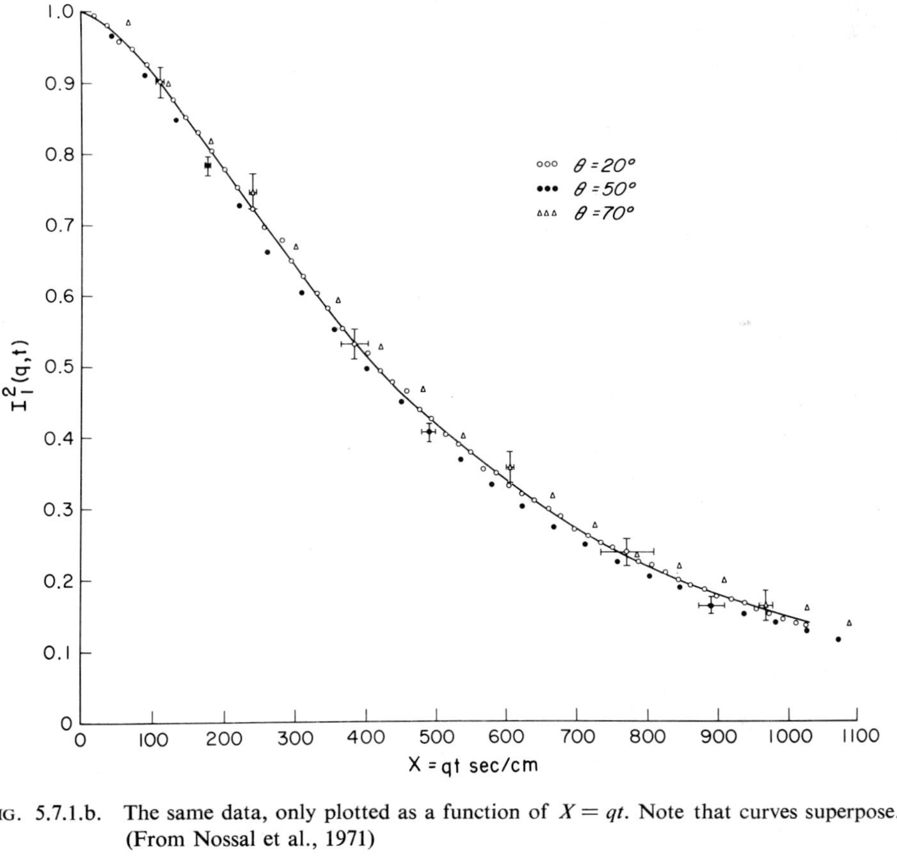

FIG. 5.7.1.b. The same data, only plotted as a function of $X = qt$. Note that curves superpose. (From Nossal et al., 1971)

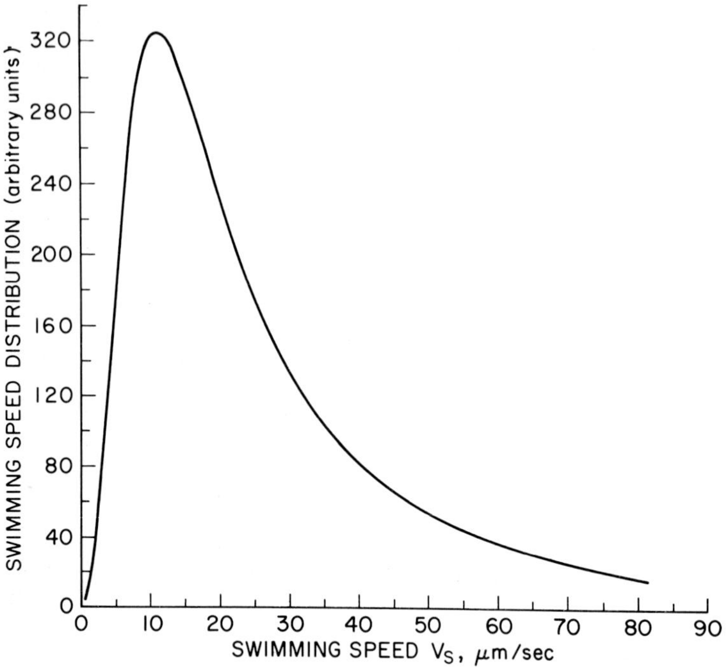

FIG. 5.7.1.c. The swimming-speed distribution function, $W(V)$ as derived from Eq. (5.7.7) for the data presented in Fig. 5.7.1*b*. (From Nossal et al., 1971.)

5.7.1 corresponds all appeared to be swimming when observed under the light microscope. When 10^{-2}M $CuCl_2$ was added to similarly prepared samples, the persistent motion gradually disappeared and after several hours the bacteria ceased to swim, assuming the characteristics of large Brownian particles (Nossal and Chen, 1972). Scattering data for a sample so treated with $CuCl_2$ are presented in Fig. 5.7.2, where they have been plotted against $y = q^2t$. The curves corresponding to different scattering angles overlap remarkably well. If we assume that the sample contains bacteria of a single radius then this motion-arrested sample would have a correlation function $F_1(q, t) \propto \exp -q^2Dt$. However as seen from the insert in Fig. 5.7.2 these data do not follow this simple exponetial decay. It is not difficult to find a possible explanation for this. A bacterial sample certainly contains scatterers of different sizes, that is, it is polydisperse. This might arise from clumping of the bacteria, dust, bacteria in different stages of growth, and so on. Then $F_1(\mathbf{q}, t)$ is a superposition of exponentials each corresponding to scattering from particles of a given size, and the resultant correlation function F_1 may deviate from a simple exponential form.

A sample of bacteria may contain both motile and nonmotile microorganisms. An analysis of the scattering data is slightly more complicated in this case, for then

$$F_1(\mathbf{q}, t) = \langle N \rangle \{X_L F_s^L(\mathbf{q}, t) + X_D F_s^D(\mathbf{q}, t)\} \tag{5.7.8}$$

where X_L and X_D are the fractions of living and dead microorganisms, and $F_s^L(\mathbf{q}, t)$ and $F_s^D(\mathbf{q}, t)$ are the functions in Eqs. (5.7.3) and (5.4.11) respectively. Measurements of $F_1(\mathbf{q}, t)$ as a function of q can give X_D and X_L. Chen and Nossal have devised a

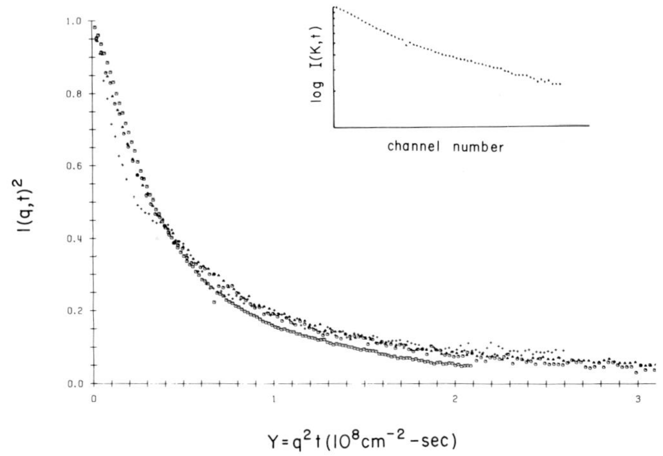

FIG. 5.7.2. The square of heterodyne time-correlation function ($F_1(q,t)$) extracted from homodyne experiment, for nonmotile bacteria. The sample was similar to that of Fig. 5.7.1, except that 10^{-2} M $CuCl_2$ was added to cause the bacteria to lose their motiliy. $F_1^2(q,t)$ is plotted as a function of $y = q^2t$ for different scattering angles θ. (key: ⊡⊡⊡ $\theta = 20$; ⊙⊙⊙ $\theta = 50°$; △△△ $\theta = 60$; +++ $\theta = 120°$). The insert shows log F_1 (**q**, t) for $\theta = 20°$. Note that the correlation function in this insert does not follow a simple exponential decay. (From Nossal and Chen, 1972.)

scheme for determining X_L and X_D in such mixtures. The interested reader should consult Nossal and Chen (1972) for the details of this method.

It might appear from the foregoing that light scattering only provides information about the speed distribution. This is not true. The bacterial samples used usually contain about 10^7 bacteria/cc. If the incident laser beam is focused down so that the scattering volume contains only a few (<100) bacteria on the average, the term $\langle\delta N(0)\,\delta N(t)\rangle$ in Eq. (5.5.8) becomes significant and can in fact be determined. Whereas q^{-1} is small compared with the free path length Λ of the bacterium L, the characteristic length of the scattering volume, can be large compared with the mean free path length. Thus $\langle\delta N(0)\,\delta N(t)\rangle$ will contain information about Λ and can be used to determine this quantity.

For example, if the displacement is a Gaussian random process, and if the scattering volume characteristics are as given in Section 6.6, the occupation number fluctuations are given by

$$\langle\delta N(0)\,\delta N(t)\rangle = \langle N\rangle\left[1 + \frac{2\langle\Delta r^2(t)\rangle}{3\sigma_1^2}\right]^{-1}\left[1 + \frac{2\langle\Delta r^2(t)\rangle}{3\sigma_2^2}\right]^{-1/2}$$

where $\langle\Delta r^2(t)\rangle$ is the mean square displacement, σ_1 is the focused incident beam diameter, and σ_2 is the slit width in the collection optics (see Section 5.6).

If the microorganisms swim along linear trajectories for distances long compared with σ_1 or σ_2 ; that is, if $\Lambda \gg \sigma_1, \sigma_2$

$$\langle\Delta r^2(t)\rangle = \langle V^2\rangle\, t^2$$

and the experimental decay $\langle\delta N(0)\,\delta N(t)\rangle$ gives $\langle V^2\rangle$. Experiments on "long-swimming" mutant strains of *E. coli* have yielded values for the root mean-square velocity (see Banks et al., 1974).

In the opposite limit, when $\Lambda \ll \sigma_1, \sigma_2$

$$\langle \Delta r^2(t) \rangle = 6Dt$$

where the "diffusion coefficient" of the motile microorganisms is

$$D = \frac{1}{3} \Lambda V$$

where V is the mean speed. Then an experimental study of $\langle \delta N(0)\ \delta N(t) \rangle$ gives D. When these results are combined with Λ determined from $F_s(\mathbf{q}, t)$, Λ can be determined. This limit has been observed (Schaefer and Berne, 1975).

The intermediate case where $\Lambda \sim \sigma_1, \sigma_2$ is much more difficult to treat. A general theory for this has been given (see Schaefer and Berne, 1975), from which it is possible to extract Λ from an analysis of the decay of $\langle \delta N(0)\ \delta N(t) \rangle$. The general decay is shown in Fig. 5.7.3.

It is important to note that occupation-number fluctuations always give information

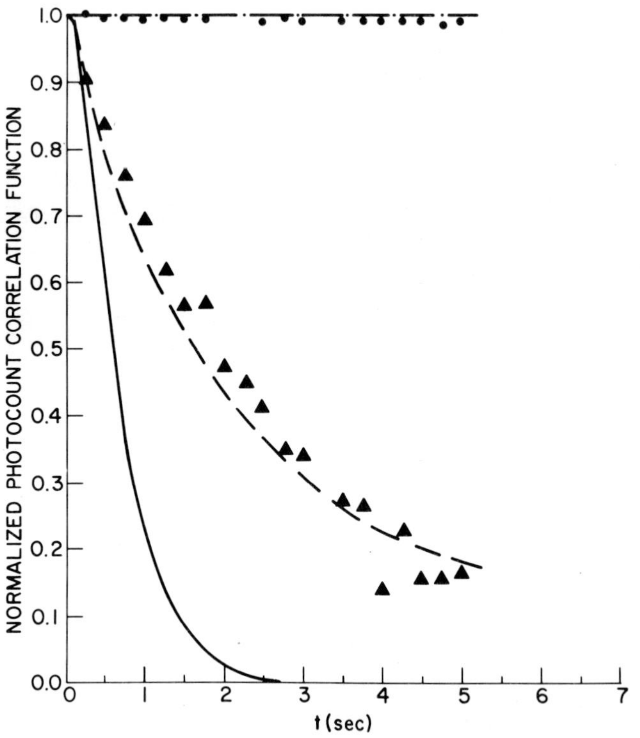

FIG. 5.7.3. $\langle \delta N(0)\ \delta N(t) \rangle$ for different systems. The solid line is the calculated curve for a freely swimming bacterium when $\Lambda \gg$ dimensions of scattering volume. The dot–dash curve is calculated for a diffusing particle. The dashed curve is the approximate curve for a random walk. The triangles and the circles correspond to measurements respectively of motile and nonmotile *E. coli*. (From Schaefer, 1974.)

about the translational motion, whereas the coherent decay embodied in $F_1(\mathbf{q}, t)$ might contain effects of the rotational motions.[18]

5 · 8 MOLECULES IN UNIFORM MOTION

If the macromolecules are forced by some external agency to flow with a velocity $\mathbf{V}$, light scattering can be used to measure this velocity. There are several possible examples of this: (a) macromolecules suspended in a fluid which is in uniform motion with velocity $\mathbf{V}$, (b) macromolecules falling at their terminal velocities in a viscous solvent under the action of gravity (sedimentation velocity), (c) macroions accelerated to a terminal velocity by an externally imposed electric field (electrophoresis), and (d) macromolecules accelerated to their terminal velocity in an ultracentrifuge (sedimentation velocity).

In ordinary particle diffusion, the flux $\mathbf{J}$ of particles at the point $\mathbf{R}$, at time t is given by Fick's first law of diffusion

$$J = -D\nabla c(\mathbf{R}, t) \tag{5.8.1}$$

where D is the diffusion coefficient, $c(R, t)$ is the concentration of particles (number per cc) at the point $\mathbf{R}$ at time t and ∇ is the gradient operator. $\mathbf{J}$ tells us the number of particles passing through a unit area (1 cm^2) perpendicular to $\mathbf{J}$ per unit time. According to this equation particles will tend to move from regions of high concentration to regions of low concentration.

In the presence of a force which accelerates the particles to a terminal velocity $\mathbf{V}$, there will be an additional flux of particles, $\mathbf{V}c$, which would exist even in the absence of particle diffusion, so that

$$\mathbf{J} = \mathbf{V}c(\mathbf{R}, t) - D\nabla c(\mathbf{R}, t) \tag{5.8.2}$$

This additional flux is due to the fact that uniformly moving particles should flow through the unit area.

In what follows, we shall assume that the number of macromolecules in the system is constant in time. This is true only if no chemical reactions involving the macromolecules take place. In Chapter 6 we study the effects of chemical reactions.

The conservation of macromolecules is expressed differentially by the continuity equation (cf. Section 10.3)

$$\frac{\partial c}{\partial t} + \nabla \cdot \mathbf{J} = 0 \tag{5.8.3}$$

Substitution of the flux $\mathbf{J}$ from Eq. (5.8.2) yields the equation

$$\frac{\partial c}{\partial t} + \mathbf{V} \cdot \nabla c = D\nabla^2 c \tag{5.8.4}$$

which describes how the particles flow and spread out (diffuse) in the system.

In Section 5.4 it was shown that it is reasonable to assume that $G_s(\mathbf{R}, t)$ should satisfy the same equation as $c(\mathbf{R}, t)$ so that $G_s(\mathbf{R}, t)$ is the solution of

$$\frac{\partial}{\partial t} G_s(\mathbf{R}, t) + \mathbf{V} \cdot \nabla G_s(\mathbf{R}, t) = D\nabla^2 G_s(\mathbf{R}, t) \tag{5.8.5}$$

subject to the initial condition,

$$G_s(\mathbf{R}, 0) = \delta(\mathbf{R}). \tag{5.8.6}$$

The spatial Fourier transform of these equations is

$$\begin{aligned} &\frac{\partial}{\partial t} F_s(\mathbf{q}, t) - i\mathbf{q} \cdot \mathbf{V} F_s(\mathbf{q}, t) = q^2 D F_s(\mathbf{q}, t) \\ &F_s(\mathbf{q}, 0) = 1 \end{aligned} \tag{5.8.7}$$

so that the corresponding solution $F_s(\mathbf{q}, t)$ is

$$F_s(\mathbf{q}, t) = \exp i\mathbf{q}\cdot\mathbf{V}t \exp -q^2 Dt \tag{5.8.8}$$

It should be noted from Eqs. (5.4.2) and (5.8.8) that the heterodyne correlation function is then proportional to (see Section 4.3)

$$Re\, F_1(\mathbf{q}, t) = \langle N\rangle\, Re\, F_s(\mathbf{q}, t) = \langle N\rangle\, [\exp -q^2 Dt] \cos(\mathbf{q} \cdot \mathbf{V}t) \tag{5.8.9}$$

with the corresponding power spectrum[19] [cf. Eq. (5.4.14)]

$$S_1(\mathbf{q}, \omega) = \frac{1}{2}\pi^{-1}\left\{\frac{q^2 D}{[\omega - \mathbf{q} \cdot \mathbf{V}]^2 + [q^2 D]^2} + \frac{q^2 D}{[\omega + \mathbf{q} \cdot \mathbf{V}]^2 + [q^2 D]^2}\right\} \tag{5.8.10}$$

The maximum in each term of the heterodyne spectrum is shifted from $\omega = 0$ to[20]

$$\omega(\mathbf{q}) = \pm\mathbf{q} \cdot \mathbf{V} = \pm 2k_i V \cos\phi \sin\frac{\theta}{2} \tag{5.8.11}$$

where we have substituted Eq. (3.2.5) for q, and where ϕ is the angle between $\mathbf{q}$ and $\mathbf{V}$. It should be noted that when $\mathbf{V}$ is $\perp$ to the scattering plane there is no frequency shift (Doppler shift), but when $\mathbf{V}$ is $\|$ to $\mathbf{q}$ there is the maximum Doppler shift.

The homodyne correlation function is [cf. Eq. (5.5.8)]

$$\begin{aligned} F_2(\mathbf{q}, t) &= \langle N\rangle^2\{1 + |F_s(\mathbf{q}, t)|^2\} + \langle \delta N(0)\, \delta N(t)\rangle \\ &= \langle N^2\rangle\{1 + \exp -2q^2 Dt\} + \langle \delta N(0)\, \delta N(t)\rangle \end{aligned} \tag{5.8.12}$$

Thus even though there is uniform motion leading to a shift in the heterodyne spectrum, there is no change in the homodyne spectrum, except for the influence uniform motion has on the decay of $\langle \delta N(0)\, \delta N(t)\rangle$ (See Eq. (6.6.11)].

The measurement of a flow was illustrated by Ben-Yosef et al. (1972), who studied convective heating in colloidal AgCl solutions. These authors noted that because about 5% of the laser beam is absorbed by the solution, there will be significant heating. They pointed out that since the flow pattern of a horizontally heated cylinder (whose axis is defined by the incident laser beam) is perpendicular to the axis, the convective velocity will have a component along **q** for 90° scattering so that there should be a shift in the heterodyne spectrum. They observed a shift (see Fig. 5.8.1) at a frequency expected by their calculations. The important thing to learn from this is that convective heating can be a source of shifts. *Caution* should therefore be exercised in interpreting light scattering experiments.

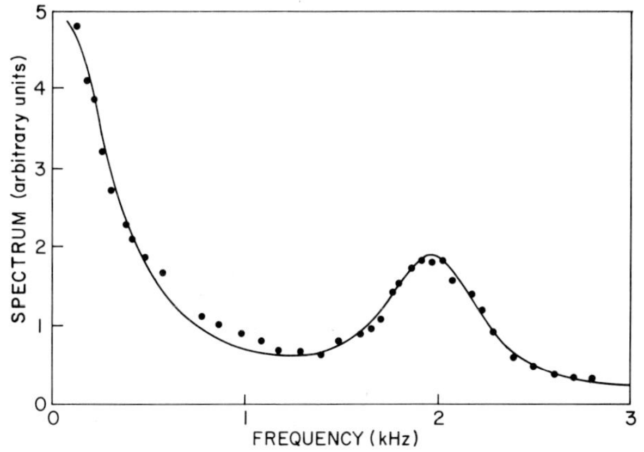

FIG. 5.8.1. Power spectrum of laser light scattered from silver chloride colloids in water. The circles are for the actual experimental data. The solid line is the sum of two Lorentzians, one centered at zero frequency, the second one at 1.97 kHz. The half-width of the first Lorentzian is 0.35 kHz and that of the second, 0.29 kHz. The instrumental width is 50 Hz. (From Ben-Yosef et al., 1972.)

In a novel application Mustacich and Ware (1974) have studied the streaming motion of protoplasm in the common alga *Nitella flexilis* by observing a Doppler-shifted heterodyne spectrum. The *Nitella* cells were positioned in the apparatus in a special holder which held them in place. The Doppler shift, [Eq. (5.8.11)] could thus be studied as a function of q at different points in the cell. Representative spectra are shown in Figs. 5.8. 2*a*–*c*. There is clearly a Doppler shift in Figs. 5.8.2*a* and 5.8.2*b*. This shift is quenched immediately after a streaming inhibitor is added, as can be seen in Fig. 5.8.2*c*. Fig. 5.8.2*d* demonstrates the validity of Eq. (5.8.11) in that it shows that $\omega(q)$ varies linearly with $\sin\theta$ for the geometry adopted. This method has certain advantages over microscopy as a method for studying flow patterns in cells. These are enumerated by Mustacich and Ware (1974).

Tanaka and Benedek (1974), in a novel and important study, have measured the velocity of blood flow in the femoral vein of rabbit by detecting (in the heterodyne mode) the Doppler shift of laser light introduced into the vein by means of a fiber optic catheter. The light is scattered from the moving erythrocytes in the blood. It is important to recognize that the blood-flow velocity is not uniform across a vein but varies from zero near the vein wall to a maximum in the center. Thus the spectrum observed should be an average over the distribution of velocities of the illuminated erythrocytes, approximately

$$I_1(\omega) \propto \frac{1}{2\pi}\int d\mathbf{V}P(\mathbf{V})\left\{\frac{\Gamma}{(\omega-\mathbf{q}\cdot\mathbf{V})^2+\Gamma^2}+\frac{\Gamma}{(\omega-\mathbf{q}\cdot\mathbf{V})^2+\Gamma^2}\right\}$$

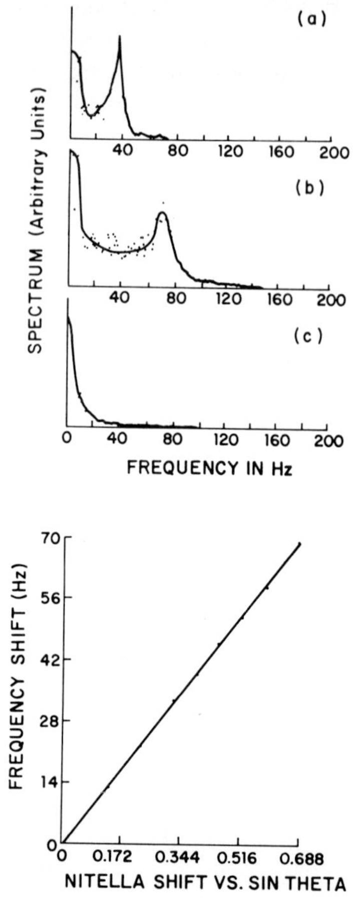

FIG. 5.8.2. Spectra of light scattered from the protoplasm of *Nitella*. The horizontal axis is frequency in Hz and the vertical axis is relative intensity. Spectrum (*a*) was taken at a scattering angle of 19.5 deg. Spectra (*b*) and (*c*) were taken at a scattering angle of 36.1 deg. Spectrum (*c*) was taken from the same point on the cell as spectrum (*b*), immediately after addition of parachloromercuribenzoate, a streaming inhibitor. (Each of these spectra was collected in about 30 sec. The points are the output of the spectrum analyzer, and the dark lines have been drawn merely to make the data more perspicuous and to emphasize the reproducible features of the data.) Part (*d*) is a plot of the magnitude of the Doppler shift from a fixed point on a *Nitella* cell as a function of the sine of the scattering angle θ. The line is the best least-square fit to the points. The predicted linear dependence is verified, and the deviations from the line provide an estimate of the experimental precision. (From Mustacich and Ware, 1974.)

where Γ is a width (due to diffusion). This average can give a spectrum which does not exhibit shifted peaks (or equivalently a time-correlation function that displays no oscillations). The situation is actually more complicated because absorption and multiple scattering of the laser light cannot be neglected. Tanaka and Benedek (1974) consider all of these complications and derive an expression which allows them to deduce from experiment the average blood-flow velocity at different points in the cross section of the vein. They show that the faster the flow velocity, the faster is the decay of the correlation function. Much useful information can be gained by this kind of *light-scat-*

tering velocimetry. A comparison between the time-correlation function of the blood flow at a fixed point of the vein prior to and after destroying the rabbit is shown in Fig. 5.8.3. Note that the blood flow decreases drastically in the dead rabbit; this observation should not escape even our most confused reader. The paper by Tanaka and Benedek (1974) contains an exhaustive discussion of this experiment. In our view this application of light scattering should have important clinical applications.

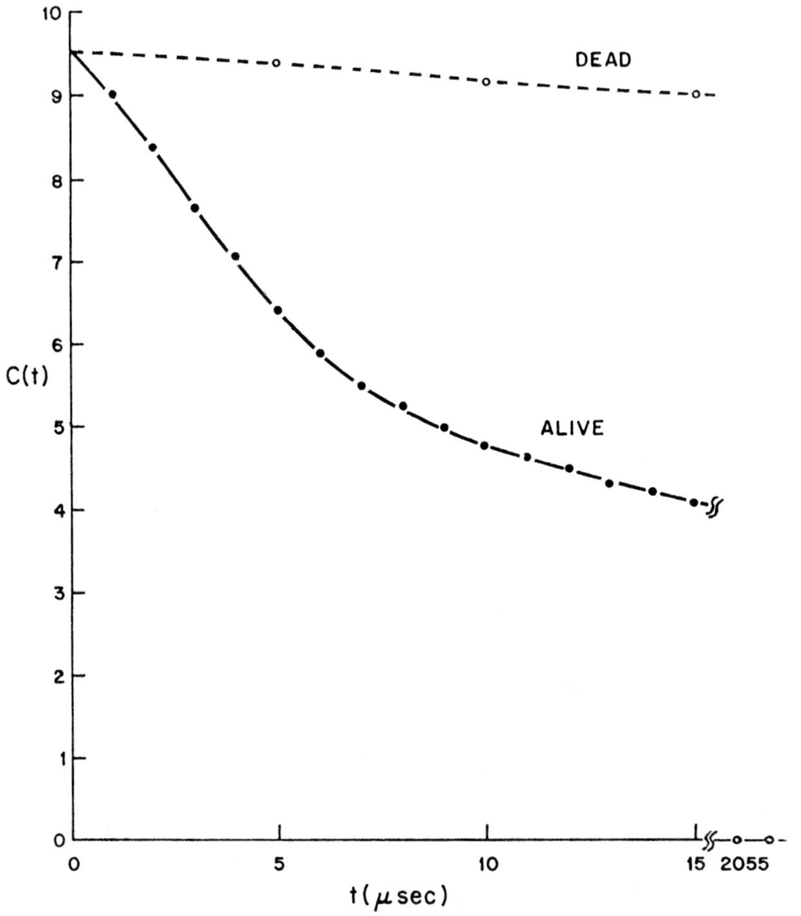

FIG. 5.8.3. The heterodyne correlation function of the scattered light from the blood flow in the femoral vein of an albino rabbit. The lower curve was taken prior to killing the rabbit, and the upper curve was measured just after its death. (From Tanaka and Benedek, (1974.)

Another application of the foregoing has been developed by Ware and Flygare (1971) and Uzgiris (1972). A detailed discussion of experimental techniques has been given by Ware (1974). These authors study macromolecules of charge Q in the presence of an externally applied electric field. It is well known that the macromolecule will be accelerated by the force $Q\mathbf{E}$ it experiences in the electric field, and will reach a terminal velocity $\mathbf{V}$ given by

$$\mathbf{V} = \mu \mathbf{E} \tag{5.8.13}$$

where μ is the electrophoretic mobility of the molecule. In the simplest case

$$\mu = \frac{Q}{\xi} \tag{5.8.14}$$

where ζ is the friction constant.[21] Substituting Eq. (5.8.13) into Eq. (5.8.14) yields the frequency shift

$$\omega(\mathbf{q}) = \mu(\mathbf{q} \cdot \mathbf{E}) = \mu q E \cos \phi \tag{5.8.15}$$

where ϕ is the angle between **q** and the applied field. Thus, one experiment can give both the mobility of an ion and its diffusion coefficient. Since in many cases the Stokes approximation $\zeta = 6\pi\eta a$ is valid

$$D = \frac{k_B T}{6\pi\eta a}\ ;\ \mu = \frac{Q}{6\pi\eta a} \tag{5.8.16}$$

the effective charge on the ion can be determined.

In highly charged macromolecules Eq. (5.8.16) does not apply. It is then necessary to generalize these arguments to include hydration and deviations from spherical shape. However there are even more formidable complications to consider first. In aqueous solution, the macroion is surrounded by an ion atmosphere composed mainly of ions of opposite charge. This means that the local field —the field felt by the macroion— will be considerably different than the applied field. Corrections for this effect can be made if one uses the Debye–Huckel–Henry theory to calculate the properties of the ion atmosphere (see Chapters 9 and 13). An estimate (Tanford, 1961) gives

$$\mu = \frac{Q}{\zeta} \left\{ \frac{X(\kappa a)}{1 + \kappa a} \right\}$$

where

$$\kappa = \left(\frac{8\pi N_0 e^2}{1000\ \varepsilon k_B T} \right)^{1/2} I^{\frac{1}{2}}$$

In these expressions N_0 is Avogadro's number, ε is the dielectric constant of the solution, a is the ionic randius, I is the ionic strength, and κ^{-1} is the radius of the ion atmosphere. The function $X(\kappa a)$ is called *Henry's function*. It varies between 1.0 and 1.5 as κa goes from zero to infinity so that the mobility might appear smaller than that predicted in Eq. (5.8.14). Even this theory is a gross oversimplification. Nevertheless, without knowing the explicit form of μ we can proceed using Eq. (5.8.13).

By applying an electric field to a solution of macromolecules, the molecules are accelerated to a terminal velocity determined by their electrophoretic mobilities.[22] The spectrum of scattered light is thereby Doppler-shifted to a frequency determined by Eq. (5.8.15). From the Doppler shift the mobility can be measured, whereas from the width of the Doppler shifted line the diffusion coefficient can be measured [cf. Eq. (5.8.10)]. In this way Ware and Flygare (1971) have measured the electrophoretic mobility and diffusion coefficient of bovine serum albumin (BSA) at a variety of pH values and ionic strengths. In their experiments the field was applied in the scattering plane but perpendicular to $\mathbf{k}_i$, the direction of the incident light. From Eq. (5.8.15) it then follows that $\omega(q) = \mu Eq \cos \theta/2$ which reduces to $\omega(q) = \mu E k_i \sin \theta$. The Doppler shift is better resolved the larger the shift compared to the width ($q^2 D$). The ratio of the shift $\omega(q) = \mu Eq \cos \theta/2$ to the half-width $\Gamma(q) = q^2 D$ is

$$R = \frac{\omega(q)}{\Gamma(q)} = \frac{\mu E \cos \theta/2}{qD} = \frac{\mu E}{2\,D\,k_i} \cdot \left[\frac{\cos \theta/2}{\sin \theta/2}\right]$$

According to this the highest resolution will be achieved at low scattering angles where R reduces to

$$R \simeq \frac{\mu E}{k_i\,D\,\theta}$$

R increases without bound in the limit of low scattering angle and the experiment is therefore limited by limitations on observing low scattering angles.

Figure 5.8.4 shows the heterodyne correlation function $ReF_1(q, t)$ for a variety of electric fields. Note the cosinusoidal oscillations at frequency $\omega(q)$. In Fig. 5.8.5 $\omega(q)/$

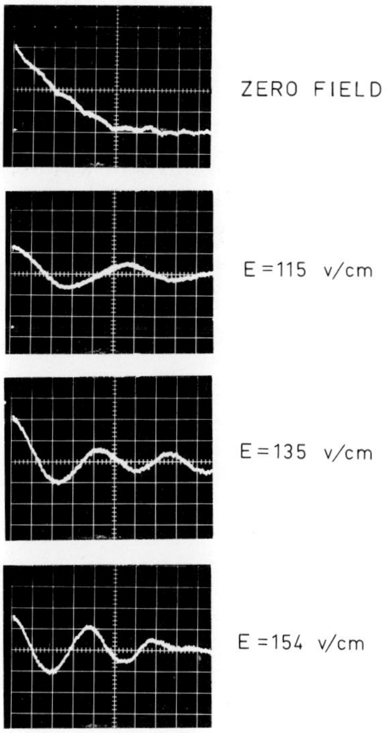

FIG. 5.8.4. The heterodyne correlation function at zero field and fields of 115, 135, and 154 V/cm respectively in a 5% solution of BSA in 0.004 M NaCl titrated to pH = 9.2 with n-butylamine. $\lambda_0 = 5154\text{A}°$, $T = 10°\text{C}$, and $\theta = 0.079$ rad $= 4.5°$. The average mobility from these data is 18×10^{-5} cm /sec • V. (From Ware and Flygare, 1971, Fig. 3.)

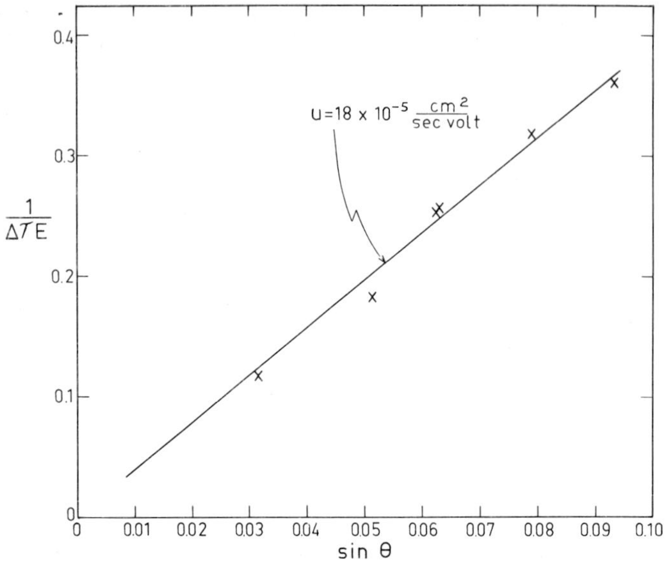

FIG. 5.8.5. A plot of $\omega(q)/2\pi E$ as a function of $\sin\theta$ where $\Delta\tau = 2\pi/\omega(q)$ is the period of oscillation of the heterodyne time-correlation function. The expected linear dependence on $\sin\theta$ is observed The slope of the line leads to an average value of the mobility shown. All data in this figure were taken in solutions under conditions identical to those described in Fig. 5.8.4 (From Ware and Flygare, 1971, Fig. 3.)

$2\pi E$ is plotted against $\sin\theta$. The slope of this plot was used to determine the mobility of BSA.

Ware and Flygare (1972) have also applied this method to the analysis of a mixture. Suppose that there is a mixture containing two species of macroions of charges Q_1 and Q_2, mobilities μ_1 and μ_2, diffusion coefficients D_1 and D_2, and polarizabilities α_1 and α_2. Furthermore, if these macroions are independent, then[23]

$$F_1(\mathbf{q}, t) = Re\,[\alpha_1^2\langle N_1\rangle F_{s_1}(\mathbf{q}, t) + \alpha_2^2\langle N_2\rangle F_{s_2}(\mathbf{q}, t)] \tag{5.8.17}$$

where $F_{s_1}(\mathbf{q}, t)$ and $F_{s_2}(\mathbf{q}, t)$ are self-intermediate scattering functions for the two species,

$$\begin{aligned} F_{s_1}(\mathbf{q}, t) &= \exp i\mu_1(\mathbf{q}\cdot\mathbf{E})t \exp -q^2D_1t \\ F_{s_2}(\mathbf{q}, t) &= \exp i\mu_2(\mathbf{q}\cdot\mathbf{E})t \exp -q^2D_2t \end{aligned} \tag{5.8.18}$$

and the heterodyne spectrum is

$$S_1(\mathbf{q}, \omega) = \pi^{-1}\left\{\langle N_1\rangle\alpha_1^2 \frac{q^2D_1}{(\omega - \omega_1)^2 + [q^2D_1]^2} + \langle N_2\rangle\alpha_2^2 \frac{q^2D_2}{(\omega - \omega_2)^2 + (q^2D_2)^2}\right\} \tag{5.8.19}$$

where the shifts are

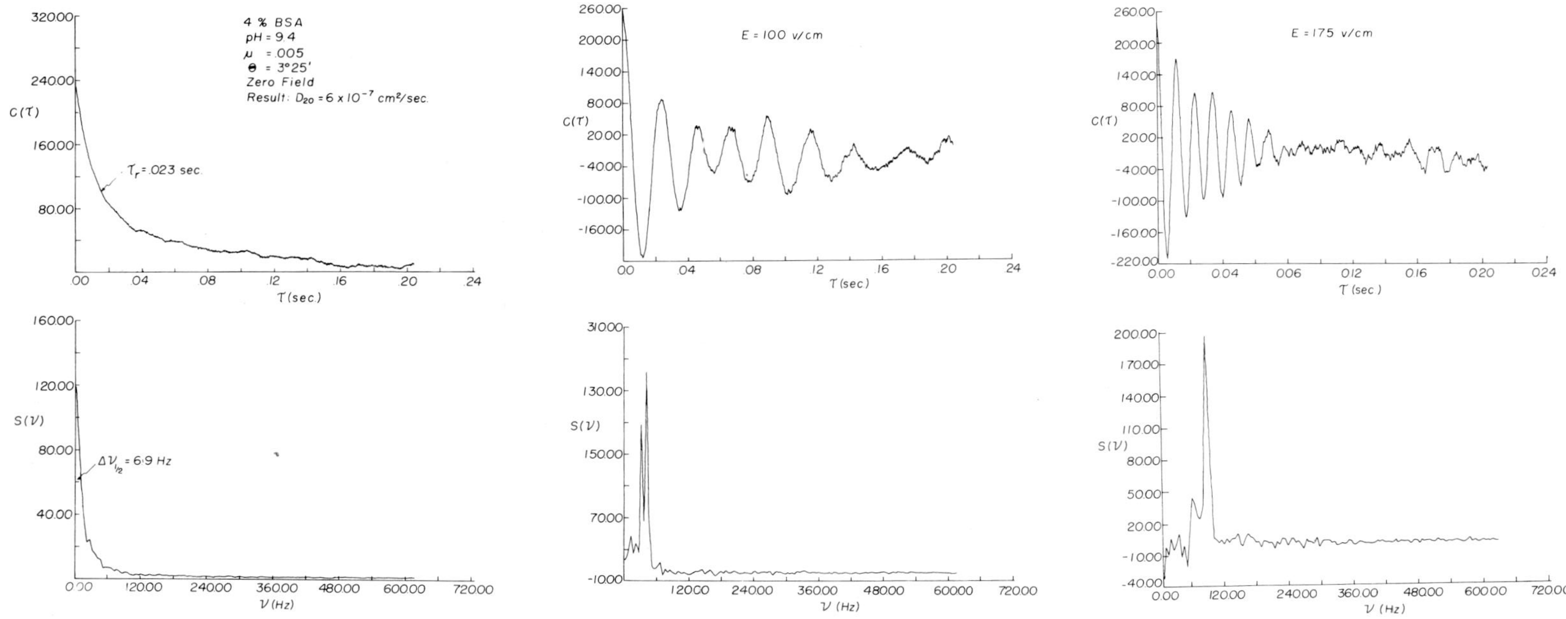

FIG. 5.8.6. Representative data for an experiment on a BSA solution at $E = 0$, 100, and 175 V/cm under conditions indicated above. (From Ware and Flygare, 1972, Fig. 3.)

$$\begin{aligned} \omega_1(\mathbf{q}) &= \mu_1(\mathbf{q} \cdot \mathbf{E}) \\ \omega_2(\mathbf{q}) &= \mu_2(\mathbf{q} \cdot \mathbf{E}) \end{aligned} \tag{5.8.19}$$

Ware and Flygare have studied two different solutions. One solution is simply a commercial preparation of BSA which contains 10–20% BSA-dimer and 80–90% BSA-monomer. The results are shown in Fig. 5.8.6. Note that for $E = 100$ V/cm, the splitting expected from Eq. (5.8.19) is already apparent. Furthermore this splitting increases as E increases. The second solution contains BSA and fibrinogen. This is shown in Fig. 5.8.7. Ware and Flygare assign the leading doublet to BSA-monomer and BSA-dimer, and they assign the trailing peak to fibrinogen.

The foregoing represents a rather naive approach to electrophoresis. In dilute electrolyte solutions, different ionic species interact through long-ranged Coulomb forces. Each ion is surrounded by an atmosphere of oppositely charged ions which influences

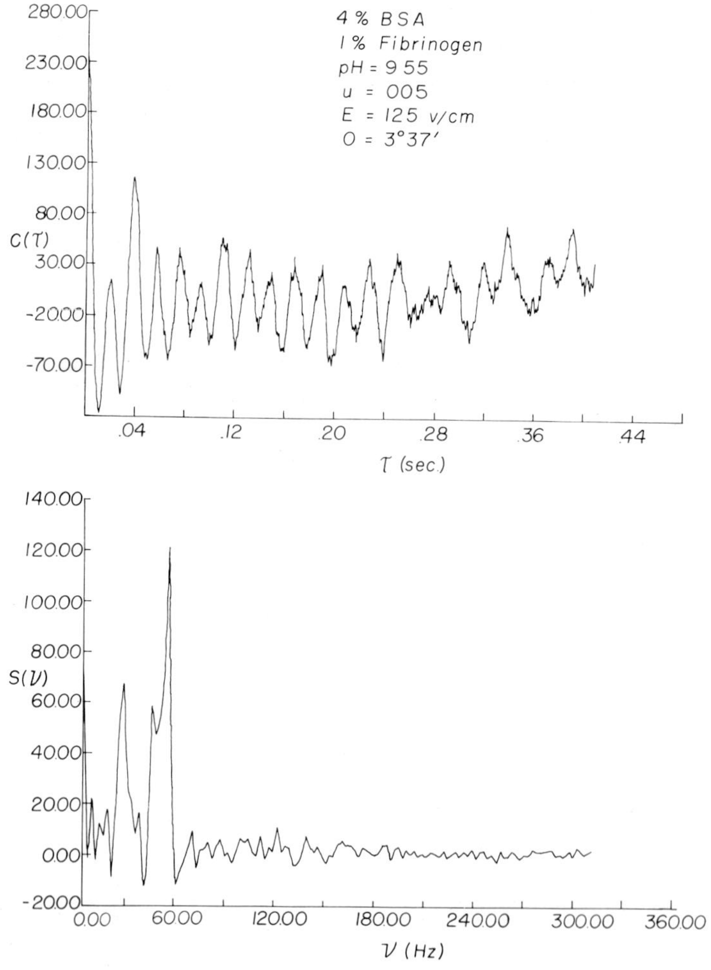

FIG. 5.8.7. Data for electrophoretic light-scattering experiment on a solution containing BSA, BSA dimers, and fibrinogen. (From Ware and Flygare, 1972, Fig. 4.)

its motion so dramatically that the above results may under certain conditions be totally invalid. In Section 9.4 a simple theory of these effects is presented. In Section 13.8 electrophoresis is reconsidered using the theory of nonequilibrium thermodynamics, and the Doppler shift is given in terms of the conductance and the transference numbers of the solution.

5 · 9 BROWNIAN MOTION

Suppose the motion of a molecule in a gas or liquid is followed for a time t. The path followed by the molecule will appear quite erratic due to the many collisions that it experiences during the time of observation. If the time is divided into very small intervals Δt, then the total displacement of the particle in the time t, $\Delta\mathbf{R}(t)$, is the resultant of all of the small steps. The displacement of the particle during the time increment Δt will, in general, vary from one interval to the next so that there will be a distribution of jump lengths. Suppose we know the distribution of jump lengths and suppose further that this distribution is not pathological.[24] Then according to the central limit theorem of probability theory, the probability for a particle to suffer a displacement in the neighborhood d^3R of $\mathbf{R}$ should be the Gaussian distribution function

$$G_s(\mathbf{R}, t) = \left[\frac{2\pi}{3}\langle \Delta R^2(t)\rangle\right]^{-3/2} \exp\left[-3R^2/2\langle \Delta R^2(t)\rangle\right] \tag{5.9.1}$$

where $\langle \Delta R^2(t)\rangle$ is the mean-square displacement of the particle in the time t.

It is therefore often assumed that G_s is Gaussian. The spatial Fourier transform of this function is the intermediate scattering function [cf. Eq. (5.4.5)]

$$F_s(\mathbf{q}, t) = \int d^3R G_s(\mathbf{R}, t)\, e^{i\mathbf{q}\cdot\mathbf{R}} \tag{5.9.2}$$

which when evaluated by completing the squares is

$$F_s(\mathbf{q}, t) = \exp -q^2 \langle \Delta R^2(t)\rangle/6 \tag{5.9.3}$$

For a freely moving molecule

$$\Delta\mathbf{R}(t) = \mathbf{V}t \tag{5.9.4}$$

$$\langle \Delta\mathbf{R}^2(t)\rangle = \langle V^2\rangle t^2$$

and the Gaussian assumption yields

$$F_s(\mathbf{q}, t) = \exp -q^2\langle V^2\rangle t^2/6 \tag{5.9.5}$$

which is precisely what we found before [cf. Eq. (5.6.9)]. For a diffusing particle

$$\langle \Delta R^2(t)\rangle = 6Dt \tag{5.9.6}$$

and

$$F_s(\mathbf{q}, t) = \exp - q^2 Dt \tag{5.9.7}$$

which is precisely what we found before [cf. Eq. (5.4.11)].

In general the displacement is

$$\Delta\mathbf{R}(t) = \int_0^t d\tau\, \mathbf{V}(\tau) \tag{5.9.8}$$

where $\mathbf{V}(t)$ is the velocity at time t. The mean-square displacement is therefore

$$\langle \Delta R^2(t) \rangle = \int_0^t dt_2 \int_0^t dt_1 \langle \mathbf{V}(t_1) \cdot \mathbf{V}(t_2) \rangle \tag{5.9.9}$$

This expression can be simplified to (cf. Appendix 5.A)

$$\langle \Delta R^2(t) \rangle = 2 \int_0^t d\tau (t - \tau) \langle \mathbf{V}(0) \cdot \mathbf{V}(\tau) \rangle \tag{5.9.10}$$

where $\langle \mathbf{V}(0) \cdot \mathbf{V}(t) \rangle$ is the velocity-autocorrelation function of the molecule. It follows from Eqs. (5.9.3) and (5.9.10) that

$$F_s(\mathbf{q}, t) = \exp - \frac{1}{3} q^2 \int_0^t d\tau (t - \tau) \langle \mathbf{V}(0) \cdot \mathbf{V}(\tau) \rangle \tag{5.9.11}$$

The velocity-correlation function decays to zero such that the integral has the asymptotic long-time form

$$\int_0^t d\tau (t - \tau) \langle \mathbf{V}(0) \cdot \mathbf{V}(\tau) \rangle \xrightarrow[t \to \infty]{} t \int_0^\infty d\tau \langle \mathbf{V}(0) \cdot \mathbf{V}(\tau) \rangle \tag{5.9.12}$$

Thus for times long compared to the velocity-correlation time

$$F_s(\mathbf{q}, t) = \exp - q^2 \frac{1}{3} \int_0^\infty d\tau \langle V(0) \cdot V(\tau) \rangle\, t \tag{5.9.13}$$

Comparison with the result of the diffusion equation [Eq. (5.4.11)] leads us to associate the self-diffusion coefficient D with the time integral of the velocity-correlation function,

$$D = \frac{1}{3} \int_0^\infty d\tau \langle \mathbf{V}(0) \cdot \mathbf{V}(\tau) \rangle \tag{5.9.14}$$

This is called a Green–Kubo relation[25] (see Chapters 10 and 11, and Zwanzig, 1965). Thus we expect the results of the diffusion equation to be valid for times long compared to the velocity-correlation time, and the coefficient of self diffusion to be proportional to the area under the velocity correlation function.

To proceed we must determine the velocity-correlation function of a macromolecule. Because macromolecules are much more massive than solvent molecules, they have a much lower velocity on the average[26] than the solvent molecules. The motion of massive molecules in solvents consisting of small molecules has received much attention. The theory that describes this situation is Brownian motion theory. In Brownian motion theory it is assumed that the force on a Brownian particle, $M(d\mathbf{V}/dt)$, consists of a systematic frictional component, $-\zeta\mathbf{V}$, and a random or fluctuating component, $\mathbf{F}(t)$. The equation of motion for the Brownian particle (B-particle) is consequently

$$M\frac{d\mathbf{V}}{dt} = -\zeta\mathbf{V} + \mathbf{F}(t) \tag{5.9.15}$$

where ζ is called the *friction constant*. This equation is called the *Langevin equation*. Hydrodynamic studies of the uniform motion of fluids past large spherical bodies show that the friction coefficient is related to the shear viscosity of the solvent η and to the radius of the sphere a by the Stokes formula $\zeta = 6\pi\eta a$ (stick-boundary conditions) and $\zeta = 4\pi\eta a$ (slip-boundary conditions). The systematic frictional force represents the tendency of a moving B-particle to be slowed down because, on the average, there will be more collisions on its front side than on the back side. The *random force* $\mathbf{F}(t)$ arises from occasional fluctuations in which the particle actually experiences collisions that accelerate and decelerate it beyond the systematic frictional force. The Langevin equation would be meaningless without the fluctuating force, since without it the B-particle would be systematically slowed down until it stops moving, and there would be no mechanism by which it could start moving again. Yet we know that a B-particle is in perpetual thermal motion.

The random force is not a prescribed function of the time, but a *random* function of the time. The Langevin equation is consequently a stochastic differential equation. In order to solve such an equation, the statistical properties of the random force must be specified. Of course, the Langevin equation can be solved formally to give

$$\mathbf{V}(t) = \mathbf{V}(0)\exp-\left(\frac{\zeta}{M}\right)t + \int_0^t d\tau\exp-\left(\frac{\zeta}{M}\right)(t-\tau)\,\mathbf{F}(\tau) \tag{5.9.17}$$

The velocity correlation function can be determined from Eq. (5.9.17) by taking the dot product of $\mathbf{V}(0)$ with each term in the equation followed by averaging over a Maxwell distribution of initial velocities. Then

$$\langle\mathbf{V}(0)\cdot\mathbf{V}(t)\rangle = \langle\mathbf{V}(0)\cdot\mathbf{V}(0)\rangle\exp-\left(\frac{\zeta}{M}\right)t$$

$$+\int_0^t d\tau\left[\exp-\left(\frac{\zeta}{M}\right)(t-\tau)\right]\langle\mathbf{V}(0)\cdot\mathbf{F}(\tau)\rangle \tag{5.9.18}$$

It is usually assumed that the random force $\mathbf{F}(\tau)$ is uncorrelated with the initial velocity, that is,

$$\langle\mathbf{V}(0)\cdot\mathbf{F}(\tau)\rangle = 0 \tag{5.9.19}$$

This actually can be proved (cf. Chapter 11). The physical reason is that $\mathbf{F}(t)$ depends only on fluctuations in the solvent and not on the initial velocity of the B-particle.

The velocity-correlation function for a Brownian particle thus turns out to be

$$\phi(t) \equiv \langle\mathbf{V}(0)\cdot\mathbf{V}(t)\rangle = \frac{3k_BT}{M}\exp-\left(\frac{\zeta}{M}\right)t \tag{5.9.20}$$

where we have used the equipartition theorem

$$\frac{1}{2} M \langle V^2 \rangle = \frac{3}{2} k_B T \tag{5.9.21}$$

to find $\langle \mathbf{V}(0) \cdot \mathbf{V}(0) \rangle = \langle V^2 \rangle$.

Substitution of $\phi(t)$ into the Kubo relation [Eq. (5.9.14)] yields

$$D = \frac{k_B T}{\zeta} \tag{5.9.22}$$

This is called the *Einstein relation* [cf. Eq. (5.4.12)].

Substitution of $\phi(t)$ into Eq. (5.9.10) followed by the required integrations leads to

$$\langle \Delta R^2(t) \rangle = 6Dt - 6D(M/\zeta)\left[1 - \exp - \left(\frac{\zeta}{M}\right) t\right] \tag{5.9.23}$$

For times such that $(\zeta/M)t$ is small; that is, for $t \ll M/\zeta$, the exponential can be expanded so that to lowest order in t

$$\langle \Delta R^2(t) \rangle = \frac{3k_B T}{M} t^2 = \langle V^2 \rangle t^2 \tag{5.9.24}$$

This is the result for a freely moving particle. Initially the mean-square displacement behaves as it would for a free particle because for very short times the solvent molecules are essentially stationary, and the particle experiences a constant force.

For times $t \gg M/\zeta$

$$\langle \Delta R^2(t) \rangle = 6Dt - \frac{6Mk_B T}{\zeta^2} = 6Dt - \text{constant} \tag{5.9.25}$$

as we expected.

In order to understand the situation that pertains to macromolecules we note that for polystyrene spheres of radius 0.01 μ in aqueous solution at room temperature $M/\zeta = 2.2 \times 10^{-9}$ sec, $D = 2.2 \times 10^{-7}$ cm^2/sec, $\zeta = 1.9 \times 10^{-7}$ gm/sec, and $6Mk_BT/\zeta^2 = 2.88 \times 10^{-15}$ cm^2.
In typical light-scattering experiments $\langle \Delta R^2(t) \rangle$ must be of order $q^{-2} \sim 10^{-10}$ cm^2. This is very much larger than $6Mk_BT/\zeta^2 \sim 10^{-15}$ cm^2 so that the constant term in Eq. (5.9.25) can be ignored. The general condition for omission of the constant term is that $\dfrac{6Mk_BT\, q^2}{\zeta^2} \ll 1$.

APPENDIX 5.A THE CALCULATION OF THE MEAN-SQUARE DISPLACEMENT

The integral

$$\langle \Delta R^2(t) \rangle = \int_0^t dt_2 \int_0^t dt_1 \langle \mathbf{V}(t_1) \cdot \mathbf{V}(t_2) \rangle$$

simplifies considerably because of the following properties:

a. $\mathbf{V}$ is a stationary random process; that is, $\langle \mathbf{V}(t_1) \cdot \mathbf{V}(t_2) \rangle$ is independent of the origin of time

$$\langle \mathbf{V}(t_1) \cdot \mathbf{V}(t_2) \rangle = \langle \mathbf{V}(0) \cdot \mathbf{V}(t_2 - t_1) \rangle$$

b. $\langle \mathbf{V}(0) \cdot \mathbf{V}(t) \rangle \to 0$ as $t \to \infty$

c. $\langle \mathbf{V}(0) \cdot \mathbf{V}(t) \rangle$ is an even function of time

$$\langle \mathbf{V}(0) \cdot \mathbf{V}(-t) \rangle = \langle \mathbf{V}(0) \cdot \mathbf{V}(t) \rangle$$

The region of integration is represented in Fig. 5.A.1. Subregion A is defined by $t_2 < t_1$ and subregion B defined by $t_1 > t_2$. Integration of $\langle \mathbf{V}(t_1) \cdot \mathbf{V}(t_2) \rangle$ over A and B must give the same result because of the preceding properties of the integrand. Thus the total integral is simply twice the integral over region A, or

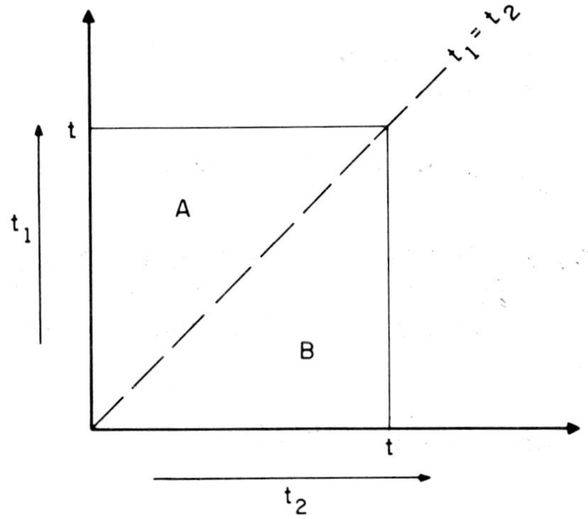

FIG. 5.A.1. The region of integration of Eq. (5.9.9) is the square. Subregion A corresponds to $t_1 > t_2$ and Subregion B corresponds to $t_2 > t_1$.

$$\langle \Delta R^2(t) \rangle = 2 \int_0^t dt_2 \int_0^{t_2} dt_1 \langle \mathbf{V}(0) \cdot \mathbf{V}(t_2 - t_1) \rangle$$

Changing variables to $y = (t_2 - t_1)$ which has limits $(t_2, 0)$ and $x = t_2$ gives

$$\langle \Delta R^2(t) \rangle = 2 \int_0^t dx \int_0^x dy \langle \mathbf{V}(0) \cdot \mathbf{V}(y) \rangle$$

Integration by parts with

$$u = \int_0^x dy \langle \mathbf{V}(0) \cdot \mathbf{V}(y) \rangle, \quad dv = dy$$

$$du = \langle \mathbf{V}(0) \cdot \mathbf{V}(y) \rangle\, dy, \qquad v = y$$

gives

$$\langle \Delta R^2(t)\rangle = 2y \int_0^x dy \langle \mathbf{V}(0) \cdot \mathbf{V}(y)\rangle \Big|_0^t - 2\int_0^t dy\, y \langle \mathbf{V}(0) \cdot \mathbf{V}(y)\rangle$$

$$= 2t \int_0^t dy \langle \mathbf{V}(0) \cdot \mathbf{V}(y)\rangle - 2\int_0^t dy\, y \langle \mathbf{V}(0) \cdot \mathbf{V}(y)\rangle$$

or

$$\langle \Delta R_2(t)\rangle = 2\int_0^t dy (t - y) \langle \mathbf{V}(0) \cdot \mathbf{V}(y)\rangle$$

Thus proving Eq. (5.9.10).

NOTES

1. This is called the *Einstein summation convention.*
2. A second-rank tensor quantity such as the polarizability can be written in the form of Eq. (5.2.7) for tedrahedral or higher molecular point group symmetries.
3. For $I_{if}^{\alpha}(\mathbf{q}, t) = (\mathbf{n}_i \cdot \mathbf{n}_f)^2 \alpha^2 F_1(\mathbf{q}, t)$. This follows from Eq. (3.3.3).
4. The Fourier transform of ρ_0 is proportional to $\delta(\mathbf{q})$ which is zero for $\mathbf{q} \neq 0$; that is,

$$\int_V d^3r\, e^{i\mathbf{q}\cdot\mathbf{r}} \propto \mathrm{v}(2\pi)^3 \delta(\mathbf{q})$$

5. I_{VH} and I_{HV} are called the *depolarized components* of the scattered light.
6. One might expect the solvent scattering to be comparable or larger than the solute scattering in dilute macromolecular solutions because of the much larger number of solvent molecules relative to solute molecules. However, this is usually not the case because the solvent structure gives rise to much more destructive interference than does the macromolecular structure. These effects are discussed in Chapter 8.
7. Although we ignore the solvent scattering here, we treat the general problem of scattering from binary mixtures in Sections 10.6 and 13.5. The conditions under which solvent scattering may be ignored are discussed in those sections.
8. If two variables x, y are statistically independent

$$\langle x(0)y(t)\rangle = \langle x\rangle\langle y\rangle$$

Since $\langle \exp i\mathbf{q} \cdot \mathbf{r}_j(t)\rangle \propto \delta(\mathbf{q}) = 0$ for $\mathbf{q} \neq 0$ it follows that for $j \neq l$ $\langle \exp -i\mathbf{q} \cdot \mathbf{r}_j(0) \exp i\mathbf{q} \cdot \mathbf{r}_l(t)\rangle = 0$

9. In diffusion theory the mean-square displacement of a particle along a given axis is equal to $2Dt$. Setting this equal to L^2 gives the time scale in Eq. (5.4.1a). The factor of 2 is omitted since we are interested only in the order of magnitude of the time scale. Likewise by setting the mean square displacement Dt equal to $(q^{-1})^2$ we obtain Eq. (5.4.1b).
10. It should be noted that here we are describing the phenomenon of self-diffusion. The quantity D is to be regarded as the "self-diffusion" coefficient, a quantity often measured in tracer diffusion, NMR, and incoherent neutron scattering experiments. It is the assumption of particle independence in an "infinitely dilute" solution that allows us to describe the light scattering in terms of the theory of self-diffusion. In Chapters 9, 10, and 13 we consider mutual diffusion. Light-scattering experiments determine mutual diffusion coefficients in general and the above theory is valid only in the limit of infinite dilution.
11. The Fourier transform of an exponential $e^{-\alpha|t|}$ is

$$\frac{1}{2\pi}\int_{-\infty}^{+\infty} dt\, e^{-i\omega t}\, e^{-\alpha|t|} = Re\ \exp\frac{1}{\pi}\int_0^{\infty} dt\, e^{-i\omega t}\, e^{-\alpha t} = \pi^{-1}\ Re \exp\frac{1}{i\omega + \alpha} = \pi^{-1}\left[\frac{\alpha}{\omega^2 + \alpha^2}\right]$$

12. Note that where the spectral density has a δ function singularity, the correlation function has a constant (in time) background.

13. This follows from the grand canonical ensemble.
14. It should be noted that number fluctuations do not depend on q. In fact, coherent light is not required for this experiment. The intensity of the scattered light is then proportional to $N(t)$ and $\langle i(o)i(t)\rangle \propto \langle N(o)N(t)\rangle$. It is convenient to designate the coherent effects as *interference fluctuations* and the incoherent effects as *number fluctuations*.
15. The spectrum corresponding to Eq. (5.6.9) is a Gaussian function of ω of half-width

$$\omega_q = 2\sqrt{\frac{\ln 2}{\tau_q}}$$

For a given q, τ_q, and ω_q are simply a measure of the temperature.
16. In any case, since the time scale is very fast, filter methods are normally used and this question is moot.
17. Only if the bacteria are regarded as uniform rigid, optically isotropic spheres, can these formulas apply.
18. Bacteria are usually large nonspherical particles. Internal interference effects can be very important. Recent work shows that qt scaling should not occur in these cases. In this event it will be difficult to determine the speed distribution from interference fluctuations (see Berne and Nossal, 1974).
19. One important point that should be noted is that optical mixing experiments cannot be used to find the direction of $\mathbf{V}$ since $S_1(\mathbf{q}, \omega)$ is unchanged when $\mathbf{V} \to -\mathbf{V}$. In principle, filter experiments would give the direction of $\mathbf{V}$ since the spectrum of $F_s(\mathbf{q}, t)$ not $ReF_s(\mathbf{q}, t)$ is measured in these experiments. Since the applications that we discuss involve polymers, the motions are too slow for filter methods. The direction of the velocity can be determined by observation of the sense of motion of the laser speckle patterns.
20. This shift is the change in frequency of a light wave (photon) on being scattered by a uniformly moving particle. The well-known formula for the Doppler shift is

$$\omega_f = \omega_i \left[1 + \frac{\hat{\mathbf{q}} \cdot \mathbf{V}}{c}\right]$$

where c is the velocity of light and $\hat{\mathbf{q}}$ is a unit vector in the direction of $\mathbf{q}$. Solving this equation for $\omega(\mathbf{q}) \equiv (\omega_i - \omega_f)$ gives Eq. (5.8.11).
21. This is found by balancing the frictional force against the force due to the field

$$-\zeta\mathbf{V} + Q\mathbf{E} = 0$$

22. The characteristic time required for the particle to reach its terminal velocity is M/ζ, which is usually many orders of magnitude shorter than the decay time of $F_s(\mathbf{q}, t)$.
23. This spectrum is merely the superposition of the spectra each species would have in the absence of the other.
24. A distribution function is called pathological here if it does not obey the central limit conditions.
25. All transport coefficients can be expressed in terms of time-correlation functions.
26. The root mean square (rms) velocity of a molecule is $\langle V^2\rangle^{1/2} = (3k_BT/M)^{1/2}$, so that the larger M, the slower the molecule moves.

REFERENCES

Ben-Yosef, N, Zeigenbaum, S., and Weitz, A., *Appl. Phys. Lett.* **21,** 436 (1972)

Berge, P., Volochine, B., Billard, R., and Hamelin, N. *C.R. Acad. Sci. Ser. D* **265,** 889 (1967)

Berne, B. J. and Nossal, R., *Biophys.* in press (1974).

Berne, B. J. and Pecora, R. *Ann. Rev. Phys. Chem.* (1974).

Berne, B. J. and Schaefer, D. W., Biophys, in Press (1975).

Chandrasekar, S., *Rev. Mod. Phys.* **15,** 1 (1943).

Dubin, S. B., Benedek, B. G., Bancroft, F. C., Freifelder, D., *J. Mol. Biol.* **54,** 547 (1970).

Foord, R., Jakeman, E., Oliver, C. J., Pike, E. R., Blagrove, R. J., Wood, E., and Peacocke, A. R., *Nature* (Lond.) **227,** 242 (1972).

Greytak, T. J. and Benedek, G. B., *Phys. Rev. Lett.*, **17,** 179 (1966).

Koppel. D. E., Ph. D. dissertation, Columbia University (1973).

Landau, L. D. and Lifshitz, I. M. *Fluid Mechanics* Addison-Wesley, Reading, Mass. (1959).

Lee, W. I. and Schurr, J. M. *Biopolym.* **13,** 903 (1974).

Mustacich, R. V. and Ware, B. R., *Phys. Rev. Lett.* **33,** 617 (1974).

Nelkin, M. and Yip, S., *Phys. Fluids* **9,** 380 (1966).

Newman, J., Swinney, H. L., Berkowitz, S. A. and Day, L. A. *Biochem.* **13,** 4832 (1974).

Nossal, R. and Chen, S. H., *Jour. Phys. (Paris)*. Suppl. **33,** Cl-171 (1972).

Nossal, R., Chen, S. H., and Lai, C. C., *Opt. Comm.* **4,** 35 (1971).

Raj, R. and Flygare, W. H., *Biochem.* **13,** 3336 (1974).

Schaefer, D. W., *Science* **180,** 1293 (1973).

Schaefer, D. W., "Applications of photon statistics and photon correlations," in *Laser Applications to Optics and Spectroscopy,* Jacobs, S. F., Scully, M. O. and Sargent, M. (eds.), Addison-Wesley, Reading, Mass. (1974).

Schaefer, D. W., Banks, G., Alpert, S., Biophys. J., in press (1975)

Schaefer, D. W. and Berne, B. J., *Phys. Rev. Lett.* **28,** 475 (1972).

Schleich, T. and Yeh, Y. Biopolym. **12,** 993 (1973).

Tanaka, T. and Benedek, G. B., *Applied Optics,* in press (1974).

Tanford, C., *Physical Chemistry of Macromolecules,* Wiley, New York (1961).

Uzgiris, E. E., *Opt. Comm.* **6,** 55 (1972).

Van Holde, *Physical Biochemisty,* Prentice-Hall, Englewood Cliffs, N.J. (1971).

Van Hove, *L., Phys. Rev.* **95,** 249 (1954).

Ware, B. R., *Adv. Coll. Interface Sci.* **4,** 1 (1974).

Ware, B. R., and Flygare, W. H., *Chem. Phys. Lett.* **12,** 81 (1971).

Ware, B. R. and Flygare, W. H., *J. Coll. Int. Sci.* **39,** 670 (1972).

Wilson, W. W., Luzzana, M. R., Penniston, J. T., and Johnson, C. S., *Proc. Nat. Acad. Sci.* U S A **71,** 1260 (1974).

Zwanzig, R., *Ann. Rev. Phys. Chem.,* **61,** 670 (1965).

CHAPTER 6

FLUCTUATONS IN CHEMICALLY REACTING SYSTEMS

6 · 1 INTRODUCTION

In Chapter 5 it was shown that the scattered field $E_s(\mathbf{q}, t)$ fluctuates on a time scale determined by the time it takes a particle to translate a distance q^{-1}. Actually the situation can be far more complicated, as can be seen by considering a typical term, $\alpha_j \exp i\mathbf{q} \cdot \mathbf{r}_j$ in Eq. (3.3.4). Suppose that the molecule j can undergo a chemical transformation. Then its polarizability α_j jumps from one value to another as the state changes. The instantaneous value of $\alpha_j(t)$ can be expressed as $\alpha_j(t) = \bar{\alpha}_j + \delta\alpha_j(t)$ where $\bar{\alpha}_j$ is the average value of the polarizability over all states of j and $\delta\alpha_j(t)$ is the deviation of $\alpha_j(t)$ from this average value. Thus $\alpha_j(t) \exp i\mathbf{q} \cdot \mathbf{r}_j(t) = \bar{\alpha}_j \exp i\mathbf{q} \cdot \mathbf{r}_j(t) + \delta\alpha_j(t) \exp i\mathbf{q} \cdot \mathbf{r}_j$. The first term fluctuates on a time scale τ_q determined by the time it takes a molecule to translate a distance q^{-1} whereas the second term fluctuates on two time scales, one determined by the translations (through $\exp i\mathbf{q} \cdot \mathbf{r}_j(t)$) and the other determined by the chemical reactions through $\delta\alpha_j(t)$. The time correlation function then consists of two parts: (a) a purely diffusive part $\bar{\alpha}_j^2 \exp q^2 Dt$ and (b) a reactive part $\langle\delta\alpha_j(\mathrm{o})\delta\alpha_j(t) \exp i\mathbf{q} \cdot \Delta r_j(t)\rangle$. The integrated intensities corresponding to these two parts are, respectively, $\bar{\alpha}_j^2$ and $\langle\delta\alpha_j^2\rangle$, so that our ability to resolve the reactive part is proportional to $\langle\delta\alpha^2{}_j\rangle/\bar{\alpha}^2{}_j$. Since the polarizability is not expected to change very much in a chemical reaction, $\langle\delta\alpha^2\rangle/\bar{\alpha}^2$ is expected to be small, and this experiment should be quite difficult. There are nevertheless several modifications of the experiment that lead to promising possibilities for determining the reactive term and thereby rate constants. For example, consider the case of a dilute solution of molecules that can exist in either of two states A or B. Suppose that these two states have identical polarizabilities but different diffusion coefficients and electrical mobilities. In the presence of a static electric field, the light scattered from A and B will have different Doppler shifts, so that in the absence of chemical transformation A and B will give rise to two bands located at different frequencies with diffusion widths as in Eq. (5.8.19). If A can transform to B and vice versa, the spectrum should look quite different and should contain information about the rate constants. This is similar to NMR studies of exchange rates. In NMR when a proton jumps from one environment to another its Larmour frequency changes by

---

This chapter can be regarded as optional since (a) it is not necessary for the logical development of the subject and (b) no light-scattering experiment has yet been performed which unambiguously gives rate constants. The reader's attention is directed to Section (6.6) for a discussion of Fluorescence Fluctuation Spectroscopy which has been successfully used to determine rate constants.

virtue of the different chemical shifts in the two different environments. The NMR spectrum then reflects the exchange rate, that is, the rate at which the proton jumps.

In this chapter we present the theory of concentration fluctuations in chemically reacting systems. Only the simplest cases are considered in detail. The results of this chapter find application not only in light scattering, but also in fluorescence fluctuation spectroscopy (see Section 6.6).

6 · 2 FORMULATION OF MODEL

In this section we consider the additional mechanism for the fluctuations in concentration—chemical reactions. The simplest case to consider is that of an ideal solution containing two species that are in chemical equilibrium with each other. Each of the species, can change its concentration in a given volume either by diffusing into or out of the volume (as before) or by the

$$A \underset{k_b}{\overset{k_a}{\rightleftharpoons}} B \tag{6.2.1}$$

chemical reaction where k_a and k_b are, respectively, the forward and backward rate constants. If A and B have different polarizabilities and therefore different scattering powers or have different dynamics (different diffusion coefficients or electrophoretic mobilities), we expect the scattered light spectrum to depend on the chemical rate constants. If A and B are optically and dynamically identical (or very similar), then as far as light scattering is concerned they *are* "identical" and the chemical transformation would not appear in expressions for the scattered spectrum.

In this chapter this simple model is used to illustrate all these effects. We follow the treatment of Berne and Giniger (1973). In fact, with linearization and a suitable redefinition of parameters, the results for the above reaction can be applied to the dimerization problem (see Section 6.5)

$$2A_1 \rightleftharpoons A_2$$

Let us consider then the simple reaction Eq. (6.2.1) with equilibrium constant, K_{eq}

$$K_{eq} = \frac{k_a}{k_b} = \frac{c_2^\circ}{c_1^\circ} \tag{6.2.2}$$

where c_1 and c_2° are, respectively, the equilibrium concentrations of A and B. The equilibrium takes place in dilute solution. Conditions are assumed to be such that scattering from the solutes is separable from that of the solvent (cf. Section 10.B).

The polarizability fluctuation given by Eq. (3.3.4) is now the sum over molecules in the state A with polarizability α_1 and molecules in the state B with polarizability α_2 so that

$$\alpha_{if}(\mathbf{q}, t) = (\mathbf{n}_i \cdot \mathbf{n}_f)\{\alpha_1 \delta c_1(\mathbf{q}, t) + \alpha_2 \delta c_2(\mathbf{q}, t)\} \tag{6.2.3}$$

where $\delta c_1(\mathbf{q}, t)$ and $\delta c_2(\mathbf{q}, t)$ are, respectively the spatial Fourier transforms of the concentration fluctuations[1] $\delta c_1(\mathbf{r}, t)$ and $\delta c_2(\mathbf{r}, t)$. Substitution of Eq. (6.2.3) into Eq. (3.3.3b) then shows that $I_{if}^{\alpha}(\mathbf{q}, t)$ is proportional to

$$S(\mathbf{q}, t) = Re \sum_{i,j=1}^{2} \alpha_i \alpha_j F_{ij}(\mathbf{q}, t) \tag{6.2.4}$$

where

$$F_{ij}(\mathbf{q}, t) \equiv \langle \delta c_i^*(\mathbf{q}, 0)\, \delta c_j(\mathbf{q}, t) \rangle \tag{6.2.5}$$

or the corresponding spectrum $I_{if}^{\alpha}(\mathbf{q}, \omega)$ is proportional to

$$S(\mathbf{q}, \omega) = \pi^{-1}\, Re\, \tilde{S}(\mathbf{q}, s = i\omega) \tag{6.2.6}$$

where $\tilde{S}(\mathbf{q}, s = i\omega)$ is the Laplace transform[2] of $S(\mathbf{q}, t)$.

In order to proceed we must calculate $F_{ij}(\mathbf{q}, t)$. Thus $\delta c_1(\mathbf{r}, t)$ and $\delta c_2(\mathbf{r}, t)$ must be found. From the discussion given in Section 6.1 it is clear that c_1 and c_2 can change by: (1) diffusion, (2) electrophoretic translation, and (3) chemical reaction.

It is a simple matter to generalize Eq. (5.8.3) to include the chemical reaction (Eq. (6.2.1))

$$\frac{\partial c_1}{\partial t} + \nabla \cdot \mathbf{J}_1 = k_b c_2 - k_a c_1$$

$$\frac{\partial c_2}{\partial t} + \nabla \cdot \mathbf{J}_2 = k_a c_1 - k_b c_2 \tag{6.2.7}$$

where the terms on the right-hand side include the effects of the chemical reaction and $\mathbf{J}_1$ and $\mathbf{J}_2$ are the fluxes of A and B, respectively. In the presence of an external electric field $\mathbf{E}$ the fluxes become[3]

$$\mathbf{J}_1 = \mu_1 \mathbf{E} c_1 - D_1 \boldsymbol{\nabla} c_1$$

and

$$\mathbf{J}_2 = \mu_2 \mathbf{E} c_2 - D_2 \boldsymbol{\nabla} c_2$$

where μ_1, μ_2, D_1, and D_2 are the respective electrophoretic mobilities and diffusion coefficients of A and B. If the external field is zero $\mathbf{J}_1$ and $\mathbf{J}_2$ reduce to the usual Fick's law result (Eq. 5.8.1).

It remains to solve the linearized Eq. (6.2.7) and to evaluate $F_{ij}(\mathbf{q}, t)$. This is most easily accomplished by first finding the spatial Fourier component of Eq. (6.2.7). This is

$$\left(\frac{\partial}{\partial t} + \lambda_1\right) c_1(\mathbf{q}, t) = k_b c_2(\mathbf{q}, t)$$

$$\left(\frac{\partial}{\partial t} + \lambda_2\right) c_2(\mathbf{q}, t) = k_a c_1(\mathbf{q}, t) \tag{6.2.9}$$

where $c_i(\mathbf{q}, t) \equiv \delta c_i(\mathbf{q}, t)$ is the spatial Fourier transform of $\delta c_i(\mathbf{r}, t)$ and

$$\lambda_1 \equiv q^2 D_1 + k_a - i\omega_1(\mathbf{q}) \equiv \gamma_1 - i\omega_1(\mathbf{q})$$
$$\lambda_2 \equiv q^2 D_2 + k_b - i\omega_2(\mathbf{q}) \equiv \gamma_2 - i\omega_2(\mathbf{q}) \qquad (6.2.10)$$

Here

$$\omega_i(\mathbf{q}) \equiv \mu_i(\mathbf{q} \cdot \mathbf{E}) \qquad i = 1, 2$$

are the Doppler shifts and γ_1 and γ_2 are defined by Eq. (6.2.10)

The simplest procedure for evaluating $F_{ij}(\mathbf{q}, t)$ is as follows[4]

a. Take the Laplace transform[5] of Eq. (6.2.9).
b. Solve the resulting set of linear algebraic equations for $c_i(\mathbf{q}, s)$.
c. Multiply $\mathbf{c}_j(\mathbf{q}, s)$ by $c_i^*(\mathbf{q}, 0)$ and average over the equilibrium ensemble to get $\tilde{\mathbf{F}}_{ij}(\mathbf{q}, s)$, the Laplace transforms of $F_{ij}(\mathbf{q}, t)$.
d. Laplace invert the resulting $\tilde{\mathbf{F}}_{ij}(\mathbf{q}, s)$ to find $F_{ij}(\mathbf{q}, t)$. Actually, step d can be bypassed altogether since we can evaluate the spectrum from the Laplace transforms using Eq. (6.2.6). Nevertheless, we will determine $F_{ij}(\mathbf{q}, t)$.

Steps a–c then give

$$\tilde{\mathbf{F}}_{11}(\mathbf{q}, s) = \frac{(s + \lambda_2)\, F_{11}(\mathbf{q}) + k_b F_{12}(\mathbf{q})}{\Delta(s)} \qquad (6.2.11a)$$

$$\tilde{\mathbf{F}}_{21}(\mathbf{q}, s) = \frac{(s + \lambda_2)\, F_{21}(\mathbf{q}) + k_b F_{22}(\mathbf{q})}{\Delta(s)} \qquad (6.2.11b)$$

$$\tilde{\mathbf{F}}_{12}(\mathbf{q}, s) = \frac{(s + \lambda_1)\, F_{12}(\mathbf{q}) + k_a F_{11}(\mathbf{q})}{\Delta(s)} \qquad (6.2.11c)$$

$$\tilde{\mathbf{F}}_{22}(\mathbf{q}, s) = \frac{(s + \lambda_1)\, F_{22}(\mathbf{q}) + k_a F_{21}(\mathbf{q})}{\Delta(s)} \qquad (6.2.11d)$$

where

$$\Delta(s) = (s + \lambda_1)(s + \lambda_2) - k_a k_b \qquad (6.2.11e)$$

and

$$\tilde{F}_{ij}(\mathbf{q}) \equiv F_{ij}(\mathbf{q}, t = 0) \equiv \langle \delta c_i^*(\mathbf{q}) \delta c_j(\mathbf{q}) \rangle \qquad (6.2.11f)$$

are the equilibrium second moments of the fluctuations.

Equations (6.2.11) present the general results for this model. They do not yet assume solution ideality but do ignore scattering from the solvent. The ideality assumption simplifies these results considerably[6]

$$F_{ij}(\mathbf{q}) = c_j^\circ \delta_{ij}$$

Then

$$\tilde{\mathbf{F}}_{11}(\mathbf{q}, s) = c_1^\circ \frac{[s + \lambda_2]}{\Delta(s)}$$

$$\tilde{\mathbf{F}}_{21}(\mathbf{q}, s) = c_2^\circ \frac{k_b}{\Delta(s)}$$

$$\tilde{\mathbf{F}}_{12}(\mathbf{q}, s) = c_1^\circ \frac{k_a}{\Delta(s)} \tag{6.2.12}$$

$$\tilde{\mathbf{F}}_{22}(\mathbf{q}, s) = c_2^\circ \frac{(s + \lambda_1)}{\Delta(s)}$$

Let us first consider some general methods of rewriting these expressions in forms such that all relevant spectra can be calculated with relative ease in limiting cases.

The quantity $\Delta(s)$ may be rewritten as

$$\Delta(s) = (s - s_+)(s - s_-) \tag{6.2.13}$$

where $s_\pm$ are the roots of the dispersion equation,[7] $\Delta(s) = 0$

$$s_\pm = -\tfrac{1}{2}[(\gamma_1 + \gamma_2) - i(\omega_1 + \omega_2)] \pm \tfrac{1}{2}[[(\gamma_1 - \gamma_2) - i(\omega_1 - \omega_2)]^2 + 4k_a k_b]^{1/2} \tag{6.2.14}$$

Laplace inversion of Eqs. (6.2.12) then gives

$$F_{11}(\mathbf{q}, t) = c_1^\circ \left\{\frac{(s_+ + \lambda_2) \exp s_+|t| - (s_- + \lambda_2) \exp s_-|t|}{(s_+ - s_-)}\right\} \tag{6.2.15a}$$

$$F_{21}(\mathbf{q}, t) = c_2^\circ k_b \left\{\frac{\exp s_+|t| - \exp s_-|t|}{(s_+ - s_-)}\right\} \tag{6.2.15b}$$

$$F_{12}(\mathbf{q}, t) = c_1^\circ k_a \left\{\frac{\exp s_+|t| - \exp s_-|t|}{(s_+ - s_-)}\right\} \tag{6.2.15c}$$

$$F_{22}(\mathbf{q}, t) = c_2^\circ \left\{\frac{(s_+ + \lambda_1) \exp s_+|t| - (s_- + \lambda_1) \exp s_-|t|}{(s_+ - s_-)}\right\} \tag{6.2.15d}$$

Now we note from Eq. (6.2.14) that $s_\pm$ and thereby $F_{ij}(\mathbf{q}, t)$ are complex functions. The presence of imaginary parts in $s_\pm$ is entirely due to the Doppler frequencies ω_1 and ω_2. Thus if we denote by $F_{ij}^{(+)}(\mathbf{q}, t)$ the above functions and by $F_{ij}^{(-)}(\mathbf{q}, t)$ the above functions with ω_1 and ω_2 replaced by $-\omega_1$ and $-\omega_2$, respectively then $F_{ij}^{(-)}(\mathbf{q}, t)$ is the complex conjugate of $F_{ij}^{(+)}(\mathbf{q}, t)$ so that

$$Re\ F_{ij}(\mathbf{q}, t) = 1/2\ \{F_{ij}^{(+)}(\mathbf{q}, t) + F_{ij}^{(-)}(\mathbf{q}, t)\} \tag{6.2.16}$$

It follows from Eq. (6.2.4) that

$$S(\mathbf{q}, t) = 1/2 \sum_{ij} \alpha_i \alpha_j \{F_{ij}^{(+)}(\mathbf{q}, t) + F_{ij}^{(-)}(\mathbf{q}, t)\} \tag{6.2.17}$$

Substitution of this into Eq. (6.2.6) then gives the spectrum as

$$S(\mathbf{q}, \omega) = \tfrac{1}{2}[S^{(+)}(\mathbf{q}, \omega) + S^{(-)}(\mathbf{q}, \omega)] \tag{6.2.18}$$

where

$$S^{(\pm)}(\mathbf{q}, \omega) \equiv \pi^{-1} \sum_{ij} \alpha_i \alpha_j \tilde{F}_{ij}^{(\pm)}(\mathbf{q}, s = i\omega) \tag{6.2.19}$$

Here $\tilde{F}_{ij}^{(\pm)}(\mathbf{q}, s)$ is the Laplace transform of $F_{ij}^{(\pm)}(\mathbf{q}, t)$. Substitution of Eqs. (6.2.15) into Eq. (6.2.18) then gives

$$S^{(\pm)}(\mathbf{q}, \omega) = \pi^{-1} \, Re \left\{ \frac{N^{(\pm)}(\mathbf{q}, \omega)}{D^{(\pm)}(\mathbf{q}, \omega)} \right\} \tag{6.2.20}$$

where

$$\begin{aligned} N^{(\pm)}(\mathbf{q}, \omega) = {} & \alpha_1^2 F_{11}(\mathbf{q})[\gamma_2 + i(\omega \mp \omega_2)] + \alpha_2^2 F_{22}(\mathbf{q})[\gamma_1 + i(\omega \mp \omega_1)] \\ & + \alpha_1\alpha_2[k_a F_{11}(\mathbf{q}) + k_b F_{22}(\mathbf{q})] \end{aligned} \tag{6.2.21}$$

$$D^{(\pm)}(\mathbf{q}, \omega) = [\gamma_1\gamma_2 - k_a k_b - (\omega \mp \omega_1)(\omega \mp \omega_2)] + i[\gamma_2(\omega + \omega_1) + \gamma_1(\omega + \omega_2)] \tag{6.2.22}$$

This is a complicated spectrum. In many cases the spectrum should simplify considerably. In the following sections we consider several limits and introduce perturbation techniques for the investigation of these limits.

6 · 3 ELECTROPHORESIS: THE FAST AND SLOW EXCHANGE LIMITS

The spectrum of a chemically reacting system is in general a very complicated function of the frequency. In this section we explore two limits in which the form of the spectrum simplifies considerably. These are the *slow exchange* and *fast exchange* limits. In the slow exchange limit the rates k_a and k_b are small compared to the Doppler frequencies ω_1 and ω_2, whereas in the fast exchange limit, the contrary is valid; that is, the rates k_a and k_b are large compared to the Doppler frequencies. We now consider each of these limits.

THE SLOW EXCHANGE LIMIT

The slow exchange limit can be investigated using a perturbation solution of the dispersion equation

$$\Delta(s) = (s + \lambda_1)(s + \lambda_2) - k_a k_b = 0 \tag{6.3.1}$$

which has the explicit form

$$\begin{aligned} \Delta(s) = {} & s^2 + [(\gamma_1 + \gamma_2) - i(\omega_1 + \omega_2)]s - i(\omega_1\gamma_2 + \omega_2\gamma_1) + \gamma_1\gamma_2 \\ & - k_a k_b - \omega_1\omega_2 = 0 \end{aligned} \tag{6.3.2}$$

In the slow exchange limit ω_1 and ω_2 are large compared to q^2D_1, q^2D_2, k_a, and k_b, and consequently to γ_1 and γ_2 also. The solution of Eq. (6.3.2) can be expressed as $s = s^{(0)} + s^{(1)} + s^{(2)} + \ldots$ where $s^{(n)}$ is the term of order n in any of the small quantities[8] q^2D_1 q^2D_2, k_a, k_b. This allows us to write a series of perturbation equations for the zeroth, first, second, . . ., nth order term. These are

$$[s^{(0)}]^2 - i(\omega_1 + \omega_2)s^{(0)} - \omega_1\omega_2 = 0 \tag{6.3.3a}$$

$$2s^{(0)}s^{(1)} + (\gamma_1 + \gamma_2)s^{(0)} - i(\omega_1 + \omega_2)s^{(1)} - i(\omega_1\gamma_2 + \omega_2\gamma_1) = 0 \tag{6.3.3b}$$

The solution of the zeroth-order equation is

$$s^{(0)}{}_{\pm} = \begin{cases} i\omega_1 \\ i\omega_2 \end{cases} \tag{6.3.3c}$$

Substituting $s^{(0)}{}_+$ and $s^{(0)}{}_-$ separately into the first-order equation (6.3.3b) gives

$$s^{(1)}{}_{\pm} = \begin{cases} -\gamma_1 \\ -\gamma_2 \end{cases} \tag{6.3.3d}$$

Combining Eqs. (6.3.3c) and (6.3.3c) gives the roots of the dispersion equation to first order in the small quantities q^2D_1, q^2D_2, k_a, and k_b

$$s_{\pm} = \begin{cases} i\omega_1 - \gamma_1 \\ i\omega_2 - \gamma_2 \end{cases} \tag{6.3.3e}$$

Substitution of this into Eq. (6.2.12) then gives $F_{ij}{}^{(+)}(\mathbf{q}, t)$. When these functions are then substituted into Eqs. (6.2.4) and (6.2.6) the results are

$$\begin{aligned} S(\mathbf{q}, t) \equiv{} & A_1 \exp -\gamma_1|t| \cos \omega_1 |t| + A_2 \exp -\gamma_2|t| \cos \omega_2|t| \\ & + B_1 \exp -\gamma_1|t| \sin \omega_1 |t| + B_2 \exp -\gamma_2|t| \sin \omega_2|t| \end{aligned} \tag{6.3.4}$$

and the corresponding spectrum

$$\begin{aligned} S(\mathbf{q},\omega) ={} & A_1\left[\frac{\gamma_1}{(\omega-\omega_1)^2+\gamma_1{}^2}+\frac{\gamma_1}{(\omega+\omega_1)^2+\gamma_1{}^2}\right] \\ & + A_2\left[\frac{\gamma_2}{(\omega-\omega_2)^2+\gamma_2{}^2}+\frac{\gamma_2}{(\omega+\omega_2)^2+\gamma_2{}^2}\right] \\ & + B_1\left[\frac{\omega+\omega_1}{(\omega+\omega_1)^2+\gamma_1{}^2}-\frac{\omega-\omega_1}{(\omega-\omega_1)^2+\gamma_1{}^2}\right] \\ & + B_2\left[\frac{\omega+\omega_2}{(\omega+\omega_2)^2+\gamma_2{}^2}-\frac{\omega-\omega_2}{(\omega-\omega_2)^2+\gamma_2{}^2}\right] \end{aligned} \tag{6.3.5}$$

where the last two terms in brackets are non-Lorentzian and

$$\begin{aligned} A_1 &\equiv \alpha_1{}^2 c_1^\circ + (\gamma_2 - \gamma_1)\, G\, \alpha_1\alpha_2 \\ A_2 &\equiv \alpha_2{}^2 c_2^\circ - (\gamma_2 - \gamma_1)\, G\, \alpha_1\alpha_2 \\ B_1 &\equiv -(\omega_2 - \omega_1)\, G\, \alpha_1\alpha_2 \\ B_2 &\equiv (\omega_2 - \omega_1)\, G\, \alpha_1\alpha_2 \end{aligned} \tag{6.3.6}$$

and to first order in the small quantities

$$G = \frac{(k_a c_1^\circ + k_b c_2^\circ)}{[(\gamma_2 - \gamma_1)^2 + (\omega_2 - \omega_1)^2]} \simeq \frac{(k_a c_1^\circ + k_b c_2^\circ)}{(\omega_2 - \omega_1)^2}$$

If the reaction is turned off ($k_a = k_b = 0$), $G = 0$ and

$$\begin{aligned} A_1 &= c_1\alpha_1{}^2 \\ A_2 &= c_2\alpha_2{}^2 \\ B_1 &= B_2 = 0 \end{aligned} \tag{6.3.7}$$

and the spectrum reduces to the electrophoretic spectrum obtained from a two-component system, that is, two Lorenztians, one centered at ω_1 with diffusive width q^2D_1 and the other centered at ω_2 with width q^2D_2 (cf. Section 5.8). When the reaction is turned on, the pair of Lorentzians at ω_1 and ω_2 are slightly shifted toward each other and become slightly skewed (due to the terms multiplied by B_1 and B_2). It is interesting to note that in the slow exchange limit the rate constants k_a and k_b separately contribute to the widths γ_1 and γ_2, and can in principle be separately determined, along with the coefficients D_1 and D_2. This could be accomplished by determining γ_1 and γ_2 as a function of q^2 (or equivalently $\sin^2 \theta/2$). The slopes of γ_1 and γ_2 versus q^2 give the diffusion constants D_1 and D_2, whereas the intercepts give the corresponding rate constants (k_a for γ_1 and k_b for γ_2).

THE FAST EXCHANGE LIMIT

In the fast exchange limit k_a and k_b are large compared to the frequencies ω_1 and ω_2 and to the diffusion rates q^2D_1 and q^2D_2 The spectrum then simplifies considerably.

Again we use a perturbation solution of the dispersion equation, Eq. (6.3.2). Now, however, we expand $s = s^{(0)} + s^{(1)} + \ . \ . \ . + s^{(n)}$, where $s^{(n)}$ is n^{th} order in any combination of the small quantities q^2D_1, q^2D_2, ω_1 or ω_2. The zero and first-order equations are

$$[s^{(0)}]^2 + (k_a + k_b)\, s^{(0)} = 0 \tag{6.3.8a}$$

$$\begin{aligned} 2s^{(0)}s^{(1)} + (k_a + k_b)\, s^{(1)} + [q^2(D_1 + D_1) - i(\omega_1 + \omega_2)]\, s^{(0)} \\ + k_a[q^2D_2 - i\omega_2] + k_b[q^2D_1 - i\omega_1] = 0 \end{aligned} \tag{6.3.8b}$$

The solution of the zero-order equation (6.3.8a) is

$$s^{(0)}{}_\pm = \begin{cases} 0 \\ -(k_a + k_b) \end{cases} \tag{6.3.8c}$$

Substitution of $s_+^{(0)}$ and then $s_-^{(0)}$ into the first-order equation (6.3.8b) gives

$$s^{(1)}{}_\pm = \begin{cases} -X_1(q^2D_1 - i\omega_1) - X_2(q^2D_2 - i\omega_2) = -q^2D_s + i\omega_s \\ -X_1(q^2D_2 - i\omega_2) - X_2(q^2D_1 - i\omega_1) = -q^2D_b + i\omega_b \end{cases} \tag{6.3.8d}$$

where

$$X_1 \equiv \frac{k_b}{k_a + k_b} \qquad X_2 \equiv \frac{k_a}{k_a + k_b} \tag{6.3.8e}$$

are the mole fractions[9] of A and B, respectively, and

$$\begin{aligned} D_s &\equiv X_1 D_1 + X_2 D_2 \\ \omega_s &\equiv X_1 \omega_1 + X_2 \omega_2 \\ D_b &\equiv X_1 D_2 + X_2 D_1 = D_1 + D_2 - D_s \\ \omega_b &\equiv X_1 \omega_2 + X_2 \omega_1 = \omega_1 + \omega_2 - \omega_s \end{aligned} \tag{6.3.9}$$

It should be noted that ω_s and D_s are the "average' frequency and diffusion coefficients, respectively. From the definition, D_s is of the same order of magnitude as D_b and likewise for ω_s and ω_b. Combining Eqs. (6.3.8c) and (6.3.8d) we obtain the roots to first order in the small quantities

$$s_{\pm} = \begin{cases} -q^2 D_s + i\omega_s \\ -\Gamma_b + i\omega_b \end{cases} \tag{6.3.10}$$

where Γ_b is a damping coefficient

$$\Gamma_b = (k_a + k_b) + q^2 D_b \tag{6.3.11}$$

In the fast exchange limit Γ_b is clearly much larger than $q^2 D_s$ and $q^2 D_b$; that is, $\Gamma_b \gg q^2 D_s, q^2 D_b$.

Substitution of these roots into Eq. (6.2.12) gives $F_{ij}^{(+)}(\mathbf{q}, t)$ as a linear combination of the terms

$$\exp -\Gamma_b|t| \exp i\omega_b|t| \text{ and } \exp -q^2 D_s|t| \exp i\omega_s|t|$$

The first is a very rapidly decaying function compared with the second because $\Gamma_b \gg q^2 D_s$. Thus the spectrum which is related to the Fourier transform of these functions should consist of a very sharp band located at ω_s with half-width $q^2 D_s$ superposed on a very broad band at ω_b with half-width Γ_b. For sufficiently large Γ_b all that will be seen is the sharp band.[10] Because this depends only on D_s and ω_s it does not give any information about the rate constants per se. The roots in Eq(6.3.10) give the spectrum

$$\begin{aligned} S(q, \omega) = A &\left\{\frac{q^2 D_s}{(\omega - \omega_s)^2 + [q^2 D_s]^2} + \frac{q^2 D_s}{(\omega + \omega_s)^2 + [q^2 D_s]^2}\right\} \\ + B &\left\{\frac{\Gamma_b}{(\omega - \omega_b)^2 + \Gamma_b^2} + \frac{\Gamma_b}{(\omega + \omega_b)^2 + \Gamma_b^2}\right\} \end{aligned} \tag{6.3.12}$$

where A and B are constants (cf. Berne and Giniger, 1973) and where we have ignored non-Lorentzian terms like the latter terms (B_1, B_2) in Eq. (6.3.5). Since the second term in Eq. (6.3.12) is small compared to the first, the observed spectrum will consist largely of the sharp term at ω_s with width $q^2 D_s$. It is important to note that this sharp band depends only on the equilibrium concentrations X_1 and X_2 and consequently gives no information about the kinetics. The broad band, which unfortunately may be difficult to observe, contains the rate information. Light scattering will probably not be useful for the measurement of rate constants in this limit.

The results for both the fast and slow exchange limits are physically reasonable. In the fast exchange limit one "sees" an "average" scattering line, whereas in the slow exchange limit one "sees" the separate Doppler lines of A and B and their respective relaxation times. The results are analogous to those in nuclear magnetic resonance spectra (see Carrington and McLachlan, 1967).

A spectrum illustrating the fast, slow, and "intermediate" exchange cases is given in Fig. 6.3.1.

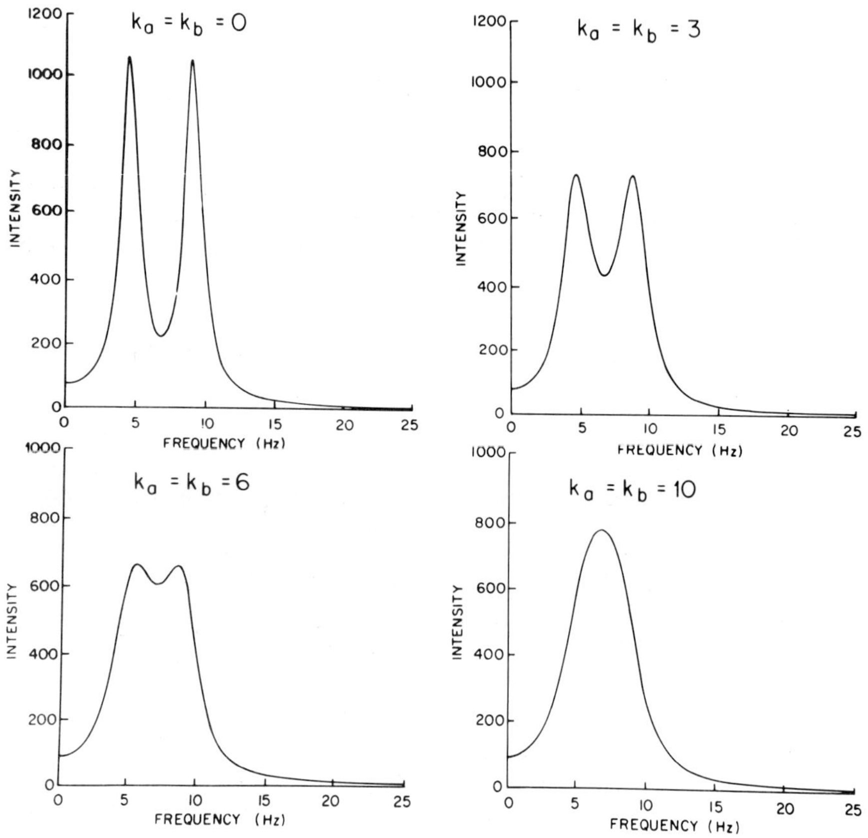

FIG. 6.3.1. A simulated electrophoretic light-scattering spectrum of a chemically reacting system

$$A \underset{k_b}{\overset{k_a}{\rightleftharpoons}} B$$

where for convenience we have taken $k_a = k_b$. This spectrum is based on Eq. (6.2.20). (From Ware, 1974.)

6 · 4 NO EXTERNAL FIELDS

When there are no external fields $\omega_1 = \omega_2 = 0$ and the roots, Eq. (6.2.14) simplify to

$$s_{\pm} = -\tfrac{1}{2}(\gamma_1 + \gamma_2) \pm \tfrac{1}{2}[(\gamma_1 - \gamma_2)^2 + 4k_a k_b]^{1/2} \tag{6.4.1}$$

where, as before, $\gamma_1 \equiv q^2 D_1 + k_a$ and $\gamma_2 \equiv q^2 D_2 + k_b$. Substitution of these roots into Eqs. (6.2.15) and (6.2.4) gives

$$S(\mathbf{q}, t) = B_+ \exp s_+|t| + B_- \exp s_-|t| \tag{6.4.2}$$

where the coefficients $B_\pm$ are

$$B_{\pm} = \pm(s_+ - s_-)^{-1}[\alpha_1{}^2 \mathring{c}_1(s_\pm + \gamma_2) + \alpha_1\alpha_2(\mathring{c}_1 k_a + \mathring{c}_2 k_b) + \alpha_2{}^2 \mathring{c}_2(s_\pm + \gamma_1)] \tag{6.4.3}$$

These represent the relative strengths of the two exponentials $\exp s_+|t|$ and $\exp s_-|t|$.

Although there are only two exponentials in the final correlation function, the reciprocal relaxation times s_+ and s_- and the factors $B_\pm$ governing the relative contributions of the two terms are rather complicated functions of rate constants and diffusion coefficients.

Several special cases follow directly from the form given above.

1. No reaction, $k_a = k_b = 0$. Then from Eqs. (6.4.1) and (6.4.3)

$$\begin{aligned} s_+ &= -q^2 D_2, \quad B^+ = \alpha_2{}^2 \mathring{c}_2 \\ s_- &= -q^2 D_1, \quad B_- = \alpha_1{}^2 \mathring{c}_1 \end{aligned} \tag{6.4.4}$$

and

$$S(\mathbf{q}, t) = [\mathring{c}_1 \alpha_1{}^2 \exp -q^2 D_1 t + \mathring{c}_2 \alpha_2{}^2 \exp -q_2 D_2 t] \tag{6.4.5}$$

which is just the correlation function expected from two independent solutes in a dilute solution.

2. Equal diffusion coefficients, $D_1 = D_2 \equiv D$. Then

$$\begin{aligned} s_+ &= -q^2 D \quad ; \quad B_+ = \bar{\alpha}^2 c \\ s_- &= -(q^2 D + 1/\tau_R); \quad B_- = X_1 X_2(\alpha_1 - \alpha_2)^2\, c \end{aligned} \tag{6.4.6}$$

where $\bar{\alpha}$ is the average polarizability, c is the total concentration of the two components A and B, and τ_R is the kinetic "relaxation time."

$$\begin{aligned} \bar{\alpha} &\equiv X_1\alpha_1 + X_2\alpha_2 \\ c &\equiv \mathring{c}_1 + \mathring{c}_2 \\ \tau_R &= (k_a + k_b)^{-1} \end{aligned}$$

Eq. (6.4.2) thus becomes

$$S(\mathbf{q}, t) = \bar{\alpha}^2 c \exp(-q^2 Dt) + X_1 X_2(\alpha_1 - \alpha_2)^2 c \exp\left(-\left[q^2 D + \frac{1}{\tau_R}\right]t\right) \tag{6.4.7}$$

This function is the sum of two exponentials, one whose decay rate depends solely on the translational diffusion coefficient and another whose decay rate depends on both the translational diffusion coefficient and the kinetic–relaxation rate $1/\tau_R$. The "strength" of the term containing the kinetic relaxation time depends on the *difference* between the polarizabilities of the molecule in states 1 and 2, as we surmised in Section (6.1). In most cases the purely diffusive contribution which contains the square of the average polarizability increment will contribute much more strongly than the second term. This result should be contrasted with the electrophoretic case in the slow exchange limit [Eq. (6.3.4)]. In the electrophoretic case k_a and k_b separately appear in the expressions, whereas in the zero field case only the combination τ_R appears.

3. Now allow both $D_1 = D_2 \equiv D$ and $\alpha_1 = \alpha_2 \equiv \alpha$. Then the previous result leads to

$$S(\mathbf{q}, t) = c\alpha^2 \exp -q^2Dt \tag{6.4.8}$$

Thus the chemical reaction is invisible to light scattering if both the dielectric constant increments and diffusion coefficients of the two species are equal. As far as light scattering is concerned, the medium consists of c molecules with polarizability increment α and diffusion coefficient D. This should be compared with electrophoresis, where even when $\alpha_1 = \alpha_2$ and $D_1 = D_2$, the reaction is seen by light scattering [cf. Eq. (6.3.4)].

4. For the fast-reaction limit, let the reaction rate be much faster than the diffusion rates,[12]

$$\frac{1}{\tau_R} \gg \begin{cases} q^2D_1 \\ q^2D_2 \end{cases} \tag{6.4.9}$$

then from Eqs. (6.4.1), (6.4.2), and (6.4.3), we obtain[13]

$$s_+ \cong -q^2D_s \quad ; \quad B_+ = \bar{\alpha}^2c$$
$$s_- \cong -1/\tau_R \quad ; \quad B_- = X_1X_2(\alpha_1 - \alpha_2)^2c$$

and

$$S(\mathbf{q}, t) = \left\{c\bar{\alpha}^2 \exp -q^2D_st + X_1X_2(\alpha_1 - \alpha_2)^2c \exp -\frac{1}{\tau_R}t\right\} \tag{6.4.11}$$

In this case the reaction is so fast that only the "average" diffusion coefficient appears in the first exponential. Note that if $\alpha_1 = \alpha_2$ the "reactive" term disappears completely, as above. In Eq. (6.4.11) the second term is usually much smaller than the first and, in addition, has a much faster decay constant. It would be very difficult in these circumstances to detect the term which contains the rate information.

A more favorable case for observing the rate constant (in the absence of an external field) is the case where although $\alpha_1 \sim \alpha_2$, D_1 is very different from D_2. It is an easy matter to find the correlation function [see Eq. (6.4.2)]. Bloomfield and Benbasat (1971) have performed such calculations. They cite cases where macromolecular interconversions exhibit large changes in D (25% ~ 40%) and show that with present equipment a slowly reacting mixture can in principle be distinguished from a rapidly reacting mixture.

6 · 5 DIMERIZATION KINETICS

The kinetic model of Section 6.2 may be easily modified to apply to dimerization reactions

$$2A_1 \underset{k_b}{\overset{k'_a}{\rightleftarrows}} A_2 \tag{6.5.1}$$

The analog of Eq. (6.2.7) for this is

$$\begin{aligned}\frac{\partial c_1}{\partial t} + \nabla \cdot \mathbf{J}_1 &= 2k_b c_2 - 2k'_a c_1^2 \\ \frac{\partial c_2}{\partial t} + \nabla \cdot \mathbf{J}_2 &= k'_a c_1^2 - k_b c_2\end{aligned} \tag{6.5.2}$$

Since fluctuations about equilibrium are small, we may set

$$c_1 = c_1^\circ + \delta c_1$$

and ignore term of order $\delta c_1{}^2$. Thus Eqs. (6.5.2) become

$$\begin{aligned}\frac{\partial(\delta c_1)}{\partial t} + \nabla \cdot (\delta J_1) &= 2k_b \delta c_2 - 4k'_a c_1^\circ \delta c_1 \\ \frac{\partial(\delta c_2)}{\partial t} + \nabla \cdot (\delta J_2) &= 2k'_a c_1^\circ \delta c_1 - k_b \delta c_2\end{aligned} \tag{6.5.3}$$

Equation (6.5.3) should be compared with Eqs. (6.2.7) for the $A \rightleftharpoons B$ reaction. These equations may be made equivalent by multiplying both sides of the first of Eqs. (6.5.3) by m_1, the mass of molecule A, and the second equation by $2m_1$, the mass of A_2. Then let $\delta c'_i$ and $\delta J'_i$ be defined as the fluctuations of the mass concentrations and mass flows per unit volume of component i. We find

$$\begin{aligned}\frac{\partial \delta c'_1}{\partial t} + \nabla \cdot (\delta \mathbf{J}'_1) &= k_b \delta c'_2 - 4k'_a c_1^\circ \delta c'_1 \\ \frac{\partial \delta c'_2}{\partial t} + \nabla \cdot (\delta \mathbf{J}'_2) &= -k_b \delta c'_2 + 4k'_a c_1^\circ \delta c'_1\end{aligned} \tag{6.5.4}$$

Thus, if

$$k_a \equiv 4k'_a c_1^\circ,$$

Equations (6.5.4) are formally the same as Eqs. (6.2.7) and the solutions for the correlation functions and spectra have similar forms (see Berne and Giniger, 1973). An example of an electrophoretic spectrum for this dimerization case is given in Fig. (6.5.1).

More complicated reactions can be easily treated by the methods outlined in the preceeding sections, that is: (a) determine the coupled diffusion-chemical reaction equations, (b) linearize the equations in the concentration fluctuations, (c) solve the linearized rate equations by Fourier-Laplace transforms, (d) solve the dispersion equation

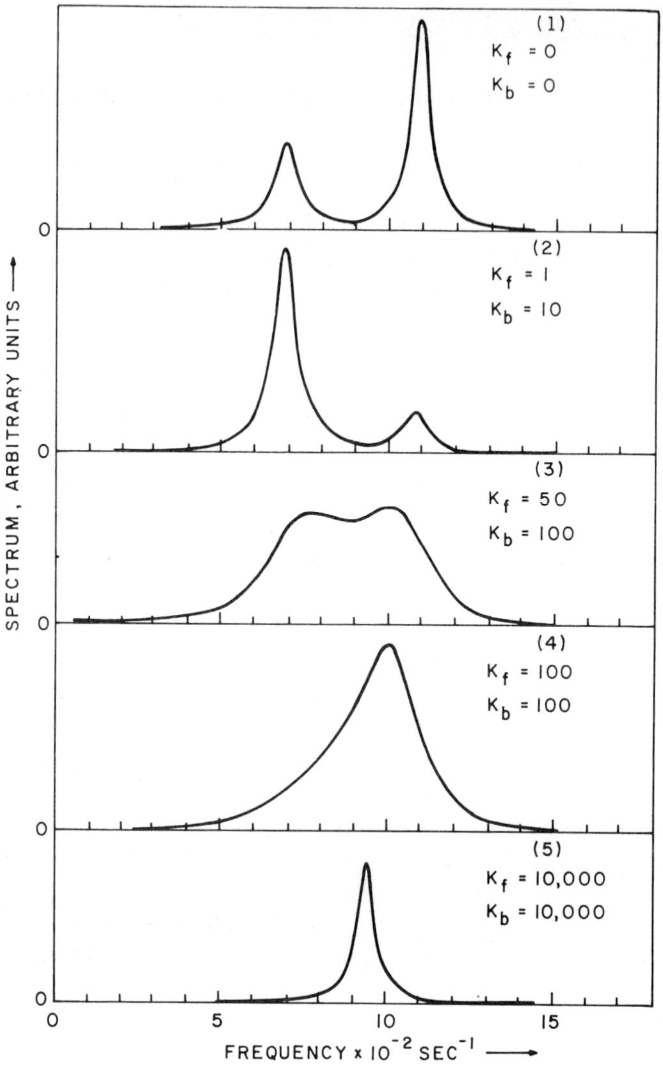

FIG. 6.5.1. The light-scattering spectrum of a chemically reacting system. In this case the spectrum is computed for association–dissociation equilibrium of chymotrpysin $2A_1 \underset{k_b}{\overset{k_f}{\rightleftharpoons}} A_2$ where k_f and k_b are the forward and backward rate constants. The constant in the figure is k_f $2k_f\, c_1{}^{\circ}$ where $c_1{}^0$ is the concentration of monomer. (For details see Berne and Giniger, 1973.)

either exactly, or by perturbation techniques as the case warrants, (e) invert the Laplace transforms, expanding the preexponential factors to the same order in the small parameters as were retained in calculating the roots. This gives the correlation function. Then perform the Fourier transforms to find the spectrum.

Chemical kinetics also influence the depolarized spectrum. Rather than treat this complicated subject in this book we refer the reader to the literature (e.g., see Berne and Pecora, 1969).

6 · 6 FLUORESCENCE CORRELATIONS

Light scattering from liquids measures the dynamics of dielectric constant fluctuations. In the dilute solutions that we have considered in this chapter these dielectric constant fluctuations are caused mainly by concentration fluctuations. Concentration fluctuations are then shown to decay by various mechanisms—mainly diffusion and chemical relaxation. It is possible to measure fluctuations in concentration and, hence, diffusion coefficients and chemical relaxation times by other techniques. In this section we discuss fluorescence fluctuation spectroscopy,[13] first introduced by Magde et al. (1972). The intensity of the fluorescent signal from a small steadily illuminated volume[14] V of solution should depend on the number of fluorescing molecules in the volume at a given time. Thus the fluorescence intensity fluctuates and its correlation function should reveal how the concentration fluctuation decays. If $I(\mathbf{r})$ is the intensity of incident radiation at position $\mathbf{r}$ in the scattering volume and $c(\mathbf{r}, t)$ is the concentration of molecules containing fluorescent tags, the fluorescent intensity is

$$I_f(t) = Q\varepsilon \int I(\mathbf{r})c(\mathbf{r}, t)\, d^3r \tag{6.6.1}$$

where Q and ε are, respectively, the fluorescence quantum efficiency and extinction coefficient for the fluorescent species. The correlation function of the fluctuations $\delta I_f(t)$ in $I_f(t)$ (which can be measured with an autocorrelator) is given by

$$\langle \delta I_f(t)\, \delta I_f(0)\rangle = (\varepsilon Q)^2 \int\int I(\mathbf{r})I(\mathbf{r}')\langle \delta c(\mathbf{r}, t)\delta c(\mathbf{r}', 0)\rangle d^3r\, d^3r' \tag{6.6.2}$$

The signal-to-noise ratio in such a fluorescence fluctuation experiment is of the order $[\langle \delta I_f^2\rangle/\langle I_f\rangle^2]^{1/2}$. This is obviously of order $\langle \delta N^2\rangle^{1/2}/\langle N\rangle$ where $\langle \delta N^2\rangle$ is the mean-square fluctuation in the number of fluorescent molecules in the illuminated volume and $\langle N\rangle$ is the average number of molecules in this same volume. From fluctuation theory (cf. Section 5.5) it follows that this fraction is of order $1/\sqrt{\langle N\rangle}$ or $1/\sqrt{Vc}$ where c is the concentration of fluorescing molecules. Thus in order to study the concentration fluctuations it is necessary to either look at a very dilute solution or to focus the laser light so that V is very small.

Since we are interested in the time-dependence of $\langle \delta I_f(0)\delta I_f(t)\rangle$ and not in its absolute value, it is convenient to work with the normalized correlation function. In Appendix 6.A we show that this can be written as

$$\hat{C}_f(t) \equiv \frac{\langle \delta I_f(0)\delta I_f(t)\rangle}{\langle \delta I_f(0)\delta I_f(0)\rangle} = \frac{\int d^3q\, |I(\mathbf{q})|^2 F(\mathbf{q}, t)}{\int d^3q\, |I(\mathbf{q})|^2 F(\mathbf{q})} \tag{6.6.3}$$

where $F(\mathbf{q}, t)$ is the correlation function of the spatial Fourier transform of the concentration fluctuations, $I(\mathbf{q})$ is the spatial Fourier transform of $I(\mathbf{r})$, and $F(\mathbf{q}) \equiv F(\mathbf{q}, 0)$.

In order to evaluate $\hat{C}_f(t)$ it is necessary to specify $I(\mathbf{q})$. In this chapter we restrict ourselves to a simple but physically meaningful form of $I(\mathbf{q})$ that enables us to derive analytical results. It should be clear that in any experiment $I(\mathbf{r})$ has a form fixed by the experimental conditions and this must be known in order to proceed. For our purposes then let the incident beam which propagates along the z-axis have a Gaussian spatial profile in the x and y directions of width σ_1 and let us observe the fluorescent beam along the x-direction through a slit which admits a Gaussian spatial profile of width σ_2 These Gaussian functions effectively describe the illuminated volume. We may set

$$I(\mathbf{r}) = I_0 \exp \frac{-(x^2 + y^2)}{2\sigma_1^2} \exp \frac{-z^2}{2\sigma_2^2} \tag{6.6.4}$$

The spatial Fourier transform of this function is easily obtained and is

$$I(\mathbf{q}) = (2\pi)^{3/2}(\sigma_1{}^2\sigma_2)I_0 \exp - \frac{1}{2} (q_x{}^2 + q_y{}^2)\, \sigma_1^2 \exp - \frac{1}{2}\, q_z{}^2\sigma_2{}^2 \tag{6.6.5}$$

This function decreases very rapidly as a function of q and is nonzero in effect for values of q of the order $q \sim 1/\sigma_1$ or $1/\sigma_2$, whichever is larger. The quantities σ_1 and σ_2 should be chosen as small as possible so that the signal-to-noise ratio ($\propto 1/\sqrt{c\sigma_1{}^2\sigma_2}$) is large. There is a lower limit on the size of σ_1 called the *diffraction limit*. In the usual applications σ_1 can be made as small as 5×10^{-4} cm, whereas σ_2 is usually larger. In this case $I(\mathbf{q})$ is nonzero for q smaller than $1/\sigma_1 \sim 10^3\ \text{cm}^{-1}$. Thus it is easy to convince ourselves that the q's that contribute nonzero values of $I(\mathbf{q})$ in Eq. (6.6.5) are small compared with the q's ($\sim 10^5\ \text{cm}^{-1}$) that are usually involved in light scattering.

If the concentration fluctuations arise and decay only because of diffusion into and out of the illuminated volume (cf. Chapter 5)

$$\langle \delta c^*(\mathbf{q}, 0)\, \delta c(\mathbf{q}, t)\rangle = \langle |\delta c(\mathbf{q})|^2\rangle \exp -q^2Dt \tag{6.6.6}$$

Away from critical points or phase transitions, $\langle |\delta c(\mathbf{q})|^2\rangle$ should be independent of q for the values of q discussed above.[15] Substitution of Eqs. (6.6.5) and (6.6.6) into Eq. (6.6.3) followed by an integration over $dq_x dq_y dq_z$ gives

$$\hat{C}_f(t) = \left[1 + \frac{t}{\tau_1}\right]^{-1} \left[1 + \frac{t}{\tau_2}\right]^{-1/2} \tag{6.6.7}$$

where

$$\tau_1 = \frac{\sigma_1{}^2}{D} \tag{6.6.8}$$

is essentially the characteristic time it takes a particle to diffuse out of a two-dimensional region the size of the incoming beam diameter and

$$\tau_2 = \frac{\sigma_2{}^2}{D} \tag{6.6.9}$$

is the characteristic time it takes a particle to traverse a distance equal to the one-dimensional distance defined by the slit width.

Thus, if τ_1 is in a time range accessible to autocorrelation (roughly $1\text{–}10^{-7}$ sec), fluorescence fluctuations may be used to measure macromolecular translational diffusion coefficients. The presence of a fluorescent label enables this method to measure the translational diffusion coefficient of a molecule in a complex mixture. Such a measurement would be very difficult in an ordinary light-scattering experiment because all components of the mixture contribute.[17] The advantage of fluorescent probes is that they allow particular species to be labeled and thereby separately studied. For σ_1 of the order of 10^{-4} cm and for a particle with a diffusion coefficient of the order 10^{-5} cm^2/sec, $\tau_1 = 10^{-3}$ sec, well within our ability to measure. This leads to the interesting possibility of measuring diffusion coefficients of labeled molecules in membranes, and in cells *in vivo*.

If the solution is flowing with velocity $\mathbf{V}$ through the illuminated volume,

$$\langle \delta c^*(\mathbf{q}, 0)\, \delta c(\mathbf{q}, t)\rangle = \langle |\delta c(\mathbf{q})|^2\rangle \exp i\mathbf{q}\cdot\mathbf{V}t \exp -q^2Dt \tag{6.6.10}$$

which follows from Eq. (5.8.8). Substitution of Eqs. (6.6.10) and (6.6.5) into Eq. (6.6.3), followed by integration over $dq_x dq_y dq_z$ gives

$$\hat{C}_f(t) = \left[1 + \frac{t}{\tau_1}\right]^{-1}\left[1 + \frac{t}{\tau_2}\right]^{-1/2} \exp -\frac{t^2}{4}\left[\frac{(V_x^2 + V_y^2)}{\sigma_1^2\left[1 + \frac{t}{\tau_1}\right]} + \frac{V_z^2}{\sigma_2^2\left[1 + \frac{t}{\tau_2}\right]}\right] \tag{6.6.11}$$

If either $\dfrac{\sigma_1}{(V_x^2 + V_y^2)^{1/2}} \ll \tau_1$ or $\dfrac{\sigma_2}{V_z} \ll \tau_2$, $\hat{C}_f(t)$ simplifies to

$$\hat{C}_f(t) = \exp -\frac{t^2}{4}\left\{\frac{(V_x^2 + V_y^2)}{\sigma_1^2} + \frac{(V_z)^2}{\sigma_2^2}\right\} \tag{6.6.12}$$

so that fluorescence-fluctuation spectroscopy can in principle be used to measure the flow velocity of a labeled component. This should have applications in the study of turbulence, in the study of flow through capillaries and microtubules, and so on. It should be recognized that the decay times, $\sigma_1/\sqrt{V_x^2 + V_y^2}$, σ_2/V_z are the times it takes the fluid to flow a distance σ_1 and σ_2 respectively. It should also be clear that the same considerations apply to electrophoresis. Another possible application is to bacterial motility. Of course Eqs. (6.6.7) and (6.6.11) can be directly applied to number fluctuation studies by light scattering for these same processes.

Now consider the simple chemical reaction

$$A \underset{k_b}{\overset{k_a}{\rightleftharpoons}} B$$

where only B fluoresces when the system is steadily irradiated at a given frequency. Then clearly the fluctuations in the fluorescence will occur not only because of diffusion but also because of chemical reaction which produces fluctuations in the concentration of B. We must now substitute the correlation function,

$$F_{22}(\mathbf{q}, t) = \langle \delta c_B^*(\mathbf{q}, 0)\, \delta c_B(\mathbf{q}, t)\rangle,$$

[which has already been evaluated (cf. Eq. (6.2.15))] into Eq. (6.6.3) and must evaluate the integral

$$\hat{C}_f(t) = \frac{\int d^3q\,|I(\mathbf{q})|^2 F_{22}(\mathbf{q}, t)}{\int d^3q\,|I(\mathbf{q})|^2 F_{22}(\mathbf{q})} \tag{6.6.13}$$

This would be quite difficult to do generally. Fortunately it is possible in most cases to closely estimate $\hat{C}_f(t)$ as follows. The presence of $I(\mathbf{q})$ in Eq. (6.6.13) insures that only small values of q contribute to the integral. Now we note that if these values of q are such that

$$k_b,\ k_a \gg q^2 D_1,\ q^2 D_2 \tag{6.6.14}$$

then this is the fast exchange limit (Sec. 6.4). Since we are considering the case where there is no electric field, the roots of the dispersion equation $\Delta(s) = 0$ are given by Eq. (6.3.10) with $\omega_s = \omega_b = 0$; that is,

$$s_\pm = \begin{cases} -q^2 D_s \\ -(k_a + k_b) - q^2 D_b \end{cases} \tag{6.6.15}$$

The simplest case occurs when $D_1 = D_2 \equiv D$. Then $D_s = D$, $D_b = D$, and[18]

$$s_\pm = \begin{cases} -q^2 D \\ -\dfrac{1}{\tau_R} - q^2 D \end{cases} \tag{6.6.16}$$

where τ_R is the reaction relaxation time $\left(\dfrac{1}{\tau_R} = k_a + k_b\right)$

Substitution of Eq. (6.6.16) into Eq. (6.2.15) then gives

$$F_{22}(\mathbf{q}, t) = \exp -q^2 Dt \frac{c_B}{c}\left[c_B + c_A \exp\frac{-t}{\tau}R\right] \tag{6.6.17}$$

When this is substituted into Eq. (6.6.13) and the integrals are evaluated we find

$$\hat{C}_f(t) = \left[1 + \frac{t}{\tau_1}\right]^{-1}\left[1 + \frac{t}{\tau_2}\right]^{-1}\frac{1}{c}\left[c_B + c_A \exp\frac{-t}{\tau}R\right] \tag{6.6.18}$$

where τ_1 and τ_2 are given by Eqs. (6.6.8) and (6.6.9).

If the reaction is turned off, that is, $k_a = k_b = 0$, Eq. (6.6.8) becomes identical to Eq. (6.6.7) for a single fluorescent species diffusing through the observation volume. Note that the term containing the chemical rate information is proportional to c_A/c so that the contribution of the reaction rate to the spectrum is a maximum when $c_A = c_B$.

The expression $\langle \delta I_f(t)\delta I_f(0)\rangle$ is more complicated than the simple expression given in Eq. (6.6.18) if A and B have different diffusion coefficients, for then the roots [cf. Eq. (6.4.1)] must be substituted into Eq. (6.2.11d) and the preexponential factors expanded to first order in the small parameters q^2D_s, q^2D_b. This is a simple but tedious task.

The above model shows the main physical features of this type of experiment. The same methods may be applied to fluctuations of other optical quantities—for instance, Raman and infrared line intensities. However, no one has as yet performed optical experiments on other than the fluorescence fluctuations.

Fluorescence fluctuation measurements have been used to obtain a chemical rate constant by Magde et al. (1972) for a binding reaction of the form

$$A + B \rightleftharpoons C$$

In their experiment A represents DNA, B is ethidium bromide (a dye that inhibits nucleic acid synthesis), and C is the DNA–EtBr complex. The complex is strongly fluorescent so that fluctuations in complex concentration are measured. Magde et al. (1972) have worked out the theoretical expressions for $\langle \delta I_f(t)\delta I_f(0)\rangle$ for the complexing reaction and have obtained the reaction relaxation times under various conditions.

6 · 7 PROSPECTS

There have as yet been no unequivocal measurements of chemical reaction rates by light scattering, although much theoretical work has been done and several experimental attempts made.

The best prospect for this field appears to be the electrophoretic technique. Variants of the light-scattering method such as fluorescence and Raman and infrared intensity fluctuations are also promising, although only the fluorescence technique has been used so far to measure reaction rate constants. Feher (1973) has introduced a technique which directly measures fluctuations of electrical conductivity of dilute electrolyte solutions. These fluctuations may be related to diffusion coefficients and reaction rate constants by methods similar to those described in this chapter.

APPENDIX 6.A. THE DERIVATION OF EQ. (6.6.3)

Consider an infinite system divided into macroscopic cubic cells of volume $\Omega = L^3$ that are open to particle flux. Assume periodic boundary conditions on each cell and expand the concentration fluctuation $\delta c(\mathbf{r}, t)$ in a Fourier series

$$\delta c(\mathbf{r}, t) = \frac{1}{\Omega} \sum_{\mathbf{q}} \delta c(\mathbf{q}, t)\, e^{i\mathbf{q}\cdot\mathbf{r}} \tag{6.A.1}$$

where

$$\delta c(\mathbf{q}, t) = \int_{\Omega} d^3r \delta c(\mathbf{r}, t)\, e^{-i\mathbf{q}\cdot\mathbf{r}} \tag{6.A.2}$$

Here the wave vectors are constrained by the periodic boundary conditions to be $\mathbf{q} = ((2\pi/L)n_x, (2\pi/L)n_y, (2\pi/L)n_z)$ where n_x, n_y, n_z are integers, and in the limit $\Omega \to \infty$, $\sum_{q} \to [\Omega/(2\pi)^3] \int d^3q$, that is, the sum can be replaced by the integral. Substitution of Eq. (6.A.1) into (6.6.2) then gives

$$\langle \delta I_f(0)\delta I_f(t)\rangle = (Q\varepsilon)^2 \frac{1}{\Omega^2} \sum_{q} \sum_{\mathbf{q}'} I(\mathbf{q}')I(\mathbf{q})\langle \delta c(\mathbf{q}', 0)\delta c(\mathbf{q}, t)\rangle \tag{6.A.3}$$

where $I(\mathbf{q})$ is

$$I(\mathbf{q}) \equiv \int_{\Omega} d^3r\, I(\mathbf{r})\, e^{i\mathbf{q}\cdot\mathbf{r}} \tag{6.A.4}$$

and likewise for $I(\mathbf{q}')$. Now we note that

$$\langle \delta c(\mathbf{q}', 0)\, \delta c(\mathbf{q}, t)\rangle = \int_{\Omega} d^3r \int_{\Omega} d^3r' \langle \delta c(\mathbf{r}', 0)\, \delta c(\mathbf{r}, t)\rangle \exp -i[\mathbf{q}' \cdot \mathbf{r}' + \mathbf{q} \cdot \mathbf{r}] \tag{6.A.5}$$

In an isotropic homogeneous system we expect that $\langle \delta c(\mathbf{r}', 0)\ \delta c(\mathbf{r}, t)\rangle$ can only depend on the separation between R and R', so that

$$\langle \delta c(\mathbf{r}', 0)\delta c(\mathbf{r}, t)\rangle = G(\mathbf{r} - \mathbf{r}', t) \tag{6.A.6}$$

Then transforming the variables of integration in Eq. (6.A.5) from $(\mathbf{r}', \mathbf{r})$ to $(\mathbf{r}', \mathbf{R} = \mathbf{r}' - \mathbf{r})$ we find

$$\langle \delta c(\mathbf{q}', 0)\, \delta c(\mathbf{q}, t)\rangle = \int_{\Omega} d^3r' \exp(\mathbf{q}' + \mathbf{q}) \cdot \mathbf{r}' \int d^3R \exp i\mathbf{q}\cdot\mathbf{R}\, G(\mathbf{R}, t) \tag{6.A.7}$$

We note that

$$\langle \delta c(-\mathbf{q}, 0)\, \delta c(\mathbf{q}, t)\rangle = \Omega \int d^3R \exp i\mathbf{q}\cdot\mathbf{R}\, G(\mathbf{R}, t) \tag{6.A.8}$$

Eliminating the last integral between Eqs. (6.A.7) and (6.A.8) gives[18]

$$\langle \delta c(\mathbf{q}', 0)\, \delta c(\mathbf{q}, t)\rangle = \delta_{q,'-q}{}' \langle \delta c^*(\mathbf{q}, 0)\, \delta c(\mathbf{q}, t)\rangle \tag{6.A.9}$$

Substituting Eq. (6.A.9) into Eq. (6.A.3), recognizing that because $I(\mathbf{r})$ is real $I^*(\mathbf{q}) = I(-\mathbf{q})$, and converting the sum over $\mathbf{q}$ to an integral as specified above, we obtain[19]

$$\langle \delta I_f(0)\, \delta I_f(t)\rangle = (2\pi)^{-3}(Q\varepsilon)^2\, \Omega^{-1} \int d^3q\, |I(\mathbf{q})|^2 \langle \delta c^*(\mathbf{q}, 0)\, \delta c(\mathbf{q}, t)\rangle \tag{6.A.10}$$

and the normalized fluorescence correlation function is

$$\hat{C}_f(t) \equiv \frac{\langle \delta I_f(0)\,\delta I_f(t)\rangle}{\langle |\delta I_f|^2\rangle} = \frac{\int d^3q\,|I(\mathbf{q})|^2 F(\mathbf{q}, t)}{\int d^3q\,|I(\mathbf{q})|^2 F(\mathbf{q})} \tag{6.A.11}$$

where

$$F(\mathbf{q}, t) \equiv \langle \delta c^*(\mathbf{q}, 0)\,\delta c(\mathbf{q}, t)\rangle \tag{6.A.12}$$

and

$$F(\mathbf{q}) \equiv F(\mathbf{q}, 0) \tag{6.A.13}$$

NOTES

1. $\delta c_1(\mathbf{q}, t) \equiv \sum_{j\varepsilon A} e^{i\mathbf{q}\cdot\mathbf{r}_j(t)} = \int d^3r\, e^{i\mathbf{q}\cdot\mathbf{r}} \sum_{j\varepsilon A} \delta(\mathbf{r} - \mathbf{r}_j(t)) = \int d^3r\, e^{i\mathbf{q}\cdot\mathbf{r}}\, \delta c_1(\mathbf{r}, t)$ and likewise for $\delta c_2(\mathbf{q}, t)$.
2. This follows because $S(\mathbf{q}, t)$ must be a real even function of the time (see Section 11.5, Theorem 3). The spectrum is

$$S(\mathbf{q}, \omega) \equiv (2\pi)^{-1} \int_{-\infty}^{+\infty} dt\, e^{-i\omega t} S(\mathbf{q}, t)$$

The integral is the sum of an integral from $(0, \infty)$ and of one from $(-\infty, 0)$. Transforming the second integral from t to $-t$ and using $S(\mathbf{q}, -t) = S(\mathbf{q}, t)$ gives

$$S(\mathbf{q}, \omega) = (2\pi)^{-1} \int_0^\infty dt\, [e^{i\omega t} + e^{-i\omega t}]\, S(\mathbf{q}, t) = \pi^{-1}\, Re \int_0^\infty dt\, e^{-i\omega t}\, S(\mathbf{q}, t)$$
$$= \pi^{-1}\, Re\, \tilde{S}(\mathbf{q}, s = i\omega)$$

where

$$\tilde{S}(\mathbf{q}, s) = \int_0^\infty dt\, e^{-st}\, S(\mathbf{q}, t)$$

is the Laplace transform of $S(\mathbf{q}, t)$.
3. In general, the fluxes $\mathbf{J}_1$ and $\mathbf{J}_2$ should contain cross terms. At low concentrations these terms are expected to be small. We are omitting these effects here in the interest of simplicity. Inclusion of these cross terms leads to simple but algebraically tedious results. See Chapter 13 for a discussion of these terms.
4. The use of macroscopic transport equations for the determination of time-correlation functions of fluctuating quantities is equivalent to the Onsager regression hypothesis outlined in Section 10.2 and discussed in Chapter 11.
5. The Laplace transform of $c_i(\mathbf{q}, t)$ is defined as

$$\mathbf{c}_i(\mathbf{q}, s) \equiv \int_0^\infty dt\, e^{-st}\, c_i(\mathbf{q}, t)$$

The Laplace transform of $\dfrac{\partial c_i(\mathbf{q}, t)}{\partial t}$ is

$$\int_0^\infty dt\, e^{-st}\, \frac{\partial c_i(\mathbf{q}, t)}{\partial t} = s\tilde{c}_i(\mathbf{q}, s) - \mathbf{c}_i(\mathbf{q}, 0)$$

where $c_i(\mathbf{q}, 0) = c_i(\mathbf{q}, t = 0)$ is the initial value of the concentration fluctuation. The Laplace transform of Eqs. (6.2.9) is a set of two coupled algebraic equations which are easily solved for $\tilde{c}_1(\mathbf{q}, s)$ and $\tilde{c}_2(\mathbf{q}, s)$.

6. Theoretical methods for treating the zero time-correlation functions in the nonideal multicomponent case have been given by Stockmayer (1950) and by Kirkwood and Goldberg (1950).
7. These are simply the roots of the quadratic equation [Eq. (6.2.11e)].
8. For example, $k_a q^2 D_2$ is a second-order term.
9. From the definition of X_1 and X_2 and from the law of mass action which gives

$$\frac{c_2^\circ}{c_1^\circ} = \frac{k_a}{k_b}$$

So that

$$\frac{c_1^\circ}{c_1^\circ + c_2^\circ} = \frac{k_b}{k_a + k_b} = X_1 \text{ and } \frac{c_2^\circ}{c_2^\circ + c_2^\circ} = \frac{k_a}{k_a + k_b} = X_2$$

10. The broad band will probably be buried in the noise.
11. Here we allow D_1 and D_2 to differ, and define D_s to be the average diffusion coefficient $D_s = X_1 D_1 + X_2 D_2$.
12. This can be shown by a perturbation solution of the dispersion equation.
13. This technique involves the same kind of consideration as discussed in Section 5.5 in connection with occupation number fluctuations $\langle \delta N(0) \delta N(t) \rangle$.
14. The illuminated volume is defined by the intersection of the incident beam and the angle of acceptance of the detector.
15. Equation (6.6.6) follows from the diffusion equation and $\langle |\delta c(\mathbf{q})|^2 \rangle$ depends on q only for values of q such that two particles separated by a distance q^{-1} are still correlated. These values are typically of order 10^7cm^{-1} unless a phase transition is approached.
16. This is also an exact result for this particular case.
17. The scattering of light at wavelengths near an electronic absorbtion band of a given species in a complex mixture gives rise to "resonance enhancement" of the light scattered by this species, and allows one do label molecules in such a way that light scattering can be used to plobe dynamic processes of one molecular moiety in a complex mixture (See Bauer et al., 1975).
18. Here we have introduced the Kronecker delta symbol

$$\delta_{\mathbf{q},-\mathbf{q}'} \equiv \frac{1}{\Omega} \int_\Omega d^3 r' e^{i(\mathbf{q}+\mathbf{q}') \cdot \mathbf{r}'} = \begin{cases} 1 & \mathbf{q} = -\mathbf{q}' \\ 0 & \mathbf{q} \neq -\mathbf{q}' \end{cases}$$

and have also used $\delta c(-\mathbf{q}, 0) = \delta c^*(\mathbf{q}, 0)$.

19. Note that Eq. (6.A.8) depends on the size of the cell Ω but that $\Omega \langle \delta c(\mathbf{q}, 0)\, \delta c(\mathbf{q}, t) \rangle$ should be independent of the cell size. For example, for small q, the initial value is

$$\lim_{q \to 0} \frac{1}{\Omega} \langle |\delta c(\mathbf{q})|^2 \rangle = kT \left(\frac{\partial \bar{c}}{\partial \mu} \right)_{T,V}$$

where $\bar{c}$ is the average concentration.

REFERENCES

Bauer, D. R, Hudson, B. and Pecora, R., *J. Chem. Phys.* **63,** 588 (1975).

Berne, B. J. and Giniger, R., *Biopolym.* **12,** 1161 (1973).

Berne, B. J. and Pecora, R., *J. Chem. Phys.* **50,** 783 (1969).

Bloomfield, V. and Benbasat, J. A., *Macromolecul.,* **4,** 609 (1971), and references cited, particularly the work of Berne, Frisch, Blum, Salsburg, and co-workers.

Carrington, A., and McLachlan, A., *Introduction to Magnetic Resonance,* Chapter 12, Harper and Row, New York (1967).

Feher, G. and Weissman, M., *Proc. Natl. Acad. Sci.* **40,** 870 (1973).

Khatchalsky, A. and Curran. P. F., *Nonequilibrium Thermodynamics in Biophysics,* Harvard University Press (1965)

Kirkwood, J. G. and Goldberg, R. J., *J. Chem. Phys.* **18,** 54 (1950).

Magde, D., Elson, E., and Webb, W. W., *Phys. Rev. Lett.* **29,** 705 (1972). [See also Biopolymers **13,** 29 (1974).]

Stockmayer, W. H., *J. Chem. Phys.* **18,** 58 (1950).

CHAPTER 7

MODEL SYSTEMS CONTAINING OPTICALLY ANISOTROPIC MOLECULES

7·1 INTRODUCTION

In light-scattering light of a given polarization impinges on a molecule, inducing a dipole moment that subsequently radiates. The magnitude and the direction of the induced dipole moment depend in general on the orientation of the molecule with respect to the incident electric field of the light (see Section 5.1). Because a molecule continually reorients, the magnitude and direction of its induced dipole moment fluctuates. This leads to a change in the polarization and the electric field strength of the light emitted by the fluctuating induced dipole moment. The light scattered from an assembly of molecules therefore contains information about molecular tumbling.

According to Eqs. (3.3.3) and (3.3.4) the spectral density of the scattered field is determined by the autocorrelation function of $\delta\alpha_{if}(\mathbf{q}, t) = \sum_j' \alpha_{if}^{(j)}(t) \exp^{i\mathbf{q}.\mathbf{r}j(t)}$ where $\alpha_{if}^{j}(t)$ is given by Eq. (3.3.2). This formula contains the projection of the polarizability tensor $\boldsymbol{\alpha}^j$ of molecule j onto the initial and final polarization directions of the light wave

$$\alpha_{if}{}^j = \mathbf{n}_i \cdot \boldsymbol{\alpha}^j \cdot \mathbf{n}_f = (\mathbf{n}_i)_\alpha \, \alpha_{\alpha\beta}{}^j (\mathbf{n}_f)_\beta$$

In general, molecules are optically anisotropic; that is, the polarizability tensor $\alpha_{\alpha\beta}$ generally has off-diagonal elements. This means that when such a molecule is placed in an applied electric field, the components of the dipole moment induced by the field

$$\mu_\alpha = \alpha_{\alpha\beta} E_\beta$$

will not generally be parallel to the applied field. A set of axes in the molecule can always be found such that with these axes as basis vectors the polarizability tensor is diagonal. These axes are called the principal axes of the polarizability. Along these axes, $\boldsymbol{\mu}$ and $\mathbf{E}$ have the same direction, while for other choices of the body fixed axes this is generally not the case. These axes define an ellipsoid with the principal axes acting as the major and minor axes. This polarizability ellipsoid will have the same symmetry as the charge distribution, which as a rule has the same symmetry as the nuclear frame of the rigid molecule. Consequently, any axis of symmetry of the molecular point group is a principal axis of the ellipsoid and any plane of symmetry contains two axes of the ellipsoid. When all three axes are equal, as in spherical top molecules, the polarizability is isotropic [cf. Eq. (5.2.7)]. When two or more axes are different the polarizability is anisotropic.

This chapter is restricted to a considerration of dilute systems. It follows that only self-correlations need be considered.[1] From the discussion in Section 5.3, we may write

$$I^{\alpha}_{if}(\mathbf{q}, t) = \sum_{j=1}^{N} \langle \alpha^{j}_{if}(0)\, \alpha^{j}_{if}(t) \exp i\mathbf{q} \cdot [\mathbf{r}_j(t) - \mathbf{r}_j(0)] \rangle \tag{7.1.1}$$

where $\mathbf{r}_j(t) - \mathbf{r}_j(0)$ is the displacement of particle j in time t.

In this chapter it is assumed that the center of mass position and the orientation of a molecule are statistically independent. This assumption is not completely justified since the interaction potential between any two molecules is not separable in the relative position and orientations of the two molecules. The problem may, however, be treated in the special case of the translational and rotational diffusion approximations where one considers both the rotational and translational diffusion coefficients to be tensors (see Appendix 7.A). With the assumption of statistical independence of molecular rotation and translation Eq. (7.1.1) becomes

$$I^{\alpha}_{if}(\mathbf{q}, t) = \sum_{j=1}^{N}{}' \langle \alpha^{j}_{if}(0)\alpha^{j}_{if}(t) \rangle F_s(\mathbf{q}, t) \tag{7.1.2}$$

Since the correlation function $\langle \alpha^{j}_{if}(0)\; \alpha^{j}_{if}(t) \rangle$ involves an ensemble average, it is the same for every equivalent molecule in the system. Consequently Eq. (7.1.1) becomes

$$I^{\alpha}_{if}(\mathbf{q}, t) = \langle N \rangle \langle \alpha_{if}(0)\, \alpha_{if}(t) \rangle F_s(\mathbf{q}, t) \tag{7.1.3}$$

The correlation function $\langle \alpha_{if}(0)\, \alpha_{if}(t) \rangle$ involves the elements $\alpha_{\alpha\beta}$ of the molecular polarizability tensor in the laboratory fixed coordinate system. The α_{if} change with time because of molecular reorientation. Note that the only q dependence on the right-hand side of Eq. (7.1.3) is in the "translational" factor $F_s(\mathbf{q}, t)$. The $\langle \alpha_{if}(0)\, \alpha_{if}(t) \rangle$ is purely local in character and hence does not depend on q. In the remaining sections of this chapter we evaluate this correlation function for various combinations of molecular symmetries and models of reorientation in fluids.

7·2 SCATTERING FROM CYLINDRICALLY SYMMETRIC MOLECULES

In order to calculate $\langle \alpha_{if}(0)\, \alpha_{if}(t) \rangle$, the laboratory-fixed components of the molecular polarizability tensor must be expressed in terms of the molecule-fixed components and functions of the molecular orientation angles. There are quite general techniques for doing this. In Appendix 7.B we discuss a Cartesian tensor technique and in Section 7.4, a spherical tensor technique. The spherical tensor methods are especially useful in dealing with asymmetric molecules. However, in order to illustrate the basic physics of the problem we treat the simplest nontrivial case—light scattering from a dilute solution of cylindrically symmetric molecules—using simple geometrical methods.

Consider geometry II of Section (3.4). In this geometry $\mathbf{k}_i$ is somewhere in the x–y plane and $\mathbf{k}_f$ is in the x direction. Polarizers and analyzers then select out the components of the scattered electric field for ($\mathbf{n}_i = \mathbf{n}_f = \hat{\mathbf{z}}$), yielding $I^{\alpha}_{VV}(\mathbf{q}, t)$ or the components ($\mathbf{n}_i = \hat{\mathbf{z}}$; $\mathbf{n}_f = \hat{\mathbf{y}}$) yielding $I^{\alpha}_{VH}(\mathbf{q}, t)$. Thus from Eqs. (3.4.2) or simply using the definition of α_{if} and the polarization directions noted

$$I^{\alpha}{}_{VV}(\mathbf{q}, t) = \langle N\rangle \langle \alpha_{zz}(0)\, \alpha_{zz}(t)\rangle F_s(\mathbf{q}, t)$$

and (7.2.1)

$$I^{\alpha}{}_{VH}(\mathbf{q}, t) = \langle N\rangle \langle \alpha_{yz}(0)\, \alpha_{yz}(t)\rangle F_s(\mathbf{q}, t)$$

Now let the molecule have molecule-fixed polarizability component $\alpha_{\parallel}$ parallel to its symmetry axis and $\alpha_{\perp}$ in any direction perpendicular to this axis. The spherical polar coordinates specifying the orientation of the molecular symmetry axis in the laboratory-fixed coordinate system are (θ, ϕ) as defined in Fig. 7.2.1.

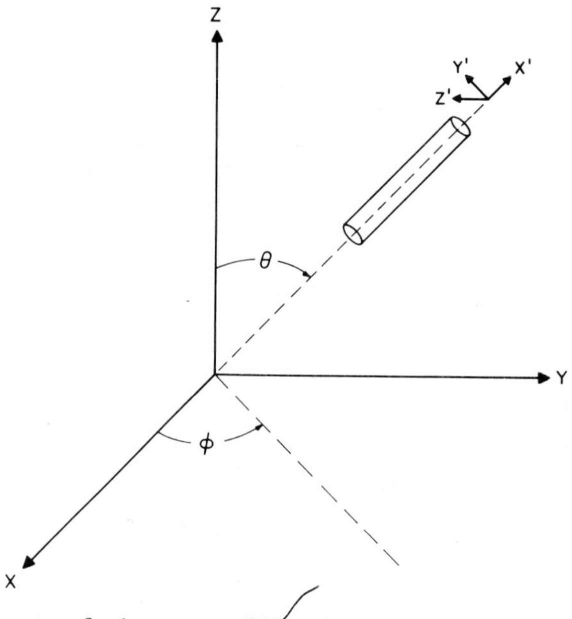

FIG. 7.2.1. The laboratory-fixed axes are XYZ and the molecule-fixed axes are $X'Y'Z'$. The orientation angles of the symmetry axis (X' axis) of the cylindrical molecule are given by θ and ϕ.

The laboratory-fixed quantity $\alpha_{zz}(t)$ may be thought of as the z component of the dipole moment induced in the molecule by a unit field in the z direction, since in this case $\boldsymbol{\mu} = \boldsymbol{\alpha} \cdot \hat{\mathbf{z}}$, it follows that

$$\mu_z = \hat{z} \cdot \mathbf{u} = \hat{z} \cdot \boldsymbol{\alpha} \cdot \hat{z} \tag{7.2.2}$$

A similar interpretation, of course, holds for $\alpha_{yz}(t)$.

In order to completely specify the orientation of the cylinder we must specify two orthogonal unit vectors in a plane perpendicular to the cylindrical axis. For convenience we chose these vectors such that one (the y' axis) lies in the plane formed by the cylindrical axis (the x' axis) and the space-fixed Z axis, and the other, the z' axis, is perpendicular to this plane (see Fig. (7.2.1.)). The former makes an angle $(\pi/2-\theta)$ with the space-fixed z axis and the latter is perpendicular to this axis. Since the molecule is symmetric with respect to rotations of the $y'z'$ plane about the x' axis, rotations about x' will not give rise to any changes in the laboratory-fixed polarizability components and therefore will not affect the spectrum. Consequently, for the purposes of this calculation we ignore rotations of the molecule about this axis.

The projections of the unit vector $\hat{\mathbf{z}}$ along the x', y', z' axes gives

$$\hat{\mathbf{z}} = \begin{pmatrix} \cos\theta \\ \sin\theta \\ 0 \end{pmatrix}$$

and the projection of the unit vector $\hat{\mathbf{y}}$ along the x', y', z' axes

$$\hat{\mathbf{y}} = \begin{pmatrix} \sin\theta\,\sin\phi \\ -\cos\theta\,\sin\phi \\ -\cos\phi \end{pmatrix}$$

Then

$$\alpha_{zz} = (\cos\theta,\, \sin\theta,\, 0)\begin{pmatrix} \alpha_{\parallel} & 0 & 0 \\ 0 & \alpha_{\perp} & 0 \\ 0 & 0 & \alpha_{\perp} \end{pmatrix}\begin{pmatrix} \cos\theta \\ \sin\theta \\ 0 \end{pmatrix}$$

$$= \alpha_{\parallel}\cos^2\theta + \alpha_{\perp}\sin^2\theta$$

and

$$\alpha_{yz} = (\sin\theta\,\sin\phi,\; -\cos\theta\,\sin\phi,\; -\cos\phi)\begin{pmatrix} \alpha_{\parallel} & 0 & 0 \\ 0 & \alpha_{\perp} & 0 \\ 0 & 0 & \alpha_{\perp} \end{pmatrix}\begin{pmatrix} \cos\theta \\ \sin\theta \\ 0 \end{pmatrix}$$

$$= (\alpha_{\parallel} - \alpha_{\perp})\sin\theta\,\cos\theta\,\sin\phi$$

These components may be expressed in terms of the spherical harmonics of order 2, $Y_{2,m}(\theta, \phi)$

$$\begin{aligned} Y_{2,0}(\theta, \phi) &= \sqrt{\frac{5}{16\pi}}\,(3\cos^2\theta - 1) \\ Y_{2,\pm 1}(\theta, \phi) &= \mp\sqrt{\frac{15}{8\pi}}\,\sin\theta\,\cos\theta\,\exp \pm i\phi \end{aligned} \tag{7.2.3}$$

Solving for the geometrical factors and substituting into Eq. (7.2.2) yields

$$\begin{aligned} \alpha_{zz} &= \alpha + \left(\frac{16\pi}{45}\right)^{1/2}\beta\, Y_{2,0}(\theta, \phi) \\ \alpha_{yz} &= i\left(\frac{2\pi}{15}\right)^{1/2}\beta[Y_{2,1}(\theta, \phi) + Y_{2,-1}(\theta, \phi)] \end{aligned} \tag{7.2.4}$$

The only molecular parameters that appear in these space-fixed polarizabilities are

$$\alpha \equiv \frac{1}{3}(\alpha_{\parallel} + 2\alpha_{\perp})$$

and (7.2.5)

$$\beta \equiv (\alpha_{\parallel} - \alpha_{\perp})$$

where α is 1/3 the trace of the molecule-fixed polarizability tensor. It is called the *isotropic part* of the polarizability tensor since it is independent of molecular orientation; that is it has the same value in the laboratory-fixed system as it does in the molecule-fixed system. The parameter β, however, measures the optical anisotropy of the molecule. (For spherical molecules $\alpha_{\parallel} = \alpha_{\perp}$ and $\beta = 0$). Consequently, β is called the molecular *optical anisotropy* or the anisotropic part of the polarizability. These two parameters determine the intensities of the different components of the light-scattering spectrum from this system.

Substituting Eqs. (7.2.4) into Eqs. (7.2.1) we obtain

$$I^{\alpha}{}_{VV}(\mathbf{q}, t) = \langle N \rangle \left[\alpha^2 F_s(\mathbf{q}, t) + \left(\frac{16\pi}{45}\right) \beta^2 F^{2}_{2,0}(t) F_s(\mathbf{q}, t)\right] \tag{7.2.6a}$$

$$I^{\alpha}{}_{VH}(\mathbf{q}, t) = \langle N \rangle \left(\frac{2\pi}{15}\right) \beta^2 [F^{(2)}_{1,1}(t) + F^{(2)}_{1,-1}(t) + F^{(2)}_{-1,1}(t) + F^{(2)}_{-1,-1}(t)] F_s(\mathbf{q}, t) \tag{7.2.6b}$$

where

$$F^{(l)}_{m,m'}(t) \equiv \langle Y^*{}_{lm'}(\theta(0), \phi(0))\, Y_{lm}(\theta(t), \phi(t)) \rangle \tag{7.2.7}$$

are orientational correlation functions which reflect how the angles $\phi(t)$ and $\theta(t)$ specifying the orientation of the symmetry axis (see Fig. 7.2.1) change in time. In arriving at Eqs. (7.2.5) we made use of the fact that $\langle Y_{2,0}(\phi(t), \theta(t)) \rangle = 0$. This follows from the observation that in an equilibrium system there is a random distribution of orientations. This is clarified in successive sections.

It is important to note that the first term on the right-hand side of Eq. (7.2.6a) involves the isotropic part of the polarizability tensor and is hence independent of the rotations. This term would appear even for spherical molecules (when $\beta = 0$), whereas the other terms would be zero. This term gives rise to "isotropic scattering" and we consequently define

$$I^{\alpha}{}_{ISO}(\mathbf{q}, t) = \langle N \rangle \alpha^2 F_s(\mathbf{q}, t) \tag{7.2.8}$$

Although in Chapter 5 we mentioned only spherical molecules, it should be obvious that many of the concepts of Chapter 5 apply to the isotropic scattering from nonspherical molecules as well.

In the following sections we evaluate the orientational correlation functions [Eq. (7.2.6)] for the rotational diffusion model.

7·3 ROTATIONAL DIFFUSION OF LINEAR MOLECULES

The orientation of a rod can be specified by a unit vector $\mathbf{u}$ directed along the axis of the rod with spherical polar coordinates $\boldsymbol{\Omega} = (\theta, \phi)$. The orientation of the rod $\mathbf{u}$ can therefore be regarded as a point on a sphere of unit radius (a unit sphere). Since in a liquid the rod is expected to suffer many reorienting collisions per second, **u** should execute a kind of "random walk" on the surface of the unit sphere (see Fig. (7.3.1).

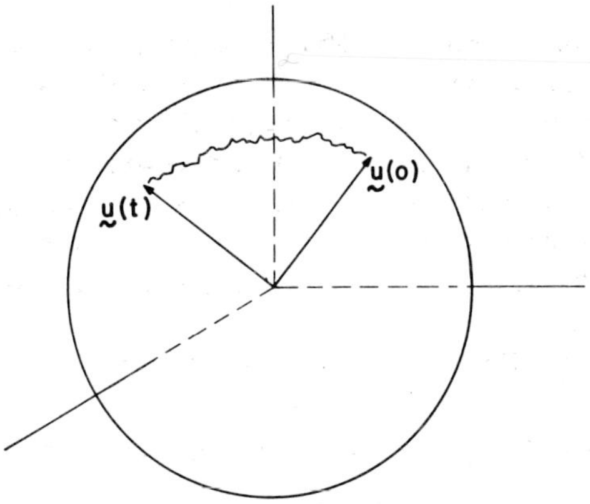

FIG. 7.3.1. **u**(0) and **u**(t) are unit vectors representing the orientation angles of the symmetry axis of a cylindrically symmetric molecule at times 0 and t, respectively. The locus of all the possible vectors **u**(t) is the surface of a sphere of unit radius (a unit sphere). The reorientation of the molecule can be regarded as a trajectory on the surface of the unit sphere. A random walk trajectory gives rise to rotational diffusion.

Debye (1929) developed a model for the reorientation processes based on the assumption that collisions are so frequent in a liquid that a molecule can only rotate through a very small angle before suffering a reorienting collision (small-step diffusion). We give here a heuristic treatment of the Debye model.

Any assembly of molecules initially oriented along some direction, say $\mathbf{u}_0$, behaves such that each molecule follows a different trajectory—made up of small steps—on the surface of the unit sphere. Initially this assembly is represented by a cloud of points which is very intense in the direction $\mathbf{u}_0$, but as time progresses, molecules reorient and the cloud spreads out, finally covering the unit sphere uniformly. It is the basic assumption of the Debye theory that the cloud of points simply diffuses on the surface of the unit sphere. We begin, therefore, with the diffusion equation, and ask the question how particles diffuse if they are constrained to remain on the sphere of unit radius? The equation that governs this motion is the diffusion equation

$$\frac{\partial c(\mathbf{r}, t)}{\partial t} = D\nabla^2 c(\mathbf{r}, t) \qquad (7.3.1)$$

where $|\mathbf{r}|$ is constrained to be 1. $c(\mathbf{r}, t)$ is then simply the concentration of rods at the point $\mathbf{r} = \mathbf{u}$ on the surface of the unit sphere at time t.

Because of the spherical symmetry—the points diffuse on the surface of a sphere—it is most convenient to solve Eq. (7.3.1) in spherical polar coordinates (r, θ, ϕ) where $r = 1$. The Laplacian ∇_r^2 in spherical polar coordinates is

$$\nabla_r^2 = \left(\frac{1}{r}\frac{\partial}{\partial r} r\right)^2 + \frac{1}{r^2 \sin^2\theta}\left[\sin\theta \frac{\partial}{\partial\theta}\left(\sin\theta \frac{\partial}{\partial\theta}\right) + \frac{\partial^2}{\partial\phi^2}\right]$$

For fixed $r = 1$, all the derivatives with respect to r vanish and

$$\nabla_r^2 = \frac{1}{\sin^2\theta}\left[\sin\theta\frac{\partial}{\partial\theta}\left(\sin\theta\frac{\partial}{\partial\theta}\right) + \frac{\partial^2}{\partial\phi^2}\right] \tag{7.3.2}$$

Letting $c(\mathbf{u}, t)d^2u$ be the fraction of rods[2] with orientation $\mathbf{u}$ in the solid angle, d^2u $(=\sin\theta\, d\theta\, d\phi)$ at time t, and substituting Eq. (7.3.2) into Eq. (7.3.1) we obtain the *rotational diffusion equation* (or Debye equation)

$$\frac{\partial c(\mathbf{u},t)}{\partial t} = \Theta\frac{1}{\sin^2\theta}\left[\sin\theta\frac{\partial}{\partial\theta}\left(\sin\theta\frac{\partial}{\partial\theta}\right) + \frac{\partial^2}{\partial\phi^2}\right] c(\mathbf{u}, t) \tag{7.3.3}$$

where Θ is called the *rotational diffusion coefficient.*

Those who are familiar with elementary quantum mechanics should recognize that the differential operator in Eq. (7.3.3), that is, the angular part of the Laplacian operator, is $-\hat{I}^2$ where $\hat{I}$ is the dimensionless orbital angular momentum operator of quantum mechanics (see Dicke and Witke, 1960). Thus the *rotational diffusion equation* can be written as

$$\frac{\partial c(\mathbf{u},t)}{\partial t} = -\Theta\,\hat{I}^2\, c(\mathbf{u}, t) \tag{7.3.4}$$

This simplifies the solution of the rotational diffusion equation considerably. As is well known, the spherical harmonics $Y_{lm}(\theta,\phi) \equiv Y_{lm}(\mathbf{u})$ are eigenfunctions of $\hat{I}^2$ and $\hat{I}_z$ corresponding to the eigenvalues $l(l+1)$ and m_l respectively, that is,

$$\begin{aligned} \hat{I}^2\, Y_{lm}(\mathbf{u}) &= l(l+1)\, Y_{lm}(\mathbf{u}) \qquad l = 0, 1, 2, \ldots \infty \\ \hat{I}_z\, Y_{lm}(\mathbf{u}) &= m_l\, Y_{lm}(\mathbf{u}) \qquad \mathrm{m}_l = -l, \ldots, 0, \ldots +l \end{aligned} \tag{7.3.5}$$

These functions form a complete orthonormal set spanning the space of functions of $\mathbf{u}$, so that

$$\int d^2u\, Y_{l'm'}(\mathbf{u})\, Y^*_{lm}(\mathbf{u}) = \delta_{l'l}\delta_{m,m'} \tag{7.3.6}$$

with the closure relation

$$\delta(\mathbf{u} - \mathbf{u}_0) = \sum_{l=0}^{\infty}\sum_{lm=-l}^{+l} Y_{lm}(\mathbf{u}_0)\, Y^*_{lm}(\mathbf{u}) \tag{7.3.7}$$

The formal solution of Eq. (7.3.4) is

$$c(u, t) = \exp(-t\Theta\hat{I}^2)\, c(\mathbf{u}, 0)$$

where $\hat{I}^2$ is an operator acting only on $\mathbf{u}$. The particular solution of Eq. (7.3.4) subject to the initial condition

$$c(\mathbf{u}, 0) = \delta(\mathbf{u} - \mathbf{u}_0) = \sum_{lm} Y_{lm}(\mathbf{u}_0)\, Y^*_{lm}(\mathbf{u}) \tag{7.3.8}$$

is therefore

$$c(\mathbf{u}, t) = \exp(-t\Theta\hat{I}^2) \sum_{lm} Y_{lm}(\mathbf{u}_0)\, Y_{lm}{}^*(\mathbf{u}) \tag{7.3.9a}$$

From Eq. (7.3.5) it may be seen that $\exp(-\, t\Theta\hat{I}^2)\; Y_{lm}(\mathbf{u}) = \exp -\; l(l+1)\Theta t\; Y_{lm}(\mathbf{u})$ Thus Eq. (7.3.9a) may be written

$$c(\mathbf{u}, t) = \sum_{lm} \exp - l(l+1)\Theta t Y^*_{lm}(\mathbf{u})\; Y_{lm}(\mathbf{u}_0) \tag{7.3.9b}$$

This particular solution of the diffusion equation can be interpreted as the transition probability; that is, the probability density for a rod to have orientation $\mathbf{u}$ at time t given that it had orientation $\mathbf{u}_0$ initially. Let us therefore take

$$K_s(\mathbf{u}, t \,|\, \mathbf{u}_0, 0) = \sum_{lm} Y_{lm}(\mathbf{u}_0)\; Y^*_{lm}(\mathbf{u}) \exp -\; l(l+1)\Theta t \tag{7.3.10}$$

where K_s is this transition probability. Note that[3]

$$\lim_{t\to\infty} K_s\,(\mathbf{u}, t \,|\, \mathbf{u}_0, 0) = Y^*_{00}(\mathbf{u})\;\; Y_{00}(\mathbf{u}_0) = \frac{1}{4\pi} \tag{7.3.11}$$

This simply means that the rods eventually become uniformly distributed on the surface of the unit sphere [of area (4π)].

The correlation functions required in light scattering (cf. Eqs. (7.2.1) and (7.2.4) are of the form $\langle Y^*_{l'm'}(\mathbf{u}(0)\, Y_{lm}\,(\mathbf{u}(t))\rangle$. These may be written as

$$\langle Y^*_{l'm'}(\mathbf{u}(0)\;\, Y_{lm}(\mathbf{u}(t))\rangle = \int d^2u_0 \int d^2u\;\, Y_{lm}\,(\mathbf{u}) G_s(\mathbf{u},t;\mathbf{u}_0, 0)\;\, Y^*_{l'm'}(\mathbf{u}_0) \tag{7.3.12}$$

where $G_s(\mathbf{u}, t;\, \mathbf{u}_0, 0) d^2u_0\, d^2u$ is the joint probability of finding a rod with orientation $\mathbf{u}_0$ in d^2u_0 initially and $\mathbf{u}$ in d^2u at time t. G_s can be expressed in terms of K_s and the probability distribution function $p(\mathbf{u}_0)$ of the initial orientation as,

$$G_s(\mathbf{u}, t;\, \mathbf{u}_0, 0) = K_s(\mathbf{u}, t \,|\, \mathbf{u}_0, 0)\, p(\mathbf{u}_0) \tag{7.3.13}$$

In an equilibrium ensemble of rods, we expect a uniform distribution[4] of molecular orientations so that $p(\mathbf{u}_0) = 1/4\pi$. Combining Eqs. (7.3.12), (7.3.13), and (7.3.10) and evaluating these integrals using Eq.(7.3.6) gives[5]

$$\langle Y^*_{l'm'}\,(\mathbf{u}(0))\;\, Y_{lm}(\mathbf{u}(t))\rangle = F_l(t)\;\; \delta_{l,l'}\; \delta_{m,m'} \tag{7.3.14}$$

with

$$F_l(t) = \frac{1}{4\pi} \exp\left(-l\,(l+1)\Theta t\right) \tag{7.3.15}$$

Note that the correlation functions are zero unless the indices $l = l'$ and $m = m'$. Moreover, they are independent[5] of m.

Returning now to Eq. (7.2.7) we see that the orientational correlation functions required are

$$F^{(2)}_{m\, m'} = F_2(t)\, \delta_{m,m'} = \frac{1}{4\pi} \exp -\;\, 6\Theta t\, \delta_{m,m'} \tag{7.3.16}$$

In Chapter 5 we found $F_s(\mathbf{q}, t) = \exp - q^2Dt$ for translational diffusion. Combining this with Eqs. (7.2.16) and (7.2.6) gives for combined rotational and translational diffusion

$$I^{\alpha}{}_{VV}(\mathbf{q}, t) = \langle N \rangle \left\{\alpha^2 + \frac{4}{45}\beta^2 \exp - 6\Theta t\right\} \exp - q^2Dt \tag{7.3.17}$$

$$I^{\alpha}{}_{VH}(\mathbf{q}, t) = \frac{1}{15}\langle N \rangle \beta^2 \exp - 6\Theta t \exp - q^2Dt \tag{7.3.18}$$

We note that $I^{\alpha}{}_{VV}(\mathbf{q}, t)$ can also be expressed as

$$I^{\alpha}{}_{VV}(\mathbf{q}, t) = I^{\alpha}_{\text{ISO}}(\mathbf{q}, t) + \frac{4}{3} I^{\alpha}{}_{VH}(\mathbf{q}, t) \tag{7.3.19}$$

where $I^{\alpha}_{ISO}(\mathbf{q}, t)$ is given by Eq. (7.2.8). Thus the scattering is characterized by two functions, $I^{\alpha}_{\text{ISO}}(\mathbf{q}, t)$ and $I^{\alpha}{}_{VH}(\mathbf{q}, t)$ where the former gives information only about the translational motion, and the latter gives in addition information about the rotational motion. The corresponding spectra are

$$\begin{aligned} I^{\alpha}{}_{VV}(\mathbf{q}, \omega) &= \langle N \rangle \pi^{-1} \left\{\alpha^2 \frac{q^2D}{\omega^2 + [q^2D]^2} + \frac{4}{45}\beta^2 \frac{[6\Theta + q^2D]}{\omega^2 + [6\Theta + q^2D]^2}\right\} \\ I^{\alpha}{}_{VH}(\mathbf{q}, \omega) &= \langle N \rangle \pi^{-1} \frac{\beta^2}{15} \left\{\frac{[6\Theta + q^2D]}{\omega^2 + [6\Theta + q^2D]^2}\right\} \end{aligned} \tag{7.3.20}$$

Equations (7.3.19) and (7.3.20) are applicable to scattering from very dilute solutions of cylindrically symmetric macromolecules. Such systems usually satisfy the assumptions in the derivation of these equations: (a) dilute solutions, (b) independence of molecular rotation and translation, (c) translational motions described by the translational diffusion equation, and (d) rotational motions described by the rotational diffusion (Debye) equation.

These equations may also be applicable to the depolarized Rayleigh scattering of cylindrically symmetric small molecules in solution. In such cases one must usually use interferometric detection since the rotation is generally fast enough to give frequency shifts of the order of wavenumbers (see Section 7.8). In such cases the contribution of the translational diffusion coefficient to the I^{α}_{VH} spectrum is usually negligible.

As we have seen in Section 3.2, the integrated intensity of a spectrum $I_{if}(\mathbf{q})$ is determined by the initial value of the time-correlation function

$$I_{if}(\mathbf{q}) = \int_{-\infty}^{+\infty} d\omega I_{if}(\mathbf{q}, \omega) = I_{if}(\mathbf{q}, t = 0) \tag{7.3.21}$$

It is a simple matter to determine the integrated intensities for the preceding geometry. It follows from Eqs. (7.3.17), (7.3.18) and (7.2.8) that

$$I^{\alpha}_{\text{ISO}} = \langle N \rangle \alpha^2 \tag{7.3.22a}$$

$$I^{\alpha}_{VH} = \langle N \rangle \frac{1}{15}\beta^2 \tag{7.3.22b}$$

$$I^{\alpha}_{VV} = I_{\text{ISO}} + \frac{4}{3} I_{VH} \tag{7.3.22c}$$

This leads to the depolarization ratio

$$\frac{I_{VH}^{\alpha}}{I_{\rm ISO}^{\alpha}} = \frac{1}{15}\frac{\beta^2}{\alpha^2} = \frac{3}{5}\left[\frac{\alpha_{\parallel} - \alpha_{\perp}}{\alpha_{\parallel} + 2\alpha_{\perp}}\right]^2 \tag{7.3.22d}$$

Consequently the intensity of depolarized scattering becomes relatively more important the larger the optical anisotropy of the scattering center.

7 · 4 SCATTERING FROM ANISOTROPIC MOLECULES

In this section we treat the scattering from molecules of arbitrary shape. This problem requires the use of mathematical methods which are common in atomic and molecular spectroscopy (Shore and Menzel, 1968). The uninitiated reader will find these techniques difficult to digest. Consequently in this section we omit the unpalatable details, leaving them for Appendix 7.C, and only present an outline of the relevant arguments.

According to Eq. (7.2.1) we require the polarizability components $\alpha_{ZZ}(t)$ and $\alpha_{YZ}(t)$ in the laboratory-fixed coordinate system. For compactness, we use the spherical tensor formulation outlined in Appendix 7.C. Accordingly, the nine spherical components of the polarizability tensor can be expressed in terms of the nine Cartesian components according to

$$\begin{aligned}
\alpha_0^{(0)} &= \frac{1}{\sqrt{3}}\,[\alpha_{XX} + \alpha_{YY} + \alpha_{ZZ}] \\
\alpha_0^{(1)} &= \frac{1}{2}\,(\alpha_{XY} - \alpha_{YX}) \\
\alpha_{\pm 1}^{(1)} &= \pm\frac{1}{2\sqrt{2}}\,[(\alpha_{YZ} - \alpha_{ZY}) \pm i(\alpha_{ZX} - \alpha_{XZ})] \\
\alpha_0^{(2)} &= \frac{1}{\sqrt{6}}\,[3\alpha_{ZZ} - (\alpha_{XX} + \alpha_{YY} + \alpha_{ZZ})] \\
\alpha_{\pm 1}^{(2)} &= \pm\frac{1}{2}\,[(\alpha_{ZX} + \alpha_{XZ}) \pm i(\alpha_{ZY} + \alpha_{YZ})] \\
\alpha_{\pm 2}^{(2)} &= \frac{1}{2}\,[(\alpha_{XX} - \alpha_{YY}) \pm i(\alpha_{XY} + \alpha_{YX})]
\end{aligned} \tag{7.4.1}$$

Where XYZ stands for a specified Cartesian coordinate frame. Thus once a Cartesian coordinate frame is chosen, the nine spherical components $\alpha_M^{(J)}$ can be determined using Eq. (7.4.1) from the nine Cartesian components. Clearly the spherical components and the Cartesian components change if the coordinate axes are rotated. Suppose we know the values of the Cartesian components of the polarizability tensor in a coordinate frame rigidly fixed within the molecule[6] (the body-fixed frame OXY Z). Then the problem confronting us is to determine the Cartesian components of the polarizability tensor in a coordinate system rigidly fixed in the laboratory (the laboratory frame $OX'Y'Z'$). The relative orientations of the molecular and laboratory-fixed frames is specified by a set of Euler angles $\Omega \equiv (\alpha, \beta, \gamma)$ defined in Fig. 7.4.1. As the molecule tumbles $\boldsymbol{\Omega}$ changes and can thus be regarded as a function of time $\boldsymbol{\Omega}(t)$. Thus the laboratory-fixed polarizability should be some function of $\boldsymbol{\Omega}(t)$. The spherical representation of the polarizability tensor enables us to express this function in a compact form. Let $\alpha_M^{(J)}(L, t)$ and $\alpha_M^{(J)}(B)$ be, respectively, the spherical components of the polarizability tensor in the laboratory-fixed(L) and body-fixed (B) frames at time t.[7] In appendix 7.C we show that these two sets of components are related by Eq. (7.C.15)

$$\alpha_M^{(J)}(L, t) = \sum_{m'} \alpha_{M'}^{(J)}(B) D_{M'M}^{(J)}(\boldsymbol{\Omega}(t)) \tag{7.4.2}$$

where $\{D_{M'M}^{(J)}(\boldsymbol{\Omega}(t))\}$ are functions of the Euler angles known as Wigner rotation functions. These form a complete orthogonal set of functions of the Euler angles and will be useful later when we discuss the rotational dynamics.

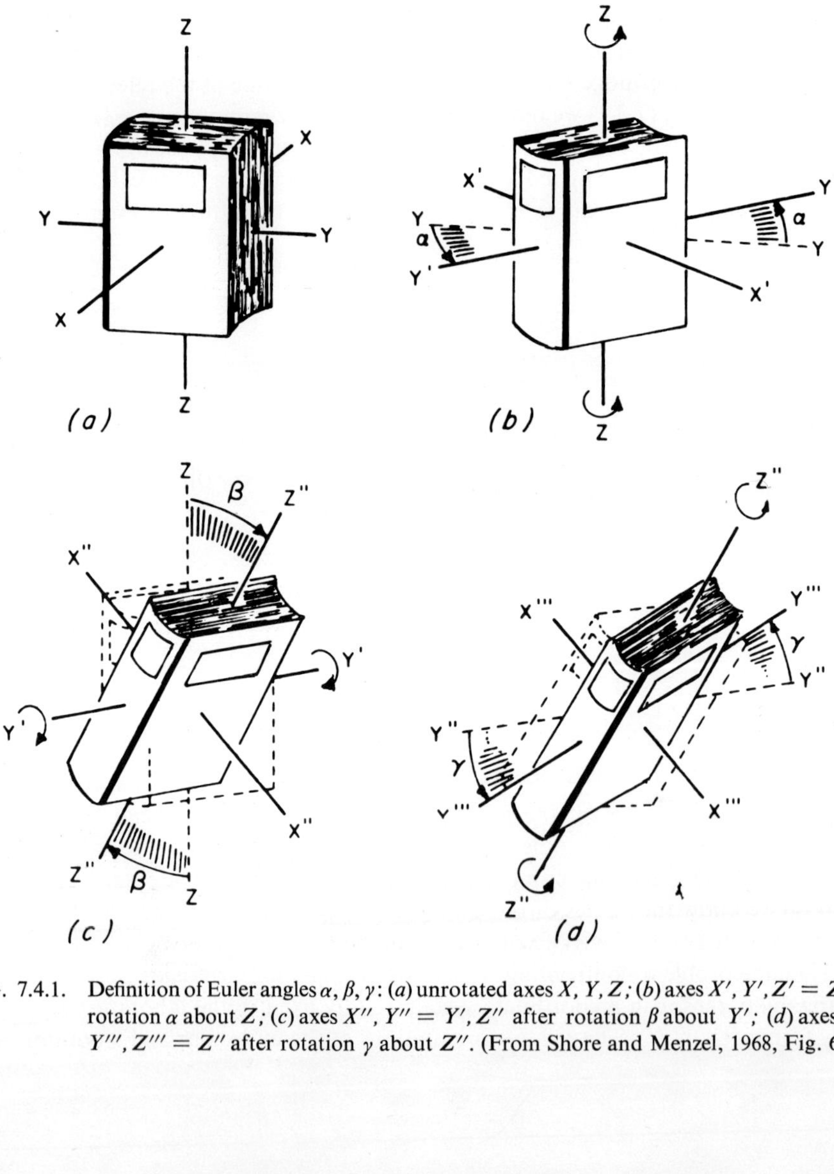

FIG. 7.4.1. Definition of Euler angles α, β, γ: (a) unrotated axes X, Y, Z; (b) axes $X', Y', Z' = Z$ after rotation α about Z; (c) axes $X'', Y'' = Y', Z''$ after rotation β about Y'; (d) axes $X''', Y''', Z''' = Z''$ after rotation γ about Z''. (From Shore and Menzel, 1968, Fig. 6.2.)

The relationship Eq. (7.C.13) between the spherical components and the Cartesian components in any coordinate frame can be solved for the Cartesian components in terms of the spherical components. Then using Eq. (7.C.14), the Cartesian components in frame (L) can be expressed in terms of the components in frame (B). For example, when these simple algebraic operations are carried out we find that

$$\alpha_{ZZ}(L, t) = \frac{1}{\sqrt{3}}\,\alpha_0^{(0)}(B) + \sqrt{\frac{2}{3}} \sum_{M'=-2}^{+2} \alpha_{M'}^{(2)}(B)\, D_{M'0}^{(2)}(\boldsymbol{\Omega}(t)) \tag{7.4.3a}$$

$$\alpha_{ZY}(L, t) = -\frac{1}{2}\Big\{\sqrt{2}\sum_{M'=1}^{1} \alpha_{M'}^{(1)}(B)\,[D_{M'1}^{(2)}(\Omega(t)) - D_{M'-1}^{(2)}(\Omega(t))]$$

$$+\, i \sum_{M'=-2}^{+2} \alpha_{M'}^{(2)}(B)\,[D_{M,1}^{(2)}(\Omega(t)) + D_{M,-1}^{(2)}(\Omega(t))^{1}\} \tag{7.4.3b}$$

If the molecule has a center of inversion or a plane of symmetry (i.e., if it is optically inactive), the Cartesian polarizability tensor in frame (B) must be symmetric; that is, $\alpha_{ij}(B) = \alpha_{ji}(B)$. Referring now to Eq. (7.4.1) we see that in this case $\alpha_0^{(1)} = \alpha_{+1}^{(1)} = 0$, and Eq. (7.4.3b) simplifies to

$$\alpha_{YZ}(L, t) = -\frac{i}{2} \sum_{m'=-2}^{+2} \alpha_{M'}^{(2)}(B)\,[D_{M',1}^{(2)}(\boldsymbol{\Omega}(t)) + D_{M',-1}^{(2)}(\boldsymbol{\Omega}(t))] \tag{7.4.4}$$

We shall assume henceforth that we are dealing with optically inactive molecules. It is a simple matter to include the optically active scattering[8]

Substitution of Eqs. (7.4.3a) and (7.4.4) into Eq. (7.2.1) shows that the spectrum depends on time-correlation functions of the form

$$\langle D_{MK}^{(J)*}(\boldsymbol{\Omega}(0))\, D_{M'K'}^{(J')}(\boldsymbol{\Omega}(t))\rangle \tag{7.4.5}$$

We shall write the explicit form of these functions in Section 7.5, where we consider rotational diffusion.

7 · 5 ROTATIONAL DIFFUSION OF ANISOTROPIC MOLECULES

The orientation of a general anisotropic molecule (not necessarily a rigid rod) is given by the Euler angles $\boldsymbol{\Omega} = (\alpha, \beta, \gamma)$ which specify the orientation of the molecular body-fixed axes with respect to the space-fixed axes (see Fig. 7.4.1).[9] What is required is the conditional probability distribution $K_s(\boldsymbol{\Omega}, t \,|\, \boldsymbol{\Omega}_0)$ which specifies the probability distribution for the molecule to have an orientation $\boldsymbol{\Omega}$ at time t given that it had an orientation $\boldsymbol{\Omega}_0$ at time 0. This conditional distribution must satisfy the initial condition

$$\lim_{t \to 0} K_s(\boldsymbol{\Omega}, t \,|\, \boldsymbol{\Omega}_0, 0) = \delta(\boldsymbol{\Omega} - \boldsymbol{\Omega}_0) \tag{7.5.1}$$

The same kind of treatment can be given for this conditional probability distribution function as was given for the rigid rod case (Section 7.3). The functions $\{C_{K,M}^{(J)}(\boldsymbol{\Omega})\}$

form a complete orthonormal set of functions of the Euler angles[10] and are eigenfunctions of the operators $\hat{I}^2$, $\hat{I}_{Z'}$, $\hat{I}_Z$.

$$\begin{aligned} \hat{I}^2 C^{(J)}_{K,M}(\boldsymbol{\Omega}) &= J(J+1)\, C^{(J)}_{K,M}(\boldsymbol{\Omega}) \\ \hat{I}_{Z'}\, C^{(J)}_{K,M}(\boldsymbol{\Omega}) &= K C^{(J)}_{K,M}(\boldsymbol{\Omega}) \\ \hat{I}_Z\, C^{(J)}_{K,M}(\boldsymbol{\Omega}) &= -M C^{(J)}_{K,M}(\boldsymbol{\Omega}) \end{aligned} \tag{7.5.2}$$

These operators have the same form as the angular momentum operators of an anisotropic rigid rotor in quantum mechanics (see Edmunds, 1957). $\hat{\boldsymbol{I}}$ is the total angular momentum operator of the molecule; $\hat{\boldsymbol{I}}_{Z'}$; is the angular momentum operator about the spacefixed Z' axis; and $\hat{I}_Z$ is the angular momentum about the body-fixed Z axis (usually chosen as the axis of highest symmetry) in the molecule.

The Debye model for rotational diffusion, when applied to the general case of an anisotropic diffusor, is (see Favro, 1960)

$$\frac{\partial}{\partial t} K_s(\boldsymbol{\Omega}, t \mid \boldsymbol{\Omega}_0) = -\sum_{i,j=x,y,Z} \hat{I}_i\, \Theta_{ij} \hat{I}_j K_s(\boldsymbol{\Omega}, t \mid \boldsymbol{\Omega}_0) \tag{7.5.3}$$

(body-fixed axes)

where $\hat{I}_X$, $\hat{I}_Y$, and $\hat{I}_Z$ are angular momentum operators about some axes stuck in the molecules and Θ_{ij} is the rotational diffusion tensor (rotational diffusion will be different around different axes in an anisotropic molecule.) The body-fixed axes may be chosen for convenience as those for which Θ_{ij} is diagonal. As we have noted previously, these axes are called the principal axes of $\boldsymbol{\Theta}$. Then

$$\boldsymbol{\Theta} = \begin{pmatrix} \Theta_{XX} & 0 & 0 \\ 0 & \Theta_{YY} & 0 \\ 0 & 0 & \Theta_{ZZ} \end{pmatrix} \tag{7.5.4}$$

For this, choice of the body-fixed axes Eq. (7.5.3) simplifies to

$$\frac{\partial}{\partial t} K_s = -(\Theta_{XX}\hat{I}_X^2 + \Theta_{YY}\hat{I}_Y^2 + \Theta_{ZZ}\hat{I}_Z^2)\, K_1 \tag{7.5.5}$$

For a completely *anisotropic diffusor* $\Theta_{XX} \neq \Theta_{YY} \neq \Theta_{ZZ}$; for a *symmetric diffusor* $\Theta_{XX} = \Theta_{YY} \equiv \Theta_\perp$ and $\Theta_{ZZ} \equiv \Theta_\parallel$ $(\Theta_\parallel \neq \Theta_\perp)$; and for a *spherical diffusor* $\Theta_{XX} = \Theta_{YY} = \Theta_{ZZ} \equiv \Theta$.

The spherical diffusion[11] case is trivial to solve. The solutions and theory for this case are presented in Section 7.3, although the context used there is slightly different. There we consider rotational Brownian motion of the symmetry axis of a cylindrical molecule. If an axis of this sort is rigidly embedded in a spherical molecule (passing through the molecular center) and if one wishes to study the rotational diffusion of this system, then the theory of Section 7.3 applies.

The case of the symmetric diffusor is also easily solved, in an analytic form, but the solutions for the asymmetric diffusor may only be found from perturbation theory on the symmetric diffusor solution. We therefore first solve the symmetric diffusor problem and then present the asymmetric diffusor solutions.

For the symmetric rotor, Eq. (7.5.3) becomes

$$\frac{\partial}{\partial t} K_s = -[\Theta_\perp(\hat{I}_X^2 + \hat{I}_Y^2) + \Theta_\parallel \hat{I}_Z^2]\, K_s \tag{7.5.6}$$

adding and subtracting $\Theta_\perp, \hat{I}_Z^2$ on the left side yields

$$\frac{\partial}{\partial t} K_s = -(\Theta_\perp \hat{I}^2 + (\Theta_\parallel - \Theta_\perp)\hat{I}_Z^2)\, K_s \tag{7.5.7}$$

Now the $C_{K,M}^{(J)}$ functions are eigenfunctions of this operator [see Eq. (7.5.2)]

$$[\Theta_\perp \hat{I}^2 + (\Theta_\parallel - \Theta_\perp)\hat{I}_Z^2]\, C_{K,M}^{(J)}(\boldsymbol{\Omega}) = (J(J+1)\,\Theta_\perp + M^2(\Theta_\parallel - \Theta_\perp))\, C_{K,M}^{(J)}(\boldsymbol{\Omega}) \tag{7.5.8}$$

Following the logic developed in Section 7.3 it follows that

$$K_s(\boldsymbol{\Omega}, t \,|\, \boldsymbol{\Omega}_0, 0) = \sum_{JKM} C_{K,M}^{(J)}(\boldsymbol{\Omega}_0)\, C_{K,M}^{(J)*}(\boldsymbol{\Omega}) \exp\overline{\ }[\Theta_\perp \hat{I}^2 + (\Theta_\parallel - \Theta_\perp)\hat{I}_Z^2]t \tag{7.5.9}$$

At $t = 0$, we see that the boundary condition Eq. (7.5.1) applies since the functions $\{C_{KM}^{(J)}\}$ form a complete orthonormal set

$$\delta(\boldsymbol{\Omega} - \boldsymbol{\Omega}_0) = \sum_{JKM} C_{K,M}^{(J)}(\boldsymbol{\Omega}_0)\, C_{K,M}^{(J)*}(\boldsymbol{\Omega}) \tag{7.5.10}$$

Now the initial orientational distribution function for an anisotropic molecule in an equilibrium ensemble is[12]

$$p(\boldsymbol{\Omega}) = \frac{1}{8\pi^2}$$

Thus the joint probability distribution function is

$$G_s(\boldsymbol{\Omega}, t; \boldsymbol{\Omega}_0, 0) = \frac{1}{8\pi^2} \sum_{JKM} C_{K,M}^{(J)}(\boldsymbol{\Omega}_0)\, C_{K,M}^{(J)*}(\boldsymbol{\Omega}) \exp -[\Theta_\perp \hat{I}^2 + (\Theta_\parallel - \Theta_\perp)\hat{I}_Z^2]t \tag{7.5.11}$$

Using the orthonormality condition

$$\int d\Omega\, C_{K,M}^{(J)'*}(\boldsymbol{\Omega})\, C_{K,M}^{(J)}(\boldsymbol{\Omega}) = \delta_{JJ'}\, \delta_{KK'}\, \delta_{MM'} \tag{7.5.12}$$

we can determine the time-correlation functions

$$F_{K,M}^{J}(t) = \langle C_{K,M}^{(J)*}(\boldsymbol{\Omega}(0))\, C_{K,M}^{(J)}(\boldsymbol{\Omega}(t))\rangle \tag{7.5.13}$$

where $\boldsymbol{\Omega}(t)$ specifies the orientation of the molecule at time t. Moreover we can show that

$$\langle C_{K',M'}^{(J)'*}(\boldsymbol{\Omega}(0))\, C_{K,M}^{(J)}(\boldsymbol{\Omega}(t))\rangle = F_{K,M}^{J}(t)\, \delta_{JJ'}\, \delta_{MM'}\, \delta_{KK'} \tag{7.5.14}$$

The correlation functions can be evaluated using G_s

$$F_{K,M}^{J}(t) = \int d\Omega_0 \int d\Omega\, G_s(\boldsymbol{\Omega}; t; \boldsymbol{\Omega}_0, 0)\, C_{K,M}^{(J)*}(\boldsymbol{\Omega}_0)\, C_{K,M}^{(J)}(\boldsymbol{\Omega})$$

We find[13]

$$F_{K,M}^J(t) = \frac{1}{8\pi^2} \exp - [J(J+1)\,\Theta_\perp + M^2\,(\Theta_\parallel - \Theta_\perp)] \tag{7.5.15}$$

We now apply these results to the calculation of the light-scattering correlation functions for symmetric diffusors. From Eqs. (7.4.3a), (7.4.4), and (7.2.1) we find that

$$I_{VV}^\alpha(\mathbf{q}, t) = \langle N \rangle \Big[\frac{1}{3}\,\alpha_0^{(0)\,2} + \frac{1}{\sqrt{3}}\,\alpha_0^{(0)} \Big\langle \sum_{M=-2}^{+2} \alpha_M^{(2)}\,(B)\,D_{M,0}^{(2)}\,(\Omega(t)) \Big\rangle +$$

$$\frac{\sqrt{2}}{\sqrt{3}}\,\alpha_0^{(0)} \Big\langle \sum_{M=-2}^{+2} \alpha_M^{(2)*}\,(B)\,D_{M,0}^{(2)*}\,(\Omega(0)) \Big\rangle +$$

$$\frac{2}{3} \sum_{M,M'} \alpha_{M}^{(2)*}\,(B)\alpha_{M'}^{(2)}\,(B) \big\langle D_{M,0}^{(2)*}\,(\Omega(0))\,D_{M',0}^{(2)}\,(\Omega(t)) \big\rangle \Big]\,F_s(\mathbf{q}, t) \tag{7.5.16}$$

and

$$I_{VH}^\alpha(\mathbf{q}, t) = \langle N \rangle \frac{1}{4} \Big\{ \sum_{M,M'} \alpha_M^{(2)*}\,(B)\alpha_{M'}^{(2)}\,(B)\big[\big\langle D_{M,1}^{(2)*}\,(\Omega,(0))\,D_{M',1}^{(2)}\,(\Omega(t)) \big\rangle$$

$$+ \big\langle D_{M,+1}^{(2)*}\,(\Omega(0))D_{M',-1}^{(2)}\,(\Omega(t)) \big\rangle$$

$$+ \big\langle D_{M,-1}^{(2)*}\,(\Omega(0))\,D_{M',+1}^{(2)}\,(\Omega(t) \big\rangle$$

$$+ \big\langle D_{M,-1}^{(2)*}\,(\Omega(0))\,D_{M',-1}^{(2)}\,(\Omega(t)) \big\rangle \big]\Big\}\,F_s(\mathbf{q}, t) \tag{7.5.17}$$

Using Eqs. (7.5.14) and (7.5.15) we find that Eqs. (7.5.16) and (7.5.17) simplify to[14]

$$I_{VV}^\alpha(\mathbf{q}, t) = \langle N \rangle\,\alpha^2 F_s(\mathbf{q}, t) + \frac{2}{15} \sum_{M=-2}^{+2} |\alpha_M^{(2)}\,(B)|^2 \{\exp[6\Theta_\perp$$

$$+ M^2(\Theta_\parallel - \Theta_\perp)]t\}\,F_s(\mathbf{q}, t) \tag{7.5.18}$$

and

$$I_{VH}^\alpha(\mathbf{q}, t) = \langle N \rangle \frac{1}{10} \sum_M |\alpha_M^{(2)}\,(B)|^2 \{\exp-[6\Theta_\perp + M^2(\Theta_\parallel - \Theta_\perp)]t\}\,F_s(q, t) \tag{7.5.19}$$

We note that

$$I_{\text{ISO}}^\alpha(\mathbf{q}, t) \equiv \langle N \rangle\,\alpha^2 F_s(\mathbf{q}, t)$$

and $$\tag{7.5.20}$$

$$I_{VV}^\alpha(\mathbf{q}, t) = I_{\text{ISO}}^\alpha(\mathbf{q}, t) + \frac{4}{3}\,I_{VH}^\alpha(\mathbf{q}, t)$$

as in Eqs. (7.2.8) and (7.3.19). From Eq. (7.5.19) it is easily seen that $I_{VH}^\alpha(q, t)$ consists of three exponentials, the first corresponding to $M = 0$, the second corresponding to $M = \pm 1$, and the third to $M = \pm 2$.

The frame (B) was chosen such that the rotational diffusion tensor Θ is diagonal. In general, the polarizability tensor α will not be diagonal in the same body fixed frame that diagonalizes Θ. In the special case when α and Θ are simultaneously diagonalized in the frame (B); that is, when the molecule is a *true symmetric top*, then $\alpha_{ij}(B) = 0$ for $i \neq j$. Referring back to Eq. (7.4.1) we see that in this eventuality $\alpha_{ZZ}(B) = \alpha_\parallel$, and $\alpha_{XX}(B) = \alpha_{YY}(B) = \alpha_\perp$, and

$$\begin{aligned} \alpha_0^{(2)}(B) &= \frac{2}{3}(\alpha_\parallel - \alpha_\perp) = \frac{2}{3}\beta \\ \alpha_{\pm 1}^{(2)}(B) &= 0 \\ \alpha_{\pm 2}^{(2)}(B) &= 0 \end{aligned} \tag{7.5.21}$$

Eq. (7.5.19) thus simplifies for a true symmetric top to

$$I_{VH}^{\alpha}(q, t) = \frac{1}{15}\langle N\rangle \beta^2 \exp - [6\Theta_\perp + q^2 D]|t| \tag{7.5.22}$$

which is identical to the result (Eq. (7.3.18)) that we found for a symmetric top in Section 7.3 by a simpler method.

Equation (7.5.22) applies rigorously if the Z axis is a four fold or more symmetry axis of rotation. The more general result given in Eq. (7.5.19) holds if the polarizability tensor does not have cylindrical symmetry about the Z axis in the body-fixed coordinate system while the rotational diffusion tensor does.

Asymmmetric Diffusor

The asymmetric diffusor problem is much more complicated than that of the symmetric diffusor. We give here only a brief indication of how the results are obtained—referring the reader to the literature for the details (Favro, 1960). Since the results are important in the interpretation of light-scattering experiments we present them in full.

The solution proceeds by analogy with the symmetric diffusor problem. The eigenfunctions and eigenvalues of the operator on the right-hand side of Eq. (7.5.6) must be found. However, the eigenfunctions of this operator are no longer simply the angular momentum eigenfunctions given in Eq. (7.5.8), but may be expressed as linear combinations of them. Thus

$$\Psi_{\tau M}^{(J)} = \sum_K a_K^{(J)}(\tau)\, C_{K,M}^{(J)} \tag{7.5.23}$$

where the $a_K^{(J)}(\tau)$ are the expansion coefficients. The $a_K^{(J)}(\tau)$ and the eigenvalues of $\Psi_{\tau M}^{(J)}$ cannot be written in a simple closed form except for small values of J. Fortunately, in order to solve the light-scattering problem we need only the eigenvalues and eigenfunctions for $J = 2$. These quantities are given in Table 7.5.1.

The solution of Eq. (7.5.3) may then be written as

$$K_1(\Omega, t|\Omega_0) = \sum_{J,\tau,M} \Psi_{\tau M}^{(J)}(\Omega_0)^*\, \Psi_{\tau M}^{(J)} \exp - f_\tau^{(J)} t \tag{7.5.24}$$

where the eigenvalues $f_\tau^{(J)}$ depend on J and τ. By a conceptually simple but algebraicly tedious procedure one may substitute the various $\Psi_{\tau M}^{(J)}$ and $f_\tau^{(J)}$ into Eq. (7.5.24), use the definition of the $C_{K,M}^{(J)}$ in terms of the $D_{K,M}^{(J)}$ and then evaluate the $F_{M,M'}^{(J)}(t)$ as defined by Eq. (7.5.14). The results are shown in Table 7.5.2.

The $F_{M,M'}^{J}$ of Table 7.5.2 may then be substituted into Eq. (7.5.17) to obtain $I_{VH}^{\alpha}(t)$. We find

$$I_{VH}^{\alpha}(\mathbf{q}, t) = \frac{1}{15}\langle N\rangle \sum_{i=1}^{5} A_i \exp - (q^2 D + f_i)t \tag{7.5.25}$$

TABLE 7.5.1

Eigenfunctions and Eigenvalues for the Asymmetric Rotor for J = 2

$\Psi_{\tau M}^{(J)}$	$f_{\tau}^{(2)}$
$\Psi_{2,M}^{(2)} = \frac{1}{N}\left[aC_{0,M}^{(2)} + \frac{b}{\sqrt{2}}(C_{2,M}^{(2)} + C_{-2,M}^{(2)})\right]$	$6\,\Theta_I + 2\Delta$
$\Psi_{1,M}^{(2)} = \frac{1}{\sqrt{2}}(C_{1,M}^{(2)} + C_{-1,M}^{(2)})$	$3\,(\Theta_{XX} - \Theta_I)$
$\Psi_{0,M}^{(2)} = \frac{1}{N}\left[bC_{0,M}^{(2)} - \frac{a}{\sqrt{2}}(C_{2,M}^{(2)} + C_{-2,M}^{(2)})\right]$	$6\,\Theta_I - 2\Delta$
$\Psi_{-1,M}^{(2)} = \frac{1}{\sqrt{2}}(C_{1,M}^{(2)} - C_{-1,M}^{(2)})$	$3\,(\Theta_{YY} - \Theta_I)$
$\Psi_{-2,M}^{(2)} = \frac{1}{\sqrt{2}}(C_{2,M}^{(2)} - C_{-2,M}^{(2)})$	$3\,(\Theta_{ZZ} - \Theta_I)$

where $a = \sqrt{3}\,(\Theta_{XX} - \Theta_{YY})$

$b = (2\,\Theta_{ZZ} - \Theta_{XX} - \Theta_{YY} + 2\Delta)$

$N = 2\Delta^{\frac{1}{2}}\,b^{\frac{1}{2}}$

$\Delta = [(\Theta_{XX} - \Theta_{YY})^2 + (\Theta_{ZZ} - \Theta_{XX})\,(\Theta_{ZZ} - \Theta_{YY})]^{\frac{1}{2}}$

$\Theta_I = 1/3\,(\Theta_{XX} + \Theta_{YY} + \Theta_{ZZ})$

TABLE 7.5.2

Components of the rotational correlation function for J = 2

$$F_{0,0}^{2}(t) = [(a^2/N^2)\exp(-f_1 t) + (b^2/N^2)\exp(-f_2 t)]$$

$$F_{1,\pm 1}{}^{2}(t) = (1/2)[\exp(-f_4 t) \pm \exp(-f_3 t)]$$

$$F_{2,\pm 2}{}^{2}(t) = [(b^2/2N^2)\exp(-f_1 t) \pm 1/2\exp(-f_5 t) + (a^2/2N^2)\exp(-f_2 t)]$$

$$F_{2,0}{}^{2}(t) = (ab/N^2\sqrt{2})\,[\exp(-f_1 t) - \exp(-f_2 t)]$$

$$F_{n,n'}{}^{2}(t) = F_{-n,-n'}{}^{2}(t) = F_{-n',-n}{}^{2}(t) = F_{n',n}(t)$$

$a = \sqrt{3}\,(\Theta_{XX} - \Theta_{YY})$	$f_1 = 6\Theta_I + 2\Delta$
$b = (2\Theta_{ZZ} - \Theta_{XX} - \Theta_{YY} + 2\Delta)$	$f_2 = 6\Theta_I - 2\Delta$
$N = 2\Delta^{\frac{1}{2}}b^{\frac{1}{2}}$	$f_3 = 3\,(\Theta_{ZZ} + \Theta_I)$
$\Delta = [(\Theta_{XX} - \Theta_{YY})^2 + (\Theta_{ZZ} - \Theta_{XX})\,(\Theta_{ZZ} - \Theta_{YY})]^{\frac{1}{2}}$	$f_4 = 3\,(\Theta_{YY} + \Theta_I)$
$\Theta_I = \frac{1}{3}(\Theta_{XX} + \Theta_{YY} + \Theta_{ZZ})$	$f_5 = 3\,(\Theta_{ZZ} + \Theta_I)$

where the f_i are defined in Table 5.7.2 and the coefficients depend on quantities given in the table as well as the body-fixed polarizability components:

$$A_1 = \frac{1}{3}\left(\frac{a^2}{N^2}\right)[\alpha_{ZZ}\,(B) - \frac{1}{2}\,(\alpha_{XX}(B) + \alpha_{YY}(B))]^2$$

$$+ \frac{ab}{N^2\sqrt{3}}\,[\alpha_{ZZ}(B) - \frac{1}{2}\,(\alpha_{XX}(B) + \alpha_{YY}(B)] \cdot [\alpha_{XX}(B) - \alpha_{YY}(B)]$$

$$+ \frac{b^2}{2N^2}[\alpha_{XX}(B) - \alpha_{YY}(B)]^2; \quad (7.5.26)$$

A_2 is the same as A_1 except that a and b are interchanged and the ab term has a negative sign and

$$\begin{aligned} A_3 &= \alpha_{ZX}^2(B) \\ A_4 &= \alpha_{ZY}^2(B) \\ A_5 &= \alpha_{XY}^2(B) \end{aligned} \quad (7.5.27)$$

Thus the correlation function I_{VH}^{α} of an asymmetric diffusor consists in general of five exponentials. The spectrum consists of five Lorentzians. It is clear that it would be difficult to extract the time constants of these five Lorentzians from an experimental spectrum.

The results simplify considerably if the body-fixed axis system is a principal axis system for the polarizability tensor as well as for the rotational diffusion tensor. In this case $A_3 = A_4 = A_5 = 0$ in Eq. (7.5.27). Then the spectrum consists of only two Lorentzians. Many asymmetric diffusors do have enough symmetry to rigorously satisfy this condition-for instance, planar molecules with at least one two fold rotation axis in the molecular plane. Others may have these axes so close together that $A_3 \cong A_4 \cong A_5 \cong 0$ and the spectrum effectively consists of only two Lorentzians. In any particular application, it must be kept in mind that the spectrum might very well be the five-Lorentzian form given by the Fourier transform of Eq. (7.5.25).

Note that these general results reduce to those for symmetric diffusor molecules when the molecule fixed $\boldsymbol{\Theta}$ and $\boldsymbol{\alpha}$ are given the appropriate symmetry.

7 · 6 EXTENDED DIFFUSION MODELS OF MOLECULAR REORIENTATION

The rotational motion of molecules is generally very complicated. The rotational diffusion model discussed in Section 7.5 is applicable to the case in which molecules undergo many collisions before reorienting through an appreciable angle. It is, in fact, as we point out in Section 7.3, the rotational analog of the small-step translational diffusion model. It is clear, however, that there are many cases in which this model does not apply. For instance, in a very dilute gas the rotational motion is essentially that of a quantum-mechanical rotor. The rotational motion is not strongly affected by intermolecular collisions since there are very few collisions. This is called the *inertial* limit. As the system becomes more dense, however, one expects collisions to play a more important role—until the rotational diffusion limit is attained. Of course, if the intermolecular forces are very weakly dependent on the relative orientation of collision partners, then we expect reorientation to be determined by inertial effects rather than by diffusion regardless of the fluid density. Thus for liquids of almost spherical molecules such as CH_4 and CCl_4 experiments show that the reorientation is predominantly inertial.

What are the consequences of these considerations for depolarized light scattering? In a dilute gas where reorientation is predominantly inertial, we expect the spectrum to be what is normally called the pure rotational Raman spectrum of the molecule. As higher densities are approached, the discrete spectral lines broaden and overlap to form a continuous band. We show how the band shape can be computed for freely rotating linear molecules and spherical top molecules and then indicate the assumptions that have been used by several authors to include collisions in the theory.

First consider a gas of spherical top molecules that is so dilute that intermolecular interactions can be ignored. Although the motions of the individual molecules are clearly quantized, we present here a classical calculation of the spectrum, leaving the quantum mechanical calculation as an exercise for the reader.

In a freely rotating spherical top the angular velocity, $\boldsymbol{\omega}_0$, is conserved. The unit vector $\hat{\boldsymbol{\omega}}_0$ parallel to $\boldsymbol{\omega}_0$ specifies the orientation of the axis of rotation, whereas the magnitude ω_0 of $\boldsymbol{\omega}_0$ specifies the rotational speed.

Consider the body-fixed axes xyz defined such that $\hat{\mathbf{z}}$ is parallel to $\boldsymbol{\omega}_0$, and $\hat{\mathbf{x}}$ and $\hat{\mathbf{y}}$ lie in a plane perpendicular to $\boldsymbol{\omega}_0$. If $\hat{\mathbf{x}}(0)$, $\hat{\mathbf{y}}(0)$, and $\hat{\mathbf{z}}(0)$ are unit vectors specifying the orientation of this frame at time $t = 0$, then at time t the frame is specified by

$$\hat{\mathbf{x}}(t) = \hat{\mathbf{x}}(0) \cos \omega_0 t + \hat{\mathbf{y}}(0) \sin \omega_0 t$$

$$\hat{\mathbf{y}}(t) = - \hat{\mathbf{x}}(0) \sin \omega_0 t + \hat{\mathbf{y}}(0) \cos \omega_0 t$$

$$\hat{\mathbf{z}}(t) = \hat{\mathbf{z}}(0)$$

Thus $\hat{\mathbf{x}}(t)$ and $\hat{\mathbf{y}}(t)$ rotate as the molecule rotates whereas $\hat{\mathbf{z}}(t)$ does not rotate because $\boldsymbol{\omega}_0$ is fixed.

As in Sec 7.3, let $\mathbf{u}$ be a unit vector imbedded rigidly in the sphere and rotating with it. $\mathbf{u}(t)$ can be expressed in terms of u_x, u_y, u_z, its projections on the orthogonal axes $\hat{\mathbf{x}}(t)$, $\hat{\mathbf{y}}(t)$ and $\hat{\mathbf{z}}(t)$ of the body-fixed frame, as

$$\mathbf{u}(t) = u_x\hat{\mathbf{x}}(t) + u_y\hat{\mathbf{y}}(t) + u_z\hat{\mathbf{z}}(t)$$

It follows that

$$\mathbf{u}(0) \cdot \mathbf{u}(t) = (1 - u_z^2) \cos \omega_0 t + u_z^2 \tag{7.6.1}$$

where we have used $(u_x^2 + u_y^2 + u_z^2 = 1)$ to eliminate $(u_x^2 + u_y^2)$. Thus $\mathbf{u}(0) \cdot \mathbf{u}(t)$ contains a time-invariant part u_z^2 and a part that varies with t. Now $u_z = \mathbf{u}(0) \cdot \hat{\boldsymbol{\omega}}_0$ can either be regarded as the projection of $\mathbf{u}(0)$ on $\hat{\boldsymbol{\omega}}_0$ or the projection of $\hat{\boldsymbol{\omega}}_0$ on $\mathbf{u}(0)$. In the remaining analysis the latter point of view is adopted, that is,

$$(\hat{\boldsymbol{\omega}}_0)_z \equiv u_z$$

Our object is to determine the correlation functions

$$C_l^{(0)}(t) \equiv \langle P_l(\mathbf{u}(0) \cdot \mathbf{u}(t))\rangle \tag{7.6.2a}$$

where the superscript zero denotes an average over an ensemble of freely rotating molecules. Equation (7.6.2a) can be expressed as

$$C_l^{(0)}(t) = \int d^3\omega_0\, p(\omega_0)\, P_l((1 - u_z^2) \cos \omega_0 t + u_z^2) \tag{7.6.2b}$$

where $p(\omega_0)$ is the Maxwell distribution of rotational velocities

$$p(\omega_0) = \left[\frac{2\pi}{3}\langle\omega_0^2\rangle\right]^{-3/2} \exp - \left[\omega_0^2 \Big/ \frac{2}{3}\langle\omega_0^2\rangle\right] \tag{7.6.2c}$$

$\langle\omega_0^2\rangle$ is the mean-square angular velocity, $\omega_0^2 = 3k_BT/I$, and I is the moment of inertia. The integral can be expressed in terms of spherical polar coordinates (ω_0, θ, ϕ) where the polar axis is defined by $\mathbf{u}(0)$. Then $u_z = \cos\theta$ and Eq. (7.6.2b) becomes

$$C_l^{(0)}(t) = 4\pi \int_0^\infty d\omega_0 \omega_0^2 \, p(\omega_0) \int_{-1}^{1} \frac{dx}{2} P_l((1 - x^2)\cos\omega_0 t + x^2)$$

where we have taken $x \equiv \cos\theta$. The x integral is evaluated first, then the ω_0 integral is evaluated by completing the squares. This results in

$$C_1^{(0)}(\tau) = \frac{1}{3} + \frac{2}{3}[1 - \tau^2] \exp \frac{-\tau^2}{2} \tag{7.6.3a}$$

$$C_2^{(0)}(\tau) = \frac{1}{5} + \frac{2}{5}[1 - 4\tau^2] \exp - 2\tau^2 + \frac{2}{5}[1 - \tau^2] \exp \frac{-\tau^2}{2} \tag{7.6.3b}$$

and in general

$$C_l^{(0)}(\tau) = \frac{1}{2l+1} \sum_{m=-l}^{+l} (1 - m^2\tau^2) \exp - m^2\tau^2/2 \tag{7.6.3c}$$

where τ is a reduced time defined by

$$\tau = \left[\frac{1}{3}\langle\omega_0^2\rangle\right]^{1/2} t = \left[\frac{k_BT}{I}\right]^{1/2} t \tag{7.6.3c}$$

$C_1^{(0)}(\tau)$ and $C_2^{(0)}(\tau)$ are plotted in Figs. 7.6.1a, b. It should be noted that neither of these two functions decay to zero as $\tau \to \infty$. This is because there are always molecules for which $\mathbf{u}(t)$ has a time-invariant projection along the axis of rotation. These functions decay on the time scale defined by $t_0 = (I/k_BT)^{1/2}$, the "average" rotational period.

The spectral densities $I_1^{(0)}(\omega)$ and $I_2^{(0)}(\omega)$ corresponding to these functions are easily evaluated. In reduced units these are

$$I_1^{(0)}(\tilde{\omega}) = \frac{1}{3}\delta(\tilde{\omega}) + \frac{1}{3}\left(\frac{2}{\pi}\right)^{1/2} \tilde{\omega}^2 \exp - \frac{1}{2}\tilde{\omega}^2$$

$$I_2^{(0)}(\tilde{\omega}) = \frac{1}{5}\delta(\tilde{\omega}) + \frac{1}{40}\left(\frac{2}{\pi}\right)^{1/2} \tilde{\omega}^2 \exp - \frac{1}{8}\tilde{\omega}^2 + \frac{1}{5}\left(\frac{2}{\pi}\right)^{1/2} \tilde{\omega}^2 \exp - \frac{1}{2}\tilde{\omega}^2$$

where $\tilde{\omega}$ is the reduced frequency $\tilde{\omega} = \omega t_0$. The delta functions arise from the 1/3 and 1/5 in the correlation functions, Eqs. (7.6.3a) and (7.6.3b).

The depolarized spectrum is the spectral density of the product $C_2^{(0)}(t)F_s(\mathbf{q}, t)$ where $F_s(\mathbf{q}, t)$ is given by Eq. (5.6.9). Because $F_s(\mathbf{q}, t)$ decays on a much slower time scale[15] than the time dependent quantities in Eq. (7.6.3) it can be replaced by $F_s(\mathbf{q}, t = 0)$ whenever it appears in a product with these functions so that

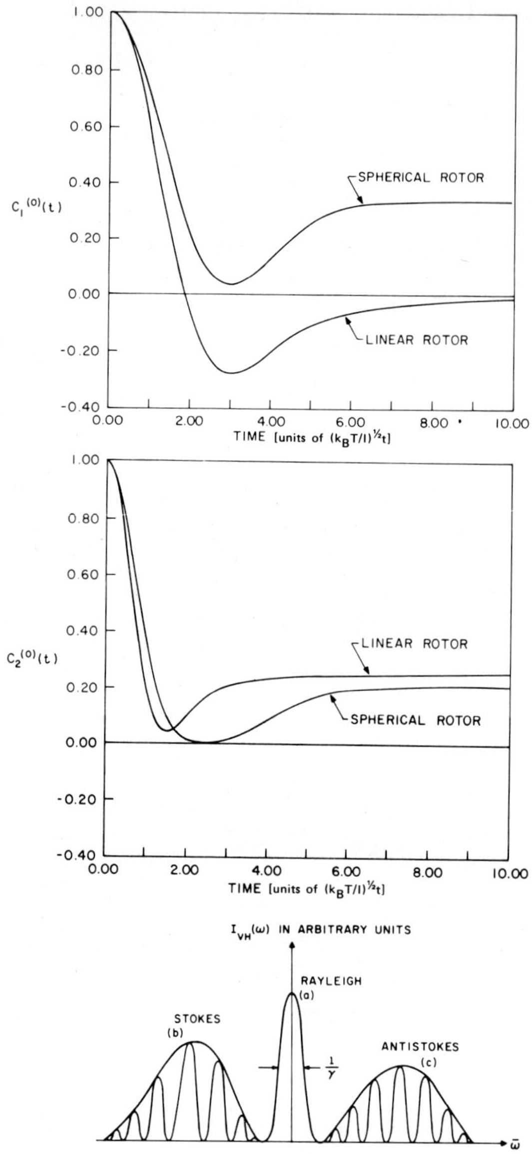

FIG. 7.6.1(a). The free-rotor time-correlation function

$$C_1^{(0)}(t) \equiv \langle P(\mathbf{u}(0) \cdot \mathbf{u}(t)) \rangle$$

for linear and spherical rotors. (*b*) The free-rotor time correlation function

$$C_2^{(0)}(t) \equiv \langle P_2(\mathbf{u}(0) \cdot \mathbf{u}(t)) \rangle$$

for linear and spherical rotors. (*c*) A schematic rotational Raman spectrum for linear molecules. Classically the spectrum consists of: (1) a Rayleigh band fo relative width $1/\gamma$ and two symmetric doublets, (2) (Stokes) and (3) (anti-Stokes) of relative width 1. A quantum-mechanical calculation gives rise to a fine structure in the rotational bands (2) and (3) corresponding to a selection rule of $\Delta J = -2$ in the transitions between the discrete rotational states. Note that in the quantum-mechanical spectrum the doublets are not symmetric.

$$I_{VH}^{\alpha}(t) \simeq \frac{\langle N\rangle\beta^2}{15}\left\{\frac{1}{5}\exp - \gamma^2\tau^2 + \frac{2}{5}[1 - 4\tau^2]\exp - 2\tau^2 + \frac{2}{5}[1 - \tau^2]\exp - \frac{1}{2}\tau^2\right\}$$

where

$$\gamma(q) \equiv \left[\frac{6M}{Iq^2}\right]^{1/2}$$

γ is a dimensionless parameter that measures the ratio of the translational to the rotational relaxation times. Thus the spectral density of the depolarized spectrum in reduced units is proportional to

$$I_{VH}(\mathbf{q}, \tilde{\omega}) = \left[\underbrace{\frac{5\pi}{\gamma(q)}\exp - \tilde{\omega}^2/4\gamma^2}_{\text{central}} + \underbrace{\frac{1}{8}\tilde{\omega}^2\exp - \tilde{\omega}^2/8}_{\text{shifted}} + \underbrace{\tilde{\omega}^2\exp - \tilde{\omega}^2/2}_{\text{shifted}}\right]$$

Usually $\gamma \ll 1$. Then the spectrum consists of a very narrow central band due entirely to the translational motion (this is called the *Rayleigh band*), and a broad symmetrical doublet shifted from the central line due entirely to the rotations.

Consider now the case of a gas of linear molecules that is sufficiently dilute that intermolecular interactions can be ignored. The free rotor correlation functions for a rigid linear molecule are quite different from those for spherical tops. This is now demonstrated. As in Section 7.3, let **u** be a unit vector pointing along the axis of the rotor. Then for a free rotor

$$\mathbf{u}(0) \cdot \mathbf{u}(t) = \cos \omega_0 t$$

where ω_0 is the rotational speed about an axis of rotation perpendicular to the symmetry axis.[16] Then, the average of the second-order Legendre polynomial of $\mathbf{u}(0) \cdot \mathbf{u}(t)$ that is needed in light scattering [see Appendix 7.B, Eq. (7.B.19)] is

$$C_2^{(0)}(t) \equiv \langle P_2(\mathbf{u}(0) \cdot \mathbf{u}(t))\rangle = \int d\omega_0 p(\omega_0)\, P_2(\cos \omega_0 t) \tag{7.6.4a}$$

where $p(\omega_0)$ is the distribution function of molecular speeds. In an equilibrium canonical ensemble

$$p(\omega_0) = \left[\frac{I}{k_B T}\right]\omega_0 \exp\frac{-I\omega_0^2}{2k_B T} \tag{7.6.4b}$$

and where I is the moment of inertia of the rod.

In order to proceed with the computation of the average it is convenient to use the trigonometric identity $\cos^2 \omega_0 t = \frac{1}{2}(1 + \cos 2\omega_0 t)$ in the definition[16] of $P_2(\cos \omega_0 t)$. This gives

$$P_2(\cos \omega_0 t) = \frac{1}{4} + \frac{3}{4}\cos 2\omega_0 t$$

so that

$$\langle P_2(\mathbf{u}(0) \cdot \mathbf{u}(t))\rangle = \frac{1}{4} + \frac{3}{4}\int_0^\infty d\omega_0 p(\omega_0)\cos 2\omega_0 t \tag{7.6.4c}$$

Although we cannot evaluate this latter integral analytically we have presented a numerical calculation in Figs. 7.6.1a, b. This should be compared with the results for the spherical tops.

Equation (7.B.17b) shows that the I_{VH}^{α} spectrum of a system composed of linear molecules is given by

$$I_{VH}^{\alpha}(\mathbf{q}, \omega) = \frac{\langle N \rangle \beta^2}{15} \left(\frac{1}{2\pi}\right) \int_{-\infty}^{+\infty} dt \, \langle P_2(\mathbf{u}(0) \cdot \mathbf{u}(t)) \rangle \, F_s(\mathbf{q}, t) \exp -i\omega t \qquad (7.6.5a)$$

where $F_s(\mathbf{q}, t)$ is given for a perfect gas by Eq. (5.6.6) and $\langle P_2(\mathbf{u}(0) \cdot \mathbf{u}(t)) \rangle$ is given by Eq. (7.6.4c). Because of the small values of q used in light scattering experiments, $F_s(\mathbf{q}, t)$ varies slowly compared with the time-dependent part of $\langle P_2 \rangle$.[12] Thus $F_s(\mathbf{q}, t)$ may be replaced by one whenever it multiplies $\langle P_2 \rangle$.

Upon carrying out the calculation we find the dimensionless normalized spectrum

$$I_{VH}(\mathbf{q}, \tilde{\omega}) = \left\{ \left(\frac{\gamma^2}{3\pi}\right)^{1/2} \exp - \left[\frac{16}{3} \gamma^2 \tilde{\omega}^2\right] + 6|\omega| \exp - 8\tilde{\omega}^2 \right\} \qquad (7.6.5b)$$

Where γ was defined previously. Usually $\gamma \gg 1$. Thus the spectrum consists of a narrow central band resulting from the first term on the right-hand side of Eqs. (7.6.5), due entirely to translational motion[17] (the *Rayleigh band*) and a broad symmetrical doublet arising from the second term on the right-hand side of Eqs. (7.6.4c) due entirely to the rotational motion. This spectrum is shown schematically in Fig. 7.6.1.

This is the classical rotational Raman spectrum of a gas. Light is inelastically scattered by a molecule either exchanging translational energy (central band) or rotational energy (doublet) and in the process suffering a frequency shift. Since translational energies are smaller than rotational energies, the central band (Q-band) is much narrower than the doublets. The left-most band in the figure (P-branch) is the rotational Stokes line—here the scattered light is shifted to lower frequencies because it excites a rotation. The right-most band is the rotational anti-Stokes line—here the scattered light is shifted to higher frequencies because it gains energy. If a purely quantum-mechanical calculation were carried out, the Stokes and anti-Stokes bands would break up into sharp lines, since only discrete transitions would be observed. Moreover, the Stokes side would be more intense than the anti-Stokes side. Since we have decided to confine our attention to Rayleigh–Brillouin scattering, we shall not dwell further on this subject. The interested reader should consult the references on Raman scattering (e.g., Szmansky, 1970 and references therein).

Gordon (1966) has proposed what he calls "extended diffusion" models to take into account the effect of collisions. The $\langle P_2(\mathbf{u}(0) \cdot \mathbf{u}(t)) \rangle$ calculated from these models cannot be expressed in simple analytical form, and we must content ourselves here with a description of the assumptions in the model and a description of results for a few particular systems.

Gordon's model assumes that molecules in a liquid are undergoing collision-interrupted free rotation. A "collision" is defined as an event which changes the angular momentum of a molecule. It is furthermore assumed that: (a) collisions are of zero duration, (b) collisions change the molecule's rotational velocity but do not change its orientation, (c) successive hard-core collisions are uncorrelated; that is, the instants at which the collisions occur are a random process and the angular velocity changes produced by the collisions are uncorrelated, and (d) each collision randomizes the direction of the angular momentum.

Condition (d) can be split into two cases which Gordon calls *J*-diffusion and *M*-diffusion. In *J*-diffusion, the angular momentum is randomized in both magnitude and direction at every collision while in *M*-diffusion, only the direction of the angular momentum vector is randomized. In the following only *J*-diffusion is considered in detail.

The correlation functions required are

$$C_l(t) \equiv \langle P_l(\mathbf{u}(0) \cdot \mathbf{u}(t))\rangle \tag{7.6.6}$$

Consider those molecules in the fluid which during time t have suffered n-1 collisions. Each of these molecules will execute n free rotations between collisions. For each molecule the n-1 collisions will occur at different times. Let us now define

$$C_l(n, t) \equiv \langle P_l(\mathbf{u}(0) \cdot \mathbf{u}(t))\rangle_{(n)} \tag{7.6.7}$$

as the average orientational correlation function for molecules that have executed n free rotation steps (punctuated by n-1 collisions) in time t. Accordingly

$$C_l(1, t) \equiv \left[\exp\frac{-t}{\tau_c}\right] C_l^{(0)}(t) = \exp\frac{-t}{\tau_c}\left[\int d\omega_0 p(\omega_0)\, P_l\,(\cos \omega_0 t)\right] \tag{7.6.8}$$

where $C_l^{(0)}(t)$ is the free-rotor correlation function which can be evaluated either analytically or numerically and $\exp\frac{-t}{\tau_c}$ is the probability that a molecule will not suffer a collision in time t. τ_c is obviously the mean-free-time.

It is possible to evaluate $C_l(n, t)$ in terms of the free-rotor correlation function $C_l^{(0)}(t)$. First we note that

$$C_l(n, t) = \int_0^t \frac{dt'}{\tau_c} C_l(1, t - t')\, C_l(n - 1, t') \tag{7.6.9}$$

where $C_l(n - 1, t')$ is the correlation function for those molecules that have suffered $n - 2$ collisions in the time t', dt'/τ_c is the fraction of those molecules that undergo one collision between times t' and $t' + dt'$, and $C_l(1, t - t')$ describes the evolution of those molecules that do not collide in the remaining time $t - t'$. To find $C_l(n, t)$ the product $(dt'/\tau_c)C_l(1, t - t')C_l(n - 1, t')$ is integrated over t' from 0 to t in order to count all possible sequences of the $n - 1$ collisions that contribute to $C_l(n, t)$. The factorization of $C_l(n, t)$ into two parts is a consequence of assumption (3).

Equation (7.6.9) is a convolution integral. The Laplace transform of Eq. (7.6.9) is consequently

$$\tilde{C}_l(n, s) = \frac{1}{\tau_c} \tilde{C}_l(1, s)\, \tilde{C}_l(n - 1, s) \tag{7.6.10a}$$

or

$$\tilde{C}_l(n, s) = \frac{1}{\tau_c} \tilde{C}_l(n - 1, s)\, \tilde{C}_l^{(0)}(s + 1/\tau_c) \tag{7.6.10b}$$

where s is the Laplace variable and where we have substituted $\tilde{C}_l(1, s) \equiv \tilde{C}_l^{(0)}(s + 1/\tau_c)$ (which follows from the Laplace transform of Eq. (7.6.8)). Iteration of Eq. (7.6.10b) gives

$$\tilde{C}_l(n, s) = \left[\frac{1}{\tau_c}\tilde{C}_l^{(0)}\left(s + \frac{1}{\tau_c}\right)\right]^{n-1}\tilde{C}_l^{(0)}\left(s + \frac{1}{\tau_c}\right) \tag{7.6.11}$$

The Laplace transform of Eq. (7.6.6) can be obtained by summing $\tilde{C}_l(n, s)$ from $n = 1$ to $n = \infty$, so that

$$\tilde{C}_l(s) = \tilde{C}_l^{(0)}\left(s + \frac{1}{\tau_c}\right)\sum_{n=1}^{\infty}\left(\frac{1}{\tau_c}\tilde{C}_l^{(0)}\left(s + \frac{1}{\tau_c}\right)\right)^{n-1} \tag{7.6.12a}$$

or

$$\tilde{C}_l(s) = \frac{\tilde{C}_l^{(0)}\left(s + \frac{1}{\tau_c}\right)}{1 - \frac{1}{\tau_c}\tilde{C}_l^{(0)}\left(s + \frac{1}{\tau_c}\right)} \tag{7.6.12b}$$

where in the last step we have summed the geometric series.[18]

Equation (7.6.12b) is much more useful than it may at first sight appear. First we note that since

$$\tilde{C}_l(s) = \int_0^\infty dt \exp(-st)\, C_l(t) \tag{7.6.13}$$

the area under the orientational correlation functions $C_l(t)$; that is, its time integral, is

$$\tau_l \equiv \lim_{s\to 0}\tilde{C}_l(s) = \int_0^\infty dt\, C_l(t) \tag{7.6.14}$$

These are called the *orientational correlation times.* From Eq. (7.6.12b) we thus find

$$\tau_l = \frac{\tilde{C}_l^{(0)}\left(\frac{1}{\tau_c}\right)}{1 - \frac{1}{\tau_c}\tilde{C}_l^{(0)}\left(\frac{1}{\tau_c}\right)} \tag{7.6.15}$$

Thus if we know the free rotor functions $C_l^{(0)}(t)$ which is always possible we can evaluate through their Laplace transforms the orientational correlation times τ_l.

Since $C_l(t)$ is a real even function of the time (see Section 11.5) its spectral density $I_l(\omega)$ (see Eq. (6.2.6) and footnote 2 of chapter 6) can be expressed as

$$I_l(\omega) = \frac{1}{\pi}\, Re\, \tilde{C}_l\,(s = i\omega) \tag{7.6.16}$$

Substitution of Eq. (7.6.12b) then gives

$$I_l(\omega) = \frac{1}{\pi} \frac{F_l'(\omega)[1 - \frac{1}{\tau_c} F_l'(\omega)] - [F_l''(\omega)]^2}{\left[1 - \frac{1}{\tau_c} F_l'(\omega)\right]^2 + [F_l''(\omega)]^2} \tag{7.6.17}$$

where

$$F_l'(\omega) \equiv \int_0^\infty dt \exp \frac{-t}{\tau_c} C_l^{(0)}(t) \cos \omega t \tag{7.6.18a}$$

$$F_l''(\omega) \equiv \int_0^\infty dt \exp \frac{-t}{\tau_c} C_l^{(0)}(t) \sin \omega t \tag{7.6.18b}$$

are functions easily evaluated from the free-rotor functions. The time-correlation function $C_l(t)$ can be obtained from the spectral density function by an inverse Fourier transformation; that is,

$$C_l(t) \equiv 2 \int_0^\infty d\omega I_l(\omega) \cos \omega t \tag{7.6.19}$$

It should be noted that τ_l, $I_l(\omega)$, and $C_l(t)$ depend on a parameter τ_c and on the free-rotor function $C_l^{(0)}(t)$. This latter function decays on a time scale determined by the root-mean square angular speed which is proportional to $[k_BT/I]^{1/2}$. Thus we have two characteristic times in this model.

$$\tau_f \equiv \left[\frac{I}{k_BT}\right]^{1/2} = \text{(free-rotor time)}$$

$$\tau_c = \text{(collision time)}$$

It is useful to define a reduced time $\tau = t/\tau_f$. In terms of this reduced time it is easy to show that only one parameter enters the theory, and that is

$$\beta = \left(\frac{\tau_f}{\tau_c}\right) \tag{7.6.20}$$

When β is small the rotors execute many free rotation cycles between collisions ($\tau_c \gg \tau_f$) and the theory then gives for $C_l(t)$, the free-rotor function $C_l^{(0)}(t)$ for times shorter than τ_c. On the other hand when β is large ($\tau_c \ll \tau_f$), the rotors can execute only a small fraction of a full cycle before they are interrupted by a collision and should perform a kind of random walk in angle space. The theory then gives a "rotational diffusion limit" albeit not necessarily the Debye limit.

So far no independent method has been given for determining the parameter β as a function of density and temperature. Without this the theory is somewhat phenomenological. The procedure usually followed is the following.

a. The $C_l(t)$ obtained from experiment is used in Eq. (7.6.14) to find an experimental τ_l for a particular l.
b. Given the experimental τ_l Eq. (7.6.15) is solved for τ_c and β is evaluated.
c. The value of β so determined is used in the theory to evaluate $C_l(t)$ and $I_l(\omega)$ and these "theoretical" results are compared with experiment.

If $C_1(t)$ and $C_2(t)$ are determined for the same thermodynamic state of a system it is easy to check the internal consistency of the model. Steps a and b can be used to predict the two parameters β required to fit $C_1(t)$ and $C_2(t)$. If the model is internally consistent these two values should be identical. Although we do not present detailed tests of this model here it is important to note that when the model is tested against molecular dynamics studies of rough spheres and ellipsoids of revolution, different values of β are obtained (O'Dell and Berne, 1975). These authors also show that the best fit values of β bear no relation to the collision dimes, τ_c, in the rough sphere fluid. Jonas, (1974) has analyzed experimental data on CH_3I and CD_3I and come to the same conclusions.

It is interesting to note that the "Debye model" gives for τ_l

$$\tau_l = \frac{1}{l(l+1)\,\Theta} \tag{7.6.21}$$

so that $\tau_2/\tau_1 = \frac{1}{3}$, whereas the extended diffusion model gives Eq. (7.6.15). Thus we do not expect the two models to always give the same results. In fact it is found from the molecular dynamic studies that τ_2/τ_1 deviates significantly from $\frac{1}{3}$ over the whole gas-liquid range of the rough sphere system so that the Debye model cannot apply to this model system.

Comparisons of the extended diffusion model with experimental spectra on small molecules have been performed by Steele, who concludes that the model, although not very successful, is still superior to any other model studied (van Koynenberg and Steele, 1974; Jones, 1974).

Chandler (1974) has recently provided an interpretation of τ_c. Based on the rough sphere model, Chandler shows that $\tau_c = \tau_\omega$, where τ_ω is the angular velocity correlation time. This interpretation has recently been subjected to test by comparison with rough-sphere molecular dynamics where it deviates considerably from the data (see O'Dell and Berne, 1974).

Fixman and Rider (1969) have generalized the original extended diffusion models to cover the situations intermediate between the *M*-diffusion and the *J*-diffusion models. They conclude that none of the models reduce to ordinary diffusional relaxation for long times. They make the interesting point that "the fact that angular space is finite has the qualitative consequence that diffusional relaxation is recovered only in the simultaneous limit of long times and large collision frequencies." It is interesting to note that this generalized extended diffusion model is not consistent with the rough sphere molecular dynamics of O'Dell and Berne, (1975) The only theory that appears to be consistent with these experiments is a Fokker-Planck theory,

Gordon (1966) has used these assumptions to treat $\langle P_2(\mathbf{u}(0) \cdot \mathbf{u}(t))\rangle$ as well as $\langle P_1(\mathbf{u}(0) \cdot \mathbf{u}(t))\rangle$ for linear molecules. Some $\langle P_2(\mathbf{u}(0) \cdot \mathbf{u}(t))\rangle$ functions for linear molecules as calculated by Gordon are shown in Figs. 7.6.2 and 7.6.3. Both figures show a plot of a normalized $\langle P_2(\mathbf{u}(0) \cdot \mathbf{u}(t))\rangle$ versus reduced time, $\tau \equiv t(k_BT/I)^{1/2}$. The parameter τ^{-1} is the rate at which collisions (as defined above) terminate the "free rotation." Note that as τ becomes very large the correlation functions in both cases approach those of the freely rotating molecule.

The extended diffusion model has been solved for linear molecules (Gordon, 1966), spherical top molecules (McClung, 1969, Fixman and Rider, 1969), and symmetric top molecules (Fixman and Rider, 1969).

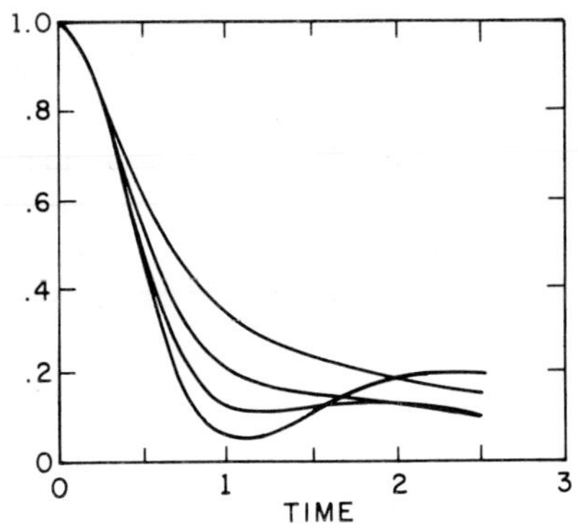

FIG. 7.6.2. Raman correlation functions for the *m*-diffusion model, with $\tau = 30, 1.75, 0.8$, and 0.4. (From Gordon, 1966, Fig. 4.)

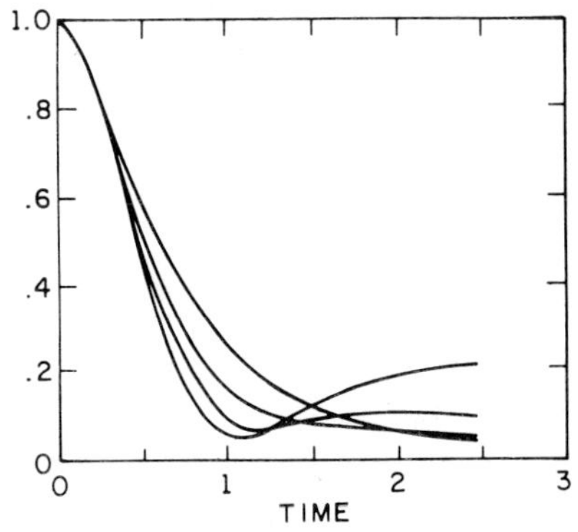

FIG. 7.6.3. Raman correlation functions for the *J*-diffusion model, with $\tau = 30, 1.75, 0.8$, and 0.4. (From Gordon, 1966, Fig. 6.)

Another model of rotational reorientation is the jump-diffusion model first described by Ivanov (1964). In this model the molecule reorients by a series of discontinuous jumps (with an arbitrary distribution of jump angles). This should be contrasted with the Debye model, which involves infinitesimal jumps, and the Gordon model, which involves continuous free rotations between collisions. This model is probably applicable to the situation where the molecular orientation is "frozen" until a volume fluctuation occurs, at which time the molecular orientation jumps to a new "frozen value." We present our own version of the jump model here. It is assumed that: (a) the jump takes place instantaneously, (b) successive jumps are uncorrelated in time with an average time τ_v between jumps, and (c) the dihedral angle between the two planes defined by the orientation vector **u** in two successive jumps is randomized.

Now let $C_l(n, t)$ denote the correlation function for that assembly of molecules that have suffered n jumps in the time t. Then

$$C\,(n,\,t) = \int_0^t \frac{dt'}{\tau_v} \exp -(t - t')\,\tau_v \langle P_l(\cos\chi)\rangle\, C_l(n - 1,\,t') \tag{7.6.22}$$

where dt'/τ_v is the probability that a jump occurs between t' and $t' + dt'$; $\exp -(t-t')/\tau_v$

is the probability that no jump occurs in the remaining time $t - t'$ and $\langle P_l(\cos\chi)\rangle$ is the average of $P_l(\cos\chi)$ over the distribution $p(\chi)$ of jump angles χ; that is,

$$\langle P_l(\cos\chi)\rangle = \frac{1}{2}\int_0^\pi d\chi \sin\chi\, p(\chi)\, P_l(\cos\chi) \tag{7.6.23a}$$

and

$$\frac{1}{2}\int_0^\pi d\chi \sin\chi\, p(\chi) = 1 \tag{7.6.23b}$$

The Laplace transform of Eq. (7.6.22) is

$$\tilde{C}_l(n, s) = \left(\frac{1}{\tau_v}\frac{\langle P_l(\cos\chi)\rangle}{\left(s + \dfrac{1}{\tau_v}\right)}\right)\tilde{C}_l(n-1, s) \tag{7.6.24}$$

Now note that $C(0, t)$ is the correlation function for those particles which have executed no jump in time t. Clearly this is

$$C_l(0, t) = \exp\frac{-t}{\tau_v}\, C_l^{(0)}(t) = \exp\frac{-t}{\tau_v} \tag{7.6.25a}$$

with Laplace transform

$$\tilde{C}_l(0, s) = \left(s + \frac{1}{\tau_v}\right)^{-1} \tag{7.6.25b}$$

The factor $\exp -t/\tau_v$ represents the probability of no jump in time t and $C_l^{(0)}(t)$ is the orientational correlation function if no jump occurs; that is, $C_l^{(0)}(t) = 1$. Iteration of Eq. (7.6.24) gives

$$\tilde{C}_l(n, s) = \left(\frac{1}{\tau_v}\left(\frac{\langle P_l(\cos\chi)\rangle}{\left(s + \dfrac{1}{\tau_v}\right)}\right)\right)^n \frac{1}{\left(s + \dfrac{1}{\tau_v}\right)} \tag{7.6.26}$$

The correlation function $\tilde{C}_l(s)$ is found by summing Eq. (7.6.26) from $n = 0$ to ∞, that is, over all numbers of jumps and taking note that this is a geometric series. This gives

$$\tilde{C}_l(s) = \frac{1}{s + \dfrac{1}{\tau_v}(1 - \langle P_l(\cos\chi)\rangle)} \tag{7.6.27}$$

The inverse Laplace transform of $\tilde{C}_l(s)$ is

$$C_l(t) = \exp - \left[\frac{|t|}{\tau_l(v)}\right] \tag{7.6.28a}$$

where

$$\tau_l(v) = \frac{\tau_v}{1 - \langle P_l(\cos\chi)\rangle} \tag{7.6.28b}$$

Now if the distribution function of jump angles is very peaked for small angles, then we can expand $P_l(\cos\chi)$ around $\chi = 0$. This gives

$$\frac{\tau_2(\nu)}{\tau_1(\nu)} \simeq \frac{\left(\frac{1}{2}\right)\langle\chi^2\rangle}{\left(\frac{3}{2}\right)\langle\chi^2\rangle} = \frac{1}{3} \tag{7.6.29a}$$

and we recover the rotational diffusion result. This is as it should be. The Debye model is after all a small-step diffusion model. If, on the other hand, this distribution is broad, say, it is uniform—all angles appearing with equal probability—then $\langle P_l(\cos\chi)\rangle = \delta_{l,0}$ and

$$\frac{\tau_2(\nu)}{\tau_1(\nu)} = 1 \tag{7.6.29b}$$

which is quite different from the rotational diffusion result. Thus one can, in principle, get large deviations from the Debye result. Other angle distributions could obviously be used.

This model is similar to a model proposed by Ivanov (1964). The main weakness of this model is that it ignores inertial effects; that is, when the constraint is released the molecule should rotate freely and not jump discontinuously to a new orientation.

7 · 7 MACROMOLECULES IN SOLUTION

The depolarized light-scattering methods discussed in this chapter have been applied to the study of the rotational motion of macromolecules in solution. Thus the light-scattering technique may take its place alongside fluorescence depolarization, dielectric relaxations, spin label, and birefringence decay techniques in the study of macromolecular rotation. Such studies are usually of interest because of the relations of tumbling times to macromolecular dimensions. Perrin (1934, 1936) has put forth a hydrodynamical model of molecular rotational motion. This theory assumes that the macromolecule may be treated as a macroscopic particle immersed in a continuum fluid. It furthermore assumes "stick boundary conditions," which stipulate that at the surface of the macromolecule (particle) the fluid (solvent) velocity is zero relative to the particle velocity. For an ellipsoid of revolution of major semiaxis a and minor semiaxis b Perrin has shown that the rotational diffusion coefficient characterizing the rotation of the symmetry axis is

$$\Theta = \frac{3k_BT}{16\pi\eta a^3}\left\{\frac{(2-\rho)\,G(\rho)-1}{(1-\rho^2)}\right\} \tag{7.7.1}$$

where ρ is the axial ratio, $\rho \equiv b/a$, η is the shear viscosity of the solution and $G(\rho)$ is a function of the axial ratio which for prolate ellipsoids, $\rho < 1$, has the form

$$G(\rho) = (1-\rho^2)^{-1/2}\ln\left\{\frac{1+(1-\rho^2)^{1/2}}{\rho}\right\};\ \rho < 1 \tag{7.7.2a}$$

and for oblate ellipsoids (plates), $\rho > 1$, has the form

$$G(\rho) = (\rho^2 - 1)^{1/2}\, \rho \ \text{arc tan}\, [(\rho^2 - 1)^{1/2}];\ \rho > 1 \qquad (7.7.2b)$$

This should be compared to the result $\Theta = k_B T/8\pi\eta a^3$ for the sticky sphere of radius a. Perrin also determined the translational diffusion coefficients for ellipsoidal molecules.[19] His result is

$$D = \frac{k_B T}{6\pi\eta a}\, G(\rho) \qquad (7.7.3)$$

where $G(\rho)$ is specified in Eqs. (7.7.2). Measurement of D and Θ by light scattering can be used in conjunction with Eq. (7.7.2) to detemine the dimensions a and b.

An interesting application of these ideas has been made by Dubin et al. (1971) to the study of the dimensions of the enzyme lysozyme. Lysozyme is an enzyme of approximate molecular weight 14,000. Macromolecules of this molecular weight should have depolarized spectra of half-width $\sim$10 MHz in aqueous solution.[20] Spectra of this half-width are most easily measured using high resolution Fabry–Perot interferometers. In fact, Dubin et al. (1971) used a spherical Fabry–Perot interferometer. They determine Θ and D and using Eqs. (7.7.1) and (7.7.3) conclude that lysozyme is a prolate ellipsoid with $2a = 55 \pm 1$Å and $2b = 33 \pm 1$ Å. On the basis of the molecular weight of lysozyme and its partial specfic volume in solution, the authors conclude that lysozyme in solution is a prolate ellipsoid of dimensions 48 ± 1Å by 26 ± 0.8 Å surrounded by a shell of water 3.5Å thick. This compares well with the x-ray data of unsolvated crystalline lysozyme, which is also a prolate ellipsoid of approximate dimensions 48 Å by 30 Å. Bauer et al. (1975) have repeated this experiment using ultra-pure lysozyme and do not find the intense spike at zero frequency change observed by Dubin et al. They, nevertheless, obtain the same Θ as Dubin et al. Bauer et al. have also measured the rotational diffusion coefficient of muscle calcium binding protein (molecular weight approximately symbol 12,000 Ds).

A further application of the depolarized scattering technique has been to the measurement of the rotational diffusion coefficient of a very large ($\sim$ 3000 Å in length) rodlike tobacco mosaic virus molecule. Because of the large size of the molecule (see Chapter 8), Eq. (7.3.20) applies only at very small angles. Wada et al. (1969) and more recently King et al. (1973) and Schurr and Schmitz (1973) have performed experiments of this type using optical mixing techniques.

This technique may be further applied to the study of intramolecular motions of large molecules. If there is a change in optical anisotropy (in a laboratory-fixed system) associated with an intramolecular motion (see Pecora, 1968), the relaxation rate of the motion should affect the depolarized spectrum. Schmitz and Schurr (1973) have detected time constants for motions of this type in the depolarized spectra of DNAs in solution.

7 · 8 APPLICATION TO SMALL MOLECULES IN LIQUIDS

Many techniques (e.g., see Gordon (1968)) are now used for the study of reorientational motion of small molecules in liquids. These methods include dielectric dispersion and relaxation, nuclear magnetic and nuclear quadrupole relaxation, ESR line shapes, picosecond pulse techniques, and neutron, Raman, and, as discussed above, depolarized Rayleigh scattering. Each of these technques has its own difficulties and strengths, and progress in the study of molecular reorientation in liquids will certainly require a systematic application of several of them. In the past few years much progress has been made in this field. Raman and infrared band-shape studies have, for instance, been very useful in elucidating deviations from the rotational diffusion model in liquids whose intermolecular forces are not very spherical. Electron-spin resonance studies have probed the relation between the relaxation times for the reorientational motion and for angular momentum relaxation.

Carbon-13 NMR has been used to study anisotropic rotational motion in liquids, as have combinations of techniques. Gillen and Griffiths (1972) have obtained the two reorientational relaxation times for benzene (a symmetric top) by combining reorientation relaxation times obtained from Raman band shapes and deuterium spin-lattice NMR relaxation times. The most extensive series of measurements probing anisotropic molecular reorientations have been made by Pecora and co-workers (Alms et al., (1973a, b)), who combined Carbon-13 spin lattice relaxation times with those obtained from depolarized Rayleigh spectra.

The simplest molecules studied by these latter authors were benzene and mesitylene. Both these molecules are symmetric tops and hence have only two relaxation times for reorientation. One of these reorientations is about the symmetry axis perpendicular to the plane of the ring (called $\tau_{\parallel}$). There is no change in optical anisotropy associated with this reorientation, hence it does not affect the light-scattering spectrum. Reorientation about an axis in the plane of the ring (perpendicular to the symmetry axis) does, however, affect the light-scattering spectrum. Thus the single molecule (or self) reorientational correlation function that contributes to the depolarized light-scattering spectrum should, in the rotational diffusion approximation, consist of a single Lorentzian [see Eq. (7.3.20)] with relaxation time

$$\frac{1}{\tau_{\perp}} = 6\Theta_{\perp} \tag{7.8.1}$$

The Carbon-13 spin-lattice relaxation-time measurements are related to the reorientation time of a C—H bond axis. For mesitylene the C—H bond axis observed was that of the ring carbons (not the methyl carbon). For symmetric top molecules such as benzene and mesitylene the τ_{NMR} is related to *both* $\tau_{\parallel}$ and $\tau_{\perp}$ by (Huntress, 1968)

$$\tau_{NMR} = \frac{1}{4}\tau_{\perp} + \frac{9}{4}\frac{\tau_{\perp}\tau_{\parallel}}{[\tau_{\parallel} + 2\tau_{\perp}]} \tag{7.8.2}$$

Thus a measurement of $\tau_{\perp}(=\tau_{LS})$ from light scattering, and τ_{NMR} by NMR can be used to determine $\tau_{\parallel}$ through eq. (7.8.2); that is,

$$\tau_{\parallel} = \frac{4\,\tau_{NMR} - \tau_{LS}}{5 - 2\left(\dfrac{\tau_{NMR}}{\tau_{LS}}\right)} \tag{7.8.3}$$

Alms, et al. (1973, 1974) have performed depolarized Rayleigh scattering and Carbon-13 spin-lattice relaxation-time measurements on solutions of benzene and mesitylene as a function of solvent viscosity. The solvents used were isopentane, cyclooctane, cyclohexanol, carbon tetrachloride, and their mixtures. These solvents contribute negligibly to the depolarized spectra of the solutions in the frequency range of interest.

These authors found that the light-scattering reorientation times varied linearly with solvent viscosity. It was determined, moreover, that the light-scattering reorientation times at different solute concentrations happen to have the same viscosity-dependence. This was taken as evidence that for these liquids pair correlations do not affect the light-scattering reorientation times (see Section 12.3). Thus both the light-scattering and NMR experiments measure single particle reorientation times for these liquids. The resultant $\tau_{\parallel}$ and $\tau_{\perp}$ for benzene and mesitylene are shown, respectively, in Figs. 7.8.1 and 7.8.2.

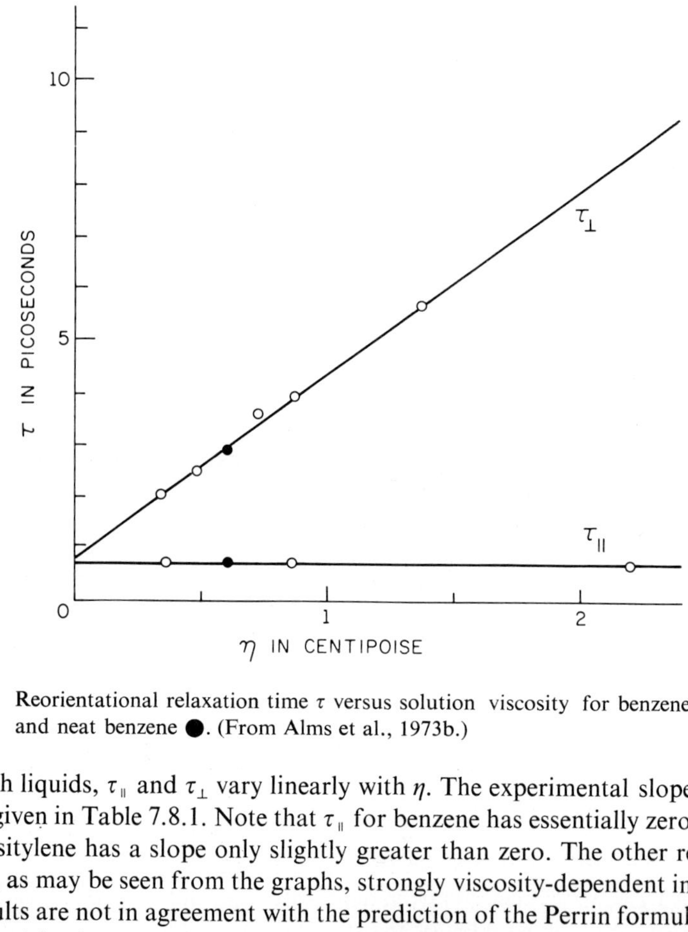

FIG. 7.8.1. Reorientational relaxation time τ versus solution viscosity for benzene solutions ○ and neat benzene ●. (From Alms et al., 1973b.)

For both liquids, $\tau_{\parallel}$ and $\tau_{\perp}$ vary linearly with η. The experimental slopes and intercepts are given in Table 7.8.1. Note that $\tau_{\parallel}$ for benzene has essentially zero slope while $\tau_{\parallel}$ for mesitylene has a slope only slightly greater than zero. The other reorientation time $\tau_{\perp}$ is, as may be seen from the graphs, strongly viscosity-dependent in both cases. These results are not in agreement with the prediction of the Perrin formulas [Eq. (7.7.1)] discussed in the preceding section. In all cases the reorientation times are much faster than predicted.

The Perrin formulas are derived on the basis of a hydrodynamic approximation. That is, the rotating system is treated as a particle reorienting in a "continuum" solvent. This is a good approximation for a macromolecule immersed in a solvent of small molecules, but one would not a priori expect it to apply to the case of molecules such as benzene and mesitylene immersed in solvents composed of molecules of approximately

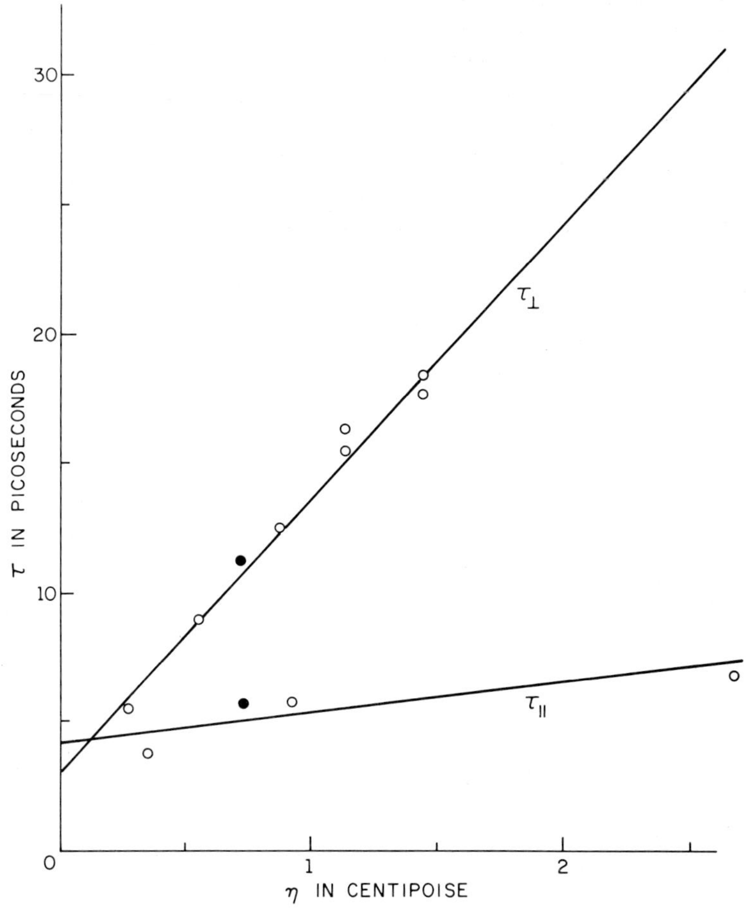

FIG. 7.8.2. Reorientational relaxation time τ versus solution viscosity for mesitylene solutions ○ and neat mesitylene ●. (From Alms et al., 1973b.)

TABLE 7.8.1

Viscosity dependence of $\tau_\perp$ and $\tau_\parallel$ for Benzene and Mesitylene

	$\tau_\perp$	$\tau_\parallel$
Benzene		
Experimental slope (psec/*cP*)	3.5 ± .1	0.0 ± .1
Experimental intercept (psec)	0.8 ± .5	0.7 ± .1
Mesitylene		
Experimental slope (psec/*cP*)	10.6 ± .8	1.0 ± .5
Experimental intercept (psec)	3.2 ± 1.0	4.4 ± 1.0

the same size. However, recent molecular dynamics calculations on perfect smooth elastic spheres (Alder, et al., 1970) give molecular dynamics results (computer simulations) that show that the hydrodynamic theory correctly predicts the *translation-*

al self-diffusion coefficient for hard-sphere fluids. However, in order to obtain agreement the "nonstick" or "slip" boundary conditions must be used in the hydrodynamic theory. The Perrin results for both translational and rotational diffusion utilize "stick" boundary conditions; that is, they assume that at the surface of the solute particle, the solvent rotates or translates with the particle. For a sphere rotating with slip boundary conditions, the particle does not have to displace or push any fluid out of its path in order to rotate. Thus one would not expect any viscosity-dependence of the reorientation time for a sphere rotating under these conditions. However, as a molecule becomes less spherical even with slip boundary conditions there should be a dependence of the orientational relaxation time on the solution viscosity. The *slopes* of plots such as those in Figs. 7.8.1 and 7.8.2 should thus be greatly dependent on molecular shape. For instance, the $\tau_{\parallel}$ reorientation of benzene looks almost "spherical" in the sense that the molecule looks pretty much the same in any orientation about this axis. From a microscopic kinetic point of view, it appears that the molecule would not have to push much solvent out of the way when it reorients about this axis. Thus, we might expect that this rotation will not be strongly viscosity-dependent. It is clear, however, from similar considerations that the reorientation described by $\tau_{\perp}$ should indeed be more strongly viscosity-dependent—as is the case.

Hu and Zwanzig (1974) have performed hydrodynamic calculations of the rotational friction coefficents of prolate and oblate ellipsoids as a function of the axial ratio using slip boundary conditions. The ratio of the friction calculated with slip to that calculated with stick boundary conditions is shown in Fig. 7.8.3 as a function of the axial ratio ρ.

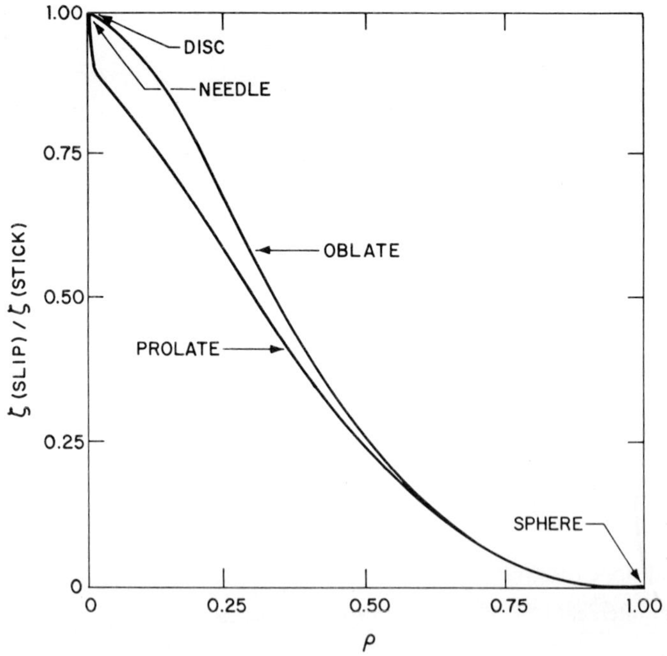

FIG. 7.8.3. The ratio of the friction coefficient ζ calculated using slip-boundary conditions to that calculated using stick-boundary conditions for prolate and oblate ellipsoids versus ρ, the ratio of the shorter to longer axis. (From Bauer, et al., 1974.)

Bauer et al. (1974) have studied reorientational relaxation times of a wide variety of molecules in organic solvents and find that the single particle rotational reorientation time about a given molecular axis is of the form

$$\tau = C\eta + \tau_0$$

where τ_0 is a constant whose value correlates well with the classical "free-rotor reorientation" time

$$\tau_{\text{FR}} = \frac{2\pi}{9}\left(\frac{I}{k_B T}\right)^{1/2}$$

where I is the moment of inertia about the axis under consideration. The values of C may be calculated by modeling molecules as ellipsoids and then applying the results of Hu and Zwanzig. These theoretical values are compared to those calculated using stick-boundary conditions and the experimental results obtained by Bauer et al. (see Table 7.8.2). Note that the "slip" results are in general agreement with the experimental results while the "stick" results are much too large.

APPENDIX 7.A THE COUPLING BETWEEN TRANSLATIONAL AND ROTATIONAL DIFFUSION IN DILUTE SOLUTION

The translational diffusion of a long rigid rod or an ellipsoid of revolution is described by two diffusion coefficients $D_{\parallel}$ and $D_{\perp}$ where $D_{\parallel}$ is the diffusion coefficient for motion parallel to the principal axis and $D_{\perp}$ is the diffusion coefficient perpendicular to this axis. These coefficients can be related to the size of the ellipsoid and to the viscosity of the solution by hydrodynamic arguments (Perrin, 1934; 1936). If **u** is a unit vector parallel to the symmetry axis, then by symmetry[21] the translational diffusion tensor is

$$\mathbf{D} = D_0\mathbf{I} + (D_{\parallel} - D_{\perp})\left[\mathbf{uu} - \frac{1}{3}\mathbf{I}\right] \tag{7.A.1}$$

where

$$D_0 = (D_{\parallel} + 2D_{\perp})/3 \tag{7.A.2}$$

is the isotropic translational diffusion coefficient.

Similar considerations apply to the rotational motion. For simplicity we ignore the rotation around the principal axis. For symmetric top molecules this is not a restriction. If the molecule suffers a displacement in the direction $\hat{\mathbf{R}}$, it will encounter more friction if it is rotating with its principal axis in a plane perpendicular to $\hat{\mathbf{R}}$ than in a plane parallel to $\hat{\mathbf{R}}$. This means that the rotational diffusion coefficient should differ depending on such factors as the orientation of $\hat{\mathbf{R}}$ relative to **u**. This can be expressed as

$$\boldsymbol{\Theta} = \Theta_0\mathbf{I} + (\Theta_{\parallel} - \Theta_{\perp})\left[\hat{\mathbf{R}}\hat{\mathbf{R}} - \frac{1}{3}\mathbf{I}\right] \tag{7.A.3}$$

where

$$\Theta_0 = \frac{1}{3}(\Theta_{\parallel} + 2\Theta_{\perp}) \tag{7.A.4}$$

TABLE 7.8.2

Viscosity-Dependences in Organic Solvents

Molecule	Axis[a]	Volume Å^3	ρ	C_{stick} psec/cP	C_{slip} psec/cP	C_{exp} psec/cP
Benzene		80	0.52 ± 0.01	22	5.0 ± 0.5	3.5 ± 0.1
	$\parallel$	80	~ 1	26	~ 0	0.0 ± 0.1
Hexafluorobenzene		108	0.44 ± 0.01	31	10.3 ± 0.5	9.5 ± 0.5
Mesitylene		133	0.46 ± 0.01	37	11.3 ± 0.5	10.6 ± 0.8
	$\parallel$	133		48		1.0 ± 0.3
CH_3I		55	0.62 ± 0.02	16	2.2 ± 0.3	1.3 ± 0.3
Toluene	x	80[b]	0.54 ± 0.01	21.5	4.6 ± 0.4	3.2 ± 0.4
	y	97	0.43 ± 0.01	30	10.5 ± 1	12.5 ± 1.5
	z	97	0.87 ± 0.04	33	0.5 ± 0.3	0.0 ± 0.3
Nitrobenzene	x	103	0.54 ± 0.01	27	5.5 ± 1	2.5 ± 0.5
	y	103	0.38 ± 0.02	33	14.5 ± 2	$26. \pm 5$
	z	103	0.8 ± 0.04	35	$1. \pm 0.5$	0.6 ± 0.4
p-Xylene		115	0.47 ± 0.01	32	$9.7 \pm 1.$	$10.5 \pm 1.$
Methylacetate		71	0.58 ± 0.03	22	3.9 ± 0.6	3.5 ± 0.8
Biphenyl		150	0.41 ± 0.01	64	$23. \pm 2$	$24. \pm 3$
Fluorene		157	0.42 ± 0.01	68	$23. \pm 2$	$27. \pm 4$
1, 4-Diphenyl 1, 3-Butadiene		204	0.26 ± 0.01	162 ± 10	$91. \pm 10$	$98. \pm 6$
Benzoic acid[c]		212	0.29 ± 0.01	143 ± 5	$74. \pm 5$	$76. \pm 5$
Acetic acid[c]		107	0.43 ± 0.01	43	$14. \pm 2$	15.7 ± 1.5
Propanoic acid[c]		141	0.40 ± 0.02	62	$22. \pm 4$	$20. \pm 2$
Valeric acid[c]		209	$0.36 - 0.45$	80–105	23–43	$29. \pm 5$

a. Unless otherwise stated, "reorientation" is reorientation about an axis normal to the symmetry axis:
- $\parallel$ — reorientation about symmetry axis
- x — reorientation about axis in ring passing through substituent
- y — reorientation about axis in ring — to x
- z — reorientation about axis $\perp$ to ring.

b. Assumes methyl group rotates independently.

c. Dimer in CCl_4.

and $\Theta_{\parallel}$ and $\Theta_{\perp}$ are the components of $\boldsymbol{\Theta}$ parallel and perpendicular to the displacement.

Now $c(\mathbf{u},\mathbf{R},t)d^3R,d^2\mathbf{u}$, the number of molecules found at $\mathbf{R}$ in d^3R with orientation $\mathbf{u}$ in d^2u at time t, is described by the diffusion equation

$$\frac{\partial c}{\partial t} = \boldsymbol{\nabla} \cdot \mathbf{D} \cdot \boldsymbol{\nabla} c - \hat{I} \cdot \Theta \cdot \hat{I} c \tag{7.A.5}$$

where $\hat{I}$ is the dimensionless orbital angular momentum vector (see Section 7.4). Substitution of Eqs. (7.A.1) and (7.A.3) gives the combined rotational and translational diffusion equation

$$\frac{\partial c}{\partial t} = D_0\nabla^2 c - \Theta_0\hat{I}^2 c + (D_{\parallel} - D_{\perp})\left[(\mathbf{u}\cdot\nabla)^2 - \frac{1}{3}\nabla^2\right]c$$

$$- (\Theta_{\parallel} - \Theta_{\perp})\left[(\hat{R}\cdot\hat{I})^2 - \frac{1}{3}\hat{I}^2\right]c \tag{7.A.6}$$

Were the anisotropies $(D_{\parallel} - D_{\perp})$ and $(\Theta_{\parallel} - \Theta_{\perp})$ small, as would be the case for short rods, the last two terms could be ignored and the equation would be separable in the rotations and translations. This is the approximation used throughout Chapter 7. On the other hand, if $(D_{\parallel} - D_{\perp})$ and/or $(\Theta_{\parallel} - \Theta_{\perp})$ are sufficiently large the full equation would apply. Then there is strong coupling between translational and rotational diffusion.

APPENDIX 7.B AN ALTERNATIVE TREATMENT OF SYMMETRIC TOP MOLECULES

There is an alternative and more general method for handling the material in Section 7.2. This method exploits the symmetry properties of the flüid more fully. Because this method requires somewhat more mathematical sophistication we treat it in this appendix.

Let **u** be a unit vector which lies along the symmetry axes of a symmetric top molecule. An applied electric field parallel to this axis will induce a dipole moment $\boldsymbol{\mu}$

$$\boldsymbol{\mu} = \alpha_{\parallel}\mathbf{E}$$

where $\alpha_{\parallel}$ designates the polarizability of the molecule along the symmetry axis. Any direction perpendicular to **u** will be such that an applied electric field along this direction will induce a dipole moment

$$\boldsymbol{\mu} = \alpha_{\perp}\mathbf{E}$$

where $\alpha_{\perp}$ designates the polarizability along any axis perpendicular to **u**.

The polarizability tensor for a symmetric top molecule can be expressed most generally as

$$\alpha_{\alpha\beta} = \alpha_{\parallel}u_\alpha u_\beta + \alpha_{\perp}(\delta_{\alpha\beta} - u_\alpha u_\beta)$$

where u_α is the αth component of the unit vector **u** which points along the symmetry axis of the top.[22]

This expression can be rearranged into the form

$$\alpha_{\alpha\beta} = \alpha\delta_{\alpha\beta} + \beta[u_\alpha u_\beta - \frac{1}{3}\delta_{\alpha\beta}] = \alpha\delta_{\alpha\beta} + \beta_{\alpha\beta} \tag{7.B.1}$$

where $\alpha \equiv \frac{1}{3}(\alpha_{\parallel} + 2\alpha_{\perp})$, $\beta \equiv (\alpha_{\parallel} - \alpha_{\perp})$, as before, and where

$$\beta_{\alpha\beta} = \beta(u_\alpha u_\beta - \frac{1}{3}\delta_{\alpha\beta}) \tag{7.B.2}$$

The trace of the matrix α is simply the sum of its diagonal elements $(\alpha_{11} + \alpha_{22} +$

α_{33}). This can be denoted as $\alpha_{\alpha\alpha}$ because we are using the Einstein convention stipulating that repeated indices be summed. Thus[23] $Tr\,\boldsymbol{\alpha} = \alpha\delta_{\alpha\alpha} + \beta[u_\alpha u_\alpha - 1/3\delta_{\alpha\alpha}] = 3\alpha$. The isotropic polarizability α is simply $\frac{1}{3}$ the trace of the polarizability tensor, that is $\alpha = \frac{1}{3}Tr\boldsymbol{\alpha}$. It should be noted that the tensor $\boldsymbol{\alpha}$ is symmetric ($\alpha_{\alpha\beta} = \alpha_{\beta\alpha}$) and that the anisotropic part ($\beta_{\alpha\beta}$) is traceless (zero trace) because $u_\alpha u_\alpha - \frac{1}{3}\delta_{\alpha\alpha} = 0$.

As the molecule rotates, the vector **u** reorients and the tensor $\boldsymbol{\alpha}$ changes in time. We note that because $\alpha\delta_{\alpha\beta}$ is independent of **u**, this part of the polarizability tensor does not change as the molecule rotates. $\alpha\delta_{\alpha\beta}$ is said to be rotationally invariant. $\beta_{\alpha\beta}$ on the other hand depends on **u**, so that this part of the polarizability changes as the molecule rotates.

The quantity α_{if} that appears in $I_{if}^{\alpha}(\mathbf{q}, t)$ is

$$\alpha_{if} = (\mathbf{n}_i)_\alpha (\mathbf{n}_f)_\beta\, \alpha_{\alpha\beta} = w_{\alpha\beta}{}^{if}\alpha_{\alpha\beta} \tag{7.B.3}$$

where

$$w_{\alpha\beta}{}^{if} = (\mathbf{n}_i)_\alpha (\mathbf{n}_f)_\beta \tag{7.B.4}$$

The time-correlation function of α_{if} can then be expressed as

$$\langle\alpha_{if}(0)\, \alpha_{if}(t)\rangle = w_{\alpha\beta}^{if}\langle\alpha_{\alpha\beta}(0)\, \alpha_{\gamma\delta}(t)\rangle\, w_{\gamma\delta}^{if} \tag{7.B.5}$$

In this appendix we compute this correlation function.

The molecules in a liquid are randomly oriented so that the probability of finding **u** in the solid angle $d\Omega = \sin\theta\, d\theta\, d\phi$ is $p(\Omega)\, d\Omega = 1/4\pi\, \sin\theta d\theta\, d\phi$. The average value of $u_\alpha u_\beta - \frac{1}{3}\delta_{\alpha\beta}$ is

$$\langle u_\alpha u_\beta - \frac{1}{3}\,\delta_{\alpha\beta}\rangle = \frac{1}{4\pi}\int_0^{2\pi} d\phi \int_0^{\pi} d\,\theta \sin\theta \left[u_\alpha u_\beta - \frac{1}{3}\,\delta_{\alpha\beta}\right]$$

To perform this average we express u_α, u_β in spherical polar coordinates, ($u_1 = \sin\theta$ $\cos\phi$; $u_2 = \sin\theta\, \sin\phi$; $u_3 = \cos\theta$) and then perform the above integration. This gives $\langle u_\alpha u_\beta - \frac{1}{3}\delta_{\alpha\beta}\rangle = 0$. The average value of the polarizability tensor is consequently $\langle\alpha_{\alpha\beta}\rangle = \alpha\delta_{\alpha\beta}$ because the anisotropic polarizability $\beta_{\alpha\beta}$ averages to zero. Because $\beta_{\alpha\beta}$ averages to zero it follows that the terms involving products of α with $\beta_{\alpha\beta}$ in Eq. 7.B.5 average to zero. Thus

$$\langle\alpha_{\alpha\beta}(0)\alpha_{\gamma\delta}(t)\rangle = \alpha^2\delta_{\alpha\beta}\delta_{\gamma\delta} + \langle\beta_{\alpha\beta}(0)\, \beta_{\gamma\delta}(t)\rangle \tag{7.B.6}$$

The quantitites $\langle\alpha_{\alpha\beta}(0)\alpha_{\gamma\delta}(t)\rangle$ and $\langle\beta_{\alpha\beta}(0)\beta_{\gamma\delta}(t)\rangle$ which involve four indices (α, β, γ, δ) are fourth-rank tensors. The angular brackets indicate an ensemble average where the ensemble represents a uniform, isotropic (rotationally invariant) liquid or gas. The second-rank tensors $\alpha_{\alpha\beta}$ and $\beta_{\alpha\beta}$ are symmetric in the indices α, β. The fourth-rank tensors in Eq. (7.B.6) are consequently symmetric in the indices (α, β) and (γ, δ) separately. It follows from very general considerations (Jeffreys, 1961) that the most general isotropic (rotationally invariant) fourth-rank tensor possessing the required symmetry[24] in (α, β) and (γ, δ) is

$$\langle\beta_{\alpha\beta}(0)\, \beta_{\gamma\delta}(t)\rangle = A(t)\, \delta_{\alpha\beta}\delta_{\gamma\delta} + B(t)\, [\delta_{\alpha\gamma}\delta_{\beta\delta} + \delta_{\alpha\delta}\delta_{\beta\gamma}] \tag{7.B.7}$$

where $A(t)$ and $B(t)$ are two independent coefficients. These coefficients can be evaluated from two independent equations. Note that

$$\langle\beta_{\alpha\alpha}(0)\beta_{\gamma\gamma}(t)\rangle = A(t)\,\delta_{\alpha\alpha}\delta_{\gamma\gamma} + B(t)[\delta_{\alpha\gamma}\delta_{\gamma\alpha} + \delta_{\alpha\gamma}\delta_{\alpha\gamma}] \tag{7.B.8}$$

$\beta_{\alpha\alpha}$ is the trace of $\beta_{\alpha\beta}$ which is zero, moreover $\delta_{\alpha\alpha} = \delta_{\gamma\gamma} = 3$ and $\delta_{\alpha\gamma}\delta_{\gamma\alpha} = \delta_{\alpha\alpha} = 3$, so that $9A(t) + 6B(t) = 0$ or

$$A(t) = -\frac{2}{3}\,\mathrm{B}(t) \tag{7.B.9}$$

Substitution of Eq. (7. B. 9) into Eq. (7.B.7) yields,

$$\langle\beta_{\alpha\beta}(0)\beta_{\gamma\delta}(t)\rangle = B(t)\left\{\delta_{\alpha\gamma}\delta_{\beta\delta} + \delta_{\alpha\delta}\delta_{\beta\gamma} - \frac{2}{3}\,\delta_{\alpha\beta}\delta_{\gamma\delta}\right\} \tag{7.B.10}$$

The coefficient $B(t)$ can be determined as follows. First we note by contraction that

$$\langle\beta_{\alpha\beta}(0)\,\beta_{\beta\alpha}(t)\rangle = B(t)\left\{\delta_{\alpha\beta}\delta_{\beta\alpha} + \delta_{\alpha\alpha}\delta_{\beta\beta} - \frac{2}{3}\,\delta_{\alpha\beta}\delta_{\beta\alpha}\right\}.$$

Since $\delta_{\alpha\beta}\delta_{\beta\alpha} = 3$; $\delta_{\alpha\alpha}\delta_{\beta\beta} = 9$, it follows that

$$\langle\beta_{\alpha\beta}(0)\,\beta_{\beta\alpha}(t)\rangle = 10B(t) \tag{7.B.11}$$

Substitution of the explicit form of $\beta_{\alpha\beta}$ from Eq. (7.B.2) into Eq. (7.B.11) yields[25]

$$\begin{aligned}\langle\beta_{\alpha\beta}(0)\,\beta_{\beta\alpha}(t)\rangle &= \beta^2\langle[u_\alpha(0)\,u_\beta(0) - \frac{1}{3}\,\delta_{\alpha\beta}\Big]\Big[u_\beta(t)\,u_\alpha(t) - \frac{1}{3}\,\delta_{\alpha\beta}\Big]\\ &= \beta^2\,\langle[\mathbf{u}(0)\cdot\mathbf{u}(t)]^2 - \frac{1}{3}\rangle\end{aligned} \tag{7.B.12}$$

This result can be expressed as[26]

$$\langle\beta_{\alpha\beta}(0)\,\beta_{\beta\alpha}(t)\rangle = \frac{2}{3}\,\beta^2\langle P_2(\mathbf{u}(0)\cdot(\mathbf{u}(t))\rangle \tag{7.B.13}$$

where $P_2(x)$ is the second-order Legendre polynomial. Combining Eqs. (7.B.13) and (7.B.11) gives the coefficient $B(t)$

$$B(t) = \frac{\beta^2}{15}\langle P_2(\mathbf{u}(0)\cdot\mathbf{u}(t))\rangle \tag{7.B.14}$$

The quantity $\mathbf{u}(0)\cdot\mathbf{u}(t)$ is the cosine of the angle $\theta(t)$ between the symmetry axis at time t and at time 0. $B(t)$ is consequently related to the angle $\theta(t)$ through which $\mathbf{u}$ turns in time t.

Substitution of Eqs. (7.B.10) and (7.B.14) into Eq. (7.B.6) results in

$$\langle\alpha_{\alpha\beta}(0)\,\alpha_{\gamma\delta}(t)\rangle = \alpha^2\delta_{\alpha\beta}\delta_{\gamma\delta} + \frac{\beta^2}{15}\langle P_2(\mathbf{u}(0)\cdot\mathbf{u}(t))\rangle\Big[\delta_{\alpha\gamma}\delta_{\beta\delta} + \delta_{\alpha\delta}\delta_{\beta\gamma} - \frac{2}{3}\,\delta_{\alpha\beta}\delta_{\gamma\delta}\Big] \tag{7.B.15}$$

Substitution of this into Eqs. (7.B.5) and (7.1.2) yields[27]

$$I_{if}^{\alpha}(\mathbf{q}, t) = \langle N \rangle w_{\alpha\beta}^{if} \langle \alpha_{\alpha\beta}(0)\, \alpha_{\gamma\delta}(t) \rangle\, w_{\gamma\delta}^{if}\, F_s(\mathbf{q}, t)$$

$$= \langle N \rangle\, F_s(\mathbf{q}, t) \left\{ \alpha^2 (\mathbf{n}_i \cdot \mathbf{n}_f)^2 + \frac{\beta^2}{15} \left[1 + \frac{1}{3} (\mathbf{n}_i \cdot \mathbf{n}_f)^2 \right] \langle P_2(\mathbf{u}(0) \cdot \mathbf{u}(t)) \rangle \right\} \quad (7.B.16)$$

For the particular scattering components described in Section 3.4 it follows from Eq. (7.B.16) that

$$I_{VV}^{\alpha}(\mathbf{q}, t) = \langle N \rangle \left[\alpha^2 + \frac{4}{45} \beta^2 \langle P_2(\mathbf{u}(0) \cdot \mathbf{u}(t)) \rangle \right] F_s(\mathbf{q}, t) \quad (7.B.17a)$$

$$I_{VH}^{\alpha}(\mathbf{q}, t) = \frac{1}{15} \langle N \rangle \beta^2 \langle P_2(\mathbf{u}(0) \cdot \mathbf{u}(t)) \rangle\, F_s(\mathbf{q}, t) \quad (7.B.17b)$$

$$I_{HH}^{\alpha}(\mathbf{q}, t) = \langle N \rangle\, \alpha^2 \cos^2\theta\, F_s(\mathbf{q}, t) + \langle N \rangle \frac{\beta^2}{15} \left[1 + \frac{1}{3} \cos^2\theta \right] \langle P_2(\mathbf{u}(0) \cdot \mathbf{u}(t)) \rangle F_s(\mathbf{q}, t) \quad (7.B.17c)$$

For particular scattering angles Eq. (7.B.17c) becomes

$$I_{HH}^{\alpha}(\mathbf{q}, t) = \begin{cases} I_{VH}^{\alpha}(\mathbf{q}, t) & \theta = \pi/2 \quad (7.B.18a) \\ I_{VV}^{\alpha}(\mathbf{q}, t) & \theta = \pi \quad (7.B.18b) \end{cases}$$

Let us define the isotropic scattering as [cf. Eq. (7.2.8)]

$$I_{\mathrm{ISO}}^{\alpha}(\mathbf{q}, t) = \langle N \rangle \alpha^2\, F_s(\mathbf{q}, t) \quad (7.B.19a)$$

Then

$$I_{\mathrm{ISO}}^{\alpha}(\mathbf{q}, t) = I_{VV}^{\alpha}(\mathbf{q}, t) - \frac{4}{3} I_{VH}^{\alpha}(\mathbf{q}, t) \quad (7.B.19b)$$

$$I_{VH}^{\alpha}(\mathbf{q}, t) = \frac{1}{15} \langle N \rangle\, \beta^2 \langle P_2(\mathbf{u}(0) \cdot \mathbf{u}(t)) \rangle\, F_s(\mathbf{q}, t) \quad (7.B.19c)$$

$$I_{HH}^{\alpha}(\mathbf{q}, t) = \cos^2\theta\, I_{\mathrm{ISO}}^{\alpha}(\mathbf{q}, t) + \left[1 + \frac{1}{3} \cos^2\theta \right] I_{VH}^{\alpha}(\mathbf{q}, t) \quad (7.B.19d)$$

The isotropic scattering involves only the translational motion, whereas $I_{VH}^{\alpha}(\mathbf{q}, t)$ depends on both the translational and rotational motion. The above equations consequently allow us to separate translational and rotational motions.

One important feature of these results is that if the molecular Hamiltonian has symmetric top symmetry, rotations of the molecule about the symmetry axis **u** do not contribute to the light-scattering spectrum. This follows from the fact that $\langle P_2(\mathbf{u}(0) \cdot \mathbf{u}(t)) \rangle$ is invariant to this rotation. Another important feature is that if the optical anisotropy β is zero, the rotations do not contribute at all to the scattering.

These formulas apply regardless of whether the rotations can be described by a rotational diffusion equation. In the event that the rotational diffusion equation applies, Eq. (7.B.16) reduces to the results found in Sections 7.3. and 7.5. If the molecules freely rotate, Eqs. (7.B.16) reduce to the results of Section 7.6.

This same type of analysis is useful in connection with other experimental methods for determining orientational correlation functions. As an example we consider *fluorescence depolarization experiments*. In a fluorescence experiment, the following steps are followed:

a. A plane-polarized pulse of light of polarization $\mathbf{n}_i$ and wavelength λ_a impinges on a molecule. (For convenience we assume the pulse is a delta function in time.)
b. The particular molecular transition which leads to the absorption of light of this wavelength is characterized by an absorption transition dipole moment $\boldsymbol{\mu}(a)$. This is a vector which may be regarded as rigidly fixed in the molecule whose magnitude determines the oscillator strength of the transition. The probability of the absorption is proportional to $|\hat{\boldsymbol{\mu}}(a) \cdot \mathbf{n}_i|^2$ where $\hat{\boldsymbol{\mu}}(a)$ is a unit vector along $\boldsymbol{\mu}(a)$.
c. The absorption process is followed by intramolecular dynamic processes that often result in fluorescence at wavelength λ_e from a different transition that is characterized by an emission transition dipole $\boldsymbol{\mu}(e)$. The probability of emission of light of wavelength λ_e and polarization $(\mathbf{n}_f)$ (measured using an analyzer) is proportional to $|\hat{\boldsymbol{\mu}}(e) \cdot \mathbf{n}_f|^2$.
d. The intensity of light observed at wavelength λ_e with polarization $\mathbf{n}_f$, given that the system is excited with light of wavelength λ_a of polarization $\mathbf{n}_i$, is proportional to

$$\hat{I}_{if}(t) = \langle |\hat{\boldsymbol{\mu}}(a,0) \cdot \mathbf{n}_i|^2 |\hat{\boldsymbol{\mu}}(e, t) \cdot \mathbf{n}_f|^2 \rangle$$

where $\hat{\boldsymbol{\mu}}(e, t)$ gives the orientation of $\boldsymbol{\mu}(e)$ at time t.

This can be expressed as

$$\hat{I}_{if}(t) = w^{ii}_{\alpha\beta} \langle \hat{\mu}_\alpha(a, 0)\, \hat{\mu}_\beta(a, 0)\, \mu_\gamma(e, t)\, \mu_\delta(e, t) \rangle\, w^{ff}_{\gamma\delta} \tag{7.B.20}$$

where

$$w^{ii}_{\alpha\beta} = (\mathbf{n}_i)_\alpha (\mathbf{n}_i)_\beta;\; w^{ff}_{\gamma\delta} = (\mathbf{n}_f)_\gamma (\mathbf{n}_f)_\delta$$

The same analysis as in the foregoing can be applied to the above correlation function. In fact, if we take[28] $\alpha = 1/3$, and $\beta = 1$ in Eqs. (7.B.1) and (7.B.15)

$$\langle \hat{\mu}_\alpha(a, 0)\, \hat{\mu}_\beta(a, 0)\, \hat{\mu}_\gamma(e, t)\, \hat{\mu}_\delta(e, t) \rangle = \frac{1}{9}\, \delta_{\alpha\beta}\delta_{\gamma\delta} + \frac{1}{15} \langle P_2(\hat{\boldsymbol{\mu}}(a, 0) \cdot \hat{\boldsymbol{\mu}}(e, t)) \rangle$$
$$\times \left[\delta_{\alpha\gamma}\, \delta_{\beta\delta} + \delta_{\alpha\delta}\delta_{\beta\gamma} - \frac{2}{3}\, \delta_{\alpha\beta}\delta_{\gamma\delta} \right] \tag{7.B.21}$$

Substitution of this into Eq. (7.B.20) then gives

$$\hat{I}_{if}(t) = \frac{1}{9} + \frac{4}{45} P_2(\mathbf{n}_i \cdot \mathbf{n}_f) \langle P_2(\hat{\boldsymbol{\mu}}(a, 0) \cdot \hat{\boldsymbol{\mu}}(e, t)) \rangle \tag{7.B.22}$$

Depolarization of fluorescence experiments can be performed with two independent configurations of polarizers and analyzers. For example, if $\mathbf{n}_i$ and $\mathbf{n}_f$ are parallel we find

$$\hat{I}_{\parallel}(t) = \frac{1}{9} + \frac{4}{45} \langle P_2(\hat{\boldsymbol{\mu}}(a, 0) \cdot \hat{\boldsymbol{\mu}}(e, t)) \rangle \tag{7.B.23}$$

If, on the other hand, $\mathbf{n}_i$ and $\mathbf{n}_f$ are perpendicular, we find

$$\hat{I}_{\perp}(t) = \frac{1}{9} - \frac{2}{45}\langle P_2(\hat{\boldsymbol{\mu}}(a, 0) \cdot \hat{\boldsymbol{\mu}}(e, t))\rangle \tag{7.B.24}$$

The depolarization ratio $R(t)$ is defined as

$$R(t) = \frac{I_{\parallel}(t) - I_{\perp}(t)}{I_{\parallel}(t) + 2I_{\perp}(t)} \tag{7.B.25}$$

From Eqs. (7.B.23) and (7.B.24) we find

$$R(t) = \frac{2}{15}\langle P_2(\hat{\boldsymbol{\mu}}(a, 0) \cdot \hat{\boldsymbol{\mu}}(e, t)\rangle \tag{7.B.26}$$

$R(t)$ is consequently determined by molecular reorientation. If $\boldsymbol{\mu}(a)$ and $\boldsymbol{\mu}(e)$ both lie along the symmetry axis $\mathbf{u}$ of a symmetric top diffusor

$$R(t) = \frac{2}{15}\langle P_2(\mathbf{u}(0) \cdot \mathbf{u}(t))\rangle = \frac{2}{15}\exp -6\Theta_{\perp}t \tag{7.B.27}$$

APPENDIX 7.C IRREDUCIBLE TENSORS IN LIGHT SCATTERING

By definition the components of the second-rank Cartesian tensor α_{xy} transform under rotation just like the product of coordinates xy (e.q., see Jeffreys, 1961) The motivation for what ensues springs from the observation that the spherical harmonics $Y_{lm}(\theta, \phi)$ (where θ, ϕ are the polar and azimuthal angles of the unit vector $(\mathbf{r}/|\mathbf{r}|)$) can be written in terms of the coordinates (x, y, z) of the vector $\mathbf{r}$, for example,

$$l = 0 \quad Y_{00}(\theta, \phi) = \frac{1}{\sqrt{4\pi}}$$

$$l = 1\begin{cases} Y_{1,0}(\theta, \phi) = \sqrt{\dfrac{3}{4\pi}}\,\dfrac{z}{r} \\[2ex] Y_{1,\pm 1}(\theta, \phi) = \pm\sqrt{\dfrac{3}{4\pi}}\,\dfrac{1}{\sqrt{2}}\left(\dfrac{x \pm iy}{r}\right) \end{cases} \tag{7.C.1}$$

$$l = 2\begin{cases} Y_{2,0}(\theta, \phi) = \left(\dfrac{5}{4\pi}\right)^{1/2}\dfrac{1}{2}\left[\dfrac{3z^2 - (x^2 + y^2 + z^2)}{r^2}\right] \\[2ex] Y_{2,\pm 1}(\theta, \phi) = \pm\left(\dfrac{5}{4\pi}\right)^{1/2}\dfrac{1}{2}\sqrt{\dfrac{3}{2}}\left[\dfrac{(zx + xz) \pm i(zy + yz)}{r^2}\right] \\[2ex] Y_{2,\pm 2}(\theta, \phi) = \left(\dfrac{5}{4\pi}\right)^{1/2}\sqrt{\dfrac{3}{8}}\left[\dfrac{(x^2 - y^2) \pm i(yx + xy)}{r^2}\right] \end{cases}$$

Consider now an arbitrary second-rank Cartesian tensor T_{ij}. Comparing the elements of this tensor with $Y_{2,m}(\theta, \phi)$ we see that, for example, the linear combination of components

$$3T_{zz} - (T_{xx} + T_{yy} + T_{zz})$$

transforms under rotation like $3z^2 - (x^2 + y^2 + z^2)$, that is, like the spherical harmonic $Y_{2,0}(\theta, \phi)$. The same type of method can be used for the components of a Cartesian tensor of arbitrary rank. Certain linear combinations of the components of the Cartesian tensor will transform like pure spherical harmonics. The utility of this observation lies in the fact that the transformation properties of the spherical harmonics under rotations are particularly simple. We first turn to a discussion of these transformation properties.

The set of all square integrable functions of the polar angles (θ, ϕ) forms a Hilbert space. This space is spanned by the spherical harmonics $Y_{lm}(\theta, \phi)$ and can be decomposed into subspaces such that the lth subspace is spanned by the $(2l + 1)$ spherical harmonics of index l.

$$Y_{l,-l}(\theta, \phi), \ldots, Y_{l,o}(\theta, \phi), \ldots, Y_{l,l}(\theta, \phi)$$

The angles (θ, ϕ) of course specify the orientation of a unit vector in a given set of Cartesian axes, say $Oxyz$. Suppose we express the orientation of this unit vector in a Cartesian frame $OXYZ$ which is obtained by rotating $Oxyz$ through the Euler angles $\Omega = (\alpha, \beta, \gamma)$. This rotation of the frame is given by the rotation matrix $R(\boldsymbol{\Omega})$ whose elements are direction cosines, and the orientation of the unit vector $(\mathbf{r}/r)$ is specified by the polar angles (θ', ϕ') in this new system. It is a standard result of the theory of angular momentum that the spherical harmonics $\{Y_{l,m}(\theta', \phi')\}$ are related to the set $\{Y_{lm}(\theta, \phi)\}$ by the relation

$$Y_{lm}(\theta', \phi') = \sum_{m'} Y_{lm}'(\theta, \phi)\, D^{(l)}_{m'm}(\boldsymbol{\Omega}) = R Y_{lm}(\theta, \phi) \tag{7.C.2}$$

where the functions $\{D^{(l)}_{m'm}(\boldsymbol{\Omega})\}$ are known as the Wigner rotation matrices

$$D^{(l)}_{m'm}(\boldsymbol{\Omega}) = \langle lm' | \hat{R}(\boldsymbol{\Omega}) | lm \rangle = \int_0^{2\pi} d\phi \int_0^{\pi} d\theta \sin\theta\, Y^*_{lm'}(\theta, \phi)\, \mathbf{R}(\Omega)\, Y_{l,m}(\theta, \phi) \tag{7.C.3}$$

and where the operator $R(\boldsymbol{\Omega})$ when applied to the function $Y_{l,m}(\theta, \phi)$ gives $Y_{lm}(\theta', \phi')$. An explicit form for R comes from the theory of angular momentum

$$R(\boldsymbol{\Omega}) = \exp -i\alpha \hat{J}_z \exp -i\beta \hat{J}_y \exp -i\gamma \hat{J}_z \tag{7.C.4}$$

with the Euler angles (α, β, γ) specifying the rotations of the coordinate frame from $Oxyz$ to $OXYZ$ and the angular momentum operators $(\hat{J}_x, \hat{J}_y, \hat{J}_z)$ about the axes (OX, OY, OZ). The operator does not change the index l.

Equation (7.C.2) is fundamental. It tells us that a rotation of a function in the lth subspace of Hilbert space produces another function which still lies entirely in the lth subspace. We say, therefore, that the lth subspace forms a $2l + 1$ dimensional invariant subspace corresponding to the group of proper rotations.

The set of all rotations, $R(\boldsymbol{\Omega})$ satisfy the postulates which define a group. For example, the product of two rotations R_1R_2 is also a rotation $R_3 = R_1R_2$. This is just the closure property. Thus we can think of any rotation as being decomposable into any number of smaller rotations. Repeated use of Eqs (7.C.2) and (7.C.3) then gives the result

$$D^{(l)}_{m''m'}(R_k) = \sum_{m'} D^{(l)}_{m''m'}(R_i)\, D^{(l)}_{m'm}(R_j) \tag{7.C.5}$$

In matrix notation this is

$$\mathbf{D}^{(l)}(R_k) = \mathbf{D}^{(l)}\,(R_i) \cdot \mathbf{D}^{(l)}(R_j) \tag{7.C.6}$$

The Wigner matrices multiply just like the rotations themselves. There is a one-to-one correspondence between the Wigner matrices of index l and the rotations R. These matrices form a representation of the rotation group. In fact, since the $2l + 1$ spherical harmonics of order l form an invariant subspace of Hilbert space with respect to all rotations, it follows that the matrices $D^{(l)}_{m'm}(R)$ form a $(2l + 1)$ dimensional irreducible representation of the rotation R. Explicit formulas for these matrices can be found in books on angular momentum (notably Edmunds, 1957).

The major result so far is that the spherical harmonics in one frame can be related to those in another frame through Eq. (7.C.2). For example,

$$Y_{lm}(\theta_L, \phi_L) = \sum_{m'} Y_{lm'}(\theta_B, \phi_B)\, D^{(l)}_{m'm}(\boldsymbol{\Omega}) \tag{7.C.7}$$

where (θ_L, ϕ_L) and (θ_B, ϕ_B) are the polar coordinates of a given vector in the laboratory- and body-fixed frames respectively, and $\boldsymbol{\Omega} = (\alpha, \beta, \gamma)$ are the Euler angles specifying the rotation R which transforms the body into the laboratory frame.[29]

As we have already seen, the components of any Cartesian tensor can be combined in such a way that the resulting sums transform like the spherical harmonics under rotations. This observation can be formalized as follows.

A standard irreducible tensorial set of rank l, $\mathbf{T}^{(l)}$ is defined as the set of $2l + l$ elements $\{\mathbf{T}^{(l)}_{\mu}\}$

$$T^{(l)}_{l}, \ldots, T^{(l)}_{-1}, T^{(l)}_{0}, T^{(l)}_{1}, \ldots, T^{(l)}_{l}$$

which transform under a rotation like the spherical harmonics $\{Y_{lm}\}$ of order l; that is, $\mathbf{T}^{(l)}$ is an irreducible tensorial set if and only if

$$T^{(l)}_{m}(L) = \sum_{m'} T^{(l)}_{m'}(B)\, D^{(l)}_{m'm}(\boldsymbol{\Omega}) \tag{7.C.8}$$

where $T^{(l)}_{m}(L)$ and $T^{(l)}_{m'}(B)$ are elements of the set in the laboratory-and body-fixed coordinate system, respectively. $\mathbf{T}^{(l)}$ is seen to belong to the $(2l + l)$ dimensional irreducible representation $\mathbf{D}^{(l)}(\boldsymbol{\Omega})$ of the rotation group.

The scalar product of two irreducible tensorial sets $\mathbf{a}^{(l)}$ and $\mathbf{b}^{(l)}$ of the same rank is defined as

$$\mathbf{a}^{(l)} \theta \mathbf{b}^{(l)} = \sum_{\mu=-l}^{+l} (-1)^{l-\mu}\, a_{-\mu}{}^{(l)}\, b^{(l)}_{\mu} \tag{7.C.9}$$

Ordinary Cartesian tensors can be reduced into irreducible tensorial sets. Take, for example, the second-rank Cartesian tensor $\mathbf{T}$

$$\mathbf{T} \equiv \begin{pmatrix} T_{xx} & T_{xy} & T_{xz} \\ T_{yx} & T_{yy} & T_{yz} \\ T_{zx} & T_{zy} & T_{zz} \end{pmatrix} \tag{7.C.10}$$

The element T_{ij} transforms under rotations just like the product of Cartesian components A_iB_j of the vectors **A** and **B**. We note that the second-rank Cartesian tensor can be decomposed into three parts

$$T_{ij} = \underbrace{\frac{1}{3}\,T\delta_{ij}}_{T_{ij}^{(0)}} + \underbrace{\frac{1}{2}\,(T_{ij} - T_{ji})}_{T_{ij}^{(1)}} + \underbrace{\left[\frac{1}{2}\,(T_{ij} + T_{ji}) - \frac{1}{3}\,T\delta_{ij}\right]}_{T_{ij}^{(2)}} \qquad (7.C.11)$$

where $T = Tr\mathbf{T} = (T_{xx} + T_{yy} + T_{zz})$. These parts are called $T_{ij}^{(0)}$, $T_{ij}^{(1)}$, and $T_{ij}^{(2)}$ respectively. We note that $T_{ij}^{(0)}$ is invariant under a rotation[30] (traces are invariant to any unitary transformation). Since the elements of T_{ij} transform like A_iB_j, it follows that $T_{ij}^{(1)} = \frac{1}{2}(T_{ij} - T_{ji})$ transforms like $(A_iB_j - A_jB_i)$. This latter quantity is just $(A \times B)_k$, the kth component of the vector cross product $(\mathbf{A} \times \mathbf{B})$. Thus $T_{ij}^{(1)}$ transforms like a vector cross product which itself is a pseudovector. We conclude that $T_{ij}^{(1)}$ transforms like a pseudovector.[31] Likewise $T_{ij}^{(2)}$ transforms like a second-rank tensor. Thus an ordinary Cartesian tensor can be decomposed into a scalar part, $T_{ij}^{(0)}$, which is invariant under rotation an antisymmetric part $T_{ij}^{(1)}$ whose components transform like the components of a pseudovector, and a traceless symmetric part $T_{ij}^{(2)}$ whose components transform like a pure second-rank tensor. $T_{ij}^{(0)}$, $T_{ij}^{(1)}$, and $T_{ij}^{(2)}$ respectively contain one, three, and five independent components. Thus we expect that irreducible tensorial sets of ranks 0, 1, and 2 can be formed from the components of $T_{ij}^{(0)}$, $T_{ij}^{(1)}$, and $T_{ij}^{(2)}$, respectively. These components, however, do not transform as pure spherical harmonics, but as linear combinations of them. We can, alternatively, find linear combinations of, say, the $T_{ij}^{(2)}$ which do transform as pure spherical harmonics. We describe this method in the following paragraph.

The relation between a given irreducible component $T_m^{(l)}$ and the Cartesian elements is identical to the relation between Y_{lm} and its Cartesian elements [Eq. (7.4.1)] except for normalization factors (like $(5/4\pi)^{1/2}$ in $Y_{2,\pm 2}$). Thus comparing Eq. (7.C.11) with Eq. (7.C.1) we obtain[32]

$$l = 0\left\{T_0^{(0)} = \frac{1}{3}\,T\right. \qquad (7.C.12a)$$

$$l = 1\begin{cases} T_0^{(1)} = T_z = (T_{xy} - T_{yz})/2 & (7.C.12b) \\ T_{\pm1}^{(1)} = \pm\dfrac{1}{\sqrt{2}}\,[T_x \pm iT_y] = \pm\dfrac{1}{\sqrt{2}}\left[\dfrac{1}{2}\,(T_{yz} - T_{zy}) \pm \dfrac{1}{2}\,(T_{zx} - T_{xz})\right] \end{cases}$$

$$l = 2\begin{cases} T_0^{(2)} = \dfrac{1}{2\sqrt{6}}\,[3T_{zz} - (T_{xx} + T_{yy} + T_{zz})] \\ T_{\pm1}^{(2)} = \pm\,\dfrac{1}{4}\,[(T_{zx} + T_{xz}) \pm i(T_{zy} + T_{yz})] \\ T_{\pm2}^{(2)} = \dfrac{1}{4}\,[(T_{xx} - T_{yy}) \pm i(T_{xy} + T_{yx})] \end{cases} \qquad (7.C.12c)$$

This construction is arbitrary up to an arbitrary multiplicative constant. The usual convention is to choose the constant such that

$$\mathbf{T}^{(l)}\theta\mathbf{T}^{(l)} = \sum_{ij\ldots} T_{ij}^{(l)}\ldots\, T_{ji}^{(l)}\ldots$$

where $T_{ij\ldots}^{(l)}$ $(l = 0, 1, 2)$ are defined in Eq. (7.C.11) and the scalar product of the irreducible tensorial set is defined in Eq. (7.C.9). The phase of this set is chosen to obey the same convention as $Y_{lm}(\theta, \phi)$ ($Y_{l,0}$, is always chosen real). With these conventions the sets turn out to be

$$l = 0 \left\{ T_0^{(0)} = \frac{1}{\sqrt{3}} T \right. \tag{7.C.13a}$$

$$l = 1 \left\{ \begin{aligned} T_0^{(1)} &= \frac{1}{2}(T_{xy} - T_{yx}) \\ T_{\pm 1}^{(1)} &= = \pm \frac{1}{\sqrt{2}} \left[\frac{1}{2}(T_{yz} - T_{zy}) \pm \frac{1}{2}(T_{zx} - T_{xz}) \right] \end{aligned} \right. \tag{7.C.13b}$$

$$l = 2 \left\{ \begin{aligned} T_0^{(2)} &= \frac{1}{\sqrt{6}} [2T_{zz} - T_{xx} - T_{yy})] \\ T_{\pm 1}^{(2)} &= -\frac{i}{2} [(T_{yz} + T_{zy}) \pm i(T_{zx} + T_{xz})] \\ T_{\pm 2}^{(2)} &= \frac{1}{2} [(T_{xx} - T_{yy}) \pm i(T_{xy} + T_{yx})] \end{aligned} \right. \tag{7.C.13c}$$

These equations can be solved for the Cartesian components in terms of the irreducible (spherical) components. For example,

$$T_{zz} = \frac{1}{\sqrt{3}} [T_0^{(0)} + \sqrt{2}\, T_0^{(2)}] \tag{7.C.14}$$

$$T_{zy} = \frac{1}{2} [\sqrt{2}\,(T_{-1}^{(1)} - T_{+1}^{(1)}) - i(T_{+1}^{(2)} + T_{-1}^{(2)})]$$

Now consider the case where T_{ij} is a tensor in the laboratory-fixed coordinate frame. Then the spherical components in Eq. (7.C.13) are also in the laboratory frame, and we denote this by writing these elements as $T_\mu^{(l)}(L)$. The elements $T_\mu^{(l)}(L)$ that appear in Eq. (7.C.13) can be related through Eq. (7.C.8) to the spherical components in the molecule or body-fixed coordinate system

$$T_m^{(l)}(L, t) = \sum_{m'} T_{m'}^{(l)}(B)\, D_{m'm}^{(l)}(\boldsymbol{\Omega}(t)) \tag{7.C.15}$$

where $\boldsymbol{\Omega}(t)$ specifies the orientation of the molecule at time t. Then for example,

$$T_{zz}(\boldsymbol{\Omega}(t)) = \frac{1}{\sqrt{3}} T_0^{(0)}(B) + \sqrt{2} \sum_{m'=2}^{2} T_{m'}^{(2)}(B)\, D_{m'0}^{(2)}(\boldsymbol{\Omega}(t)) \tag{7.C.16}$$

$$\begin{aligned} T_{zy}(\boldsymbol{\Omega}(t)) = \frac{1}{2} \Big\{ &\sqrt{2} \sum_{m'=-1}^{+1} T_{m'}^{(1)}(B)\, [D_{m',-1}^{(1)}(\boldsymbol{\Omega}(t)) - D_{m',1}^{(1)}(\boldsymbol{\Omega}(t))] \\ &- i \sum_{m'=-2}^{+2} T_{m'}^{(2)}(B)\, [D_{m',1}^{(2)}(\boldsymbol{\Omega}(t)) + D_{m',-1}^{(2)}(\boldsymbol{\Omega}(t))] \Big\} \end{aligned}$$

This can now be applied to the polarizability tensor (Sec. 7.4). This results in Eqs. (7.4.3.). In the event that $\boldsymbol{\alpha}$ is a symmetric tensor, this result simplifies to Eq. (7.4.4.).

NOTES

1. Cooperative effects in molecular reorientations in neat liquid are treateds in Sections 12.2 and 12.3.
2. We note that $c(\mathbf{r}, t) = c(\mathbf{u}, t)$ for $|r| = 1$.
3. This follows because $Y_{00}(\mathbf{u}) = \dfrac{1}{\sqrt{4\pi}}$
4. In a uniform distribution, the probability of finding $\mathbf{u}_0$ in a solid angle d^2u_0 is simply this solid angle d^2u_0 over the total solid angle (4π) accessible to the vector, that is, $d^2u_0/4\pi$. Thus

$$p(\mathbf{u}_0)\, d^2u_0 = \frac{1}{4\pi}\, d^2u_0.$$

5. Since in an isotropic liquid system every space fixed axis is equivalent there should be no dependence of $\langle Y_{lm}^*(\mathbf{u}(o))Y_{lm}(\mathbf{u}(t))\rangle$ on m. This is a general property of isotropic systems and follows from the rotational invariance of equilibrium systems regardless of the kinetic equation (there is no unique axis of quantization). This independence of m is more general than our derivation indicates. We return to this in Sections 7.4 and 7.7.
6. These are properties determined by the electronic structure of the molecule.
7. Note that the laboratory-fixed components vary in time due to molecular tumbling, whereas the body-fixed components do not vary in time because we are dealing with a rigid molecule.
8. See, for example, Perrin (1942).
9. Here $\Omega = (\alpha, \beta, \gamma)$ are not the polar angles of Sections (7.2) and (7.3). Also

$$\int d\Omega = \int_0^{2\pi} d\alpha \int_0^{2\pi} d\gamma \int_0^{\pi} d\beta \sin\beta.$$

10. The functions $C_{KM}^{(J)}(\Omega)$ are related to the Wigner functions $D_{KM}^{(J)}(\Omega)$ by

$$C_{K,M}^{(J)}(\Omega) = \left[\frac{2J+1}{8\pi^2}\right]^{\frac{1}{2}} D_{K,M}^{(J)}(\Omega).$$

11. It should be pointed out that in order to be a spherical diffusor a molecule does not have to have spherical symmetry. It is sufficient to have only tetrahedral (or, of course, higher) symmetry. Similarly, in order to be a symmetric diffusor the molecule need only have four fold rotational symmetry about the symmetry axis.
12. Note that this is normalized since $\displaystyle\int d\Omega\, \frac{1}{8\pi^2} = \int_0^{2\pi} d\alpha \int_0^{2\pi} d\gamma \int_0^{\pi} d\beta \sin\beta\, \frac{1}{8\pi^2} = 1$
13. Note that $F_{K,M}^{J}(t)$ is independent of K. This again follows from the isotropy of the equilibrium fluid since $\mathbf{K}$ refers to projection on the Z axis, but all space fixed axes are equivalent. This is quite general.
14. Where it should be noted from the definition of $C_{K,M}^{(J)}$ in terms of $D_{K,M}^{(J)}$ that

$$F_{K,M}^{J}(t) = \frac{(2J+1)}{8\pi^2} \langle D_{K,M}^{(J)*}(\Omega(0)) D_{K,M}^{(J)}(\Omega(t))\rangle$$

15. Let τ and τ_q be respectively the decay time for $C_2^{(0)}$ and F_s. Then $\tau/\tau_q \sim q\ \sqrt{I/M}$ where I is the moment of inertia and M is the mass. Thus $\tau/\tau_q \sim qa$ where a is the length of the molecule. For light scattering $q \sim 10^5$. Thus for small molecules $\tau/\tau_q \sim 10^{-3}$.
16. $P_2(\cos\omega_0 t) = \dfrac{3}{2}\cos^2\omega_0 t - \dfrac{1}{2}$
17. Note the dependence on q of the Rayleigh band.
18. This is the result for the J-diffusion model. In the M-diffusion model the molecules have a distribution of molecular speeds, $p(\omega_0)$, but each molecule retains the same speed through all collisions. Letting $C_l^{(0)}(t_0|\omega_0)$, $C_l(n,t|\omega_0)$ denote the free particle and n-free step-correlation

functions for those molecules with rotational speed ω_0, the same steps that lead from Eqs. (7.6.6) to Eq. (7.6.12b) can be followed giving $\tilde{C}_l(s)$ for the M-diffusion model as

$$\tilde{C}_l(s) = \int_0^\infty d\omega_0\, p(\omega_0) \left\{ \frac{\tilde{C}_l^0\left(s + \frac{1}{\tau_c}\middle|\, \omega_0\right)}{1 - \frac{1}{\tau_c}\tilde{C}_l^0\left(s + \frac{1}{\tau_c}\middle|\, \omega_0\right)} \right\}$$

19. Macromolecules small compared to q^{-1} rotate many times during the time they take to diffuse a distance q^{-1}, so that the measured translational diffusion coefficient should be the angle-averaged coefficient given in Eq. (7.7.3). In general, translational diffusion is characterized by two numbers, one for diffusion along the major axis $D_\parallel$ and one for diffusion along the minor axis $D_\perp$ (see Appendix (7.A). Equation (7.7.3) is $D_0 = (D_\parallel + 2D_\perp)/3$. Actually Perrin gives formulas like Eq. (7.7.3) for $D_\parallel$ and $D_\perp$. It should be possible to determine $D_\parallel$ and $D_\perp$ in molecules of size comparable to q^{-1}.
20. Equation (5.4.12) can be used for this estimate.
21. See Appendix 7.B.
22. Note that an electric field $\mathbf{E}$ induces the dipole moment $\mu_\alpha = \alpha_{\alpha\beta}E_\beta$, which has the explicit form
$\mu_\alpha = \alpha_\parallel u_\alpha(\mathbf{u} \cdot \mathbf{E}) + \alpha_\perp(E_\alpha - u_\alpha(\mathbf{u} \cdot \mathbf{E}))$
Thus if $\mathbf{E}_\perp\mathbf{u}$, $\mu_\alpha = \alpha_\perp E_\alpha$ and if $\mathbf{E}_\parallel \mathbf{u}$, $\mu_\alpha = \alpha_\parallel E_\alpha$ as required above.
23. This follows because
$\delta_{\alpha\alpha} = \delta_{11} + \delta_{22} + \delta_{33} = 3$
$u_\alpha u_\alpha = u_1u_1 + u_2u_2 + u_3u_3 = \mathbf{u} \cdot \mathbf{u} = 1$
24. $\delta_{\alpha\beta}\delta_{\gamma\delta}$, $\delta_{\alpha\gamma}\delta_{\beta\delta}$, $\delta_{\alpha\delta}\delta_{\beta\gamma}$, are the only isotropic tensors involving the indices $\alpha\beta\gamma\delta$; moreover the combination above is symmetric in the interchange of α with β and γ with δ.
25. This follows from
$u_\alpha(0)\, u_\beta(0)\, u_\beta(t)\, u_\alpha(t) = [u_\alpha(0)\, u_\alpha(t)]\, (u_\beta(0)\, u_\beta(t)]$
$= [\mathbf{u}(0) \cdot \mathbf{u}(t)]\, [\mathbf{u}(0) \cdot \mathbf{u}(t)]$
$u_\alpha(t)\, u_\beta(t)\, \delta_{\beta\alpha} = u_\alpha(t)\, u_\alpha(t) = \mathbf{u}(t) \cdot \mathbf{u}(t) = 1$
and
$\delta_{\alpha\beta}\delta_{\beta\alpha} = 3$
26. $P_2(x) = \left[\frac{3}{2}x^2 - \frac{1}{2}\right]$ where $-1 \leqslant x \leqslant 1$
27. This follows from
$(n_i)_\alpha(n_f)_\beta\delta_{\alpha\beta}\delta_{\gamma\delta}(n_i)_\gamma(n_f)_\delta = (n_i)_\alpha(n_f)_\alpha(n_i)_\gamma(n_f)_\gamma = (\mathbf{n}_i \cdot \mathbf{n}_f)^2$
$(n_i)_\alpha(n_f)_\beta\delta_{\alpha\gamma}\delta_{\beta\delta}(n_i)_\gamma(n_f)_\delta = (n_i)_\alpha(n_i)_\alpha(n_f)_\beta(n_f)_\beta = (\mathbf{n}_i \cdot \mathbf{n}_i)\ (\mathbf{n}_f \cdot \mathbf{n}_f) = 1$
28. Note that if $\alpha = 1/3$, $\beta = 1$ in Eq. (7.B.1)
$$\alpha_{\alpha\beta} = \frac{1}{3}\delta_{\alpha\beta} + [u_\alpha u_\beta - \frac{1}{3}\delta_{\alpha\beta}] = u_\alpha u_\beta$$
as required here.
29. In many papers R is defined such that the rotation brings the laboratory into the body frame. This is reciprocal to our R. If this definition is used Eq. (7.C.7) should read
$$Y_{lm}(\theta_L, \phi_L) = \sum_{m'} D_{mm'}^{(l)*}(R)\, Y_{lm'}(\theta_B, \phi_B)$$
30. See for example, the special case treated in Section 7.2.
31. A pseudovector (unlike a real vector) is even with respect to coordinate inversion. In all other respects its transformation properties are identical to those of vectors.
32. It is clear, however, that a special prescription must be given for constructing the set $\mathbf{T}^{(1)}$. This set is constructed by substituting $T_k = 1/2\, \varepsilon_{ijk}T_{ij}$ where $k = x,y,z$ in place of x,y,z in $Y_{1,m}(\theta,\phi)$. The symbol ε_{ijk} is the Levi-Civita tensor density. It has the value $+1$ if ijk are in the order 1,2,3 or any cyclic permutation, thereof and -1 for any noncyclic permutation of 1,2,3. Here 1,2,3 stand respectively, for x,y,z. If any of the indices are equal, $\varepsilon_{ijk} = 0$. Thus, for example, $T_0^{(1)} = T_z = (1/2)\varepsilon_{ijk}\, T_{ij} = 1/2(T_{xy} - T_{yx})$ where it should be remembered that we sum over i,j.

REFERENCES

Alder, B., Gass, D. M., and Wainwright, T. E., *J. Chem. Phys.* **53,** 3813 (1970).

Alms, G. R., Bauer, D. R., Brauman, J. I., and Pecora, R., *J. Chem. Phys.* **58,** 5570 (1973); **61,** 2255 (1974).

Bauer, D. R., Brauman, J. I. and Pecora, R., *J. Am. Chem. Soc., 96,* 6840, (1974).

Bauer, D. R., Opella, S. J., Nelson, D. J., and Pecora, R., *J. Am. Chem. Soc.* **97,** 2580 (1975).

Ben-Reuven, A. and Gershon, N. D. *J. Chem. Phys.* **51,** 893 (1969).

Brink, D. M. and Satchler, G. R., *Angular Momentum* Chapter IV, Oxford University Press (1968).

Chandler, D., *J. Chem. Phys.* **60,** 3508 (1974).

Debye, P., *Polar Molecules,* Dover, New York (1929).

Dicke, R. H. and Witke, J. P., *Introduction to Quantum Mechanics,* Reading, Addison-Wesley, Mass. (1960).

Dubin, S. B., Clark, N. A. and Benedek, G. B., *J. Chem. Phys.* **54,** 5158 (1971).

Edmunds, A. R. *Angular Momentum in Quantum Mechanics,* Princeton University Press (1957).

Fano, V. and Racah, G. *Irreducible Tensorial Sets,* Academic, New York (1959).

Favro, D., *Phys., Rev.* **119,** 53 (1960).

Fixman, M. and Rider, K., *J. Chem. Phys.* **51,** 2425 (1969).

Gillen, K. T. and Griffiths, J. E., *Chem. Phys. Lett.* **17,** 359 (1972).

Gordon, R. G., *Adv. Mag. Res.* **3,** 1, (1968). This is a comprehensive review of these techniques.

Gordon, R. G., *J. Chem. Phys.* **44,** 1830 (1966).

Happel, J. and Brenner, H., *Low Reynolds Number Hydrodynamics,* Prentice Hall, Englewood Cliffs, N.J. (1965).

Hu, C. and Zwanzig, R., *J. Chem. Phys.,* **60,** 4354 (1974).

Huntress, W. T., Jr., *J. Chem. Phys.* **48,** 3524 (1968).

Ivanov, E. N., *Sov. Phys. JETP* **18,** 1041 (1964).

Jeffreys, H., *Cartesian Tensors,* Cambridge, University Press (1961).

Jones, D. R., Ph. D. Thesis, Stanford University (1974).

King, T. A., Knox, A., and McAdam, J. D., *Biopolym.* **12,** 1917 (1973).

McClung, R. E. W., *J. Chem. Phys.* **51,** 3842 (1969).

O'Dell, J. and Berne B. J., *J. Chem. Phys.,* in press (1975).

Pecora, R., *J. Chem. Phys.* **49,** 1036 (1968).

Perrin, F., *J. de Phys. et Rad.* **V,** 497 (1934); **VII,** 1 (1936).

Perrin, F., *J. Chem. Phys.* **10,** 415 (1942).

Schurr, J. and Schmitz, K. S. *Biopolym.* **12,** 1021 (1973a); **12,** 1543 (1973b).

Shore, B. W. and Menzel, D. H., *Atomic Spectra,* Section 10 and Chapter VI, Wiley, New York (1968).

St. Pierre, A. G. and Steele, W. A., preprint (1974).

Szmansky, H. A. (ed.), *Raman Spectroscopy Theory and Practice,* Plenum, New York, (1970).

van Koynenberg, P. and Steele, W. A., preprint (1974).

Wada, A., Suda, N. Tsuda, K. and Soda, J., *J. Chem. Phys. 50,* **31** (1969).

CHAPTER 8

SCATTERING FROM VERY LARGE MOLECULES

8 · 1 INTRODUCTION

When molecules are very large, intramolecular interference must be taken into account in calculations of scattered intensities and spectra. We present here a discussion of these effects for large molecules in dilute solution.

Intramolecular interference is usually negligible for scattering from small molecules since the wavelets scattered from different segments of the same molecule all have essentially the same phase and hence add constructively at the point of observation. However, for large molecules observed at large values of q, the intramolecular interference depends on the distribution and time rate of change of molecular segmental positions. Thus these effects contain information about molecular shapes, shape fluctuations, and molecular rotations.

Throughout this chapter scattering from the small solvent molecules is ignored since it is usually either small in total intensity compared to the macromolecular scattering or since solvent fluctuations generally decay on a much faster time scale than do macromolecular fluctuations they are temporally separable.

Most of the work on intramolecular interference has been concerned with isotropic scattering. The isotropic scattering is large compared to the anisotropic scattering and is hence relatively easy to measure. Thus the bulk of this chapter is concerned with the isotropic scattering. A discussion of the anisotropic scattering is given in Section 8.9.

8 · 2 ANGULAR DISTRIBUTIONS OF ISOTROPIC INTEGRATED INTENSITIES

It is a convenient artifice in discussing the scattering from solutions of large particles to consider the basic scattering element to be a polymer "segment" rather than the polymer molecule (Debye, 1947). This division of the molecule into segments aids in calculating the scattering form factors for isotropic scattering. Each segment is chosen so that its maximum size l is small compared to $1/q$, that is,

$$ql \ll 1$$

This ensures that each segment can be considered as a point scatterer, that is, that there is no significant intrasegment interference.

Consider the scattering medium to be made up of a collection of polymer *segments.*

Each molecule contains n segments and the illuminated volume contains N molecules. Thus there are nN segments. The scattered field zero-time correlation function is[1]

$$I_{if}^{\alpha}(\mathbf{q}) = I_{if}^{\alpha}(\mathbf{q}, 0) = (\mathbf{n}_i \cdot \mathbf{n}_f)^2 \left\langle \sum_{i,j,l,m} \alpha_l^i \alpha_m^j \exp i\mathbf{q} \cdot [\mathbf{r}_l^i - \mathbf{r}_m^j] \right\rangle \tag{8.2.1}$$

where $\mathbf{r}_l^i$ is the position and α_l^i the polarizability of the lth segment of the ith molecule.

The sum on the right-hand side of Eq. (8.2.1) may be written as a double sum over segments on the same molecule ($i = j$) plus a double sum over segments belonging to different molecules ($i \neq j$). If the solution is sufficiently dilute, segments on different molecules are uncorrelated so that the $i \neq j$ sum is zero. Thus if the solution is dilute and all segments and molecules are identical[2] Eq. (8.2.1) may be written

$$I_{if}^{\alpha}(\mathbf{q}) = (\mathbf{n}_i \cdot \mathbf{n}_f)^2 \langle N \rangle \alpha_M^2 S(\mathbf{q}) \tag{8.2.2}$$

where

$$\alpha_M \equiv n\alpha \tag{8.2.3}$$

is the *molecular* polarizability and

$$S(\mathbf{q}) \equiv \frac{1}{n^2} \left\langle \sum_{l,m} \exp i\mathbf{q} \cdot (\mathbf{r}_l - \mathbf{r}_m) \right\rangle \tag{8.2.4}$$

is called the *molecular form* or *structure factor*. In Eq. (8.2.4) the double summation is, of course, only over segments in the same molecule.

For concentrated solutions or polymer melts the correlations of segments on different molecules must be considered. Calculation of the summation on the right-hand side of Eq. (8.2.1) is very difficult in this case since the evaluation of intermolecular form factors demand a detailed knowledge of the solution structure. For instance, for a solution of rod-shaped molecules the average relative orientation and center-of-mass positions of pairs of rods must be known (Zimm, 1946, 1948a).

For solutions in the limit of infinite dilution only the structure of a single molecule need be known. Conversely, the scattering gives information on the structure of a single molecule and contains no information about correlations in position or orientation between different molecules. Thus since the physical chemist often wants to study the structure of a single molecule in solution, the experimental results are often extrapolated to infinite dilution.

Note from Eq. (8.2.4) that $S(0) = 1$, as we expect, since there is no destructive intramolecular interference between light waves scattered from different parts of the molecule when $q \to 0$; they all travel the same distance to the point of observation and hence arrive with the same phase. At high values of q, however, there may be destructive interference between light waves scattered from different parts of the molecule, reducing $S(\mathbf{q})$ from its zero argument value.

Scattering from some simple model systems, the rigid rod and Gaussian coil are first considered. Then a discussion relating average dimensions of molecules of arbitrary shape to light scattering intensities is given.

Rigid Rod

Consider a long thin rod-like molecule. The diameter of the rod is assumed to be small compared to its length. Furthermore, the rod diameter is not large enough for light scattered from two points along a crossection diameter to produce a significant phase difference. Thus as far as light scattering is concerned, the rod is a distribution of polarizable segments along a straight line. The rod considered here has a constant polarizability per unit length. We may then apply Eq. (8.2.4) written in the form

$$S(\mathbf{q}) = \langle | \sum_l \frac{1}{n} \exp i\mathbf{q} \cdot \mathbf{u} r_l |^2 \rangle \tag{8.2.5}$$

where $\mathbf{u}$ is a unit vector along the cylindrical axis of the rod. The sum is, as before, over all segments along the rod and the average brackets denote an average over all $\mathbf{u}$.

By making n very large while keeping the length L of the rod constant, it may be seen that the sum in Eq. (8.2.5) may be replaced by the integral

$$\lim_{\substack{n\to\infty \\ L\ constant}} \sum_l \frac{1}{n} \exp i\mathbf{q} \cdot \mathbf{u} r_l = \frac{1}{L} \int_{-L/2}^{+L/2} \exp i\mathbf{q} \cdot \mathbf{u} r \, dr \tag{8.2.6}$$

$$= j_0 \left(\mathbf{q} \cdot \mathbf{u} \frac{L}{2} \right) \tag{8.2.7}$$

where $j_0(w)$ is a spherical Bessel function of order zero,

$$j_0(w) = \sin w/w$$

Choosing a coordinate system such that $\mathbf{q}$ is along the z axis and expressing $\mathbf{u}$ in spherical polar coordinates we find that $\mathbf{q} \cdot \mathbf{u} = q \cos\theta$ and

$$S(\mathbf{q}) = \langle | j_0 \left(\frac{x}{2} \cos\theta \right) |^2 \rangle \tag{8.2.8}$$

where

$$x \equiv qL \tag{8.2.9}$$

In an equilibrium ensemble[3] all orientations of the rod are equally probable so that the orientational distribution function is

$$p(\theta, \phi) = \frac{1}{4\pi} \tag{8.2.10}$$

(see Chapter 7, Note 4). The brackets in Eq. (8.2.8) denote an average over all orientations of the rod; that is, over the angle θ. Substitution of Eq. (8.2.10) into Eq. (8.2.8) then gives

$$S(\mathbf{q}) = \frac{1}{4\pi} \int_0^{2\pi} d\phi \int_0^{\pi} d\theta \sin\theta \left| j_0 \left(\frac{x}{2} \cos\theta \right) \right|^2 \tag{8.2.11}$$

Integrating over ϕ and transforming from θ to $y = \cos\theta$, we find

$$S(\mathbf{q}) = \frac{1}{2} \int_{-1}^{1} dy \left| j_0 \left(\frac{xy}{2} \right) \right|^2 \tag{8.2.12}$$

This formula can be evaluated numerically to determine how $S(\mathbf{q})$ depends on q. It can also be expressed in[4] the more common form

$$S(\mathbf{q}) = \frac{2}{x}\int_0^x dz \frac{\sin z}{z} - \left(\frac{2}{x}\sin\left(\frac{x}{2}\right)\right)^2 \tag{8.2.13}$$

The first integral on the left is a tabulated function.

A plot of $S(x)$ against x is shown in Fig. 8.2.1. Note that at small x (corresponding to small rods and/or small q), $S(x) \cong 1$. In this region there is negligible intramolecular destructive interference.

If one could do a scattering experiment over the x range from 0 to 5, the shape of the $S(x)$ curve would show that the molecule is rod-like and if q is also known, would give the length of the rod. This procedure is quite difficult in practice since unless the molecule is very long or q can be made very large (for instance, by using light of low wavelength) only a small portion of the curve can be obtained. In addition, molecules

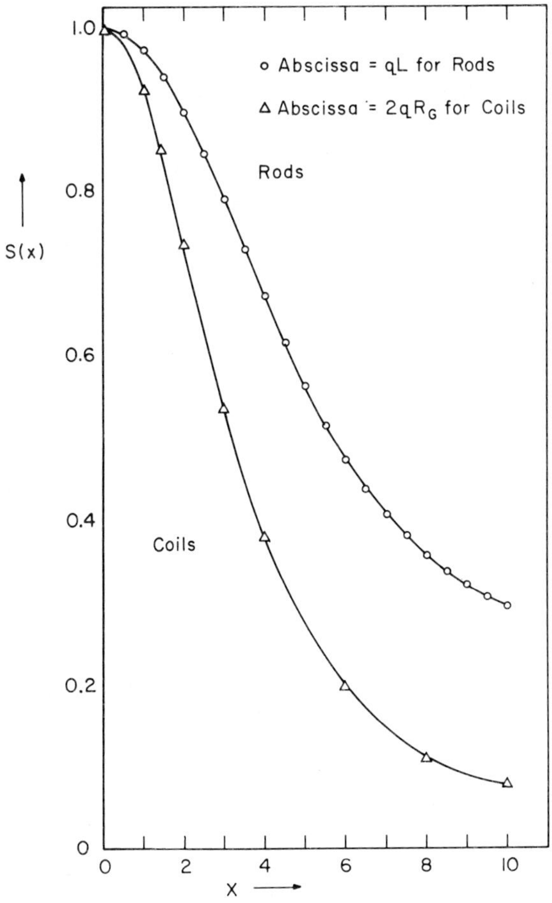

FIG. 8.2.1. The structure factor $S(x)$ of a rigid rods and of rauclom coils as a function of the dimensionless variable x where for rods $x = q$L and for coils $2qR_G$ where L is the length of the rod and R_G is the radius of gyration of the coil (Cf. Eq. 8.2.23).

of different shapes may give rise to curves that are similar to those for a rod, making the curves difficult to distinguish unless the data are very precise.

An alternative derivation of Eq. (8.2.13) is given by Zimm, Stein, and Doty (1945). This article is reprinted in McIntyre and Gornick (1964).

Gaussian Coil

By definition a Gaussian coil molecule consists of a collection of segments such that the mean-squared distance between segments m links apart is proportional to m, that is,

$$\langle r^2(m)\rangle = l^2 m \tag{8.2.14}$$

where l is a constant characteristic of the particular molecule. It may be considered to be the length of a "statistical segment." Furthermore, the probability, p of any two segments, say segments i and j (where $i - j = m$) being separated by the vector distance $\mathbf{r}_{ij}$ is given by a Gaussian distribution

$$p(\mathbf{r}_{ij})d^3r_{ij} = \left[\frac{3}{2\pi\langle r^2(m)\rangle}\right]^{3/2} \exp - \frac{3}{2}\frac{r_{ij}^2}{\langle r^2(m)\rangle} d^3r_{ij} \tag{8.2.15}$$

The probability that a particle in random flight moves a distance $\mathbf{r}_{ij}$ in a large number of steps is given by this probability distribution. Thus implicit in this model is the assumption that the polymer is very flexible and that the distance between segments along the polymer chain corresponds to a large number of elementary chemical bonds. This model is well-known and is discussed in detail in several textbooks (e.g., Flory, 1969).

From Eq. (8.2.4) the form factor for the Gaussian coil is obtained by averaging $\exp i\mathbf{q}\cdot\mathbf{r}_{ij}$ over the distribution function given in Eq. (8.2.15)[5]

$$S(\mathbf{q}) = \frac{1}{n^2}\sum_{i,j}^{n}\left[\frac{3}{2\pi\langle r^2(m)\rangle}\right]^{3/2}\int \exp i\mathbf{q}\cdot\mathbf{r}_{ij} \exp - \frac{3}{2}\frac{r_{ij}^2}{\langle r^2(m)\rangle} d^3r_{ij} \tag{8.2.16}$$

$$= \frac{1}{n^2}\sum_{i,j}^{n} \exp \frac{-q^2\langle r^2(m)\rangle}{6} \tag{8.2.17}$$

The summation over i, j may, in the limit of a large number of segments, be replaced by an integration over m

$$S(\mathbf{q}) = \frac{1}{n^2}\int_0^n n_2(m) \exp \frac{-q^2\langle r^2(m)\rangle}{6} dm \tag{8.2.18}$$

where $n_2(m)$ is the number of segment pairs separated by m links.

The calculation of $n_2(m)$ is relatively simple. Note that there are more pairs separated by a small number of links than by a large number. Furthermore it is clear that $n_2(m)$ is proportional to $n - m$

$$n_2(m) = c(n - m) \tag{8.2.19}$$

The proportionality constant c may be evaluated from the fact that the total number of pairs is n^2, that is,

$$\int_0^n n_2(m)\, dm = n^2 \tag{8.2.20}$$

Thus

$$c = \frac{n^2}{\int_0^n (n-m)\, dm} = 2$$

Combining Eqs. (8.2.14), (8.2.18), (8.2.19), and (8.2.20), we obtain

$$S(\mathbf{q}) = 2/n^2 \int_0^n (n - m) \exp - \frac{q^2 l^2 m}{6}\, dm \tag{8.2.12}$$

Upon performing the integration we find

$$S(\mathbf{q}) = \frac{2}{q^4 R_G{}^4}[\exp - q^2 R_G{}^2 - 1 + q^2 R_G{}^2] \tag{8.2.22}$$

where we have let

$$R_G{}^2 = \frac{nl^2}{6} \tag{8.2.23}$$

The quantity R_G is the "radius of gyration" of the Gaussian coil. (Radii of gyration are discussed briefly in Section 8.3). Note that nl^2 is just the mean-square distance between segments spaced n units apart along the chain; that is, it is just the mean-squared end-to-end distance of the chain, $\langle r^2(n)\rangle$. Hence if we write

$$R_G{}^2 = \frac{\langle r^2(n)\rangle}{6} \tag{8.2.24}$$

all reference to l^2 (the mean-squared segment length) may be omitted from the result.

$S(q)$ is usually written in terms of the dimensionless scattering parameter

$$y = q^2 R_G{}^2 \tag{8.2.25}$$

Then

$$S(\mathbf{q}) = \frac{2}{y^2}[e^{-y} - 1 + y] \tag{8.2.26}$$

A plot of S versus $x = 2qR_G = 2\sqrt{y}$ is shown in Fig. 8.2.1. If the shape of a particle were known to be either a Gaussian coil or a rigid rod it would be difficult to distinguish between the two if an experiment were confined to low values of the scattering parameter.

8 · 3 MOLECULES OF ARBITRARY SHAPE

Calculations of $S(\mathbf{q})$ similar to those for the rigid rod and Gaussian coil may be made for molecules of other specified shapes (e.g., see Kerker, 1969). For molecules of all shapes, however, $S(\mathbf{q})$ at low values of the scattering parameter yields information about a molecular size parameter, the radius of gyration R_G.

The radius of gyration is essentially the root mean-square radius of a macromolecule (Tanford, 1961). Let $\mathbf{R}$ be a vector locating the center of mass of a molecule and let $\mathbf{r}_i$ be a vector locating segment i with mass m_i. Then by definition

$$R_G{}^2 \equiv \frac{\langle \sum_{i=1}^{n} m_i(\mathbf{r}_i - \mathbf{R})^2 \rangle}{\sum_{i=1}^{n} m_i} \tag{8.3.1}$$

If all segments have identical masses m, then

$$R_G{}^2 = \frac{\langle \sum_{i=1}^{n} (\mathbf{r}_i - \mathbf{R})^2 \rangle}{n} \tag{8.3.2}$$

R_G may be calculated if the molecular shape is known. For example, for a rod

$$R_G{}^2 = \frac{L^2}{12} \tag{8.3.3}$$

and for a Gaussian coil $R_G{}^2$ is given by Eq. (8.2.23).

Expanding the structure factor Eq. (8.2.4) in powers of $(\mathbf{q} \cdot [\mathbf{r}_i - \mathbf{r}_j])$ retaining only terms to second order, we obtain

$$S(\mathbf{q}) \cong 1 - \frac{1}{2n^2} \langle \sum_{i,j} \{\mathbf{q} \cdot (\mathbf{r}_i - \mathbf{r}_j)\}^2 \rangle + \ldots \tag{8.3.4}$$

Averaging this form of $S(\mathbf{q})$ over all "orientation" angles of the segment displacements relative to $\mathbf{q}$ leads to

$$S(q) \cong 1 - \frac{q^2}{8\pi n^2} \langle \sum_{i,j}^{n} \int_0^{2\pi} d\phi \int_{-1}^{+1} |\mathbf{r}_i - \mathbf{r}_j|^2 \cos^2 \theta_{ij} d \cos \theta_{ij} \rangle + \ldots \tag{8.3.5}$$

$$= 1 - \frac{q^2}{6n^2} \langle \sum_{i,j} |\mathbf{r}_i - \mathbf{r}_j|^2 \rangle + \ldots \tag{8.3.6}$$

We may write[6]

$$\frac{1}{n^2} \langle \sum_{i,j=1}^{n} |\mathbf{r}_i - \mathbf{r}_j|^2 \rangle = 2 \langle \frac{1}{n} \sum_{i=1}^{n} (\mathbf{r}_i - \mathbf{R})^2 \rangle - 2 \langle \left[\sum_{i=1}^{n} \frac{\mathbf{r}_i - \mathbf{R}}{n} \right]^2 \rangle \tag{8.3.7}$$

However the position of the molecular center of mass is just

$$\mathbf{R} \equiv \frac{1}{n} \sum_{i=1}^{n} \mathbf{r}_i \tag{8.3.8}$$

Therefore

$$\langle \left[\sum_{i}^{n} \frac{\mathbf{r}_i - \mathbf{R}}{n} \right]^2 \rangle = \langle (\mathbf{R} - \mathbf{R})^2 \rangle = 0 \tag{8.3.9}$$

and the remaining term in Eq. (8.3.7) is just $2R_G^2$. Thus

$$\left\langle \frac{1}{n^2} \sum_{i,j}^{n} (\mathbf{r}_i - \mathbf{r}_j)^2 \right\rangle = 2R_G^2 \tag{8.3.10}$$

and $S(\mathbf{q})$ becomes

$$S(\mathbf{q}) = 1 - \frac{q^2}{3} R_G^2 + \ldots \tag{8.3.11}$$

From Eq. (8.3.11) we see that a plot of $S(\mathbf{q})$ as a function of q^2 at low q yields a straight line whose slope is $(-R_G^2/3)$. If the shape is known, R_G can be related to the dimensions of the molecule[7] and a measurement of R_G then gives the molecular size. Note that expansions of Eqs. (8.2.13) and (8.2.27) with the use of Eq. (8.3.3), yield Eq. (8.3.11).

It is sometimes convenient to plot $[S(\mathbf{q})]^{-1}$ against q^2 since this plot may be somewhat more sensitive to small differences in values of $S(\mathbf{q})$ for molecules of different shapes than the $S(\mathbf{q})$ against q^2 plot. In this case at *small* q (since $(1/(1 - x)) \cong 1 + x$ for small x)

$$\frac{1}{S(\mathbf{q})} \cong 1 + \frac{q^2}{3} R_G^2 + \ldots \tag{8.3.12}$$

and the slope is $\frac{R_G^2}{3}$. At higher values of q the curvature of the $[S(\mathbf{q})]^{-1}$ plot gives higher moments of the polymer segment distribution and hence information concerning the actual molecular shape.

8 · 4 MOLECULAR WEIGHT DETERMINATIONS

Perhaps the most routine use of integrated light scattering is the measurement of molecular weights of polymer samples. To determine molecular weights the constant relating I_{if} to $S(\mathbf{q})$ should be specified.

Set the scattering *per unit volume*

$$I_{if}(\mathbf{q}) = KS(\mathbf{q}) \tag{8.4.1}$$

where the structure factor $S(\mathbf{q})$ is, as described above, normalized so that

$$S(0) = 1$$

The constant is given by (see sections 3.2 and 3.3)

$$K = (B)\,(\alpha_M)^2 c \tag{8.4.2}$$

where c is the *number* of macromolecules per unit volume, α_M is the *polarizability* of a macromolecule, and

$$B = \frac{k_f^4 E_0^2}{\varepsilon^2 R^2} (\mathbf{n}_i \cdot \mathbf{n}_f)^2$$

Let α_M' be the polarizability per *unit mass* of the molecule, that is,

$$\alpha_M' = \frac{\alpha_M N_0}{M} \tag{8.4.3}$$

where M is the molecular weight of the molecule, and N_0 is Avogadro's number. Furthermore, let $c' \equiv \frac{M}{N_0} c$ be the mass density of macromolecules. Then Eq. (8.4.1) becomes

$$I_{if}(\mathbf{q}) = \frac{B}{N_0} \alpha_M'^2 c' M S(\mathbf{q}) \tag{8.4.4}$$

Equation (8.4.4) may be rearranged to read

$$\frac{Bc'\alpha_M'^2}{N_0 I_{if}(\mathbf{q})} = \frac{1}{MS(\mathbf{q})} \tag{8.4.5}$$

All quantities on the left-hand side of Eq. (8.4.5) may be determined experimentally without previous knowledge of the molecular weight of the molecule. Consider, for instance, α_M'. It is usually assumed that α_M is related to the excess dielectric constant of the solution over that of the pure solvent (ε_0) by the relation

$$\varepsilon - \varepsilon_0 = 4\pi \frac{N}{V} \alpha_M \tag{8.4.6}$$

If the solution is dilute, the left-hand side of Eq. (8.4.6) may be expanded in a power series in the mass concentration

$$\varepsilon - \varepsilon_0 \approx \left(\frac{\partial \varepsilon}{\partial c'}\right)_{c'=0} c' \tag{8.4.7}$$

From Eqs. (8.4.3), (8.4.6), and (8.4.7), we obtain

$$\alpha_M' = \left(\frac{\partial \varepsilon}{\partial c'}\right)_{c'=0} \frac{1}{4\pi} \tag{8.4.8}$$

Since the optical dielectric constant ε is equal to the square of the solution refractive index, it is evident from Eq. (8.4.8) that α_M' may be measured by differential refractometry.

Using Eq. (8.3.12) for $[S(\mathbf{q})]^{-1}$ at small q, Eq. (8.4.5) becomes

$$\frac{Bc'\alpha_M'^2}{N_0 I_{if}(\mathbf{q})} = \frac{1}{M}\left(1 + q^2 \frac{R_G^2}{3} + \cdots\right) \tag{8.4.9}$$

Thus, if the quantity on the left-hand side of Eq. (8.4.9) is plotted against q^2, the $q = 0$ intercept gives the reciprocal of the macromolecular molecular weight.

8 · 5 CORRECTIONS FOR FINITE CONCENTRATIONS AND POLYDISPERSITY

The above theory was developed for infinitely dilute, monodisperse solutions. If the solutions become concentrated the interference between scattered wavelets from different macromolecules becomes important. Consequently, the simple linear dependence of $I_{if}(\mathbf{q})$ on c' (or $[I(\mathbf{q})]^{-1}$ on c'^{-1}) is no longer valid (at low c' one might expect a first correction term proportional to c'^2) and $S(\mathbf{q})$ no longer includes only one molecule. In general, the equilibrium distribution of segments throughout the solution as a whole must be known to calculate the structure factors (see Section 8.2). Thus since measurements are made at finite q and c', a method of double extrapolation to zero q and c' must be developed. Zimm (1948b) has presented a technique for this utilizing what is now called the Zimm plot.

The Zimm plot is a plot of $[B(\alpha'_M)^2c'/N_0 I_{if}(\mathbf{q})]$ as ordinate against q^2 plus a constant times c'. (The value of the constant is chosen merely for convenience in making the plot). Values of the ordinate are experimentally determined at a fixed concentration c' for a series of q^2 values and then plotted. The procedure is then repeated for different values of c'.[8] Lines of constant c' are then extrapolated to form a $q = 0$ line and lines of constant q to form a $c' = 0$ line. This chapter has thus far been devoted to the theory of the $c' = 0$ line.

The intercept of both these lines with the ordinate axis gives M^{-1}. In addition, the initial slope of the $c' = 0$ line in accordance with Eq. (8.4.9) gives $R_G^2/3$. Furthermore at $q = 0$

$$\frac{Bc'\alpha_M^2}{N_0 I_{if}} = \frac{1}{M} + 2\beta c' + \ldots$$

where β is the solution-second virial coefficient. Thus, the initial slope of the $q = 0$ line may be used to obtain the second virial coefficient (Zimm, 1948b). This procedure is illustrated in fig (8.5.1).

For polydisperse solutions, the corrections become rather complicated at high concentrations. Let us consider only the limit of very low concentrations.

Consider a dilute solution of macromolecules differing only in molecular weight. Assume first that α'_M does not depend on the molecular weight of the molecule. Furthermore let the mass concentration, molecular weights, and structure factors of species i be, respectively, c_i', M_i, and $S_i(\mathbf{q})$. Then since each species is independent, we merely sum the scattering from each to obtain the total scattering. Thus from Eq. (8.4.4) for the scattering per unit volume

$$I_{if}(\mathbf{q}) = \frac{B\alpha'^2_M}{N_0} \sum_i M_i c'_i S_i(\mathbf{q}) \tag{8.5.1}$$

where the summation is over all species present.

At low q, $S_i(\mathbf{q}) \cong 1$ and

$$I_{if} = \frac{B\alpha'^2_M}{N_0} \sum_i M_i c'_i \tag{8.5.2}$$

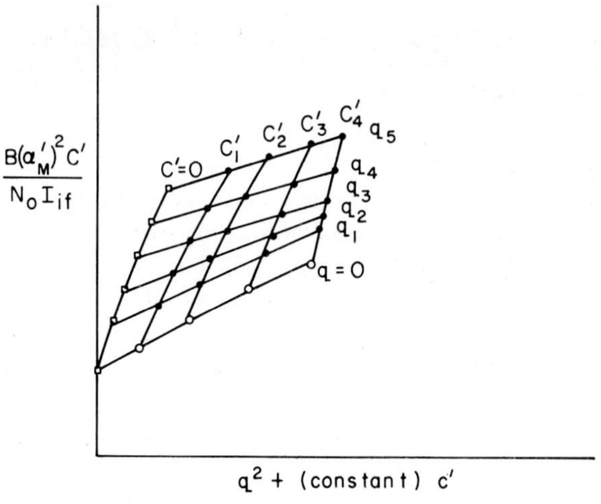

FIG. 8.5.1. Zimm plot for the double extrapolation of light-scattering data to zero concentration and zero q. The intercept is $\langle M\rangle^{-1}$ and the initial slope of the $c' = 0$ line is $R_G{}^2/3$.

$$= \frac{B\alpha'_M{}^2}{N_0} \langle M\rangle\, c' \tag{8.5.3}$$

where $\langle M\rangle$ is the weight average molecular weight of the polymer sample

$$\langle M\rangle \equiv \frac{\sum_i c'_i M_i}{\sum_i c'_i} \tag{8.5.4}$$

and

$$c' \equiv \sum_i c'_i \tag{8.5.5}$$

Thus from Eqs. (8.4.10) and (8.5.3) we see that the Zimm plot yields the weight average molecular weight of a polydisperse sample.

For large molecules observed at high values of q, the average $S(\mathbf{q})$ must be calculated. A synthetic polymer sample usually has a molecular weight distribution characterized by a continuous function. Let $f(M)dM$ be the weight fraction of molecules with molecular weights between M and $M + dM$. Then the summation in Eq. (8.5.1) may be replaced by an integration over M

$$I_{if}(\mathbf{q}) = \frac{B\alpha'_M{}^2}{N_0} \langle M\rangle c' \frac{\int_0^\infty f(M) M S(M, \mathbf{q})\, dM}{\int_0^\infty f(M) M dM} \tag{8.5.6}$$

where the dependence of the structure factor on molecular weight has been explicitly indicated. Thus if $f(M)$ is known and $S(M, \mathbf{q})$ may be found for a given model (rigid rod, coil, etc.), $I_{if}(\mathbf{q})$ can be calculated.

Further discussion of polydisperse solutions is given in Section 8.10 in connection with light-beating experiments.

8 · 6 TIME CORRELATION FUNCTIONS AND SPECTRAL DISTRIBUTIONS

In Sections 8.1–5 an outline of the theory of the frequency integrated polarized scattering from macromolecular solutions is presented. These measurements may be used to obtain information about equilibrium (static) properties of molecules in solution. Thus molecular weights, radii of gyration, molecular shapes, size and molecular weight distributions of polydisperse samples, and solution virial coefficients may be studied by observing frequency-integrated light-scattering intensities as functions of scattering angle and concentration. Performing an additional measurement of the frequency distribution of this scattered light or the corresponding time-correlation function extends the range of these experiments to the measurement of nonequilibrium (dynamic) properties of macromolecules in solution. The measurement of translational diffusion coefficients of macromolecules by this technique has already been discussed in Chapter 5. However, when the molecules are very large, other dynamic properties may also affect the spectral distribution of the scattered light (Pecora, 1964). Some of these effects are discussed in this and following sections.

Consider a dilute solution of identical polymer molecules which may be subdivided into identical segments, as in Section 8.2. The scattered-field time-correlation function is proportional to

$$I_{if}^{g}(\mathbf{q}, t) = (\mathbf{n}_i \cdot \mathbf{n}_f)^2 \alpha_M{}^2 N S(\mathbf{q}, t) \tag{8.6.1}$$

where

$$S(\mathbf{q}, t) \equiv \frac{1}{n^2} \langle \sum_{i,j} \exp i\mathbf{q} \cdot [\mathbf{r}_i(t) - \mathbf{r}_j(o)] \rangle \tag{8.6.2}$$

is the dynamic form factor for a single molecule. Note that

$$S(\mathbf{q}, o) = S(\mathbf{q}). \tag{8.6.3}$$

The summation in Eq. (8.6.2) is only over segments belonging to a single molecule. For concentrated solutions space–time correlations between segments on different molecules must also be considered.

In this section some general considerations about $S(\mathbf{q}, t)$ are given, then in Sections 8.7 and 8.8 dynamic models for scattering from rigid rods and Gaussian coils are discussed.

Equation (8.6.2) expresses $S(\mathbf{q}, t)$ in terms of the molecular segmental positions in a laboratory-fixed coordinate system. It is convenient for calculations to express these positions in terms of the position of the molecular center of mass $\mathbf{R}(t)$ and some vector giving the position of the segment relative to the center of mass. Thus,

$$\mathbf{r}_j(t) = R(\mathbf{t}) + \mathbf{b}_j(t) \tag{8.6.4}$$

Then, if we let $\mathbf{R}_t$ represent the displacement of the center of mass in the time interval t

$$\mathbf{R}_t \equiv \mathbf{R}(t) - \mathbf{R}(o) \tag{8.6.5}$$

Equation (8.6.2) may be written

$$S(\mathbf{q}, t) = \frac{1}{n^2}\langle \exp i\mathbf{q} \cdot \mathbf{R}_t \sum_{i,j} \exp i\mathbf{q} \cdot [\mathbf{b}_j(t) - \mathbf{b}_i(o)]\rangle \tag{8.6.6}$$

In cases where intramolecular interference is negligable (roughly where max $q|\mathbf{b}_j(t) - \mathbf{b}_i(o)| \ll 1$)

$$S(\mathbf{q}, t) \approx \langle \exp i\mathbf{q} \cdot \mathbf{R}_t\rangle \tag{8.6.7}$$

If it is further assumed that the molecule translates by a translational diffusion process, it is shown in Section (5.4)that

$$S(\mathbf{q}, t) \sim \exp -q^2Dt \tag{8.6.8}$$

Thus only the translational diffusion coefficient may be measured under these conditions.

If, however, intramolecular interference is important, "intramolecular" motions may, in some circumstances, affect the spectral distribution of the scattered light. The general condition for such contributions is that the terms containing the $\mathbf{b}_i(t)$ and $\mathbf{b}_j(0)$ must contribute a time-dependence to $S(\mathbf{q}, t)$. Three cases should be distinguished.

Rigid Large, Uniform Spherical Polymers.

Here the sum over segments is not time-dependent since the only relative segmental motion allowed is rotation. The sum is invariant to any rotation of the sphere. A formal mathematical argument produces the same result (Pecora, 1968), but it is clear on physical grounds that the model sphere "looks the same" to the light wave in any orientation. Thus, Eq. (8.6.6) becomes

$$S(\mathbf{q}, t) = \langle \exp i\mathbf{q} \cdot \mathbf{R}_t\rangle\, S(\mathbf{q}) \tag{8.6.9}$$

where $S(\mathbf{q})$ is the particle structure factor for a sphere

$$S(\mathbf{q}) \equiv \langle \frac{1}{n^2} \sum_{ij} \exp i\mathbf{q} \cdot (\mathbf{b}_i(0) - \mathbf{b}_j(0))\rangle \tag{8.6.10}$$

In this case $S(\mathbf{q})$ is easily evaluated. Equation (8.6.10) may be written in the form

$$S(\mathbf{q}) = \left| \frac{1}{n} \sum_{i=1}^{n} \exp i\mathbf{q} \cdot \mathbf{b}_i \right|^2$$

The sum may then be replaced by an integral,

$$S(\mathbf{q}) = \left| \left(\frac{3}{4\pi r^3}\right) \int_0^r \exp i\mathbf{q} \cdot \mathbf{b}\, 4\pi b^2 db \right|^2 \tag{8.6.11}$$

where r is the radius of the sphere. The integral in Eq. (8.6.11) is easily performed. The resulting $S(\mathbf{q})$ is

$$= \{3/(qr)^3(\sin qr - qr \cos qr)\}^2 \tag{8.6.12}$$

Rigid, Uniform, Non-Spherical Polymers.

In this case if the molecule is large enough the sum will be time-dependent. The rigid rod model, an example of this case, is discussed in detail below.

Flexible Polymers.

For flexible polymers the structural change due to intramolecular motions must be large enough for the light wave to detect the difference between the various molecular shapes. Only under these circumstances will intramolecular interference affect the light-scattering spectral distributions. An extreme example of this case, the Rouse–Zimm dynamic model of the Gaussian coil, is discussed in detail in Section 8.8.

8 · 7 TIME CORRELATION FUNCTION FOR LONG RIGID RODS

In this section, we treat a simple but important model—the rigid rod (Pecora, 1964, 1968). This model illustrates the conditions under which rotational motions of rigid, nonspherical molecules affect the isotropic spectral distributions. It is also of great practical importance, since it is applicable to a wide variety of real macromolecules such as fibrous proteins, helical polypeptides, and some viruses (e.g., tobacco mosaic virus).

The rod consists of n identical, optically isotropic segments arranged along a line of length L. The thickness of the rod is assumed to be negligible compared to the wavelength of the light.

To evaluate the average in Eq. (8.6.6), both the structure and the dynamics of the rod must be known. In general, a rod will have two independent components of the translational diffusion tensor, one of which can be taken for translations parallel to the long rod axis and one perpendicular to it. This general case yields rather complicated results. For simplicity it is assumed that only one number is necessary to characterize the translational diffusion of the molecule. This is probably a good assumption when rotation is fast compared to translation. When this is true we may use the model for combined rotational–translational diffusion discussed in Chapter 7 to calculate $S(\mathbf{q}, t)$. Since all the segments are arranged along a line, all $\mathbf{b}_j$ are either parallel or antiparallel to a given vector pointing along the rod and all we have to determine to describe the segmental motion is the joint probability distribution function $G_s(\mathbf{R}, \boldsymbol{\Omega}, t; \mathbf{0}, \boldsymbol{\Omega}_0, 0)$ for the polymer to be at position $\mathbf{R} = \mathbf{0}$ with orientation $\boldsymbol{\Omega} = \boldsymbol{\Omega}_0$ at time $t = 0$ and at position $\mathbf{R} = \mathbf{R}_t$, and orientation $\boldsymbol{\Omega} = \boldsymbol{\Omega}_t$ at time t. This probability distribution is

$$G_s(\mathbf{R}, \boldsymbol{\Omega}, t; \mathbf{0}, \boldsymbol{\Omega}_0, 0) = \frac{1}{4\pi} K_s(\mathbf{R}, \boldsymbol{\Omega}, t \,|\, \mathbf{0}, \boldsymbol{\Omega}_0, 0) \tag{8.7.1}$$

where $K_s(\mathbf{R}, \boldsymbol{\Omega}, t\,|\,\mathbf{0}, \boldsymbol{\Omega}_0, 0)$ satisfies the translation–rotation diffusion equation

$$\frac{\partial}{\partial t} K_s(\mathbf{R}, \boldsymbol{\Omega}, t\,|\,\mathbf{0}, \boldsymbol{\Omega}_0, 0) = [D\nabla_R^2 - \Theta \hat{I}^2]\, K_s(\mathbf{R}, \boldsymbol{\Omega}, t\,|\,\mathbf{0}, \boldsymbol{\Omega}_0, 0) \tag{8.7.2}$$

together with the boundary condition

$$K_s(\mathbf{R}, \boldsymbol{\Omega}, t\,|\,\mathbf{0}, \boldsymbol{\Omega}_0, 0) = \delta(\mathbf{R})\, \delta(\boldsymbol{\Omega} - \boldsymbol{\Omega}_0) \tag{8.7.3}$$

D and Θ are the translational and rotational diffusion coefficients of the polymer and $\hat{I}$ is the angular momentum operator (see Section 7.3). Equation (8.7.2) simply states that translational and rotational diffusion are independent processes (see Section 7.A). Equation (8.6.6) may be rewritten as

$$S(\mathbf{q}, t) = \frac{1}{4\pi n^2} \int d^3R \int d\boldsymbol{\Omega}_0 \int d\boldsymbol{\Omega} \exp i\mathbf{q} \cdot \mathbf{R}_t \sum_{i,j=1}^{n} \exp i\mathbf{q} \cdot [\mathbf{b}_j(t) - \mathbf{b}_i(0)]\, K_s(\mathbf{R}, \boldsymbol{\Omega}, t\,|\,\mathbf{0}, \boldsymbol{\Omega}_0, 0) \tag{8.7.4}$$

$$= \frac{1}{4\pi n^2} \int d\boldsymbol{\Omega} \int d\boldsymbol{\Omega}_0 \left(\sum_{j=1}^{n} \exp i\mathbf{q} \cdot \mathbf{b}_j(t) \right) \left(\sum_{i=1}^{n} \exp -i\mathbf{q} \cdot \mathbf{b}_i(0) \right) F_s(\mathbf{q}, \boldsymbol{\Omega}, t\,|\,\boldsymbol{\Omega}_0, 0) \tag{8.7.5}$$

where $F_s(\mathbf{q}, \boldsymbol{\Omega}, t\,|\,\boldsymbol{\Omega}_0, 0)$ is the spatial Fourier transform of K_s

$$F_s(\mathbf{q}, \boldsymbol{\Omega}, t\,|\,\boldsymbol{\Omega}_0, 0) = \int d^3R \exp i\mathbf{q} \cdot \mathbf{R}\, K_s(\mathbf{R}, \boldsymbol{\Omega}, t\,|\,\mathbf{0}, \boldsymbol{\Omega}_0, 0) \tag{8.7.6}$$

The function $F_s(\mathbf{q}, \boldsymbol{\Omega}, t\,|\,\boldsymbol{\Omega}_0, 0)$ may be determined in the same manner as in Chapter (7). It turns out to be

$$F_s(\mathbf{q}, \boldsymbol{\Omega}, t\,|\,\boldsymbol{\Omega}_0, 0) = \exp -q^2Dt \sum_l \sum_m Y_{lm}(\boldsymbol{\Omega}_0) Y_{lm}^*(\boldsymbol{\Omega}) \exp -l(l+1)\Theta t \tag{8.7.7}$$

Let us look at the integrand of Eq. (8.7.5). It should be noted that the $\exp i\mathbf{q}\cdot\mathbf{b}_j(t)$ depends on $\mathbf{b}_j(t)$ which specifies the position of the jth segment with respect to the center of mass. Since the polymer is a symmetrical rod, for every segment at $\mathbf{b}$ there is a segment at $-\mathbf{b}$. It then follows that

$$\sum_{j=1}^{n} \exp i\mathbf{q} \cdot \mathbf{b}_j(t) = 2 \sum_{j=1}^{n/2} \cos \mathbf{q} \cdot \mathbf{b}_j(t) \tag{8.7.8}$$

where the summation is over half the rod. The same is, of course, true of the zero time exponentials. Each $\mathbf{b}_i$ points at a given time in the same direction, and is therefore specified by the same orientation angles. These terms can be expanded in surface spherical harmonics as follows

$$\begin{aligned} \cos \mathbf{q} \cdot \mathbf{b}_i(0) &= 4\pi \sum_{l\ even, m} i^l Y_{lm}^*(\boldsymbol{\Omega}_0)\, Y_{lm}(\boldsymbol{\Omega}_q)\, j_l(qb_i) \\ \cos \mathbf{q} \cdot \mathbf{b}_j(t) &= 4\pi \sum_{l\ even, m} i^l Y_{lm}(\boldsymbol{\Omega})\, Y_{lm}^*(\boldsymbol{\Omega}_q)\, j_l(qb_j) \end{aligned} \tag{8.7.9}$$

where $\boldsymbol{\Omega}_q$ specifies the orientation of the vector $\mathbf{q}$ and $j_l(X)$ the spherical Bessel function (of the first kind) of order l. Substitution of Eqs. (8.7.6–9) into Eq. (8.7.5) with subsequent use of the orthonormality of the spherical harmonics yields

$$S(\mathbf{q}, t) = 16\pi \frac{1}{n^2} \sum_{l \text{ even}} \sum_{i,j=1}^{\frac{n}{2}} j_l(qb_i)\, j_l(qb_j) \exp -[q^2 D + l(l+1)\Theta]t \sum_m |Y_{lm}(\boldsymbol{\Omega}_q)|^2 \tag{8.7.10}$$

From the addition theorem of spherical harmonics

$$\sum_{m=-l}^{+l} |Y_{lm}(\boldsymbol{\Omega}_q)|^2 = \frac{(2l+1)}{4\pi} \tag{8.7.11}$$

Consequently

$$S(\mathbf{q}, t) = \sum_{l \text{ even}} \left| \frac{2}{n} \sum_{j=1}^{\frac{n}{2}} j_l(qb_j) \right|^2 (2l+1) \exp -[q^2 D + l(l+1)\,\Theta]t \tag{8.7.12}$$

$$= \left\{ \left| \frac{2}{n} \sum_{j=1}^{\frac{n}{2}} j_0(qb_j) \right|^2 \exp -q^2 Dt + 5 \left| \frac{2}{n} \sum_{j=1}^{\frac{n}{2}} j_2(qb_j) \right|^2 \exp -(q^2 D + 6\Theta)t + \ldots \right. \tag{8.7.13}$$

which may be written as

$$S(\mathbf{q}, t) = S_0(qL) \exp -q^2 Dt + S_1(qL) \exp -(q^2 D + 6\Theta t) + \ldots \tag{8.7.14}$$

where the strengths $S_i(qL)$ are defined in terms of the spherical Bessel functions

$$S_0(qL) = \left| \frac{2}{n} \sum_{j=1}^{\frac{n}{2}} j_0(qb_j) \right|^2$$

$$S_1(qL) = 5 \left| \frac{2}{n} \sum_{j=1}^{\frac{n}{2}} j_2(qb_j) \right|^2$$

$$\vdots \qquad \vdots$$

$$S_{\frac{l}{2}}(qL) = (2l+1) \left| \frac{2}{n} \sum_{j=1}^{\frac{n}{2}} j_l(qb_j) \right|^2 \tag{8.7.15}$$

Thus the scattered field time-correlation function is

$$I_{if}^{\alpha}(\mathbf{q}, t) = (\mathbf{n}_i \cdot \mathbf{n}_f)^2 N\alpha_M^2 [S_0(qL) \exp -q^2 Dt + S_1(qL) \exp -(q^2 D + 6\Theta)t + \ldots \tag{8.7.16}$$

and its Fourier transform is

$$I_{if}^{\alpha}(\mathbf{q}, \omega) = (\mathbf{n}_i \cdot \mathbf{n}_f)^2 \frac{N\alpha_M^2}{\pi} \left\{ S_0(qL) \frac{q^2 D}{\omega^2 + (q^2 D)^2} + S_1(qL) \frac{(q^2 D + 6\Theta)}{\omega^2 + (q^2 D + 6\Theta)^2} + \ldots \right. \tag{8.7.17}$$

Note that the spectra given by Eqs. (8.7.16) and (8.7.17) consist, respectively, of a sum of exponentials and Lorentzians with weights determined by the $S_{l/2}(qL)$ and time constants and widths determined by the translational and rotational diffusion coefficients of the rod. It should be emphasized that the first term in each series depends on D and not on Θ.

It remains for us to numerically calculate the $S_{l/2}$ as functions of the dimensionless argument qL and also to calculate the total integrated intensity

$$I_{if}^{\alpha}(\mathbf{q}) \equiv \int d\omega I_{if}^{\alpha}(\mathbf{q}, \omega)$$
$$= (\mathbf{n}_i \cdot \mathbf{n}_f)^2 N\alpha_M{}^2 S(qL) \tag{8.7.18}$$

where

$$S(qL) = \sum_{l\ even} S_{\frac{l}{2}}(qL) \tag{8.7.19}$$

The structure factor $S(qL)$ gives the angular (or more generally the q) dependence of the total frequency integrated scattered light intensity from a rod-like molecule. It was calculated in Section 8.2 [see (Eq. 8.2.13)].

The sums in Eqs. (8.7.15) may be numerically evaluated by dividing the molecule into a large number of segments such that each segment is small compared to $1/q$. However when these conditions are fulfilled the sums may be replaced by integrals, that is,

$$\frac{2}{n}\sum_{j=0}^{\frac{n}{2}} j_l(qb_j) \rightarrow \frac{2}{qL}\int_0^{\frac{qL}{2}} j_l(z)dz \tag{8.7.20}$$

Using the formulas for the spherical Bessel functions

$$j_0(z) = \frac{\sin z}{z}$$
$$j_1(z) = \frac{\sin z}{z^2} - \frac{\cos z}{z} \tag{8.7.21}$$
$$j_2(z) = \left[\frac{3}{z^3} - \frac{1}{z}\right]\sin z - \frac{3}{z^2}\cos z$$

we obtain for the first few terms in the series

$$S_0(qL) = \left[\frac{2}{qL}\int_0^{\frac{qL}{2}} \frac{\sin z}{z}\, dz\right]^2 \tag{8.7.22}$$

and

$$S_1(qL) = 5(qL)^{-1}\left[-3j_1\left(\frac{qL}{2}\right) + \int_0^{\frac{qL}{2}} \frac{\sin z}{z}\, dz\right]^2 \tag{8.7.23}$$

Values of the integral

$$\int_0^x \frac{\sin z}{z}\, dz \tag{8.7.24}$$

may be found in tables and the $S_0(qL)$ and $S_1(qL)$ may be calculated. These functions are plotted in Fig. 8.7.1 for values for $qL \equiv x$ from 0 to 10. Noting the form of Eq. (8.2.13) for $S(q)$ we set

$$S(qL) - (S_0(qL) + S_1(qL)) \equiv S_h(qL) \tag{8.7.25}$$

where S_h gives the contribution to the integrated intensity of all terms other than S_0 and S_1. S_h is also plotted in Fig. 8.7.1.

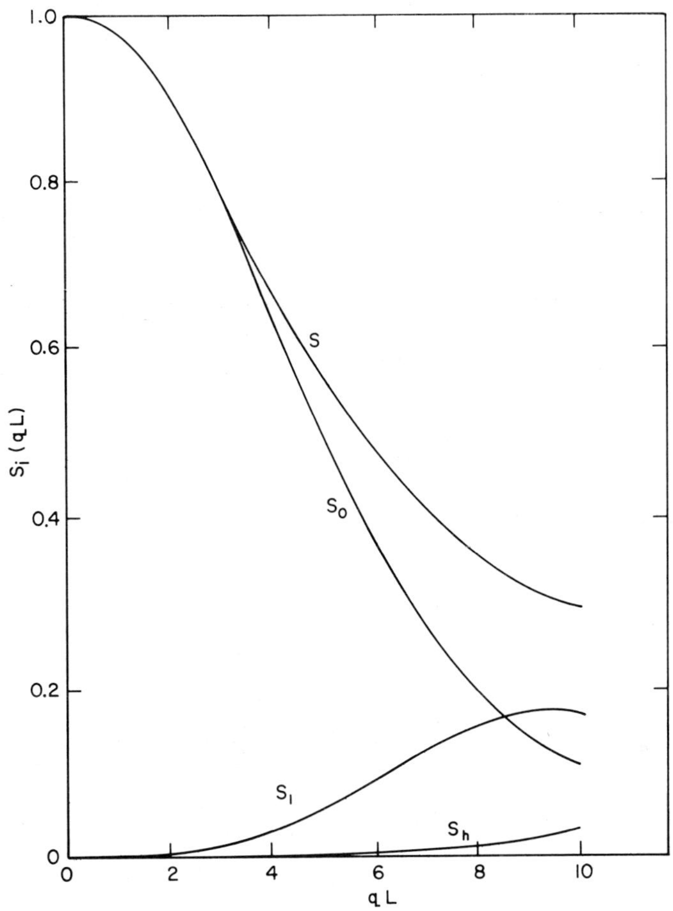

FIG. 8.7.1. Relative integrated intensities of light scattered from optically isotropic rigid rods. S is the total relative integrated intensity, S_0 the intensity of the pure translational part, S_1 the first non-zero term whose spectral width contains the rotational diffusion coefficient, and S_h the sum of intensities of all other terms.

It may be seen from Fig. 8.7.1 that for $qL \leq 3$, $S_0(qL) \sim S(qL)$. Thus $I_{if}^{\alpha}(\mathbf{q}, t)$ and $I_{if}^{\alpha}(\mathbf{q}, \omega)$ are, respectively, almost entirely given by the first terms on the right-hand sides of Eqs. (8.7.16) and (8.7.17). In this case, for instance, the width of the spectrum is determined solely by the translational diffusion coefficient. Thus the rotational diffusion coefficient could not be determined by an experiment on an *optically iso-*

tropic rod if $qL \leq 3$ (short molecule, low scattering angle, or large light wavelength in the medium). On the other hand, for $qL \geq 5$ (large molecules observed at large q) S_1 becomes an important part of the spectrum. Its contribution ranges from 12% of S at $qL = 5.2$ to 57.7% at $qL = 10$. The rest of the terms $S_h(qL)$ are negligible for $qL \leq 8$ and rise to about 10% of $S(qL)$ for $qL = 10$. Thus, for a large molecule the S_1 term on the right-hand side of Eq. (8.7.14) may be "turned off" by observing the scattering at low angles. Under these conditions, a fit of the experimental result to the remaining single Lorentzian in Eq. (8.7.17) allows easy extraction of the translational diffusion coefficient. By observation of the scattering at large angles and a fit of the spectrum to the full two-Lorentzian form given in Eq. (8.7.17) with use of the already determined D, the rotational diffusion coefficient may be obtained from the data.

A detailed experimental study of the isotropic component of light scattered from dilute solutions of tobacco mosaic virus (a rod-like molecule with $L \cong 3000$ Å and cross section diameter $\cong 180$ Å) has been perfomed by Cummins et al. (1969) using spectrum analysis techniques. These authors found that the measured spectrum fit the theory described above rather well. Wada et al. (1971) repeated these experiments using an autocorrelator with similar results.

As stated above, this theory assumes that translational diffusion is isotropic; that is, in a molecule-fixed frame, the diffusion constant parallel to the long molecular axis is the same as that perpendicular to it. For highly anisotropic large molecules this is probably not a good assumption. Maeda and Saito (1969) have calculated the spectrum taking into account the anisotropy of the translational diffusion constant. Their resulting expressions are rather complex and will not be given here. Their results are expressed as a power series in the translational diffusion coefficient anisotropy,

$$\frac{D_{\parallel} - D_{\perp}}{\bar{D}}$$

The first term in the series (independent of the anisotropy) is the same as that given above with $D \equiv \dfrac{(D_{\parallel} + 2D_{\perp})}{3}$. For tobacco mosaic virus at high qL values the extra terms dependent on the anisotropy become significant in the spectrum. Fujime (1970) and later Schaefer et al. (1971) have performed experiments on this virus and have attempted to detect the terms in the spectrum dependent on the diffusional anisotropy. The interpretation of Fujime's results was complicated by his use of polydisperse samples. Schaefer et al. (1971) used monodisperse samples but obtained essentially a zero value for the diffusion coefficient anisotropy—a result in disagreement with the usual simple hydrodynamic models.

8 · 8 GAUSSIAN COILS

In Section 8.7 a calculation was presented for a very stiff rod-like molecule. Although this model is adequate to describe light scattering from many real systems, most molecules have some degree of flexibility. When the intramolecular motions have large configurational changes associated with them, relaxation times for these motions will be present in the isotropic spectrum even if there is no accompanying change in the molecular segment-fixed polarizabilities. In order to illustrate some of the essential features of the dynamics which are likely to affect the light-scattering spectrum, we treat in some detail a dynamic version of the Gaussian coil model described in Section 8.2 (Zimm, 1956).

The model molecule consists on $n + 1$ light-scattering beads each with identical isotropic polarizabilites α. The beads are connected by "springs" (or "segments" in the previous description) which provide a restoring force linear in the displacement if some beads stray from their equilibrium separations. Each bead interacts with the surrounding medium through identical frictional coefficients ζ and, in addition, Brownian forces are exerted on the beads by solvent molecules. (see Fig. 8.8.1).

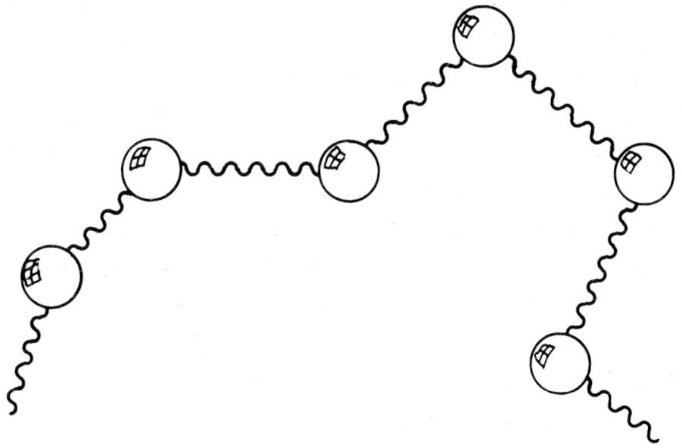

FIG. 8.8.1. Bead-spring model of a flexible macromolecule. All interactions of the macromolecule with light and the surrounding medium occurs through the beads. The "springs" (segments) merely provide an entropic restoring force to return the beads to their equilibrium separations whenever a shape fluctuation occurs.

It is convenient in analyzing the forces in this system to consider the components of the bead coordinates in a Cartesian system and to use a column vector notation. Thus let x_i be the x-coordinate of bead i and let

$$\mathbf{x} = \begin{bmatrix} x_0 \\ x_1 \\ \cdot \\ \cdot \\ \cdot \\ x_n \end{bmatrix} \tag{8.8.1}$$

The force on the beads, as stated above, consists of three parts in the absence of any external forces or hydrodynamic interactions.

1. First is a frictional part tending to slow down the bead motions. This force is proportional to the velocity of the bead. The x-components of this force may be written

$$\mathbf{F}_x^{(1)} = -\zeta \frac{d\mathbf{x}}{dt} \tag{8.8.2}$$

2. Second is an entropic force tending to restore the beads to their equilibrium separations. The x-component of this force may be written in matrix form

$$\mathbf{F}_x^{(2)} = -\sigma \mathbf{A}\mathbf{x} \tag{8.8.3}$$

where **A** is a $n + 1$ dimensional square matrix with elements 2 along the diagonal (except for the (0, 0) and $(n + 1, n + 1)$ elements which are 1) and -1 in positions adjacent to the diagonal. All other elements are zero.

$$\mathbf{A} \equiv \begin{bmatrix} 1 & -1 & 0 & 0 \ldots & 0 & 0 & 0 \\ -1 & 2 & -1 & 0 \ldots & 0 & 0 & 0 \\ 0 & -1 & 2 & 1 \ldots & 0 & 0 & 0 \\ \vdots & \vdots & \vdots & \vdots & \vdots & \vdots & \vdots \\ 0 & 0 & 0 & 0 \ldots & -1 & 2 & -1 \\ 0 & 0 & 0 & 0 \ldots & 0 & -1 & 1 \end{bmatrix} \tag{8.8.4}$$

and σ is the "entropic" force constant expressed in terms of the mean-square length of a segment $\langle l^2 \rangle$,

$$\sigma \equiv \frac{3k_B T}{\langle l^2 \rangle} \tag{8.8.5}$$

A statistical mechnical discussion of the origin of the entropic force is beyond the scope of this book.[9] Here we will merely present a heuristic derivation of Eqs. (8.8.3–5).

The probability of finding a Gaussian coil in a given configuration is

$$p(x_0 \ldots x_n) = c \exp - \frac{3}{2} \sum_{j=1}^{n} \frac{(x_j - x_{j-1})^2}{\langle l^2 \rangle}$$

where the length of the jth segment bounded by beads j and $j - 1$ is $x_j - x_{j-1}$. This can be written in the form of a Boltzmann factor with an effective potential energy

$$\mu = \frac{3k_B T}{\langle l^2 \rangle} \frac{1}{2} \sum_{j=1}^{n} (x_j - x_{j-1})^2$$

The effective force on bead j is

$$\mathbf{F}_j = - \frac{\partial \mu}{\partial x_j} = - \frac{3k_B T}{\langle l^2 \rangle} (x_{j+1} - 2x_j + x_{j-1})$$

(except, of course, for beads zero and n). This result is equivalent to Eq. (8.8.3). The restoring force in Eq. (8.8.3) is the same as that for a linear harmonic lattice with free ends.

3. Third is the stochastic Brownian force $\mathbf{F}_x^{(3)}(t)$ which plays the same role here as it does in the Langevin equation (Section 5.10).

Applying Newton's second law of motion to the beads[10]

$$\begin{aligned} m \frac{d^2\mathbf{x}}{dt^2} &= \mathbf{F}_x^{(1)} + \mathbf{F}_x^{(2)} + \mathbf{F}_x^{(3)} \\ &= -\zeta \frac{d\mathbf{x}}{dt} - \sigma \mathbf{A} \cdot \mathbf{x} + \mathbf{F}_x^{(3)} \end{aligned} \tag{8.8.6}$$

Since we are concerned only with relatively slow long-wavelength bead motions, we may ignore the term on the left-hand side of Eq. (8.8.6) (the inertial term). Consequently, the equation of motion for **x** is

$$\zeta \frac{d\mathbf{x}}{dt} = \mathbf{F}_x^{(3)}(t) - \sigma \mathbf{A}\mathbf{x} \tag{8.8.7}$$

Similar equations apply to **y** and z.

The matrix **A** couples the x components of neighboring segments. As in the problem of the harmonic lattice, it is convenient to define a set of "normal" coordinates. For Eq. (8.8.7) this set is simply the coefficients in a Fourier series expansion of **x**

$$x_i = \sum_{k=0}^{n} Q_{ik} \mu_k^{(x)} \tag{8.8.8}$$

where

$$\begin{aligned} Q_{ik} &= \left(\frac{2}{n}\right)^{1/2} \cos \pi k \left(\frac{i}{n} - \frac{1}{2}\right) && \text{for } k \text{ even} \\ &= \left(\frac{2}{n}\right)^{1/2} \sin \pi k \left(\frac{i}{n} - \frac{1}{2}\right) && \text{for } k \text{ odd} \end{aligned} \tag{8.8.9}$$

Similar equations give $\mathbf{y}_i$ and $\mathbf{z}_i$ and also define a new normal coordinate vector $\boldsymbol{\mu}_k \equiv (\mu_k^x, \mu_k^y, \mu_k^z)$.

Upon substitution of Eq. (8.8.8) into Eq. (8.8.7), we obtain the equations of motion for the normal coordinate vectors

$$\zeta \frac{d\boldsymbol{\mu}_k}{dt} = \mathbf{F}_k^{(3)}(t) - 4\sigma \sin^2 \left(\frac{\pi k}{2n}\right) \boldsymbol{\mu}_k \tag{8.8.10}$$

The last term on the right-hand side of Eq. (8.8.10) simplifies considerably if we confine our attention only to low k values (which correspond to long "wavelength" modes)

$$k \ll n$$

Then

$$4 \sin^2 \left(\frac{\pi k}{2n}\right) \sim \left(\frac{\pi k}{n}\right)^2 \tag{8.8.11}$$

and Eq. (8.8.10) becomes

$$\frac{d\boldsymbol{\mu}_k}{dt} = \frac{1}{\zeta} \mathbf{F}_k^{(3)}(t) - \frac{\boldsymbol{\mu}_k}{\tau_k} \qquad (k = 1, 2, 3 \ldots) \tag{8.8.12}$$

where τ_k, the relaxation time of the kth normal mode, is given by

$$\tau_k = \frac{\langle l^2 \rangle n^2 \zeta}{3 k_B T \pi^2 k^2} \qquad (k = 1, 2 \ldots)$$

These relaxation times may be expressed in terms of the measurable parameters R_G^2 [see Eq. (8.2.25)] and the molecular translational diffusion coefficient

$$D = \frac{k_B T}{n\zeta} \tag{8.8.13}$$

$$\tau_k = \frac{2R_G{}^2}{\pi^2 k^2 D} \tag{8.8.14}$$

As expected, the larger the radius of gyration the larger the relaxation times and the larger the diffusion constant the smaller τ_k. The $k = 0$ mode represents translational diffusion of the molecular center of resistance and is given by the Langevin equation for the diffusive motion of a single particle (see Section 5.9).

Equation (8.8.12) may be solved by passing to a Fokker–Planck equation (see Appendix 8.A) for $\Psi_k(\boldsymbol{\mu}_k,t)$ the probability distribution function for the normal coordinate $\boldsymbol{\mu}_k$

$$\frac{\partial \Psi_k}{\partial t} = \frac{1}{\tau_k}\left[\frac{\langle \mu_k^2\rangle}{3}\nabla_k^2\Psi_k + \nabla_k\cdot(\boldsymbol{\mu}_k\Psi_k)\right] \tag{8.8.15}$$

where

$$\langle \mu_k^2\rangle \equiv \frac{nR_G{}^2}{6\pi^2 k^2} \tag{8.8.16}$$

is the equilibrium mean-squared length of mode k.

The total distribution function for all modes is a product of those for the individual modes

$$\Psi\{\boldsymbol{\mu}_k(t)\} = \prod_{k=0}^{n} \Psi_k(\boldsymbol{\mu}_k(t)) \tag{8.8.17}$$

The time-dependent form factor for light scattering is given in terms of the Green function solution of Eq. (8.8.15), $\Psi(\{\mu_k(t)\} \mid \{\mu_k(0)\} \mid t)$, and the equilibrium–distribution function of the model $\Phi(\{\boldsymbol{\mu}_k(0)\}$,

$$S(\mathbf{q}, t) = \left[\frac{1}{n+1}\right]^2 \sum_{i,j}^{n} \int \cdots \int \exp i\mathbf{q}\cdot[\Sigma_k(Q_{jk}\boldsymbol{\mu}_k(t) - Q_{jk}\boldsymbol{\mu}_k(0))]\,\Psi\Phi d\boldsymbol{\mu}(t)d\boldsymbol{\mu}(0)$$

where

$$d\boldsymbol{\mu}(t)d\boldsymbol{\mu}(0) = d\boldsymbol{\mu}_0(t)\ldots d\boldsymbol{\mu}_n(t)\,d\boldsymbol{\mu}_0(0)\ldots d\boldsymbol{\mu}_n(0) \tag{8.8.18}$$

The functions Ψ and Φ may be shown to be[11]

$$\Psi\Big[\{\boldsymbol{\mu}_k(t)\} \mid \{\boldsymbol{\mu}_k(0)\} \mid t\Big] = \left[\frac{1}{4\pi Dt}\right]^{3/2} \exp \frac{-[\boldsymbol{\mu}_0(t) - \boldsymbol{\mu}_0(0)]^2}{4Dt} \times$$

$$\sum_{k=1}^{n} \left\{\frac{3}{2\pi\langle \boldsymbol{\mu}_k{}^2\rangle\left[1 - \exp\left(-\frac{2t}{\tau_k}\right)\right]}\right\}^{\frac{2}{3}} \times$$

$$\exp\left\{-\frac{3}{2\langle \mu_k{}^2\rangle_e}\,\frac{\left[\boldsymbol{\mu}_k(t) - \mu_k(0)\exp\left(-\frac{t}{\tau_k}\right)\right]^2}{\left[1 - \exp\left(-\frac{2t}{\tau_k}\right)\right]}\right. \tag{8.8.19}$$

and

$$\Phi(\{\boldsymbol{\mu}_k(0)\}) = \delta(\boldsymbol{\mu}_0(t) - \boldsymbol{\mu}_0(0)) \prod_{k=1}^{n} \left[\frac{3}{2\pi\langle\mu_k{}^2\rangle_e}\right]^{3/2} \exp\left(-\frac{3}{2}\frac{(\boldsymbol{\mu}_k(0))^2}{\langle\mu_k{}^2\rangle_e}\right) \tag{8.8.20}$$

Substituting Eqs. (8.8.19) and (8.8.20) into Eq. (8.8.18) and performing the integrations, we find

$$S(\mathbf{q}, t) = \exp -q^2Dt \frac{1}{n^2} \sum_{i,j}^{n} \prod_{k=1}^{n} \exp\left\{-\frac{q^2}{6}\langle\mu_k{}^2\rangle_e \left[Q_{jk}{}^2 + Q_{ik}{}^2 - 2Q_{ik}Q_{jk} \exp\left(-\frac{t}{\tau_k}\right)\right]\right\} \tag{8.8.21}$$

where we have set as usual $n \approx n + 1$. This expression is still not in closed form.

By expanding the exponential term containing the time in Eq. (8.8.21) and then evaluating the coefficients,[12] we obtain

$$S(\mathbf{q}, t) = S_0(x) \exp -q^2Dt + S_2(x) \exp -\left(q^2D + \frac{2}{\tau_1}\right)t + \ldots \tag{8.8.22}$$

where $x \equiv q^2R_G{}^2$ [cf. Eq. (8.2.25)], and the first coefficient is

$$S_0(x) = \frac{\pi}{x} \exp\left(-\frac{x}{6}\right)\left[\operatorname{erf}\frac{\sqrt{x}}{2}\right]^2 \tag{8.8.23}$$

where erf y is the error function of argument y

$$\operatorname{erf} y \equiv \frac{1}{\pi^{1/2}} \int_0^y \exp(-z^2)\, dz \tag{8.8.24}$$

Equation (8.8.22) should be compared with that for a rod [Eq. (8.7.13)]. Note that the $S_0(x)$ contribution to the spectrum depends only on the translational diffusion coefficient of the coil while the term labeled S_2 is the first appreciable term which depends on the intramolecular motion. Note that it depends only on the intramolecular relaxation time of the $k = 1$ normal mode.

It should be noted that setting $t = 0$ in Eq. (8.8.21) for $S(\mathbf{q},t)$ and replacing the i, j summation by an integration yields Eq. (8.2.26) for $S(\mathbf{q})$ of the Gaussian coil.

Figure 8.8.2 shows a plot of $S(x)$, $S_0(x)$, and $S_2(x)$ and

$$S_h \equiv S - S_0 - S_2 \tag{8.8.25}$$

versus $\sqrt{x}$ in the range from $\sqrt{x} = 1$ to $\sqrt{x} = 2.6$.

For small x

$$S \approx S_0$$

and the time-correlation function decay is determined solely by the coil translational diffusion coefficient. However when $x \geq 3$ ($x^{1/2} \geqslant 1.73$), the intramolecular relaxation times appreciably affect the decay. $S - S_0$ is 15% of the relative scattered intensity at $x = 3$ and is 50% at $x = 7$. The term labeled S_2 is 11.5% of S at $x = 3$ and 24% at $x = 7$.

Thus we see that as in the case of rods, the coil must be large enough so that its configurational fluctuations can be "seen" by the light wave in order for configurational dynamic constants to affect the scattered field time-correlation function.

Polystyrenes in solutions under ideal conditions (θ solutions) are expected to exhibit

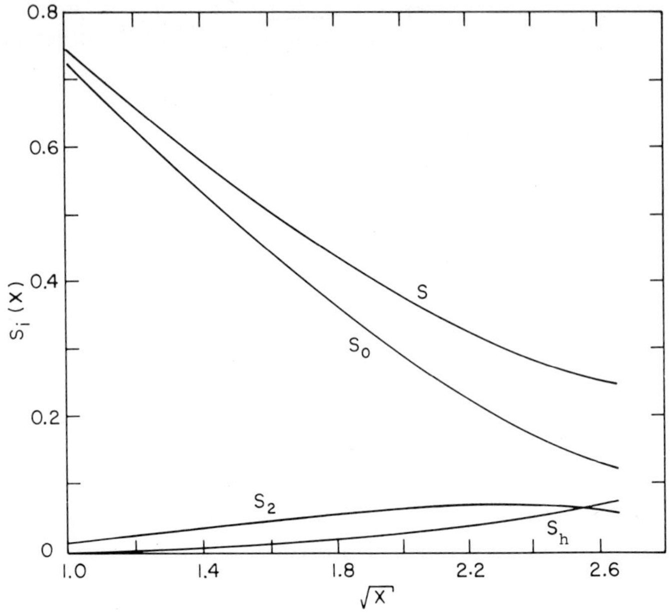

FIG. 8.8.2. Relative integrated intensities of light scattered from Gaussian coils versus $\sqrt{x} = qR_G$. S is the total relative integrated intensity, S_0 the intensity of the pure translational part, S_1 the first significant term whose time correlation function depends on intramolecular relaxation times and, $S_h \equiv S - (S_0 + S_1)$.

many of the features of the theoretical model presented above. Consequently, Reed and Frederick (1971) have studied the Rayleigh line broadening of θ solutions of polystyrenes in cyclohexane. The polystyrene molecular weights ranged from 179,000 to 4,500,000. In a later work, Kramer and Frederick (1972) studied light scattering from polystyrenes in 2-butanone for molecular weights from 8.7×10^5 to 5×10^7 with similar results. At low molecular weights the spectra are single Lorentzians as predicted by the theory. For the higher molecular weight samples studied at high scattering angle a deviation from a single Lorentzian form appears. After careful discussions of the possible causes of this non-Lorentzian spectrum (including polydispersity and critical scattering), the authors conclude that it is most likely due to intramolecular Brownian motion. These experiments are only semiquantitative. No numerical values of the long mode relaxation times were obtained. Huang and Frederick (1973), however, obtained τ_1 from the Rayleigh spectrum for a polystyrene of average molecular weight 27.3×10^6 in both cyclohexane and in 2-butanone. They conclude that the values of τ_1 obtained are in "essential agreement" with values calculated from normal coordinate theories.

8 · 9 ANISOTROPIC SCATTERING

Depolarized scattering has so far played a relatively minor role in the study of macromolecules in solution (although it has been used to a great extent in studying polymeric solids). Much of the early interest in this field arose from the need to determine how anisotropic scattering would affect molecular-weight determinations. Other workers were interested also in determining molecular optical anisotropies of chain molecules in solution and in ultimately relating these anisotropies to molecular conformations in solution. A third possibility is the use of angular studies of the integrated intensities of the depolarized component to study molecular sizes and shapes as a supplement to such studies using the "polarized" component.

The difficulties in the latter two applications stem from the fact that macromolecular optical anisotropies are usually small relative to the average molecular polarizabilities. Thus unless the polarizers in the experiment are extremely good and all other experimental factors (collimation, etc.) are correctly taken care of, the experimental results are for the most part unreliable. One difficulty is that the "polarized scattering" (very large) could "leak through" the polarizers and be measured as part of the depolarized component. Another difficulty is that since the depolarized intensity is very weak, the solution must be relatively concentrated to obtain measurable depolarized signal. This high concentration results in multiple scattering of the isotropic signal. Since polarizations change in multiple scattering (even from optically isotropic molecules), this multiply-scattered light could easily be mistaken for the single-scattered depolarized signal.

With the present use of lasers and high quality polarizers, however, these difficulties can in many cases be overcome with the result that depolarized light scattering is now becoming an important tool for the study of large molecules in solution. In addition, the extension of light-scattering techniques to measure scattered light time-correlation functions now allows measurements of the dynamics of fluctuations of molecular properties that are coupled to fluctuations of optical anisotropies.

In this section we briefly review some of the macromolecular applications of integrated depolarized scattering and then those of time-correlated depolarized light scattering.

Applications of Integrated Depolarized Scattering

Molecular Optical Anisotropies. For dilute solutions of macromolecules whose intramolecular (and intermolecular) form factors are negligible, the expression for depolarized scattering is rather simple. In the notation of Sections 3.4 and 7.2 it is easily shown that

$$I_{VV} = I_{\text{ISO}} + \frac{4}{3} I_{VH} \tag{8.9.1}$$

where

$$I_{VH} = \frac{1}{15} N\langle\gamma^2\rangle \tag{8.9.2}$$

and $\langle\gamma^2\rangle$ is the mean-squared optical anisotropy of a molecule. The quantity $\langle\gamma^2\rangle$ is formally defined as an average over all molecular conformations of

$$\begin{aligned}\langle\gamma^2\rangle &\equiv \frac{3}{2}\langle Tr(\boldsymbol{\alpha\alpha}) - \frac{1}{3}(Tr\boldsymbol{\alpha})^2\rangle \\ &= \frac{3}{2}\langle\alpha_{ij}\alpha_{ji} - \frac{1}{3}(\alpha_{ii})^2\rangle\end{aligned} \tag{8.9.3}$$

where $\boldsymbol{\alpha}$ is the molecular polarizability tensor and where in the last equation the Einstein summation convention is used.

If the molecule is optically isotropic in all conformations, that is, $\alpha_{ij} = \alpha\delta_{ij}$, then $\langle\gamma^2\rangle = 0$. If the molecules have cylindrical symmetry in all conformations, that is, in a principal axis system $\alpha_{11} = \alpha_{22}$, then

$$\langle\gamma^2\rangle = \langle(\alpha_{11} - \alpha_{33})^2\rangle. \tag{8.9.4}$$

Measurements of I_{VH} at infinite dilution yield values of $\langle\gamma^2\rangle$. These molecular optical anisotropies may be related for many macromolecular systems to the anisotropies of units comprising the chain. Since the anisotropies of individual units (or bonds) are tensor quantities, their resultant for the molecule is conformation-dependent. Thus, values of $\langle\gamma^2\rangle$ may be related to properties—such as barriers to internal rotation around chemical bonds—that determine the molecular configuration.

As noted above, experimental measurements must be very carefully performed. All measurements must be extrapolated to infinite dilution. In addition, care should be taken to subtract collision-induced scattering from the data, especially for the shorter chain molecules (see Chapter 14). For larger molecules extrapolation to zero q must be done in order to avoid intramolecular interference effects. Local field effects must also be considered when relating the experimental depolarized intensities to $\langle\gamma^2\rangle$.

Flory (1969) has discussed in detail methods for calculating $\langle\gamma^2\rangle$ from knowledge of bond polarizabilities and barriers to internal rotation. Patterson and Flory (1972) have discussed the treatment of experimental data.

Form Factors. For large macromolecules observed at large scattering angles intramolecular interference must be considered as it was in Section 8.2. Such effects arise from correlations of the orientations of macromolecular segments over large distances. These correlations lead to changes in both I_{VV} and I_{VH}. I_{ISO} is, of course, the same as computed in Sec. 8.2. In general, however, Eq. (8.9.1) is no longer valid. After writing out the basic equations for molecules composed of cylindrically symmetric segments, we consider the relatively trivial case of the freely jointed chain and in Appendix 8.B, the not-so-trivial rigid-rod case in order to illustrate the method of calculation. Pecora and Aragon (1974) have treated hollow and solid spheres. We refer the reader to their for article for further details of this case.

Let each molecule be composed of n identical, cylindrically-symmetric segments. The "average" polarizability of a segment is

$$\alpha \equiv \frac{\alpha_{\parallel} + 2\alpha_{\perp}}{3}$$

and the optical anisotropy is $\beta \equiv \alpha_{\parallel} - \alpha_{\perp}$.

The general equations for the I_{VV} and I_{VH} components of the scattered intensity may easily be written if one uses the coordinate system II of Section 3.4. In this coordinate system the incident light beam travels somewhere in the xy plane with incident polarization in the z-direction and the scattered beam travels along the x-direction. Using the results of Chapter 7, we find that

$$\begin{aligned} I_{VV}(\mathbf{q}) &= N\alpha^2\langle\sum_{i,j}^{n} \exp i\mathbf{q}\cdot[\mathbf{r}_i - \mathbf{r}_j]\rangle \\ &\quad + \frac{N}{9}\beta^2\langle\sum_{i,j}^{n}(3\cos^2\theta_i - 1)(3\cos^2\theta_j - 1)\exp i\mathbf{q}\cdot[\mathbf{r}_i - \mathbf{r}_j]\rangle \end{aligned} \tag{8.9.5}$$

$$+\frac{2N}{3}\alpha\beta\langle\sum_{i,j}^{n}(3\cos^2\theta_i - 1)\exp i\mathbf{q}\cdot[\mathbf{r}_i - \mathbf{r}_j]\rangle$$

and

$$I_{VH}(\varphi) = N\beta^2\langle\sum_{i,j}^{n}(\sin\theta_i\cos\theta_i\sin\varphi_i)(\sin\theta_j\cos\theta_j\sin\varphi_j\exp i\mathbf{q}\cdot[\mathbf{r}_i - \mathbf{r}_j]\rangle \quad (8.9.6)$$

The angles (θ_i, ϕ_i) represent the orientation angles of segment i in a laboratory-fixed spherical polar coordinate system. Note that in Eq. (8.9.5) the last term contains a "cross term" in $\beta\alpha$. For large stiff molecules this term does not, in general, vanish. In the limit $q \to 0$, however, the $\beta\alpha$ term vanishes and Eqs. (8.9.1) and (8.9.2) are valid.

Freely Jointed Chains

The case of the freely-jointed chain is trivial. I_{ISO} is, of course, the same as that given by Eq. (8.2.26). The other terms are vastly simplified by the fact that the orientation of a segment is uncorrelated with its center-or-mass position and that orientations of different segments are uncorrelated. Thus in Eqs. (8.9.5) and (8.9.6) all terms in the summations proportional to β^2 are zero unless $i = j$. Note also that the $\beta\alpha$ term in I_{VV} is zero.

Then using Eq. (8.2.26) we obtain

$$I_{VV} = N\alpha_M^2\, S(q) + \frac{4}{3} I_{VH} \quad (8.9.7)$$

and

$$I_{VH} = N\beta^2 n\langle\sin^2\theta\cos^2\theta\sin^2\varphi\rangle \quad (8.9.8)$$

Since each segmental orientation is as likely as any other in this model, the average over angles is easily computed

$$\langle\sin^2\theta\cos^2\theta\sin^2\varphi\rangle = \frac{1}{4\pi}\int_{-1}^{1} d(\cos\theta)\int_0^{2\pi} d\varphi\sin^2\theta\cos^2\theta\sin^2\varphi = \frac{1}{15} \quad (8.9.9)$$

Therefore,

$$I_{VH} = N\frac{\beta^2 n}{15} \quad (8.9.10)$$

Thus in the freely jointed chain case I_{VH} is independent of q and is proportional to the number of segments. In this model the size and number of segments into which a molecule is divided is no longer arbitrary. It must conform to a true "statistical segment." In fact, if a real chain is very flexible and the segment is taken to be a single chain unit (e.g., the CH_2 group), the total molecular optical anisotropy will, in the absence of intrachain long-range forces, vary linearly with the number of bonds in the limit of long chains.

Rigid Rods

Since rigid rod segments have strong correlations in orientation (in fact, they all have the same orientation), the calculation of the integrated intensities is more difficult than for the Gaussian coil (Horn, 1956). The I_{VH} spectrum will, in the case of long rods, have a dependence on q. We present the calculation and results for this case in Appendix 8.B.

Time-Correlation Functions for Depolarized Scattering

The theories described in Chapter 7 apply to the depolarized spectra of dilute solutions of small, rigid macromolecules. If the macromolecules are nonrigid or large (compared to $1/q$) the depolarized spectra will be modified.

Maeda and Saito (1973) have calculated the spectrum of light scattered from optically anisotropic rigid rods whose length is $\gtrsim q^{-1}$. This calculation is even more complex than that for the integrated intensity (zero-time scattered-field correlation function) for the same model given in Appendix 8.B, and will, therefore, not be given here. The interested reader should consult the article by Maeda and Saito.

The depolarized scattering for the Rouse–Zimm dynamical model of flexible polymer chains (cf. Section 8.8) may also be calculated. Ono and Okano (1971) have performed this calculation for $q = 0$ (zero scattering angle) and find that the scattered light spectral density is a series of Lorentzians each with a relaxation time characteristic of one of the Rouse–Zimm model modes. However the contribution of each mode to the spectrum is *equal*. This behavior should be contrasted with that of the isotropic spectrum where the scattering spectrum is dominated by contributions from the longest wavelength modes.

So far no experiments have been reported on depolarized scattering from systems which might correspond to these models. Schmitz and Schurr (1973), however, have examined forward depolarized scattering from the semiflexible molecule calf-thymus DNA. They report finding three relaxation times in the depolarized spectrum. The longest of these ($\tau \sim 18$ msec) is close to what one would predict on the basis of flow-dichroism results. The temperature profile of the longest relaxation time was investigated and found to exhibit a spike near the helix-coil denaturation temperature. This spike was interpreted as resulting from an increase in the molecular weight of the DNA in the denaturation transition region.

8 · 10 OTHER MODELS

Most macromolecules in solution are neither as stiff as the rigid rod nor as flexible as the Gaussian coil. For particular systems of interest a dynamical model should be made and the corresponding spectrum (or time correlation function) calculated. Measurement of the spectrum and a fit to the theoretical form then allows extraction of the model dynamic constants. These dynamic constants may then be related to equilibrium structural properties of the molecule (end-to-end distances, backbone curvature, etc.).

Models for the low-frequency motions of a semiflexible linear molecule usually treat the system as an elastic wire with force constants for bending and stretching. The model of Harris and Hearst (1966) has been extensively applied to the light-scattering problem by Fujime (1972). In this model the chain is represented by a space vector **r** which is a function of the distance along the chain s and the time t. The Langevin equation for $\mathbf{r}(s, t)$ [cf. Eq. (8.8.6)] is

$$\rho \frac{\partial^2 \mathbf{r}}{\partial t^2} + \zeta \frac{dr}{dt} + \varepsilon \frac{\partial^4 \mathbf{r}}{\partial s^4} - \kappa \frac{\partial^2 \mathbf{r}}{\partial s^2} = \mathbf{F}(s, t)$$

where ρ is the chain density, ζ the frictional constant, ε and κ are, respectively, elastic constants for bending and stretching of the chain, and $\mathbf{F}(s, t)$ is the fluctuating force. The elastic constants may be related to the persistence length (or statistical length) l of the chain. The persistence length, of course, increases as the chain becomes stiffer.

The elastic constant for bending is simply

$$\varepsilon = \frac{3k_BT}{4} l$$

while the elastic constant for stretching is for the special cases of a rod of length L

$$\kappa = \frac{3k_BT}{L} \qquad \text{(rod)}$$

and for a coil

$$\kappa = \frac{3k_BT}{l} \qquad \text{(coil)}$$

The space-curve function $\mathbf{r}(s, t)$ and $\mathbf{F}(s, t)$ may be expanded in normal coordinates and relatively simple *formal* expressions for the scattered spectrum may be found. We refer the reader to the literature for descriptions of this work (Harris and Hearst, 1966; Fujime, 1972).

For very large molecules it is very difficult to fit the data to the theoretical form. In some cases it may suffice to merely find the dependence of the time constant (or half-width) for decay of the experimental auto-correlation function (or spectral density) as a function of q^a at large q. The value of a should be related to the persistence length and hence the force constants of the chain. De Gennes (1967) has, in fact, computed the q-dependence of the spectral half-width in the limit $qR_G \gg 1$ for the Rouse–Zimm model of Gaussian coils. He finds that for the free-draining model (described in Section 8.8) $a = 4$, while DuBois-Violete and de Gennes (1967) find that strong hydrodynamic interaction leads to $a = 3$. DuBois-Violete and de Gennes (1967) and also Silbey and Deutch (1973) have included excluded volume effects in the model. The former authors find $a = \frac{8}{3}$ and the latter, $a = \frac{16}{5}$.

8 · 11 EFFECTS OF POLYDISPERSITY ON TIME-CORRELATION FUNCTIONS AND SPECTRA

Distribution-Function Method

In Section 8.5, the effects of polydispersity on integrated spectra of polydisperse solutions were discussed. It was shown, for instance, that weight-average molecular weights are obtained by double extrapolation of integrated scattering data to zero concentration and scattering vector. Equation (8.5.6) shows how the scattering factor averaged over the polymer distribution modifies the angular distribution of the composite scattered intensity. In this section, we merely mention some of the considerations that must be employed in the study of light-scattering time-correlation functions and spectral distributions from polydisperse macromolecular solutions.

In the dynamic case, Eq. (8.5.6) becomes (Pecora and Tagami, 1969)

$$I_{if}^{\alpha}(\mathbf{q}, t) = \frac{B}{N_0} \alpha_M'^2 \langle M \rangle c' \frac{\int_0^\infty f(M) M S(M, \mathbf{q}, t) \, dM}{\int_0^\infty f(M) M \, dM} \tag{8.11.1}$$

In general, $S(M, \mathbf{q}, t)$ must be calculated as a function of molecular weight and the result then averaged over the molecular weight distribution. This means that the molecular weight-dependence of both the structure *and* the dynamic constants of the macromolecule must be known.

Let us assume that the molecule is so small that all structure factors may be ignored. Hence from Eq. (8.6.8)

$$S(M, \mathbf{q}, t) \cong \exp -q^2 D(M) t \tag{8.11.2}$$

where the translational diffusion coefficient is the only quantity dependent on the molecular weight. There are many empirically based relations between D and M (some of which have theoretical justification). For instance, for many macromolecules

$$D = CM^{-\beta} \tag{8.11.3}$$

where C and β are constants over a specified molecular weight range and usually $0.3 \leq \beta \leq 1.0$. Then

$$I_{if}^{\alpha}(\mathbf{q}, t) = \frac{B}{N_0} \alpha_M'^2 \langle M \rangle c' \frac{\int_0^\infty f(M) M \exp -q^2 C M^{-\beta} t \, dM}{\int_0^\infty f(M) M \, dM} \tag{8.11.4}$$

Thus $I_{if}^{\alpha}(\mathbf{q}, t)$ is a mass-weighed superposition of exponentials. Note that molecules with higher M values contribute more to the average than those with lower M values. Since the diffusion coefficient decreases as M increases, the time correlation decays more slowly than would a monodisperse sample with $M = \langle M \rangle$. Of course if the distribution function $f(M)$ is highly skewed toward low molecular weights, this conclusion may no longer apply. For rods (Yamakawa, 1971) $\beta \cong 1$, for flexible coils $\beta \cong .5$, and for spheres $\beta \cong \frac{1}{3}$. Thus the effects of polydispersity on the spectrum should be most important for rods. It should be emphasized that a plot of $1/\tau$ (obtained by fitting the time-correlation function to a single exponential) would, in general, not vary linearly with q^2 for a polydisperse solution [cf. Eq. (5.4.11)].

A full calculation of the translational diffusion contribution using (8.11.4) including structure factors for rods and coils has been reported by Tagami and Pecora (1969) and Reed (1972) for model $f(M)$. No calculations have as yet been reported of the effects on polydispersity on terms dependent on "intramolecular" motions (see Sections 8.7 and 8.8).

Moment Methods

Analyzing spectra via the distribution-function method discussed above is practicable only if one has reason to believe a model distribution function gives a good representation of the molecular weight or size-distribution function of a particular sample. In most cases, however, especially for solutions of biological molecules, the distribution of weights or sizes is unknown. It would be particularly interesting to obtain from the sample a well-defined "average" and to be able to estimate the dispersion about this average. These quantities, which are moments of the molecular size or weight distribution function, can in certain cases be measured by studying either the homodyne or heterodyne correlation functions at small times.

Consider the homodyne correlation function, which is proportional to $|I_1(\mathbf{q}, t)|^2$ for a dilute system with negligible intramolecular interference. Then for a polydisperse system

$$|I_1(\mathbf{q}, t)|^2 = B|\sum_j g_j \exp -q^2 D_j t|^2 \tag{8.11.5}$$

where

$$g_j = \alpha_i^2 N_i \tag{8.11.6}$$

B is defined following Eq. (8.4.2), and where α_i and N_i are, respectively, the polarizability and number of molecules with diffusion coefficients D_i.

As before, we define a polarizability per unit mass

$$\alpha_i' \equiv \frac{\alpha_i}{m_i} \tag{8.11.7}$$

and *assume* that α_i' is independent of mass. Then

$$g_i = \alpha'^2 N_i m_i^2 \tag{8.11.8}$$

Then substituting Eq. (8.11.8) into Eq. (8.11.5) we obtain

$$|I_1(\mathbf{q}, t)|^2 = B^2\alpha'^4|\sum_i N_i m_i^2 \exp -q^2 D_i t|^2 \tag{8.11.9}$$

The time derivative of Eq. (8.11.9) is

$$\frac{\partial|I_1(\mathbf{q}, t)|^2}{\partial t} = -2q^2\alpha'^4 B^2|\sum_i N_i m_i^2 D_i \exp -q^2 D_i t| \times |\sum_i N_i m_i^2 \exp -q^2 D_i t| \tag{8.11.10}$$

Form the quotient

$$L(t) = -\frac{1}{|I_1(\mathbf{q}, t)|^2}\frac{\partial(|I_1(\mathbf{q}, t)|^2)}{\partial t}$$

and expand the time exponent inside the integral sign and keep only terms in the result to order t. We obtain

$$L(t) = 2q^2\langle D\rangle_z - 2q^4\langle \delta D^2\rangle_z t + \cdots \tag{8.11.11}$$

where

$$\langle D\rangle_z \equiv \frac{\sum N_i m_i^2 D_i}{\sum N_i m_i} \tag{8.11.12}$$

is the z-average diffusion coefficient, and

$$\langle(\delta D)^2\rangle_z \equiv \langle D^2\rangle_z - \langle D\rangle_z^2 \tag{8.11.13}$$

is the variance in this z-average diffusion coefficient.

Thus by analyzing the homodyne correlation function at small times and computing $L(t)$, one may obtain both $\langle D\rangle_z$ and its variance $\langle(\delta D)^2\rangle_z$. The assumptions that enter into these equations should be recalled before applying them to any set of experimental data: (a) no interactions between scatterers (infinite dilution), (b) no intramolecular interference (small scatterers and/or small q), and (c) polarizability per unit mass of a scatterer independent of the molecular mass (cf. Section 8.5).

Another technique for extracting moments of the diffusion coefficient from heterodyne and homodyne experiments has been given by Koppel (1972). First normalize the correlation function

$$N(\mathbf{q}, t) \equiv \frac{I_1(\mathbf{q}, t)}{I_1(\mathbf{q},0)} \equiv \frac{\sum g_i \exp -q^2 D_i t}{\sum g_i} \equiv \langle \exp -q^2 D t\rangle \tag{8.11.14}$$

where g_i is given by Eq. (8.11.6).

Take the logarithm of both sides of Eq. (8.11.14) and expand the right-hand side in a power series in t. This gives

$$\ln N(\mathbf{q}, t) = 1 - K_1 t + \frac{1}{2} K_2 t^2 - \frac{1}{3!} K_3 t^3 + \frac{1}{4!} K_4 t^4 \ldots$$

where

$$K_n = \left[(-1)^n \frac{d^n}{dt^n} \ln N(\mathbf{q}, t)\right]_{t=0} \tag{8.11.15}$$

is the nth cumulant of $N(\mathbf{q}, t)$. The explicit forms of the first few cummulants are

$$K_1 = \langle q^2 D\rangle \equiv q^2 \sum \frac{N_i \alpha_i^2 D_i}{\sum N_i \alpha_i^2}$$

$$K_2 = \langle(q^2 D - \langle q^2 D\rangle)^2\rangle$$

$$K_3 = \langle(q^2 D - \langle q^2 D\rangle)^3\rangle$$

$$K_4 = \langle(q^2 D - \langle q^2 D\rangle)^4\rangle - 3K_2^2$$

If $\alpha_i \equiv \alpha_i' m_i$ and α_i' is the same for all molecules in solution then

$$K_1 = q^2\langle D\rangle_z$$

$$K_2 = q^4\langle(\delta D)^2\rangle_z$$

with similar expressions for the higher coefficients. K_2 is a measure of the variance of the distribution, while K_3 and K_4 are measures of the skewness (or asymmetry) and the kurtosis (peakedness, or flatness) of the distribution. For a Gaussian distribution all K_n except K_1 and K_2 are identically zero.

The logarithm of a normalized homodyne or heterodyne correlation function may be obtained from experiment. If the data are good at small times, it is a relatively simple matter to evaluate the first few derivatives in Eq. (8.11.15) and thereby determine the coefficients K_1 and K_2.

This method of analysis may also be extended to the interpretation of electrophoretic light scattering from polydisperse systems.

Paucidisperse Systems

Many systems of interest to the macromolecular chemist contain significant concentrations of only a few macromolecular species. In systems containing only two species it may be worthwhile to fit $I_1(\mathbf{q}, t)$ to two exponentials. This is especially so when the relative concentrations of the two species can be varied. Herbert and Carlson (1971) have, for instance, combined integrated intensity and spectral measurements to study the self-association of rabbit skeletal muscle myosin.

Another way to analyze paucidisperse systems by light scattering is to use external fields such as is done in the electrophoretic light-scattering technique described in Section (5.8).

8 · 12 LARGE PARTICLES

All of the theories described in this chapter are valid only in what is variously called the Born, Rayleigh–Gans or Rayleigh–Debye approximation. These theories assume that each segment of a scattering particle "sees" the same (or nearly the same) incident light wave. This approximation breaks down when the particles become very large (size of the order of the wavelength of light) and the particle interior is very different optically from the surrounding medium. For particles with a well-defined inside and outside (e.g., as opposed to a linear chain molecule bathed in solvent), a rough criterion for the validity of the Rayleigh–Debye approximation is

$$\frac{4\pi}{\lambda} R|m - 1| \ll 1 \tag{8.12.1}$$

where R is a characteristic dimension of a particle and m is the ratio of the refractive index inside the particle to that outside. Equation (8.12.1) implies that the phase of a wave traversing the scattering body is almost the same as it would be if the particle were not there.

There is extensive literature on $I_{if}(\mathbf{q})$ for particles for which the Rayleigh–Debye criterion breaks down. Complete solutions of the problem are available for spherical particles (Mie theory) of arbitrary size and refractive index and some solutions are available for cylindrical particles. The works by van de Hulst (1957) and Kerker (1969) discuss these cases in detail.

Biological cells are among the large systems to which light scattering has been applied (Wyatt, 1973). Most of these studies use measurements of the integrated light-scattering intensity as a function of scattering angle to measure sizes and size distributions of living cell dispersions. Other experiments measure changes in $I_{if}(\mathbf{q})$ when stimuli are applied to a cell (e.g., penicillin to staphylococcus strains). This type of study should be contrasted with those on microorganism motility described in Section (5.7).

There is a possibility that spectral distribution measurements may be used to study dynamic processes in *parts* of *cells*. This could be accomplished by using a microscope to focus light on a cell region and thus to observe time-dependent fluctuations solely in that local region (Fujime, 1972).

APPENDIX 8.A THE FOKKER-PLANCK EQUATION

In this appendix we outline[13] how the equation

$$\frac{d\mu}{dt} = \frac{1}{\zeta} F(t) - \frac{\mu}{\tau} \tag{8.A.1}$$

may be used with suitable statistical assumptions to obtain the Fokker–Plank equation for the probability distribution of the $\boldsymbol{\mu}$ (Section 8.8). In Eq. (8.A.1). we have for simplicity left off the mode index and treat only one component of $\boldsymbol{\mu}$. The generalization to a vector $\boldsymbol{\mu}$ is trivial.

We start with the assumption that the random process is Markoffian, that is, that the probability for μ to take on a certain value at a time $t + \delta t$ depends only on its value at t and not on the past history of the system. The Markoffian assumption is expressed by the Smoluchowski equation for the probability function $\psi(\mu(t), \mu(0), t)$

$$\psi(\mu(t), \mu(0), t) = \int d\mu(t')\, \psi(\mu(t'), \mu(0), t')\, \psi(\mu(t), \mu(t'), t - t') \tag{8.A.2}$$

Now consider the integral

$$\int d\mu(t)\, J\,(\mu(t))\, \frac{\partial\psi}{\partial t}\,(\mu(t), \mu(0), t)$$

where $J(\mu(t))$ is an arbitrary function which goes to zero for $\mu(t) \to \pm\, \infty$ sufficiently rapidly. From the definition of the time derivative and Eq. (8.A.2), this integral may be written as

$$\begin{aligned}
&\int d\mu(t) J(\mu(t))\, \frac{\partial\psi}{\partial t}\,(\mu(t), \mu(0), t) \\
&\quad = \lim_{\Delta t \to 0} \frac{1}{\Delta t} \int d\mu(t) J(\mu(t))[\psi(\mu(t), \mu(0), t + \Delta t) - \psi(\mu(t), \mu(0), t)] \\
&\quad = \lim_{\Delta t \to 0} \frac{1}{\Delta t} \left[\int d\mu(t) J(\mu(t)) \int d\mu(t')\, \psi(\mu(t'), \mu(0), t)\, \psi(\mu(t), (t'), \Delta t) \right. \\
&\qquad\qquad \left. - \int d\mu(t') J(\mu(t')\, \psi(\mu(t'), \mu(0), t) \right]
\end{aligned} \tag{8.A.3}$$

If the order of integration over $\mu(t)$ and $\mu(t')$ in Eq. (8.A.3) is interchanged and the function $J(\mu(t))$ is expanded in a Talyor series in $\mu(t) - \mu(t')$, we obtain, upon stopping at terms of order $(\mu(t) - \mu(t'))^2$,

$$\int d\mu(t)\, J(\mu(t)) \frac{\partial \psi}{\partial t} = \int d\mu(t')\, \psi(\mu(t'), \mu(0), t) \times [J'(\mu(t'))\ A(\mu(t')) + \tfrac{1}{2} J''(\mu('t), B(\mu(t')) + \ldots] \quad (8.A.4)$$

where A and B are related to the first and second moments of the distribution function ψ.

$$A(\mu(t')) = \lim_{\Delta t \to 0} \frac{1}{\Delta t} \int d\mu(t)\, (\mu(t) - \mu(t'))\, \psi(\mu(t), \mu(t'), \Delta t) \quad (8.A.5)$$

and

$$B(\mu(t')) = \lim_{\Delta t \to 0} \frac{1}{\Delta t} \int d\mu(t)\, (\mu(t) - \mu(t'))^2 \psi(\mu(t), \mu(t'), \Delta t) \quad (8.A.6)$$

Higher-order terms in Eq. (8.A.4) are usually assumed to vanish since the integrals corresponding to those in Eq. (8.A.5) and Eq. (8.A.6) are thought to be of $0(\Delta t^n)$ where $n > 1$ and hence vanish when divided by Δt and the $\Delta t \to 0$ limit taken. Upon integrating Eq. (8.A.4) partially and relabelling the variable of integration, we obtain

$$\int d\mu(t)\, J(\mu(t)) \left[\frac{\partial \psi}{\partial t} + \frac{\partial}{\partial \mu(t)} (A\psi) - \frac{1}{2} \frac{\partial^2}{\partial \mu(t)^2} (B\psi)\right] \quad (8.A.7)$$

Since Eq. (8.A.7) must hold for any function $J(\mu(t)$, the expression in the square brackets must be zero. Thus we obtain the Fokker–Planck equation

$$\frac{\partial \psi}{\partial t} = - \frac{\partial}{\partial \mu(t)} [A(\mu(t))\, \psi] + \frac{1}{2} \frac{\partial^2}{\partial(\mu(t))^2} [B(\mu(t))\psi] \quad (8.A.8)$$

The stochastic equation for μ [Eq. (8.A.1)] is used along with suitable statistical assumptions to evaluate the moments A and B. We assume that the random force in Eq. (8.A.1) is a Gaussian random process with the average values

$$\langle F(t) \rangle = 0$$
$$\langle F(t')\, F(t) \rangle = 2 \frac{\langle \mu^2 \rangle}{\tau} \zeta^2 \delta(t_1 - t_2) \quad (8.A.9)$$

From Eq. (8.A.1)

$$\mu(t) - \mu(t') = - \frac{\mu}{\tau} \Delta t - \frac{1}{\zeta} \int_{t'}^{t} F(t) dt \quad (8.A.10)$$

Substituting Eq. (8.A.10) into Eqs. (8.A.5) and (8.A.6) and using Eqs. (8.A.9) we obtain

$$A(\mu(t)) = - \frac{\mu}{\tau}$$

and (8.A.11)

$$B(\mu(t)) = 2\frac{\langle\mu^2\rangle}{\tau}$$

Finally upon substitution of Eqs. (8.A.11) into Eq. (8.A.8), the explicit Fokker–Planck equation corresponding to Eq. (8.A.1) is obtained

$$\frac{\partial\psi}{\partial t} = \frac{1}{\tau}\left[\langle\mu^2\rangle\frac{\partial^2}{\partial\mu^2}\psi + \frac{\partial}{\partial\mu}(\mu\psi)\right] \tag{8.A.12}$$

If Eq. (8.A.1) were a vector equation we would obtain

$$\frac{\partial\psi}{\partial t} = \frac{1}{\tau}\left[\frac{\langle\mu^2\rangle}{3}\nabla^2\psi + \nabla\cdot(\boldsymbol{\mu}\psi)\right]$$

which is the same as Eq. (8.8.15).

APPENDIX 8.B FORM FACTOR FOR THE OPTICALLY ANISOTROPIC RIGID ROD

In this appendix we derive an expression for I_{VV} and I_{HV} for a rigid, optically anisotropic rod of negligible thickness (cf. Section 8.9).

First note that in coordinate system II of Section 3.4 the scattering vector may be written

$$\mathbf{q} = \frac{2\pi}{\lambda}\{(\cos\Theta - 1)\,\hat{\boldsymbol{i}} + \sin\Theta\hat{\boldsymbol{j}}\} \tag{8.B.1}$$

where Θ is the scattering angle. If we let θ and ϕ be the spherical polar orientation angles of the rod (and hence of each segment), then

$$(\mathbf{r}_i - \mathbf{r}_j) = r_{ij}\,[\sin\theta\cos\varphi\hat{\boldsymbol{i}} + \sin\theta\sin\varphi\hat{\boldsymbol{j}} + \cos\theta\,\hat{\boldsymbol{k}}] \tag{8.B.2}$$

where $r_{ij} = |\mathbf{r}_i - \mathbf{r}_j|$. Thus

$$\mathbf{q}\cdot(\mathbf{r}_i - \mathbf{r}_j) = r_{ij}\frac{2\pi}{\lambda}[(\cos\Theta - 1)\sin\theta\cos\varphi + \sin\Theta\sin\theta\sin\varphi] \tag{8.B.3}$$

which after some trigonometric substitutions becomes

$$\begin{aligned} &= \frac{4\pi}{\lambda}r_{ij}\sin\frac{\Theta}{2}\cdot\sin\theta\sin\left(\varphi - \frac{\Theta}{2}\right) \\ &= 2qr_{ij}\sin\theta\sin\left(\varphi - \frac{\Theta}{2}\right) \end{aligned} \tag{8.B.4}$$

Substituting Eq. (8.B.4) into Eqs. (8.9.5) and (8.9.6) and using the result obtained in Section 8.2 for the scattering from optically isotropic rigid rods, we obtain

$$I_{VV}(\mathbf{q}) = N\left[\alpha_M - \frac{1}{3}\beta n\right]^2 S(\mathbf{q})$$
$$+ \sum_{i,j}^{n} \left\langle \exp\left[2i\, qr_{ij} \sin\frac{\Theta}{2} \sin\theta\, sin\left(\varphi - \frac{\Theta}{2}\right)\right] \Big\{N\beta\, [\beta\cos^4\theta$$
$$- \frac{2}{3}[\beta - 3\alpha] \cos^2\theta]\Big\}\right\rangle \tag{8.B.5}$$

$$I_{VH}(\mathbf{q}) = N\beta^2 \sum_{i,j}^{n} \left\langle \exp\left[2i\, qr_{ij} \sin\frac{\Theta}{2} \sin\theta \sin\left(\varphi - \frac{\Theta}{2}\right)\right] \sin^2\theta \cos^2\theta \sin^2\varphi \right\rangle \tag{8.B.6}$$

where $\alpha_M^2 \equiv n^2\alpha^2$ is the square of the *molecular* polarizability and $S(\mathbf{q})$ is given by Eq. (8.2.10).

The averages in Eqs. (8.B.5) and (8.B.6) are easily performed. Replace the brackets by an integral

$$\frac{1}{4\pi}\int_0^{\pi} d\theta \sin\theta \int_0^{2\pi} d\varphi \ldots . \tag{8.B.7}$$

The angular integrals appearing in I_{VV} are all of the form

$$\frac{1}{4\pi}\int_0^{\pi} d\theta \sin\theta \int_0^{2\pi} d\varphi \cos^n\theta \exp\left[iqr_{ij} \sin\theta \sin\left(\varphi - \frac{\Theta}{2}\right)\right] \tag{8.B.8}$$

The integral over φ may be done immediately by noting that

$$K_1(\theta) \equiv \int_0^{2\pi} d\varphi \exp c(\theta) \sin\left(\varphi - \frac{\Theta}{2}\right) = \int_{-\frac{\theta}{2}}^{2\pi-\frac{\theta}{2}} d\varphi' \exp\left[ic(\theta) \sin\varphi'\right] \tag{8.B.9}$$

where

$$c(\theta) \equiv qr_{ij} \sin\theta \tag{8.B.10}$$

Since 2π is a complete cycle for this integrand it does not matter where the cycle begins, so that

$$K_1(\theta) = \int_0^{2\pi} d\varphi' \exp ic(\theta) \sin\varphi' \tag{8.B.11}$$

$$= \int_0^{2\pi} d\varphi' \cos\left[c(\theta) \sin\varphi'\right] - i\int_0^{2\pi} d\varphi \sin\left[c(\theta) \sin\varphi'\right] \tag{8.B.12}$$

The second integral on the right-hand side of Eq. (8.B.12) vanishes by symmetry. The integral is simply

$$K_1(\theta) = 2\int_0^{\pi} d\varphi' \cos\left(c(\theta) \sin\varphi'\right) \tag{8.B.13}$$

which may be easily written in terms of the Bessel function of order 0, $J_0(c(\theta))$

$$K_1(\theta) = 2\pi J_0(c(\theta)) \tag{8.B.14}$$

The angular integrals appearing in I_{HV} are all of the form

$$\frac{1}{4\pi}\int_0^\pi d\theta \int_0^{2\pi} d\varphi \sin^3\theta \cos^2\theta \sin^2\varphi \exp\left[2i\, qr_{ij} \sin\frac{\Theta}{2} \sin\theta \sin\left(\varphi - \frac{\Theta}{2}\right)\right] \quad (8.B.15)$$

We first perform the integral over φ. Let

$$K_2(\theta) = \int_0^{2\pi} d\varphi \sin^2\varphi \exp\left[ic(\theta)\sin\left(\varphi - \frac{\Theta}{2}\right)\right] \quad (8.B.16)$$

where $c(\theta)$ is given by Eq. (8.B.10).

Change to a new variable φ' defined as

$$\varphi' = \varphi - \frac{\Theta}{2}$$

and use the trigonometric identity

$$\sin^2\varphi = \frac{1 - \cos 2\varphi}{2}$$

Then

$$K_2(\theta) = \frac{1}{2}\int_{-\frac{\theta}{2}}^{2\pi - \frac{\theta}{2}} d\varphi'[1 - \cos(\Theta + 2\varphi')] \exp[ic(\theta)\sin\varphi'] \quad (8.B.17)$$

It does not matter for the integration over φ' where on a cycle we start the integration, so that after expanding the $\cos(\theta + 2\varphi')$ we obtain

$$\begin{aligned} K_2(\theta) = \frac{1}{2}\Bigg\{&\int_0^{2\pi} d\varphi' \exp[ic(\theta)\sin\varphi'] \\ &- \cos\Theta \int_0^{2\pi} d\varphi' \cos 2\varphi' \exp[ic(\theta)\sin\varphi'] \\ &- \sin\Theta \int_0^{2\pi} d\varphi' \sin 2\varphi' \exp[ic(\theta)\sin\varphi']\Bigg\} \end{aligned} \quad (8.B.18)$$

It is easy to see from symmetry considerations that after expanding the exponential

$$\exp[ic(\theta)\sin\varphi'] = \cos(c(\theta)\sin\varphi') + i\sin(c(\theta)\sin\varphi') \quad (8.B.19)$$

some of the integrals over φ' from $0 \to 2\pi$ are zero and others are simply twice the same integral over the integration range from $0 \to \pi$.

Thus

$$K_2(\theta) = \int_0^\pi d\varphi' \cos[c(\theta)\sin\varphi'] - \cos\Theta \int_0^\pi d\varphi' \cos 2\varphi' \cos[c(\theta)\sin\varphi'] \quad (8.B.20)$$

But

$$\int_0^{\pi} d\varphi' \cos(c(\theta) \sin \varphi') = \pi J_0(c(\theta)) \tag{8.B.21}$$

and

$$\int_0^{\pi} d\varphi' \cos 2\, \varphi' \cos(c(\theta) \sin \varphi') = \pi J_2(c(\theta)) \tag{8.B.22}$$

where J_0 and J_2 are, respectively, the Bessel functions of order zero and two.
The final expression for $K_2(\theta)$ is therefore

$$K_2(\theta) = \pi[J_0(c(\theta)) - \cos \Theta\, J_2(c(\theta))] \tag{8.B.23}$$

Using $K_1(\theta)$ and $K_2(\theta)$, we must now perform the integration over θ. These integrals may be evaluated with the aid of Sonine's first finite-integral formula

$$J_{\mu+\nu+1}(qr_{ij} \sin \theta) = \frac{(qr_{ij})^{\nu+1}}{2^{\nu}\Gamma(\nu + 1)} \times \int_0^{\frac{\pi}{2}} J_{\mu}(z \sin \theta) \sin^{\mu+1} \theta \cos^{2\nu+1}\theta d\theta \tag{8.B.24}$$

In the integrals relevant to our problem, the integration over the range $0 \to \pi$ is simply twice the integral over the range $0 \to \pi/2$.
We find using Eqs. (8.B.14), (8.B.23), and (8.B.24) that

$$\begin{aligned}
\frac{1}{4\pi}\int_0^{\pi} d\theta \sin\theta \cos^4\theta K_1(\theta) &= \frac{3j_2(z)}{z^2} \\
\frac{1}{4\pi}\int_0^{\pi} d\theta \sin\theta \cos^2\theta K_1(\theta) &= \frac{j_1(z)}{z} \\
\frac{1}{4\pi}\int_0^{\pi} d\theta \sin^3\theta\cos^2\theta\, K_2(\theta) &= \frac{1}{2}\left[\frac{j_1(z)}{z} - \frac{2j_2(z)}{z^2} - \cos\Theta \frac{j_3(z)}{z}\right]
\end{aligned} \tag{8.B.25}$$

where

$$z \equiv qr_{ij}$$

and j_l is the spherical Bessel function of order l.
Substituting Eqs. (8.B.11), (8.B.16), and (8.B.25) into Eqs. (8.B.5) and (8.B.6) we obtain

$$I_{VV}(\mathbf{q}) = N[\alpha_M - \frac{1}{3}\beta_M]^2 S(\mathbf{q}) + \sum_{i,j}^{n} \left\{N\beta^2 \frac{3j_2(z)}{z^2} - \frac{2}{3} N\beta(\beta - 3\alpha) \frac{j_1(z)}{z}\right\} \tag{8.B.26}$$

and

$$I_{VH}(\mathbf{q}) = N\frac{\beta^2}{2}\sum_{i,j}^{n}\left\{\frac{j_1(z)}{z} - \frac{3j_2(z)}{z^2} - \cos\Theta \frac{j_3(z)}{z}\right\} \tag{8.B.27}$$

where we set

$$\beta_M \equiv n\beta_0$$

The calculation may be completed by substituting the explicit forms of the spherical Bessel functions into Eqs. (8.B.26) and (8.B.27) and then performing the summation. Since for the rigid rod model there is a constant anisotropy per unit length, the summation may, as in the isotropic case in Section (8.2), be replaced by an integration, that is,

$$\sum_{i,l}^{n} f(r_{ij}) \to \frac{2n^2}{L^2} \int_0^L dr(L - r) f(r) \tag{8.B.28}$$

First we define a dimensionless variable of integration

$$z \equiv qr \tag{8.B.29}$$

and let $x \equiv qL$. Then

$$\frac{2n^2}{L^2} \int_0^L dr\,(L - r) f(r) = 2n^2 \left\{ \frac{1}{x} \int_0^x dz f(z) - \frac{1}{x^2} \int_0^x dz f(z) \right\} \tag{8.B.30}$$

Substituting Eqs. (8.B.28) and (8.B.30) into Eqs. (8.B.26) and (8.B.27), using the explicit forms for the spherical Bessel functions [cf. Eqs. (8.7.19)] and performing the integrations we obtain

$$\begin{aligned} I_{VV}(\mathbf{q}) = {} & N[\alpha_M - \frac{1}{3} \beta_M]^2 S(x) \\ & + N[\beta_M^2]^2 \frac{3}{2} X(x) - \frac{2}{3} N\beta_M^2[\beta_M - 3\alpha_M]\, Y(x) \end{aligned} \tag{8.B.31}$$

and

$$I_{VH}(\mathbf{q}) = N\beta_M^2 \left\{ \frac{1}{2} Y(x) - \frac{3}{4} X(x) - \frac{1}{4} \cos\Theta\, Z(x) \right\} \tag{8.B.32}$$

where

$$X(x) \equiv \left[\frac{\sin x}{x^5} - \frac{\cos x}{x^5} + \frac{1}{2}\frac{\sin x}{x^3} + \frac{\cos x}{2x^2} - \frac{1}{3x^2} + \frac{1}{2x} \int_0^x \frac{\sin z}{z}\, dz \right] \tag{8.B.33}$$

$$Y(x) \equiv \left[\frac{\sin x}{x^3} - \frac{2}{x^2} + \frac{\cos x}{x^2} + \frac{1}{x} \int_0^x \frac{\sin z}{z}\, dt \right. \tag{8.B.34}$$

and

$$Z(x) = \left[\frac{5 \sin x}{x^5} - \frac{5 \cos x}{x^4} + \frac{\sin x}{2x^3} + \frac{\cos x}{2x^2} - \frac{8}{3x^2} + \frac{1}{2x} \int_0^x \frac{\sin z}{z}\, dz \right] \tag{8.B.35}$$

It should be emphasized that $I_{VH}(\mathbf{q})$ depends explicitly on $\cos \Theta$, so that universal curves of I_{VH} versus $x = qL$ must be drawn for each value of θ. Alternatively, $I_{VH}/N\beta_M^2$ may be plotted against $\sin \Theta/2$ for each value of L/λ. Note that when $L/\lambda \sim 1$, I_{VH} depends strongly on $\sin \Theta/2$. All curves, of course, approach 1/15 as $\Theta \to 0$, as dictated by Eqs. (8.B.2) and (8.B.4).

An interesting feature of Eq. (8.B.31) is the term dependent on $\beta_M \alpha_M$. Measurements of $I_{VV}(\mathbf{q})$ (or perhaps some more convenient function such as I_{HH}) for long rods should yield values of the *sign* of the optical anisotropy.

Van Aartsen (1970) has generalized this theory to include cylindrically symmetric particles of any thickness. His calculation includes the Horn (1956) results discussed above, the scattering from infinitely thin discs, and the scattering from optically isotropic rods with comparable length and thickness as special cases.

NOTES

1. Chapter 3, Note 10.
2. In many cases this assumption is not appropriate (e.g., see Holtzer and Rice, 1957).
3. We are here referring to systems in which there is no long range orientational order (as in liquid crystals).
4. First note that $j_0^2\left(\frac{xy}{2}\right) = \frac{\sin^2 \frac{xy}{2}}{\left(\frac{xy}{2}\right)^2} = \frac{1}{2}\frac{(1 - \cos 2z)}{z^2}$

   where $z \equiv xy/2$. Transforming the variable of integration from y to z then gives

   $$S(\mathbf{q}) = \frac{1}{2x}\int_{-x/2}^{x/2} dz\,\frac{(1 - \cos 2z)}{z^2} = -\frac{1}{x} - \int_{-x\,2}^{+x/2} dz\,\frac{\cos 2z}{z^2}$$

   where we have evaluated the integral of $1/z^2$. The last integral is now integrated by parts and the trigonometric identity $2 \sin^2 \theta = (1 - \cos 2\theta)$ is used again. This gives Eq. (8.2.13).
5. By completing the squares in the exponential this integral may be evaluated. It is useful to remember that the Fourier transform of a Gaussian is a Gaussian.
6. Since we are interested only in the difference between $\mathbf{r}_i$ and $\mathbf{r}_j$ we may subtract the center-of-mass vector $\mathbf{R}$ from each, that is, let $\mathbf{P}_i = \mathbf{r}_i - \mathbf{R}$. Then

   $$\frac{1}{n^2}\langle\sum_{i.j}|\mathbf{r}_i - \mathbf{r}_j|^2\rangle = \frac{1}{n^2}\langle\sum_{i.j}(\mathbf{P}_i - \mathbf{P}_j)^2\rangle$$
   $$= \frac{1}{n^2}\langle\sum_{i.j}(\mathbf{P}_i^2 + \mathbf{P}_j^2 - 2\mathbf{P}_i\cdot\mathbf{P}_j)\rangle$$
   $$= 2\langle\sum_i \frac{\mathbf{P}_i^2}{n}\rangle - 2\langle\left(\sum_i \frac{\mathbf{P}_i}{n}\right)^2\rangle$$

   Upon substituting the relation between $\mathbf{P}_i$ and $\mathbf{r}_i$, we obtain Eq. (8.3.7).
7. For example Eq. (8.3.3) gives R_G in terms of the length, L, of the rod.
8. These measurements may, of course, be done in the opposite order. At fixed q^2, one may measure the ordinate for a range of c', plot the resulting values, and then repeat the procedure for a range of q^2.
9. See, for instance, Yamakawa (1971) and references cited therein.
10. It should be emphasized that the forces arising from hydrodynamic interactions are neglected in this treatment. Thus only the "free draining" or "Rouse" model is treated here. Zimm (1956) has extended this model to include hydrodynamic interactions.
11. The function ψ is the same as that calculated by Uhlenbeck and Ornstein (1930) for Brownian motion of a harmonically bound particle.
12. This is done explicitly by Pecora (1968).
13. A more complete discussion is given by Wang and Uhlenbeck (1945).

REFERENCES

Cummins, H. Z., Carlson, F. D., Herbert, T. J., and Woods, G., *Biophys. J.* **9**, 518 (1969).
Debye, P., *J. Phys. Coll. Chem.* **51,** 18 (1947).
de Gennes, P.-G., *Phys.* **3,** 37 (1967).

DuBois-Violete, E. and de Gennes, P.–G., *Phys.* **3,** 181 (1967).

Flory, P. J., *Statistical Mechanics of Chain Molecules* Wiley-Interscience, New York (1969).

Fujime, S., *J. Phys. Soc. Japan* **29,** 416 (1970).

Fujime, S., *Adv. Biophys.* **3,** 1 (1972) and references cited therein.

Fujime, S., *Rev. Sci. Instrum.* **43,** 566 (1972).

Harris, R. A. and Hearst, J. E., *J. Chem. Phys.* **44,** 2595 (1966).

Herbert, T. J. and Carlson, F. D., *Biopolym.* **10,** 2231 (1971).

Holtzer, A. and Rice, S. A., *J. Am. Chem. Soc.* **79,** 4847 (1957).

Horn, P., *Ann. de Phys.* **10,** 386 (1956).

Huang, W.–N. and Frederick, J. E. *Macromolec.* **7,** 34 (1974).

Kerker, M., *The Scattering of Light and Other Electronagnetic Radiation,* Academic New York (1969).

Koppel, D. E, *J. Chem. Phys.* **57,** 4814 (1972).

Kramer, O. and Frederick, J. E., *Macromolec.* **5,** 69 (1972).

Maeda, H. and Saito, N., *J. Phys. Soc. Japan* **27,** 4, 984 (1969).

Maeda, H., and Saito, N., *Polym. J.* **4,** 309 (1973).

McIntyre, D. and Gornick, F., *Light Scattering from Dilute Polymer Solutions,* Gordon and Breach, New York (1964).

Ono, K. and Okano, K., *Jap. J. Appl. Phys.* **9,** 1356 (1971).

Patterson, G. D. and Flory, P. J., *J. C. S. Faraday II* **68,** 1098, 1111 (1972).

Pecora, R., *J. Chem. Phys.* **40,** 1604 (1964).

Pecora, R., *J. Chem. Phys.* **48,** 4126 (1968).

Pecora, R., *J. Chem. Phys.* **49,** 1032 (1968).

Pecora, R. and Tagami, Y., *J. Chem. Phys.* **51,** 3298 (1969).

Pecora, R. and Aragon, *S. Chem. Phys.* Lipids, **13,** 1 (1974).

Reed, T. F., *Macromolec.* **5,** 771 (1972).

Reed, T. F. and Frederick, J. E., *Macromolec.* **4,** 72 (1971).

Schaefer, E. W., Benedek, G. B., Schofield, P., and Bradford, E., *J. Chem. Phys.* **55,** 3884 (1971).

Schmitz, K. S. and Schurr, J. M., *Biopolym.* **12,** 1543 (1973).

Silbey, R. and Deutch, J. M., *J. Chem. Phys.* **57,** 5010 (1973).

Tagami, Y. and Pecora, R., *J. Chem. Phys.* **51,** 3293 (1969).

Tanford, C., *Physical Chemistry of Macromolecules,* Wiley, New York (1961).

Uhlenbeck, G. E. and Ornstein, L. S., *Phys. Rev.* **36,** 823 (1930).

van Aartsen, J. J., *Eur. Polym. J.* **6,** 1095 (1970).

van de Hulst, H. C., *Light Scattering by Small Particles,* Wiley, New York, (1957).

Wada, A., Ford, N. C., Jr., and Karasz, F. E., *J. Chem. Phys.* **55,** 1798 (1971).

Wang, M. C. and Uhlenbeck, G. E. *Rev. Mod. Phys.* **17,** 323 (1945).

Wyatt, P. J., in *Methods of Microbiology,* Vol 8, Academic Press, (1973).

Yamakawa, H., *Modern Theory of Polymer Solutions,* Harper and Row, New York (1971).

Zimm, B. H. *J. Chem. Phys.* **14,** 164 (1946).

Zimm, B. H. *J. Chem. Phys.* **16,** 1093 (1948).

Zimm, B. H. *J. Chem. Phys.* **16,** 1099 (1948).

Zimm, B. H., Stein, R. S., and Doty, P., *Poly. Bull.* **1,** 90 (1945).

Zimm, B. H. *J. Chem. Phys.* **24,** 269 (1956).

CHAPTER 9

ELECTROLYTE SOLUTIONS

9 · 1 INTRODUCTION

The diffusion constants and the conductivities of ionic species in electrolyte solutions depend in general on the charges of the ions and the ionic concentrations. These strong effects arise from the long-range nature of the Coulomb forces, which give rise to long-range spatial correlations between the ions. It is clear that these effects might give rise to light-scattering spectra which differ considerably from the predictions of previous chapters where particle independence was always assumed. As we shall see, the integrated intensity, the measured diffusion coefficient, and the electrophoretic mobility will depend on the scattering angle, that is, will be q-dependent.

Our chief emphasis will be on electrolyte solutions in which one of the ionic species is a polyelectrolyte and the remaining species are small ions. The spectrum will then be proportional to the time-correlation function of the polyelectrolyte concentration fluctuation alone.

It is interesting to note that because of concentration fluctuations, the normal spherically symmetric configuration of ions around a "central" ion can become asymmetric, so that the central ion experiences a nonzero electrical force. This instantaneous force accelerates the central ion and thereby contributes to its diffusion coefficient and mobility. In a polyelectrolyte solution the relatively fast small counterions can migrate to one side or another of the macroion, so that they tend to "drag" the macroion with them and thereby give rise to larger macroion diffusion coefficients than would be expected on the basis of Stokes law. Concommitantly, the macroions tend to retard the motions of the small ions, thereby giving rise to reduced diffusion coefficients for them. These effects are expected to be larger in dilute solutions because of the large relative fluctuations there. Thus we expect strong dependence on the ionic strength of the solution.

In this chapter we present a simple model calculation that demonstrates how this cooperative motion affects the scattering spectrum. Our approach is based on the Debye–Onsager treatment of ion transport (see Falkenhagen, 1934; Stephen, 1971). This is our first discussion of cooperative effects in light scattering. In Chapter 13 this problem is reconsidered in the context of the general theory of nonequilibrium thermodynamics.

9 · 2 THE DIFFUSION EQUATION OF A STRONG ELECTROLYTE

First we consider a solution of a salt A_mB_n that is completely dissociated into ions A and B of charge z_1 and z_2, respectively[1], according to the chemical equation

$$A_mB_n \rightarrow mA + nB$$

The solution must be electrically neutral so that

$$mz_1 + nz_2 = 0$$

$$z_1c_1^\circ + z_2c_2^\circ = 0 \tag{9.2.1}$$

where c_1° and c_2° are the number densities (number per cc) of ions A and B respectively.[2] Since no chemical reactions are occurring, the number of ions of each kind is conserved so that the continuity equation applies to both ionic species (see Sections 5.8 and (10.3).

$$\frac{\partial c_i(\mathbf{r}, t)}{\partial t} + \nabla \cdot J_i(\mathbf{r}, t) = 0 \qquad i = 1, 2 \tag{9.2.2}$$

Here $\mathbf{J}_i(\mathbf{r}, t)$ is the flux of ions of type i. This flux consists of two parts:

$$\mathbf{J}_i(\mathbf{r}, t) = -D_i\nabla c_i(\mathbf{r}, t) + c_i(\mathbf{r}, t)\,\mathbf{V}_i(\mathbf{r}, t) \tag{9.2.3}$$

The first term on the right is the usual Fick law or diffusive flux and D_i is the "hydrodynamic" diffusion coefficient of the ion

$$D_i = \frac{k_BT}{\zeta_i} \tag{9.2.4}$$

where ζ_i. is the friction constant (for large spherical ions of radius $a_i\zeta_i = 6\pi\eta a_i$). The second term is simply the "convective" flux of species i. In an equilibrium solution, there is no average diffusive or convective flux. However, because of local concentration fluctuations, ions experience local electric fields and therefore experience electrostatic forces that accelerate them. For example, consider a typical ion of species i at position $\mathbf{r}$ of the fluid. The Langevin equation (cf. Section 5.9) for this ion is

$$m_i\frac{d\mathbf{V}}{dt} = -\zeta_i\mathbf{V}_i + z_i\mathbf{E} + \mathbf{F}(t) \tag{9.2.5}$$

where m_i is the mass of the ion, $\mathbf{E}$ is the electric field acting on the ion, and $\mathbf{F}(t)$ is the random force. The solution of this equation for the average velocity $\langle\mathbf{V}\rangle_{V_0}$ of the ion, given that its initial velocity is $\mathbf{V}_0$ is

$$\langle\mathbf{V}(t)\rangle_{V_0} = \mathbf{V}_0 \exp\frac{-\zeta_i t}{m_i} + \frac{z_i E}{\zeta_i}\left[1 - \exp\frac{-\zeta_i t}{m_i}\right]$$

This decays to the terminal velocity $z_i\mathbf{E}/\zeta_i$ in the very short time (m_i/ζ_i). Thus for times long compared to (m_i/ζ_i) the ion should move at its terminal velocity $(z_i\mathbf{E}/\zeta_i)$ in the electric field $\mathbf{E}$ where this field is due only to concentration fluctuations. Thus we may take

$$\mathbf{V}_i(\mathbf{r}, t) = \frac{z_i\mathbf{E}(\mathbf{r}, t)}{\zeta_i} = \beta D_i z_i\mathbf{E}(\mathbf{r}, t) \tag{9.2.6}$$

where $\beta = (k_BT)^{-1}$ and in the last equality we have substituted Eq. (9.2.4). Combining Eqs. (9.2.2), (9.2.3), and (9.2.6) gives

$$\frac{\partial c_i(\mathbf{r}, t)}{\partial t} = D_i\{\nabla^2 c_i(\mathbf{r}, t) - \beta z_i \nabla \cdot [c_i(\mathbf{r}, t)\,\mathbf{E}(\mathbf{r}, t)]\}\ ;\ i = 1, 2 \tag{9.2.7}$$

These two equations can be combined with the Poisson equation

$$\nabla \cdot \mathbf{E}(\mathbf{r}, t) = \frac{4\pi}{\varepsilon_0}\left[\sum_{i=1}^{2} z_i c_i(\mathbf{r}, t)\right] \tag{9.2.8}$$

The quantity in the square bracket is the local charge density, ε_0 is the dielectric constant of the solution, and the Poisson equation relates the local electric field to the local charge density.

In an equilibrium solution the local concentrations can be expressed as

$$c_i(\mathbf{r}, t) = c_i^0 + \delta c_i(\mathbf{r}, t) \tag{9.2.9}$$

where $\delta c_i(\mathbf{r}, t)$ is the local concentration fluctuation. $\delta c_i(\mathbf{r}, t)$ is expected to be relatively small.

Equations (9.2.7) and (9.2.8) simplify considerably when Eq. (9.2.9) is substituted and only first-order terms in $\delta c_i(\mathbf{r}, t)$ are retained. First we note that Eq. (9.2.8) becomes

$$\nabla \cdot \mathbf{E} = \frac{4\pi}{\varepsilon_0}\left[\sum_i z_i c_i^0 + \sum_i z_i \delta c_i(\mathbf{r}, t)\right] \tag{9.2.10}$$

From electroneutrality $\sum_i c_i^0 z_i = 0$ so that

$$\nabla \cdot \mathbf{E} = \frac{4\pi}{\varepsilon_0}\left[\sum_i z_i \delta c_i(\mathbf{r}, t)\right] \tag{9.2.11}$$

Equation (9.2.11) is important in that it shows that the local electric field that arises from the concentration fluctuations is first order in the concentration fluctuations. Thus the term containing $c_i(\mathbf{r}, t)\mathbf{E}(\mathbf{r}, t)$ in Eq. (9.2.7) is; to first order in the concentration fluctuations, $c_i^0\mathbf{E}(\mathbf{r}, t)$. Substitution of Eq. (9.2.9) into Eqs. (9.2.7) and (9.2.8.) leads to the linearized equations

$$\frac{\partial \delta c_i(\mathbf{r}, t)}{\partial t} = D_i[\nabla^2 \delta c_i(\mathbf{r}, t) - \beta z_i c_i^0 \nabla \cdot \mathbf{E}];\ i = 1, 2 \tag{9.2.12}$$

$$\nabla \cdot \mathbf{E} = \frac{4\pi}{\varepsilon_0} \sum_i z_i \delta c_i(\mathbf{r}, t) \tag{9.2.13}$$

Substitution of Eq. (9.2.13) into Eq. (9.2.12), followed by a spatial Fourier transformation of the resulting equation, gives the two coupled diffusion equations

$$\frac{\partial \delta c_1(\mathbf{q}, t)}{\partial t} = -(q^2 + q_1^2)\,D_1 \delta c_1(\mathbf{q}, t) + \left|\frac{z_2}{z_1}\right| q_1^2 D_1 \delta c_2(\mathbf{q}, t) \tag{9.2.14}$$

$$\frac{\partial \delta c_2(\mathbf{q}, t)}{\partial t} = -(q^2 + q_2^2)\,D_2 \delta c_2(\mathbf{q}, t) + \left|\frac{z_1}{z_2}\right| q_2^2 D_2 \delta c_1(\mathbf{q}, t) \tag{9.2.15}$$

where in these equations[3] and in the following

$$q_1^2 \equiv \left(\frac{4\pi}{\varepsilon_0}\right) \beta z_1^2 c_1{}^0; \qquad q_2^2 \equiv \left(\frac{4\pi}{\varepsilon_0}\right) \beta z_2^2 c_2{}^0; \qquad q_0^2 \equiv q_1^2 + q_2^2 \tag{9.2.16}$$

The quantities q_1, q_2 and q_0 have the dimensions cm^{-1}. They are inverse "screening" lengths. Later we shall see that q_0 also plays a role in the equilibrium structure of the ionic solution. Equation (9.2.15) can be solved by means of Laplace transformation with respect to time. When the solutions are multiplied by $\delta c_1^*(\mathbf{q}, 0)$ and $\delta c_2^*(\mathbf{q}, 0)$ and ensemble-averaged we obtain

$$\tilde{\mathbf{F}}_{11}(\mathbf{q}, s) = \frac{1}{\Delta(s)} \left\{[s + (q^2 + q_2^2) D_2] S_{11}(\mathbf{q}) + \left|\frac{z_2}{z_1}\right| q_1^2 D_1 S_{12}(\mathbf{q})\right\}$$

$$\tilde{\mathbf{F}}_{12}(\mathbf{q}, s) = \frac{1}{\Delta(s)} \left\{[s + (q^2 + q_2^2) D_2] S_{21}(\mathbf{q}) + \left|\frac{z_2}{z_1}\right| q_1^2 D_1 S_{22}(\mathbf{q})\right\}$$

$$\tilde{\mathbf{F}}_{12}(\mathbf{q}, s) = \frac{1}{\Delta(s)} \left\{[s + (q^2 + q_1^2) D_1] S_{12}(\mathbf{q}) + \left|\frac{z_1}{z_2}\right| q_2^2 D_2 S_{11}(\mathbf{q})\right\}$$

$$\tilde{\mathbf{F}}_{22}(\mathbf{q}, s) = \frac{1}{\Delta(s)} \left\{[s + (q^2 + q_1^2) D_1] S_{22}(\mathbf{q}) + \left|\frac{z_1}{z_2}\right| q_2^2 D_2 S_{21}(\mathbf{q})\right\} \tag{9.2.17}$$

where

$$\Delta(s) = [s + (q^2 + q_1^2) D_1]\,[s + (q^2 + q_2^2) D_2] - q_1^2 D_1 q_2^2 D_2 \tag{9.2.18}$$

and where $\tilde{\mathbf{F}}_{ij}(\mathbf{q}, s)$ is the Laplace transform of $F_{ij}(\mathbf{q}, t)$ where

$$F_{ij}(\mathbf{q}, t) \equiv \langle \delta c_i{}^*(\mathbf{q}, 0) \delta c_j(\mathbf{q}, t) \rangle \tag{9.2.19}$$

The quantities $S_{ji}(\mathbf{q})$ are simply,

$$S_{ji}(\mathbf{q}) = F_{ji}(\mathbf{q}, 0) = \langle \delta c_i{}^*(\mathbf{q})\, \delta c_j(\mathbf{q}) \rangle, \tag{9.2.20}$$

the equilibrium structure factors considered in the next section. By symmetry $S_{ij}(\mathbf{q}) = S_{ji}(\mathbf{q})$.

As before, we proceed by finding the roots of the dispersion equation

$$\Delta(s) = 0 \tag{9.2.21a}$$

It is interesting to note in passing that the dispersion equation can be written as

$$1 = - \sum_i \frac{q_i{}^2 D_i}{s + q^2 D_i} \tag{9.2.21b}$$

where the sum goes over all ionic species in the solution (1 and 2 in this case). Equation (9.2.21b) is also valid for solutions containing more than two species.

Before solving Eq. (9.2.21a) we note that for a 10^{-3} molar (1–1) aqueous electrolyte solution ($\varepsilon_0 = 80$), q_1 and q_2 are approximately $3 \times 10^{+8}\text{cm}^{-1}$, and for typical values of the diffusion coefficients D_1 and D_2 ($\sim 10^{-5}\text{cm}^2/\text{sec}$), $q_1^2 D_1$ and $q_2^2 D_2$ are of order 10^{12}sec^{-1}. Thus in light scattering where $q \sim 10^5\text{cm}^{-1}$.

$$q \ll q_1, q_2 \text{ and } q^2D_1, q^2D_2 \ll [(q^2 + q_1^2)D_1 + (q^2 + q_2^2)D_2]$$

Thus we should consider solutions of the dispersion equation[4]

$$\Delta(s) = s^2 + [(q^2 + q_1^2)D_1 + (q^2 + q_2^2)D_2]s + q^2(q^2 + q_0^2)D_1D_2 = 0 \qquad (9.2.21c)$$

to lowest order in q^2D_1 or q^2D_2 (compared with $[(q^2 + q_1^2)D_1 + (q^2 + q_2^2)D_2]$). We follow the procedure outlined in Section 6.3. The zeroth-order equation is

$$s^{(0)2} + [(q^2 + q_1^2)D_1 + (q^2 + q_2^2)D_2]s^{(0)} = 0 \qquad (9.2.22)$$

which has the roots

$$s_{\pm}^{(0)} = \begin{cases} 0 & \text{(slow)} \\ -[(q^2 + q_1^2)D_1 + (q^2 + q_2^2)D_2] & \text{(fast)} \end{cases} \qquad (9.2.23)$$

The first-order equation is

$$2s^{(1)}\, s^{(0)} + [q^2 + q_1^2)D_1 + (q^2 + q_2^2)D_2]\, s^{(1)} + q^2(q^2 + q_0^2)D_1D_2 = 0 \qquad (9.2.24)$$

which has the roots

$$s_{\pm}^{(1)} = \begin{cases} -q^2D_s(q) \\ +\, q^2D_s(q) \end{cases} \qquad (9.2.25)$$

where

$$D_s(q) \equiv \frac{(q^2 + q_0^2)D_1D_2}{[(q^2 + q_1^2)D_1 + (q^2 + q_2^2)D_2)} \qquad (9.2.26)$$

Thus to first order in q^2D_1 or q^2D_2, the roots are

$$s_{\pm} = \begin{cases} s_+^{(0)} + s_+^{(1)} = -\, q^2D_s(q) & \text{(slow)} \quad (9.2.27) \\ s_-^{(0)} + s_-^{(1)} = -\, [(q^2 + q_1^2)D_1 + (q^2 + q_2^2)D_2] + q^2D_s(q) & \text{(fast)} \quad (9.2.28) \end{cases}$$

and the denominator in Eq. (9.2.18) is given by

$$\Delta(s) = (s - s_+)\,(s - s_-) \qquad (9.2.29)$$

where $s_{\pm}$ is given by Eq. (9.2.28). Laplace inversion of Eqs. (9.2.17) then gives $F_{ij}(\mathbf{q}, t)$ as linear combinations of two exponential factors

$$\exp -q^2D_s(q)|t| \text{ and } \exp -[(q^2 + q_1^2)D_1 + (q^2 + q_2^2)D_2 - q^2D_s(q)]|t| \qquad (9.2.30)$$

The first exponential decays very slowly whereas the second exponential decays very rapidly. Thus the spectral densities of the correlation functions $F_{ij}(\mathbf{q}, t)$ will consist of a superposition of two Lorentzians: one relatively narrow of width $q^2D_s(q)$ and one very broad of width $[(q_1{}^2D_1 + q_2{}^2D_2) - q^2D_s(q)]$. Since $q^2D_s(q)$, q^2D_1, and q^2D_2 are usually small compared with $[q_1{}^2D_1 + q_2{}^2D_2]$, the broad line will effectively have an angle-independent width determined by τ_r, where

$$\tau_r \equiv \frac{1}{q_1^2 D_1 + q_2^2 D_2} \tag{9.2.31a}$$

is the relaxation time for the "ionic atmosphere." τ_r characterizes the time it takes for a nonequilibrium distribution of ions around a given ion to relax to equilibrium.

Three points are worth noting here.

1. The diffusion coefficient measured in light scattering is angle or q-dependent

$$D_s(q) = (q^2 + q_0^2) D_1 D_2 / [(q^2 + q_1^2) D_1 + (q^2 + q_2^2) D_2]$$

2. In the limit of q small compared to q_1, q_2 the q-dependence disappears and

$$D_s(q) \to D_s \equiv q_0^2 D_1 D_2 / [q_1^2 D_1 + q_2^2 D_2] \tag{9.2.31b}$$

Classical measurements of diffusion confirm this result.

3. In the limit of q small compared to q_1 and q_2, the broad line has an angle independent width that gives the relaxation time of the ionic atmosphere

$$\tau_r = 1/[q_1^2 D_1 + q_2^2 D_2]$$

According to point 2, the apparent diffusion coefficient of the concentration fluctuation of the macroion is related to the Stokes law diffusion coefficients D_1, D_2. Since $q_1^2/q_2^2 = |z_1/z_2|$, Eq. (9.2.31b) simplifies to

$$D_s = \frac{(1 + z) D_1 D_2}{(D_2 + z D_1)} \tag{9.2.32}$$

for a macroion of charge $z = |z_1|$ in a solution of counterions of charge $|z_2| = 1$. Thus we expect a strong charge-dependence of D_s. In particular if $D_1 \ll D_2$, then $D_1 < D_s < D_2$ and the apparent diffusion coefficient D_s of the large ion should therefore be increasingly larger than its hydrodynamic value as z increases. This has been corroborated by previous measurements of the diffusion constant.

In Eq. (6.4.7) we saw that a fast chemical reaction would give rise to a broad angle-independent background. Angle-independence might then be surmised to imply the presence of chemical reactions. Equation (9.2.30), however, shows that the relaxation of the ionic atmosphere also leads to an angle-independent width.

The terms that survive at long times are

$$\begin{aligned} F_{11}(\mathbf{q}, t) &\simeq S_{11}(\mathbf{q}) \exp -q^2 D_s(q)|t| \\ F_{12}(\mathbf{q}, t) &\simeq S_{21}(\mathbf{q}) \exp -q^2 D_s(q)|t| \\ F_{12}(\mathbf{q}, t) &\simeq S_{12}(\mathbf{q}) \exp -q^2 D_s(q)|t| \\ F_{22}(\mathbf{q}, t) &\simeq S_{22}(\mathbf{q}) \exp -q^2 D_s(q)|t| \end{aligned} \tag{9.2.33}$$

so that the time-correlation function of the dielectric fluctuations reduces at long times to[5]

$$F(\mathbf{q}, t) \cong \sum_{ij} a_i a_j F_{ij}(\mathbf{q}, t) \tag{9.2.34}$$

$$= S(\mathbf{q}) \exp -q^2 D_s(q)|t| \tag{9.2.35}$$

where the integrated intensity $S(\mathbf{q})$ is given by

$$S(\mathbf{q}) \equiv \sum_{ij} a_i a_j S_{ij}(\mathbf{q}) \tag{9.2.36}$$

The "long-time decay" of the dielectric fluctuations is characterized by an angle-dependent diffusion coefficient that decreases linearly with q^2 for values of q small compared with q_1 and q_2.

It suffices to mention recent homodyne measurements of the diffusion coefficient of R 17 virus solutions at high pH or equivalently large macroion charge, and no added salt, in which the diffusion coefficient is a rather strong function of q (see Schaefer and Berne, 1974; and Pusey, et al., 1972). In Fig. 9.2.1, $10^7 D(q)$ is plotted against $(10^{-10} q^2)$, for virus particles titrated to a charge $z = 3600$ (indicated by △△△), to a charge $z = 0$ (indicated by ○○○) and to a charge $z = 500$ (indicated by □□□). It should be noted that in the uncharged macromolecular solution, D does not vary with angle as expected (Section 5.4), whereas for large charge there is a significant q-dependence. Unfortunately Eq. (9.2.31b) is not completely consistent with these data. Nevertheless, this preliminary work should encourage future activity in this area.

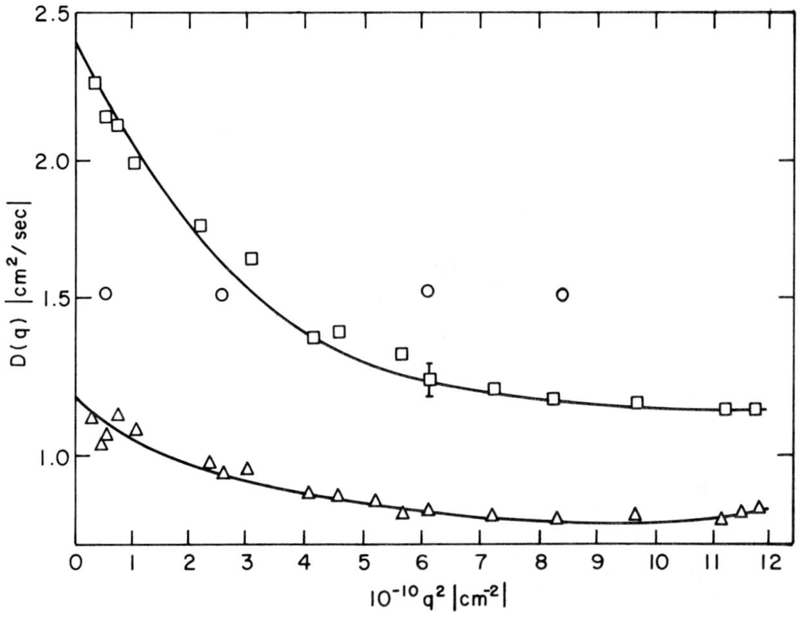

FIG. 9.2.1. The dependence of the diffusion coefficient in solutions of R-17 virus at different pH with no added salt. R-17 of charges $z = 0$, 500, and 3600 are indicated respectively by (○○○), (□□□), and (△△△). (From Schaefer and Berne, 1974).

9 · 3 EXTERNAL FIELD—ELECTROPHORESIS

When a static homogeneous external electric field $\mathbf{E}_0$ acts on the electrolyte solution, the local $\mathbf{E}_l(\mathbf{r}, t)$ electric field, that is, the field acting on an ion, consists of two parts, the external field, $\mathbf{E}_0$, and the field due to all the surrounding ions, $E(\mathbf{r}, t)$, that is,

$$\mathbf{E}_l(\mathbf{r}, t) = \mathbf{E}_0 + \mathbf{E}(\mathbf{r}, t) \tag{9.3.1}$$

Substitution of $\mathbf{E}_l(\mathbf{r}, t)$ into Eqs. (9.2.7) and (9.2.8) in place of $\mathbf{E}(\mathbf{r}, t)$ then yields, after linearizing the equations as in the previous section

$$\frac{\partial \delta c_1(\mathbf{q}, t)}{\partial t} = -[(q^2 + q_1^2)D_1 - i\omega_1(\mathbf{q})]\delta c_1(\mathbf{q}, t) + \left|\frac{z_2}{z_1}\right| q_1^2 D_1 \delta c_2(\mathbf{q}, t) \tag{9.3.2a}$$

$$\frac{\partial \delta c_2(\mathbf{q}, t)}{\partial t} = -[(q^2 + q_2^2)D_2 - i\omega_2(\mathbf{q})]\delta c_2(\mathbf{q}, t) + \left|\frac{z_1}{z_2}\right| q_2^2 D_2 \delta c_1(\mathbf{q}, t) \tag{9.3.2b}$$

where

$$\omega_i(\mathbf{q}) = \beta z_i D_i(\mathbf{q} \cdot \mathbf{E}_0) \qquad i = 1, 2 \tag{9.3.3}$$

are the Doppler shifts discussed in Section 5.8.

The only difference between Eqs. (9.2.13) and (9.2.15), and Eqs. (9.3.2) is that whenever $(q^2 + q_i^2)D_i$ appears in the latter, $[(q^2 + q_i^2)D_i - i\omega_i(\mathbf{q})]$ appears in the former. Thus Eqs. (9.2.17) (9.2.21b) apply to this case if the simple substitutions

$$[(q^2 + q_i^2)D_i] \rightarrow [(q^2 + q_i^2)D_i - i\omega_i(\mathbf{q})] \tag{9.3.4}$$

are made in these equations. The dispersion equation thus becomes

$$s^2 + \sum_{j=1}^{2} [(q^2 + q_j^2)D_j - i\omega_j(\mathbf{q})]s + [q^2(q^2 + q_0^2)D_1 D_2 - \omega_1(\mathbf{q})\omega_2(\mathbf{q})]$$
$$- i[\omega_1(\mathbf{q})(q^2 + q_2^2)D_2 + \omega_2(\mathbf{q})(q^2 + q_1^2)D_1] = 0 \tag{9.3.5}$$

or, more succinctly

$$1 = -\sum_{j=1}^{2} \frac{q_j^2 D_j}{s + q^2 D_j - i\omega_j(\mathbf{q})} \tag{9.3.6}$$

This latter form of the dispersion equation is valid for solutions in which there are more than two different ionic species.

If the quantity $\sum_{i=1}^{2} (q^2 + q_i^2)D_i$ is large compared to q^2D_1, q^2D_2, $\omega_1(\mathbf{q})$, and $\omega_2(\mathbf{q})$ the two roots of Eq. (9.3.5) to first order in these small quantities are

$$s_{\pm} = \begin{cases} s_+^{(0)} + s_+^{(1)} = -\Gamma_s(q) + i\omega_s(\mathbf{q}) & \text{(slow)} \\ s_-^{(0)} + s_-^{(1)} = -\Gamma_f(q) + i\omega_f(\mathbf{q}) & \text{(fast)} \end{cases} \tag{9.3.7}$$

where $(\omega_s(\mathbf{q}), \Gamma_s(q))$ and $(\omega_f(\mathbf{q}), \Gamma_f(q)$ are, respectively, the Doppler shift and width of the slow (s) and the fast (f) fluctuations which are explicitly

$$\omega_s(\mathbf{q}) = \left\{\frac{(q^2 + q_2^2)D_2\omega_1(\mathbf{q}) + (q^2 + q_1^2)D_1\omega_2(\mathbf{q})}{(q_2 + q_1^2)D_1 + (q^2 + q_2^2)D_2}\right\} \tag{9.3.8a}$$

$$\Gamma_s(q) = q^2 \left\{\frac{(q^2 + q_0^2)D_1 D_2}{(q^2 + q_1^2)D_1 + (q^2 + q_2^2)D_2}\right\} \equiv q^2 D_s(q) \tag{9.3.8b}$$

$$\omega_f(\mathbf{q}) = \{\omega_1(\mathbf{q}) + \omega_2(\mathbf{q}) - \omega_s(\mathbf{q})\} \tag{9.3.8c}$$

$$\Gamma_f(q) = \{(q^2 + q_1^2)D_1 + (q^2 + q_2^2)D_2 - \Gamma_s(q)\} \tag{9.3.8d}$$

Four points should be noted about these roots.

1. When $E_0 = 0$, these roots reduce to Eq. (9.2.28) as required.

2. When $q \ll q_1, q_2$; that is, when the wavelength of the fluctuation is much greater than the shielding lengths q_1^{-1}, q_2^{-1}, $\omega_s(q)$ reduces to

$$\omega_s(q) = \beta D_1 D_2(\mathbf{q} \cdot \mathbf{E}_0) \left\{\frac{z_1 q_2^2 + z_2 q_1^2}{q_1^2 D_1 + q_2^2 D_2}\right\} = 0 \tag{9.3.9}$$

which is zero because z_1 and z_2 are of opposite sign.[6] Also in this limit

$$D_s(q) = q_0^2 D_1 D_2/(q_1^2 D_1 + q_2^2 D_2) \equiv D_s \tag{9.3.10}$$

This is usually the situation in light scattering.

3. When $q \gg q_1, q_2$; that is, when the wavelength of the fluctuation is much smaller than the shielding lengths q_1^{-1}, q_2^{-1}, the two roots are exactly

$$\begin{aligned} s_+ &= -q^2 D_1 + i\omega_1(\mathbf{q}) \\ s_- &= -q^2 D_2 + i\omega_2(\mathbf{q}) \end{aligned} \tag{9.3.11}$$

In this very special case, the results even for a binary system look like the independent particle picture (cf. Section 5.8.)

4. The Doppler shift $\omega_f(\mathbf{q})$ is to first order in q, $\omega_f(\mathbf{q}) \simeq \omega_1(\mathbf{q}) + \omega_2(\mathbf{q})$, but because of the large value of $\Gamma_f(q)$, the root s_- gives rise to an overdamped decay of the fast fluctuation.

It is not difficult to understand these results physically. Fluctuations of sufficiently long wavelength ($q \ll q_1, q_2$) can be divided into two classes: (a) those that appear uncharged, and (b) those that appear to be charged. The former are not affected by an external field (hence point 2), whereas the latter are (hence point 4). Nevertheless, the latter case involves a charge separation over this large wavelength and hence a large electrostatic potential energy. This untenable situation should decay rapidly as it does in Γ_f^{-1}.

In the opposite case ($q \gg q_1, q_2$), the wavelength is small compared to the shielding length, and the fluctuations behave like an independent particle, albeit a fictitious one (hence point 3).

As before, the heterodyne correlation function is a linear combination of the exponential factors

$$[\exp \pm i\omega_s(\mathbf{q})|t|][\exp -\Gamma_s(q)|t|] \text{ and } [\exp \pm i\omega_f(\mathbf{q})|t|][\exp -\Gamma_f(q)|t|].$$

Thus the spectrum is a linear combination of the two Lorentzians

$$\frac{\Gamma_s(q)}{(\omega - \omega_s(\mathbf{q}))^2 + \Gamma_s^2(q)} \text{ and } \frac{\Gamma_f(q)}{(\omega - \omega_f(\mathbf{q}))^2 + \Gamma_f^2(q)}$$

where the first is relatively narrow, with width $\Gamma_s(q) = q^2 D_s(q)$ and the second is relatively broad, with width $\Gamma_f(q) = 1/\tau_r$. The resolutions R_s and R_f of the two frequencies are $R_s = [\omega_s(\mathbf{q})/q^2 D_s(q)]$ and $R_f = [\omega_f(\mathbf{q})/\Gamma_f(q)]$. Since for small q, $\omega_s(\mathbf{q}) \propto q^3$,

$q^2D_s(q) \propto q^2$, $\omega_f \propto q$ and $\Gamma_f(q) = \text{const}$, $R_s \to 0$ and $R_f \to 0$, as $q \to 0$ so that the Doppler frequencies should have poor resolution at low values of q. Exactly the opposite is predicted in the simple theory of Section 5.8. It is clear that the simple theory of Section 5.8 is not valid for the two-component electrolyte except for the experimentally inaccessible case when $q \gg q_1, q_2$. In the event that another salt is added in sufficient concentration, the simple theory should work, as we show in Section 9.4.

9 · 4 MACROIONS

Suppose that one of the ions, say A, is large and highly charged compared with the other ions. In this case the small ions should rapidly come to equilibrium in the presence of the big ion. Setting $\dot{c}_2(\mathbf{q}, t) = 0$ in Eq. (9.2.15) then gives

$$\delta c_2(\mathbf{q}, t) = \left|\frac{z_1}{z_2}\right| \frac{q_2^2 D_2}{q^2 + q_2^2} \delta c_1(\mathbf{q}, t) \tag{9.4.1}$$

Substitution of this into Eq.(9.2.14) gives

$$\frac{\partial \delta c_1(\mathbf{q}, t)}{\partial t} = -\, q^2 D_s(q) \delta c_1(\mathbf{q}, t) \tag{9.4.2}$$

which has the solution[7]

$$F_{11}(\mathbf{q}, t) = S_{11}(\mathbf{q}) \exp -q^2 D_s(q)|t| \tag{9.4.3}$$

where

$$D_s(q) = D_1\left[1 + \frac{q_1^2}{q^2 + q_2^2}\right] \tag{9.4.4}$$

This quantity decreases with q.

If we add salt in the form of small ions to this solution, an equation similar to Eq. (9.4.3) applies but now at sufficiently high concentration

$$D_s(q) = D_1\left[1 + \frac{q_1^2}{q^2 + (q_0^2 - q_1^2)}\right] \tag{9.4.5}$$

where, as before, the inverse Debye screening length is

$$q_0^2 = \beta\left(\frac{4\pi}{\varepsilon_0}\right) \sum_j \omega_f \tag{9.4.6}$$

At a sufficiently high concentration of added salt, $q_0 \gg q_1, q$ and Eq. (9.4.5) reduces to

$$D_s(q) \underset{\text{high ionic strength}}{\longrightarrow} D_1 \tag{9.4.7}$$

The Stokes law or hydrodynamic diffusion coefficient should then be observed. Thus at high salt concentration, the effects of ionic shielding discussed in Section 9.2 are unimportant, and the diffusion coefficient reduces to that of the isolated macroion. Physically this occurs because now a fluctuation in the counterions can be neutralized by a concentration fluctuation in ions of opposite charge, so that large electrical forces need not be exerted on the macroion.

In the presence of an external electric field, a similar approximation leads us to set $\dot{c}_2 = 0$ in Eq. (9.3.2). Then once again δc_2 can be eliminated from Eq. (9.3.2), leading to a closed equation which to first order in the applied field E_0 is

$$\frac{\partial \delta c_1(\mathbf{q}, t)}{\partial t} = -\,[q^2 D_s(q) - i\omega_s(\mathbf{q})]\delta c_1(\mathbf{q}, t) \tag{9.4.8}$$

where $D_s(q)$ is identical to Eq. (9.4.5), and where

$$\omega_s(\mathbf{q}) = \omega_1(\mathbf{q})\left[1 - \frac{q_2^2}{q^2 + q_2^2}\right] \tag{9.4.9}$$

for the two-component solution. For added salt

$$\omega_s(\mathbf{q}) = \omega_1(\mathbf{q})\left[1 - \frac{q_1^2}{(q^2 + q_0^2 - q_1^2)} \cdot \sum_{j \neq 1} q_j^2 \left|\frac{z_j}{z_1}\right|\right] \tag{9.4.10}$$

Again at high salt concentration $q_0 \gg q_1$, q, and

$$\omega_s(\mathbf{q}) \underset{\substack{\text{high ionic}\\ \text{strength}}}{\longrightarrow} \omega_1(\mathbf{q})$$

The simple theory presented in Section 5.9 should therefore be valid when the solution has a sufficiently large ionic strength. We return to a consideration of electrophoresis in Chapter 13.

9 · 5 THE EQUILIBRIUM STRUCTURE FACTORS

The structure factors defined by Eq. (9.2.20) can be expressed for[8] a solution of n ionic species as

$$S_\alpha(\mathbf{q}) = \left\langle \sum_{i\varepsilon\alpha} \sum_{j\varepsilon\beta} \exp i\mathbf{q} \cdot \mathbf{r}_{ij}^{\alpha\beta} \right\rangle; \ \alpha = 1 \ldots \text{n} \tag{9.5.1}$$

$$S_{\alpha\alpha}(\mathbf{q}) = \left\langle \sum_{i\varepsilon\alpha} \sum_{j\varepsilon\alpha} \exp i\mathbf{q} \cdot \mathbf{r}_{ij}^{\alpha\alpha} \right\rangle \tag{9.5.2}$$

where

$$\mathbf{r}_{ij}^{\alpha\beta} = \mathbf{r}_i^{\alpha} - \mathbf{r}_j^{\beta}$$

is the relative position of ion i of type α with respect to ion j of type β. This can be expressed as

$$S_{\alpha\beta}(\mathbf{q}) = \int d^3r \, \langle \sum_{i\varepsilon\alpha} \sum_{j\varepsilon\beta} \delta(\mathbf{r} - \mathbf{r}_{ij}{}^{\alpha\beta}) \rangle \, e^{i\mathbf{q}\cdot\mathbf{r}} \tag{9.5.3}$$

$$S_{\alpha\alpha}(\mathbf{q}) = \int d^3r \, \langle \sum_{i\varepsilon\alpha} \sum_{j\varepsilon\alpha} \delta(\mathbf{r} - \mathbf{r}_{ij}{}^{\alpha\alpha}) \rangle \, e^{i\mathbf{q}\cdot\mathbf{r}} \tag{9.5.4}$$

The pair-correlation functions $g_{\alpha\beta}{}^{(2)}(\mathbf{r})$ are defined by the equations

$$c_\beta g_{\alpha\beta}{}^{(2)}(\mathbf{r}) = \frac{1}{N_\alpha} \langle \sum_{i\varepsilon\alpha} \sum_{j\varepsilon\beta} \delta(\mathbf{r} - \mathbf{r}_{ij}{}^{\alpha\beta}) \rangle \tag{9.5.5}$$

$$c_\alpha g_{\alpha\alpha}{}^{(2)}(\mathbf{r}) = \frac{1}{N_\beta} \langle \sum_{i\varepsilon\alpha} \sum_{\substack{j\varepsilon\alpha \\ i\neq j}} \delta(\mathbf{r} - \mathbf{r}_{ij}{}^{\alpha\alpha}) \rangle \tag{9.5.6}$$

Eqs. (9.5.5) and (9.5.6) give, respectively, the density of ions of type β at $\mathbf{r}$, given that an ion of type α is at the origin, and the density of ions of type α at $\mathbf{r}$, given that an ion of type α is at the origin.

The structure factors are thus seen to be related to the pair-correlation functions by

$$S_{\alpha\beta}(\mathbf{q}) = N_\alpha c_\beta \int d^3r g_{\alpha\beta}{}^{(2)}(\mathbf{r}) \, e^{i\mathbf{q}\cdot\mathbf{r}} \tag{9.5.7}$$

$$S_{\alpha\alpha}(\mathbf{q}) = N_\alpha + N_\alpha c_\alpha \int d^3r g_{\alpha\alpha}{}^{(2)}(\mathbf{r}) \, e^{i\mathbf{q}\cdot\mathbf{r}} \tag{9.5.8}$$

These structure factors can be determined as follows. If an ion of type α is fixed at the origin, the other ions will distribute themselves such that the electrostatic potential at the point $\mathbf{r}$ is $\phi_\alpha(\mathbf{r})$. Thus an ion β at $\mathbf{r}$ must have an electrostatic energy $z_\beta\phi_\alpha(\mathbf{r})$. Thus the density at $\mathbf{r}$ of ions of type β and α, given that an ion of type a is at the origin, is respectively

$$c_\beta g_{\alpha\beta}(\mathbf{r}) = c_\beta \exp -\beta z_\beta \phi_\alpha(\mathbf{r}) \simeq c_\beta[1 - \beta z_\beta \phi_\alpha(\mathbf{r})] \tag{9.5.9}$$

$$c_\alpha g_{\alpha\alpha}(\mathbf{r}) = c_\alpha \exp -\beta z_\alpha \phi_\alpha(\mathbf{r}) \simeq c_\alpha[1 - \beta z_\alpha \phi_\alpha(\mathbf{r})] \tag{9.5.10}$$

where the second term is the high-temperature form. Now according to the Poisson equation [Eq. (9.2.11)] the electrical field at $\mathbf{r}$, $\mathbf{E}(\mathbf{r}) = -\nabla\phi_\alpha(\mathbf{r})$ given that an ion of type α is at the origin, is determined by the charge density at $\mathbf{r}$. This charge density is the sum of three parts: (a) the charge density at $\mathbf{r}$ due to the ion of type α fixed at the origin, $\mathbf{z}_\alpha\delta(\mathbf{r})$, (b) the charge density of ions of type β, $\mathbf{z}_\beta c_\beta g_{\alpha\beta}{}^{(2)}(\mathbf{r})$ and (c) the charge density of the remaining ions of type α, $\mathbf{z}_\alpha c_\alpha g_{\alpha\alpha}{}^{(2)}(\mathbf{r})$. Combining all these terms with Eqs. (9.5.9) and (9.5.10) gives the Poisson–Boltzmann equation

$$-\nabla^2\phi_\alpha(\mathbf{r}) = \frac{4\pi}{\varepsilon_0}\left[z_\alpha\delta(\mathbf{r}) + z_\alpha c_\alpha \exp -\beta z_\alpha\phi_\alpha(\mathbf{r}) + z_\beta c_\beta \exp -\beta z_\beta \phi_\beta(\mathbf{r})\right]; \; \beta \neq \alpha \tag{9.5.11}$$

This equation is very difficult to solve for $\phi_\alpha(\mathbf{r})$. At sufficiently high temperatures $|\beta z_\alpha \phi_\alpha| \ll 1$ and $|\beta z_\beta \phi_\alpha| \ll 1$ so that the exponentials can be expanded to first order in these quantities. Since $z_\alpha c_\alpha + z_\beta c_\beta = 0$ by electroneutrality, it follows that

$$-\nabla^2\phi_\alpha(\mathbf{r}) = \frac{4\pi}{\varepsilon_0} z_\alpha \delta(\mathbf{r}) - q_0{}^2\phi_\alpha(\mathbf{r}) \tag{9.5.12}$$

where q_0 is given by Eq. (9.2.16). Spatial Fourier transformation of this equation gives

$$\phi_\alpha(\mathbf{q}) = \frac{4\pi}{\varepsilon_0} \frac{z_\alpha}{q^2 + q_0^2} \tag{9.5.13}$$

where $\phi_\alpha(\mathbf{q})$ is the spatial Fourier transform of $\phi_\alpha(\mathbf{r})$.

Inverse Fourier transformation of Eq. (9.5.12) gives

$$\phi_\alpha(\mathbf{r}) = \frac{4\pi}{\varepsilon_0} z_\alpha \frac{\exp -q_0 r}{r} \tag{9.5.13}$$

This is a "screened" Coulomb potential. The quantity q_0^{-1} is a "screening-length." For distances much larger than q_0^{-1} the potential due to the ion α at the origin is essentially zero. This comes about because the ion α is surrounded on the average by ions of types α and β such that the ion α is neutralized.

Fourier transformation of the "high-temperature" forms of Eqs. (9.5.9) and (9.5.10), substitution of Eqs. (9.5.13) and substitution of the result into Eqs. (9.5.7) and (9.5.8) gives the Debye–Huckel structure factors

$$\begin{aligned} S_{\alpha\beta}(\mathbf{q}) &= +\, V\sqrt{c_\alpha c_\beta}\left[\frac{q_\alpha q_\beta}{q^2 + q_0^2}\right] \\ &= -\, V\left(\frac{4\pi}{\varepsilon}\,\beta\right)\left[\frac{(z_\alpha c_\alpha)\,(z_\beta c_\beta)}{q^2 + q_0^2}\right] \end{aligned} \tag{9.5.14}$$

$$S_{\alpha\alpha}(\mathbf{q}) = Vc_\alpha\left[1 - \frac{q_\alpha^2}{q^2 + q_0^2}\right] \tag{9.5.15}$$

where

$$q_\alpha^2 \equiv \beta\left(\frac{4\pi}{\varepsilon_0}\right) z_\alpha^2 c_\alpha; \quad q_\beta^2 \equiv \beta\left(\frac{4\pi}{\varepsilon_0}\right) z_\beta^2 c_\beta$$

Although there are no light scattering data that corroborate these results, it is interesting to note that the integrated intensities of R-17 virus solutions corresponding to the conditions of Fig. 9.2.1 show the same qualitative q^2 dependence as does $S_{\parallel}(\mathbf{q})$ (see Fig. 9.5.1).

It should be noted that $S_{\alpha\beta}(\mathbf{q})$ decreases as q increases whereas $S_{\alpha\alpha}(\mathbf{q})$ increases as q increases. Substitution of Eqs. (9.5.14) and (9.5.15) into Eq. (9.2.36) gives the integrated intensity

$$S(\mathbf{q}) = \sum_\alpha a_\alpha^2 c_\alpha - \beta\left(\frac{4\pi}{\varepsilon_0}\right) \frac{|\sum_\alpha a_\alpha z_\alpha c_\alpha|^2}{q^2 + q_0^2} \tag{9.5.16}$$

This function is an increasing function of q. It is usually the case in light scattering that[9] $q \ll q_0$ so that

$$S(\mathbf{q}) = \sum_\alpha a_\alpha^2 c_\alpha - \frac{|\sum_\alpha a_\alpha z_\alpha c_\alpha|^2}{\sum_\alpha z_\alpha^2 c_\alpha} \tag{9.5.17}$$

in this eventuality the integrated scattering will be angle-independent. Because in the

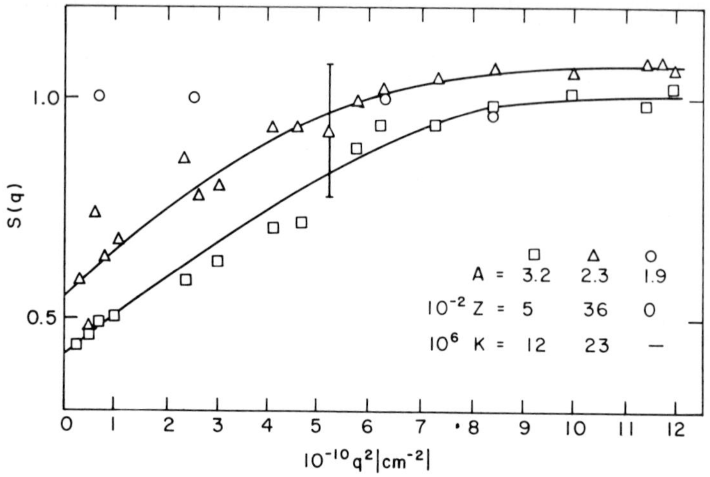

FIG. 9.5.1. The q^2-dependence of the integrated scattering from solutions of R-17 virus with no added salt. The data points are defined in Fig. 9.2.1. (From Schaefer and Berne, 1974.)

second term on the right the numerator is a quadratic function of c_α and the denominator is a linear function of c_α, this second term goes to zero as all the $c_\alpha \rightarrow 0$. Thus the integrated scattering should reduce to

$$S(q) \cong \sum_{\alpha} a_\alpha^2 c_\alpha \tag{9.5.18}$$

in the limit of infinite dilution (usually for 10^{-4}m solution of 1–1 electrolyte). This is the result to be expected for independent scatterers. For higher concentrations, this approximation is not valid and care must be taken in trying to extract molecular weights from integrated intensities in electrolyte solutions (Kerker, 1969).

It is interesting to note that when Eq. (9.2.26) is multiplied by $S(q)$ from Eq. (9.5. 15), the resulting expression is, a constant for $(q^2 + q_1^2)D_1 \ll (q^2 + q_2^2)D_2$. Thus we expect that

$$D_s(q) = D/S(q) \tag{9.5.19}$$

Although Eq. (9.5.19) has not been corroborated by experiment in a quantitative sense, Schaefer and Berne (1974) have shown that for the highly charged macroions of the R-17 virus the experimental $D_s(q)$, when multiplied by $S(q)$ determined from integrated intensities, are such that $D_s(q)S(q)$ is a very slowly varying function of q^2. This is shown in Fig. 9.5.2 for the same data points shown in Fig. (9.2.1) and Fig. (9.5.1). Actually, this observation results from a formal treatment such as that given in Chapter 11 (see Schaefer and Berne, 1974).

This section only applies to very dilute electrolytes with small charges. In fact there has been no attempt to consider the finite size of the ions. Thus the treatment here depended upon a linearization of Boltzmann factors—an approximation that is valid only if the interaction energy is small compared to k_BT. This is surely not the case for highly charged ions. Also at high concentrations, the finite sizes of the ions should be important. This section should only be regarded as pedagogical in that it points to many interesting consequences.

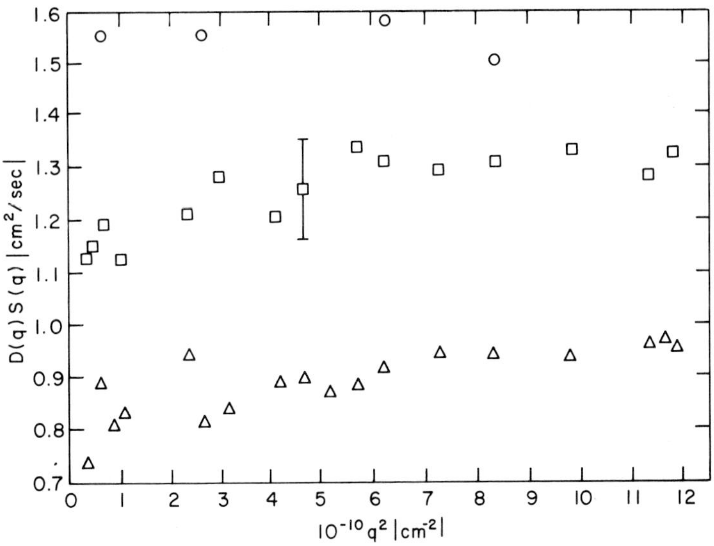

FIG. 9.5.2. The dependence of $D_s(q)\, S(q)$ on q^2 for solutions of R-17 virus specified in Fig. 9.2.1. (From Schaefer and Berne, 1974.)

NOTES

1. The charges z_i are measured in units of the charge on an electron.
2. Here we use concentrations in number per cc for convenience. Note that the concentrations in moles per liter A is $[A] = 1000\ c_i/N_0$ where N_0 is Avogadro's number.
3. As before

$$\delta c_i(\mathbf{q}, t) \equiv \int d^3r\, e^{i\mathbf{q}\cdot\mathbf{r}}\, \delta c_i(\mathbf{r}, t)$$

4. This follows from Eqs. (9.2.21a) and (9.2.18).
5. The dielectric correlation function is

$$F(\mathbf{q}, t) = \sum_{ij} a_i a_j F_{ij}(\mathbf{q}, t)$$

where $a_i \equiv (\partial\varepsilon/\partial c_i)$. This is found by expanding $\delta\varepsilon$ in terms of concentration

$$\delta\varepsilon = \sum_i \left[\frac{\partial\varepsilon}{\partial c_i}\right] \delta c_i\ {}_{c_j \neq c_i}$$

(see Section 10.1).

6. Note that $q_1^2/q_2^2 = |z_1/z_2|$. Therefore $z_1 q_1^2 + z_2 q_2^2 \propto [z_1|z_2| + z_2|z_1|] = 0$.
7. There is only one decay constant now because we have ignored all fast processes.
8. Note that

$$\delta c_\alpha(\mathbf{q}) = \sum_{i\varepsilon\alpha} \exp i\mathbf{q}\cdot\mathbf{r}_i^\alpha$$

where the sum only goes over ions of type α $(i\varepsilon\alpha)$. Thus

$$S_{\alpha\beta}(\mathbf{q}) = \langle \delta\mathbf{c}_\alpha{}^*(\mathbf{q})\delta c_\beta(\mathbf{q})\rangle$$

is given by Eq. (9.5.1).

9. For example in a 10^{-3}m solution of NaCl, $q_\alpha, q_\beta \sim 10^8$cm whereas $q \sim 10^5$cm. The q-dependence of $S(\mathbf{q})$ has actually been observed (Pusey, et al., 1972) in polyelectrolyte solutions under conditions of high ionic charge to which the above theory does not apply.

REFERENCES

Falkenhagen, H., *Electrolytes,* Oxford University Press (1934).

Kerker, M., *The Scattering of Light and Other Electromagnetic Radiation,* Section 9.23, Academic, New York (1969).

Pusey, P. N., Schaefer, D. W., Koppel, D. E., Camerini-Otero, R. D., and Franklin, R. M., *Jour. Phys.* (*Paris*) Supp. **33,** Cl–163 (1972).

Schaefer, D. W. and Berne, B. J., *Phys. Rev. Lett.,* **32,** 1110 (1974).

Stephen, M. J., *J. Chem. Phys.* **55,** 3878 (1971).

CHAPTER 10

LIGHT SCATTERING FROM HYDRODYNAMIC MODES

10 · 1 INTRODUCTION

In several of the preceding chapters it was assumed that individual molecules scatter light independently of each other. This is justified only in the case of dilute gases, or dilute macromolecular solutions, where the scattering center moves independently of its neighbors. This condition is obviously not satisfied in pure (neat) liquids or liquid mixtures where neighboring molecules are highly correlated both with respect to their spatial distribution and their relative motions. The spectrum of scattered light from such systems has fine structure, and the widths and relative intensities of the various bands give important information about their transport and thermodynamic properties. In this chapter we discuss the isotropic light scattering of liquids and liquid mixtures. We leave for subsequent chapters a discussion of collective effects on the depolarized scattering from molecular liquids.

The molecular theory of light scattering gives the spectrum[1] of light scattered from a pure monatomic liquid as

$$I_{if}^{\alpha}(\mathbf{q}, \omega) = (\mathbf{n}_i \cdot \mathbf{n}_f)^2 \alpha^2 S(\mathbf{q}, \omega) \tag{10.1.1}$$

where $S(\mathbf{q}, \omega)$ is the spectrum of density fluctuations of wave vector $\mathbf{q}$

$$S(\mathbf{q}, \omega) = \frac{1}{2\pi} \int_{-\infty}^{\infty} dt \exp - i\omega t \langle \delta\rho^*(\mathbf{q}, 0)\, \delta\rho(\mathbf{q}, t) \rangle \tag{10.1.2}$$

which appears frequently in the literature of many-body theory (e.g., see Van Hove, 1954; Pines and Nozieres, 1966) where it is called the *dynamic form factor*. The spectra corresponding to different scattering geometries is specified in Eq. (5.2.21). If the polarizability is spherical there should be no I_{VH} component. Thus a simple experimental test of this assumption is to see whether or not $I_{VH} \neq 0$. Experimentally it is found that $I_{VH} \neq 0$ even for the inert gases. The nonzero VH scattering arises because colliding molecules distort each other's charge clouds, producing an optical anisotropy which exists for the duration of the collision (see Chapter 14). This collision-induced scattering contributes appreciably to the spectrum only for frequencies that are of the order of the inverse duration of a collision, which is around 10^{13} sec^{-1}. For our purposes in this chapter it suffices to say that the contribution is weak and can be ignored compared with the contributions of the I_{VV} spectrum in the spectral range less than 10^{10} sec^{-1}.

The preceding formulas are based on an approximate molecular theory of light scattering. It is more precise to use Eq. (3.2.13), which involves the fluctuations of the spatial Fourier components of the dielectric fluctuations. Although the dielectric constant of a fluid is a scalar, fluctuations may spontaneously arise in which molecules are partially aligned in the neighborhood of any point $\mathbf{r}$. Thermal fluctuations will generally give rise to optical anisotropies characterized by off-diagonal elements in the dielectric fluctuations. Because of the spherical symmetry of atoms, the dielectric fluctuations in monatomic liquids should have no off-diagonal elements, so that $\delta\varepsilon_{\alpha\beta}(\mathbf{q}, t) = \delta\varepsilon(\mathbf{q}, t)\delta_{\alpha\beta}$. Substitution of this into Eq. (3.2.13) leads to the scattering formula

$$I_{if}^{\varepsilon}(\mathbf{q}, \omega) = (\mathbf{n}_i \cdot \mathbf{n}_f)^2 \frac{1}{2\pi} \int_{-\infty}^{+\infty} dt\, e^{-i\omega t} \langle \delta\varepsilon^*(\mathbf{q}, 0)\, \delta\varepsilon(\mathbf{q}, t)\rangle \tag{10.1.3}$$

Thus we must study dielectric fluctuations of wavelengths ($q^{-1} \sim 1000$ Å) appreciably larger than intermolecular separations. These fluctuations involve the collective motions of large numbers of molecules, and consequently can be described by the laws of macroscopic physics—thermodynamics and hydrodynamics. For this purpose it is useful to regard the fluid as a continuous medium. This means that any "small region" of the fluid is still sufficiently large to contain a great number of molecules. Thus when we talk about an infinitesimal volume element d^3r we mean a volume element that is very small compared with the volume of the whole system but large compared to the distances between atoms. Hence it makes sense to speak about the local values of such macroscopic concepts as entropy, enthalpy, and pressure. For example, to every point $\mathbf{r}$ in the fluid at time t we can ascribe values of the entropy density $s(\mathbf{r}, t)$, number density $\rho(\mathbf{r}, t)$, energy density $e(\mathbf{r}, t)$, pressure $p(\mathbf{r}, t)$, and dielectric constant $\varepsilon(\mathbf{r}, t)$. When we say that the system is in *local equilibrium*, we mean that the local values of the thermodynamic and optical properties are related to each other just as they would be if the material were in a state of overall thermodynamic equilibrium. Thus for example if a dilute gas is in local equilibrium, the local pressure would be related to the local temperature and density by the ideal-gas equation of state $p(\mathbf{r}, t) = \rho(\mathbf{r}, t)k_BT(\mathbf{r}, t)$. This assumption of local equilibrium lies at the foundations of fluid mechanics (e.g., see Landau and Lifshitz, 1960) and of irreversible thermodynamics (e.g., see DeGroot and Mazur, 1962).

The dielectric constant ε_0 of a pure fluid in total equilibrium is in general a function of the density ρ_0 and temperature T_0; that is, $\varepsilon_0 = \varepsilon(\rho_0, T_0)$. This is called the *dielectric equation of state*. Clearly on the local level, there are small fluctuations in the local density and temperature so that we can write $\rho(\mathbf{r}, t) = \rho_0 + \delta\rho(\mathbf{r}, t)$ and $T(\mathbf{r}, t) = T_0 + \delta T(\mathbf{r}, t)$. Thus if we assume local equilibrium, the dielectric equation of state can be used to determine the local value of the dielectric constant. Accordingly we write $\varepsilon(\mathbf{r}, t) = \varepsilon(\rho_0 + \delta\rho(\mathbf{r}, t); T_0 + \delta T(\mathbf{r}, t))$. Since these fluctuations are expected to be quite small, this can be expanded in a (rapidly convergent) power series in these fluctuations. The local dielectric fluctuation $\delta\varepsilon(\mathbf{r}, t) = \varepsilon(\mathbf{r}, t) - \varepsilon(\rho_0, T_0)$ is then to first order in the fluctuations $\delta\rho$, δT,

$$\delta\varepsilon(\mathbf{r}, t) = \left(\frac{\partial\varepsilon}{\partial\rho}\right)_T \delta\rho(\mathbf{r}, t) + \left(\frac{\partial\varepsilon}{\partial T}\right)_\rho \delta T(\mathbf{r}, t)) \tag{10.1.4}$$

where the derivatives are found experimentally from the temperature and density variation of ε in an equilibrium fluid. Substitution of Eq. (10.1.4) into Eq. (10.1.3) then gives the spectrum

$$I_{if}^{\varepsilon}(\mathbf{q}, \omega) = (\mathbf{n}_i \cdot \mathbf{n}_f)^2 \left\{\left(\frac{\partial\varepsilon}{\partial\rho}\right)_{T_0}^2 S_{\rho\rho}(\mathbf{q}, \omega) + \left(\frac{\partial\varepsilon}{\partial\rho}\right)_{T_0} \left(\frac{\partial\varepsilon}{\partial T}\right)_{\rho_0} [S_{\rho T}(\mathbf{q}, \omega)\right.$$

$$+S_{T\rho}(\mathbf{q},\omega)] + \left(\frac{\partial\varepsilon}{\partial T}\right)^2_{\rho_0} S_{TT}(\mathbf{q},\omega)\} \tag{10.1.5}$$

where $S_{\rho\rho}(\mathbf{q},\omega)$, $S_{\rho,T}(\mathbf{q},\omega)$, $S_{T,\rho}(\mathbf{q},\omega)$, and $S_{TT}(\mathbf{q},\omega)$ are, respectively, the spectral densities of the correlation functions $\langle\delta\rho^*(\mathbf{q},0)\delta\rho(\mathbf{q},t)\rangle$, $\langle\delta\rho^*(\mathbf{q},0)\delta T(\mathbf{q},t)\rangle$, $\langle\delta T^*(\mathbf{q},0)\delta\rho(\mathbf{q},t)\rangle$ and $\langle\delta T^*(\mathbf{q},0)\,\delta T(\mathbf{q},t)\rangle$,. This is markedly different from Eq. (10.1.1), which only involves $S_{\rho\rho}(\mathbf{q},\omega)$. The simple microscopic theory does not involve temperature fluctuations. To our knowledge no microscopic theory has yet been devised that gives anything like temperature fluctuations. In our opinion Eq. (10.1.5) should be used since its predictions accord well with experiments. Fortunately, in many simple liquids it is experimentally found that ε is only a function of the density; that is, that $(\partial\varepsilon/\partial T)_{\rho_0} \cong 0$. Equation (10.1.5) then simplifies to

$$I^{\varepsilon}_{if}(\mathbf{q},\omega) = (\mathbf{n}_i \cdot \mathbf{n}_f)^2 \left(\frac{\partial\varepsilon}{\partial\rho}\right)^2_{T_0} S_{\rho\rho}(\mathbf{q},\omega) \tag{10.1.6}$$

Thus in most applications $I^{\alpha}_{if}(\mathbf{q},\omega)$ and $I^{\varepsilon}_{if}(\mathbf{q},\omega)$ are both proportional to $S_{\rho\rho}(\mathbf{q},\omega)$. The only difference is that α in Eq. (10.1.1) is replaced by $(\partial\varepsilon/\partial\rho)_{T_0}$ in Eq. (10.1.6). In fact α in the molecular theory should be interpreted as an effective polarizability in solution and not the vacuum polarizability.[2]

$S_{\rho\rho}(\mathbf{q},\omega)$ is the spectral density of the autocorrelation function of the density fluctuation $\delta\rho(\mathbf{q},t)$. Thus from Eqs. (3.2.16) and (10.1.6) it follows that the integrated intensity is proportional[3] to the *structure factor*

$$S(\mathbf{q}) = \langle|\delta\rho(\mathbf{q})|^2\rangle, \tag{10.1.7}$$

which is simply the mean-square fluctuation of the qth Fourier component of the density fluctuation $\delta\rho(\mathbf{q})$. Light scattering probes values of q, such that $q^{-1}(\sim 1000\text{ Å})$ is much greater than the range of intermolecular interactions. It is possible to ignore the q dependence in Eq. (10.1.7). In this case

$$\lim_{q\to 0}\delta\rho(\mathbf{q}) = \lim_{q\to 0}\int_v d^3r\, e^{i\mathbf{q}\cdot\mathbf{r}}\,\delta\rho(\mathbf{r}) = \int_v d^3r\delta\rho(\mathbf{r}) = \delta N$$

where δN is the fluctuation in the number of particles in the scattering volume so that the integrated scattering is proportional to the mean-square fluctuation in N (see Appendix 10.A)

$$\lim_{\mathbf{q}\to 0} S(\mathbf{q}) = \langle\delta N^2\rangle = V\rho^2 k_B T\chi_T \tag{10.1.8}$$

where the second equality follows from statistical fluctuation theory; ρ is the mean number density, and χ_T is the isothermal compressibility $\chi_T = (1/\rho)\,[\partial\rho/\partial p]_T$. The integrated scattering is consequently expected to be

$$I_{if}(\mathbf{q}) = (\mathbf{n}_i \cdot \mathbf{n}_f)^2 \left(\frac{\partial\varepsilon}{\partial\rho}\right)^2_{T_0} V\rho^2 k_B T\chi_T \tag{10.1.9}$$

which is independent of the scattering angle except insofar as V depends on the scattering angle. This formula was first derived by Einstein (1910). Extensive discussion of the validity of Eq. (10.1.9) (or more precisely, of the equivalent expression for the turbidity) is given in Kerker (1969), Section 9.1. It has been confirmed by numerous experiments. A very important consequence of Eq. (10.1.9) is the following. At the gas–liquid critical point the fluid becomes infinitely compressible, that is, $\chi_T \to \infty$ and $I_{if}^{g}(\mathbf{q}) \to \infty$. At this point the fluid scatters so much light that it becomes opalescent—hence the term *critical opalescence*. Unfortunately, Eq. (10.1.9) does not describe the observed features of the scattering in the critical region where particles are correlated over distances of the order of q^{-1} so that the q dependence of $S(\mathbf{q})$ cannot be ignored. We return to a brief consideration of this in Section 10.7.

The major consideration thus far is that to describe light scattering from fluids we must compute time correlations of the density and temperature fluctuations. It is important to note that the time-correlation function $\langle\delta\rho^*(\mathbf{q}, 0)\delta\rho(\mathbf{q}, t)\rangle$ is proportional to the spatial Fourier transform of the integral $\int d^3r' \langle\delta\rho(\mathbf{r} - \mathbf{r}', t)\delta\rho(\mathbf{r}', 0)\rangle$. This follows from the convolution theorem of Fourier analysis. Clearly what is required for a prediction of the light-scattering spectrum is knowledge of how density fluctuations at two different points of the fluid at two different times are correlated. If a density fluctuation spontaneously occurs at $\mathbf{r}'$ at $t = 0$, how does it correlate with a density fluctuation at a different point $\mathbf{r} - \mathbf{r}'$ at sometime t later? If the distance between these two points is as "large" as the wavelength of visible light, the fluctuations should be correlated because the density fluctuation at the first point at $t = 0$ can propagate or diffuse to the second point in time t. Thus in some limit we expect that we will be able to calculate this correlation function on the basis of the "hydrodynamic equations." Similar considerations are involved in the computation of the correlation functions of the temperature fluctuations.

Hence this chapter is devoted to an investigation of hydrodynamic fluctuation theory. Much of what we use in this chapter is developed more formally in Chapter 11. In order to use hydrodynamic fluctuation theory it is necessary to discuss the derivation of the usual equations of fluid mechanics. Unfortunately this task would require the writing of a separate monograph. Space does not permit us to present a detailed account of the equations of fluid dynamics. We refer the reader to the excellent monograph on this subject by Landau and Lifshitz (1960), and will only highlight the important points here.

In subsequent sections we study the dynamic aspects of the scattering from pure fluids and binary mixtures.

10 · 2 RELAXATION EQUATIONS AND THE REGRESSION OF FLUCTUATIONS

A typical relaxation experiment consists of the following steps.

1. The system is first brought to equilibrium in the presence of certain "macroscopic constraints" so that the property to be measured, $\langle A(\mathbf{r}, t)\rangle_0$, is constant in time but depends on position. This quantity $\langle A(\mathbf{r}, t)\rangle_0$ is the ensemble-average of a property A in the ensemble defined by the constraints.

2. The constraints are removed. The system then relaxes to some new equilibrium state, during the process of which $\langle A(\mathbf{r}, t)\rangle$ relaxes from its initial constrained value to

the value it assumes in the new equilibrium state. The detailed time-dependence is then analyzed for the relaxation times. The quantity $\langle A(\mathbf{r}, t)\rangle$ is an average over the unconstrained ensemble.

The relaxation of certain properties of the system can often be described by simple phenomenological equations called *relaxation equations.* In chemical kinetics, for example, the constrained state may be a mixture of gases in metastable equilibrium—for example, hydrogen and oxygen. A spark is then introduced and the gas mixture reacts. The concentration of the reactants and products change with time until a new equilibrium state is achieved. The relaxation equations are the familiar phenomenological equations of chemical kinetics and the relaxation times are related to the chemical rate constants.

Diffusion provides another good example. A small impermeable sack containing a solute is placed in an infinite container full of solvent. At $t = 0$ the sack is broken (the constraint is removed) and the solute diffuses through the solvent until a new state of equilibrium is reached in which the solute is uniformly distributed throughout the solution. The property that is measured is the concentration of solute $\langle c(\mathbf{r}, t)\rangle$ as a function of position and time. Initially the solute is found only in the sack so that the sack geometry defines the initial concentration distribution $\langle c(\mathbf{r})\rangle_0$. This experiment is completely described by the diffusion equation the solution of which gives[4]

$$\langle c(\mathbf{r}, t)\rangle = \left(\frac{1}{2\pi}\right)^3 \int d^3q e^{-i\mathbf{q}\cdot\mathbf{r}} \exp -q^2 D|t| \, \langle c(\mathbf{q})\rangle_0 \tag{10.2.1}$$

where D is the diffusion coefficient and $\langle c(\mathbf{q})\rangle_0$ is the spatial Fourier transform of the initial concentration, $\langle c(\mathbf{r})\rangle_0$. An important thing to note about Eq. (10.2.1) is that the long-time solution is dominated by the small q-components of the concentration[5]

$$\langle c(\mathbf{q}, t)\rangle = \langle c(\mathbf{q})\rangle_0 \exp -q^2 D|t| \tag{10.2.2}$$

This is the case because as $t \to \infty$, $\exp -q^2 D|t|$ goes rapidly to zero unless q is very small.

Actually, the diffusion equation and other relaxation equations as well are macroscopic equations. They only describe phenomena over macroscopic distances and long times. In fact, as we shall see, these equations are derived on the assumption that the properties $\langle A(\mathbf{r}, t)\rangle$ do not vary much over microscopic distances (interparticle distances) and microscopic times (collision times). The relaxation equations are valid only for times such that many collisions have occurred and only for distances large compared to interparticle separations. To describe phenomena on a microscopic scale, it is necessary to apply the kinetic theory of gases and liquids. Fortunately light-scattering experiments involve long distances ($\sim q^{-1} \sim 1000$ Å) and long times, so that we can apply the phenomenological relaxation equations (except to rarefied systems) (Section 14.3). In Section 10.4 we indicate how such equations are derived; in Chapter 11 we indicate how these ideas can be extended to microscopic distances and times; and in Chapter 14 we show where the macroscopic equations do not suffice.

Light-scattering experiments are not relaxation experiments like the foregoing but instead, as we have seen, involve fluctuation phenomena and time-correlation functions. In connection with the development of the thermodynamics of irreversible processes, Onsager (1931) proposed the principle that spontaneous fluctuations in $A(\mathbf{r}, t)$ "regress" back to equilibrium according to the same relaxation equations that describe the macroscopic relaxation processes. This is known as the *Onsager regression hypothesis*. The ultimate validity of this hypothesis is based on experimental evidence and has been corroborated by light-scattering experiments, among others. The regression hypothesis enables us to compute time-correlation functions of properties involving large distances and long times (compared to molecular distances and times). A discussion of the microscopic basis of the Onsager regression hypothesis is given in Section 11.3.

We illustrate this using the example of diffusion. According to Onsager the concentration fluctuations in an equilibrium solution should satisfy the same equation that applies to macroscopic diffusion, that is,

$$\frac{\partial}{\partial t}\,\delta c(\mathbf{q}, t) = -q^2 D\delta c(\mathbf{q}, t) \tag{10.2.3}$$

where $\delta c(\mathbf{q}, t)$ is the spatial Fourier transform of the concentration fluctuation.[6] The solution of Eq. (10.2.10) is $\delta c(\mathbf{q}, t) = \exp - q^2 Dt \delta c(\mathbf{q}, 0)$, where $\delta c(\mathbf{q}, 0)$ is the initial fluctuation. Multiplication by $\delta c^*(\mathbf{q}, 0)$ followed by averaging over the equilibrium distribution function gives the correlation function

$$\langle \delta c^*(\mathbf{q}, 0)\delta c(\mathbf{q}, t)\rangle = \langle |\delta c(\mathbf{q})|^2\rangle \exp -q^2 Dt \tag{10.2.4}$$

where $\langle |\delta c(\mathbf{q})|^2\rangle$ is the mean-square concentration fluctuation to be determined from thermodynamic fluctuation theory (see Appendix 10.A). The spectral density of Eq. (10.2.4) is the Lorentzian

$$S_{cc}(\mathbf{q}, \omega) = \pi^{-1}\langle |\delta c(\mathbf{q})|^2\rangle \left\{\frac{q^2 D}{\omega^2 + [q^2 D]^2}\right\} \tag{10.2.5}$$

It is important at this juncture to note that Eq. (10.2.3) gives $\delta c(\mathbf{q}, t)$, which decays to zero as $t \to \infty$. However we know that concentration fluctuations are always arising and decaying. Equation (10.2.3) therefore gives an erroneous result. In Chapter 11 it is shown that the correct equation has the form

$$\frac{\partial}{\partial t}\,\delta c(\mathbf{q}, t) = -q^2 D\delta c(\mathbf{q}, t) + F(\mathbf{q}, t) \tag{10.2.6}$$

where $F(\mathbf{q}, t)$ is a fluctuating quantity. Physically it arises because the instantaneous flux deviates from the systematic locally averaged Fickian form $-D\nabla\delta c$. This revised equation is a stochastic differential equation like the Langevin equation in Section 5.9. As we show in Chapter 11, the solution of this equation still leads to Eq. (10.2.4), so that our conclusions are valid even though Eq. (10.2.3) is erroneous. These conclusions apply to all of the phenomenology used in this chapter.

In subsequent sections, we see that time-correlation functions often have the form

$$S_{AA}(\mathbf{q}, t) = \langle \delta A^*(\mathbf{q}, 0)\, \delta A(\mathbf{q}, t)\rangle = \langle |\delta A(\mathbf{q})|^2\rangle\, e^{-\Gamma(q)|t|} \cos \omega(q)t \tag{10.2.7}$$

with a corresponding spectral density, which is a superposition of Lorentzians

$$S_{AA}(\mathbf{q}, \omega) = \frac{1}{\pi}\, S_{AA}(\mathbf{q}) \left[\frac{\Gamma(q)}{[\omega - \omega(q)]^2 + [\Gamma(q)]^2} + \frac{\Gamma(q)}{[\omega + \omega(q)]^2 + [\Gamma(q)]^2}\right] \tag{10.2.8}$$

where $\delta A(\mathbf{q}, t)$ is the qth Fourier component of $\delta A(\mathbf{r},t)$, the fluctuation in the density of the conserved variable A; $S_{AA}(\mathbf{q}) \equiv \langle|\delta A(\mathbf{q})|^2\rangle$ is the mean-square fluctuation[7] of $\delta A(\mathbf{q}, t)$; $\Gamma(q)$ is the width; and $\omega(q)$ is the shift of the spectrum. In many applications $\Gamma(q) = \gamma q^2$ and $\omega(q) = Cq$, where γ is an attenuation coefficient and C is a "velocity of propagation." γ is usually a sum of transport coefficients. By studying the widths and shifts as a function of q, γ and C can be determined. This provides useful information about the collective modes and transport properties of a system. $\Gamma^{-1}(q)$ can be regarded as the *lifetime* $\tau(q)$ of the fluctuation. Equation (10.2.8) is plotted schematically in Fig. 10.2.1.

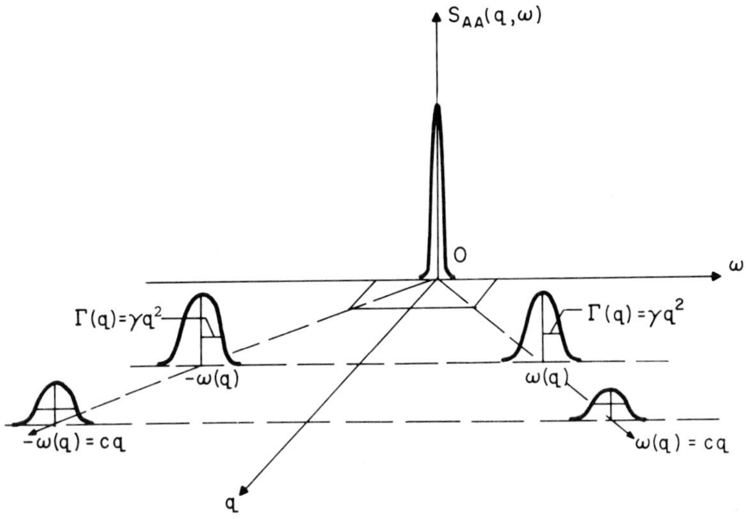

FIG. 10.2.1. A schematic spectral density corresponding to Eq. (10.2.8).

10 · 3 CONSERVATION EQUATIONS AND HYDRODYNAMIC MODES

In Section 10.2 we saw that the macroscopic relaxation equations can be used to determine correlation functions. In this section we summarize the traditional methods for deducing the macroscopic relaxation equations of *fluid mechanics*. In subsequent sections these equations are used to determine the *Rayleigh–Brillouin spectrum*. The first step in the derivation of the relaxation equation involves a discussion of *conservation laws*.

Associated with any extensive property $\mathscr{A}$ (like energy, and mass) is a specific property $A(\mathbf{r}, t)$ defined as the quantity of $\mathscr{A}$ per unit volume (or the density of $\mathscr{A}$) at the space–time point $(\mathbf{r}, t)$. For a system removed from equilibrium or for an equilibrium system undergoing spontaneous thermal fluctuations, $A(\mathbf{r}, t)$ varies with space and time. In an *arbitrary volume* V, fixed with respect to the "laboratory-fixed" coordinate axes, the total "quantity" $\mathscr{A}$ in this volume at time t is

$$\mathscr{A}(t) = \int_V d^3r A(\mathbf{r}, t) \tag{10.3.1}$$

where $A(\mathbf{r}, t)d^3r$ is the quantity of $\mathscr{A}$ in the infinitesmal volume d^3r located at $\mathbf{r}$ at time t. The quantity $A(\mathbf{r}, t)$ often has the microscopic form

$$A(\mathbf{r}, t) = \sum_j A_j(t)\delta(\mathbf{r} - \mathbf{r}_j(t))$$

where $A_j(t)$ is the property to be associated with molecule j.

The time rate of change of $\mathscr{A}(t)$ in the volume V is

$$\frac{d\mathscr{A}}{dt}(t) = \int_V d^3r \frac{\partial A}{\partial t}(\mathbf{r}, t) \tag{10.3.2}$$

Since the volume V is fixed, the order of the differentiation and the integration may be interchanged. The derivative of the integrand is a partial derivative because it is evaluated at the fixed position $\mathbf{r}$.

Another expression for $d\mathscr{A}/dt$ can be obtained as follows. The surface enclosing the volume V is divided into infinitesmal elements of area $d\mathbf{S}$. The vector $d\mathbf{S}$, of magnitude dS, is defined for each element. The direction of $d\mathbf{S}$ is taken perpendicular to the element of surface dS and pointing outward from the volume V. The "flux" or current density $\mathbf{J}_A(\mathbf{r}, t)$ of $\mathscr{A}$ is by definition the quantity of $\mathscr{A}$ passing per unit time through a surface of unit area perpendicular to the direction of $\mathbf{J}_A(\mathbf{r}, t)$. Thus $\mathbf{J}_A(\mathbf{r}, t) \cdot d\mathbf{S}$ is the quantity of $\mathscr{A}$ passing per unit time out of V through the surface element $d\mathbf{S}$. It follows that the rate of change of $\mathscr{A}$ within V due to a flow of $\mathscr{A}$ through the boundary surface of V is

$$-\int_S d\mathbf{S} \cdot \mathbf{J}_A(\mathbf{r}, t) \tag{10.3.3}$$

where the integration is taken over the entire surface S bounding V.

If $\sigma_A(\mathbf{r}, t)$ is the *internal source* of $\mathscr{A}$ per unit volume per unit time (due to chemical reactions, for example), then the rate of change of $\mathscr{A}$ within V due to the internal production of $\mathscr{A}$ is

$$\int_V d^3r \sigma_A(\mathbf{r}, t) \tag{10.3.4}$$

Adding the rates of change of $\mathscr{A}$ within V due to flow through S and due to internal production of $\mathscr{A}$ gives

$$\frac{d\mathscr{A}(t)}{dt} = -\int_S d\mathbf{S} \cdot \mathbf{J}_A(\mathbf{r}, t) + \int_V d^3r \sigma_A(\mathbf{r}, t) \tag{10.3.5}$$

The surface integral can be transformed to a volume integral by means of Gauss' theorem

$$\int_S d\mathbf{S} \cdot \mathbf{J}_A(\mathbf{r}, t) = \int_V d^3r \boldsymbol{\nabla} \cdot \mathbf{J}_A(\mathbf{r}, t) \tag{10.3.6}$$

Combining Eqs. (10.3.2), (10.3.5), and (10.3.6) gives

$$\int_V d^3r \left\{\frac{\partial A}{\partial t} + \nabla \cdot \mathbf{J}_A - \sigma_A\right\} = 0 \tag{10.3.7}$$

Since the volume V under consideration is arbitrary, the only way that Eq. (10.3.7) can hold for all volumes is for the integrand to vanish so that

$$\frac{\partial A}{\partial t} + \nabla \cdot \mathbf{J}_A = \sigma_A \tag{10.3.8}$$

Now if $\mathscr{A}$ is a conserved quantity like the total mass, the total linear momentum, or the total energy, there can be no net "creation" or "destruction" of $\mathscr{A}$ in V; that is, $\sigma_A \equiv 0$ and in this case

$$\frac{\partial A}{\partial t} + \nabla \cdot \mathbf{J}_A = 0 \tag{10.3.9}$$

According to this equation, $\mathscr{A}$ can only change by flow into or out of V. Equation (10.3.9) is the differential form of a macroscopic conservation law. It expresses locally the overall conservation of $\mathscr{A}$. *Thus if $\mathscr{A}$ is conserved, then locally, that is, at each point in the fluid, its density $A(\mathbf{r}, t)$ must satisfy Eq. (10.3.9).* Of course Eq. (10.3.9) applies equally well to a system removed from equilibrium as to a system in equilibrium undergoing spontaneous thermal fluctuations. In this latter case $A(\mathbf{r}, t) = \langle A \rangle + \delta A(\mathbf{r}, t)$, $\mathbf{J}_A(\mathbf{r}, t) = \langle \mathbf{J}_A \rangle + \delta \mathbf{J}_A(\mathbf{r}, t)$ and $\sigma_A(\mathbf{r}, t) = \langle \sigma_A \rangle + \delta\sigma_A(\mathbf{r}, t)$ where $\langle A \rangle$, $\langle \mathbf{J}_A \rangle$ and $\langle \sigma_A \rangle$ are equilibrium values and δA, $\delta \mathbf{J}_A$, and $\delta\sigma_A$ are fluctuations. In equilibrium, the average flux and source vanish ($\langle J_A \rangle = \langle \sigma_A \rangle = 0$) and $\langle A \rangle$ is independent of $(\mathbf{r}, t)$ so that the fluctuations also satisfy Eq. (10.3.8); that is,

$$\frac{\partial}{\partial t} \delta A(\mathbf{r}, t) + \nabla \cdot \delta \mathbf{J}_A(\mathbf{r}, t) = \delta\sigma_A(\mathbf{r}, t) \tag{10.3.10}$$

An interesting feature of these "conservation equations" emerges from a consideration of the spatial Fourier transform of Eq. (10.3.10), which is

$$\frac{\partial}{\partial t} \delta A(\mathbf{q}, t) = i\mathbf{q} \cdot \delta \mathbf{J}_A(\mathbf{q}, t) + \delta\sigma_A(\mathbf{q}, t) \tag{10.3.11}$$

where $\delta A(\mathbf{q}, t)$, $\delta \mathbf{J}_A(\mathbf{q}, t)$ and $\delta\sigma_A(\mathbf{q}, t)$ are the Fourier components of $\delta A(\mathbf{r}, t)$, $\delta \mathbf{J}_A(\mathbf{r}, t)$ and $\delta\sigma_A(\mathbf{r}, t)$ respectively. When $A(\mathbf{r}, t)$ corresponds to a conserved density, Eq. (10.3.11) applies with $\delta\sigma_A(\mathbf{q}, t) \equiv 0$, and the resulting equation $[\delta\dot{A}(\mathbf{q}, t) = i\mathbf{q} \cdot \delta \mathbf{J}_A]$ implies that as $q \to 0$, the rate of change of $\delta A(\mathbf{q}, t)$ [that is, $\delta\dot{A}(\mathbf{q}, t)$] approaches zero. This means that fluctuations in the densities of conserved variables become infinitely long-lived as $q \to 0$, or equivalently that the lifetime $\tau(q)$ of these fluctuation is such that $\tau(q) \to \infty$ as $q \to 0$. Any fluctuation with this property we shall call a *hydrodynamic mode*. Hydrodynamic modes can be classified according to their behavior as $q \to 0$. If there is no propagation frequency $\omega(q)$ and the width is γq^n, the mode is called a purely *diffusive mode*. Otherwise it is called a *propagating-mode*. In the ensuing chapters we study how such modes contribute to the light-scattering spectrum. If A is not a conserved property $\delta\sigma_A(\mathbf{q}, t) \neq 0$, it follows that $\delta\dot{A}(\mathbf{q}, t) \neq 0$ as $q \to 0$. Thus "fluctuations" in nonconserved densities have finite lifetimes,[8] $\tau(q)$, as $q \to 0$.

For a given finite value of q the lifetime of a nonconserved density may nevertheless be comparable to that of the conserved densities. In this eventuality, for this finite value of q, the nonconserved density can be regarded as relatively long-lived. We call such modes *quasihydrodynamic modes.* These latter modes are distinguishable from the former because their lifetimes are finite in the limit $q \to 0$.

In the special case where A is carried only by the actual convective flow of the fluid, the current is given by

$$\mathbf{J}_A(\mathbf{r}, t) = A(\mathbf{r}, t)\, \mathbf{u}(\mathbf{r}, t) \tag{10.3.12}$$

where $\mathbf{u}(\mathbf{r}, t)$ is the velocity of the fluid at $(\mathbf{r}, t)$. This is the case for the mass density $m\rho$. The conservation equation for the mass is therefore

$$\frac{\partial \rho(\mathbf{r}, t)}{\partial t} + \boldsymbol{\nabla} \cdot [\rho(\mathbf{r}, t)\, \mathbf{u}(\mathbf{r}, t)] = 0 \tag{10.3.13}$$

This is called the *equation of continuity.* It is important to note that Eq. (10.3.12) is not the general form of $\mathbf{J}_A$ but only applies to special conserved densities such as the mass density.

An example of a nonconserved density is the concentration $c(\mathbf{r}, t)$ of a component undergoing chemical reactions. Then

$$\frac{\partial c(\mathbf{r}, t)}{\partial t} + \boldsymbol{\nabla} \cdot \mathbf{J}(\mathbf{r}, t) = \sigma_c(\mathbf{r}, t) \tag{10.3.14}$$

where $\mathbf{J}$ is the flux of this component and σ_c is the increase of the component per unit volume due to all chemical reactions. σ can be related to the concentrations of all the components by applying the theory of chemical kinetics.

The conservation equations do not by themselves constitute a closed set of relaxation equations. In order to "close" Eq. (10.3.8) we must specify a "constitutive relation" relating the flux $\mathbf{J}_A(\mathbf{r}, t)$ to the "density" $A(\mathbf{r}, t)$. As an example we consider the simple example of diffusion in a binary mixture, in which there are no chemical reactions. Then Eq. (10.3.8) applies with no source term. According to Fick's second law the local average current of the solute is

$$\langle \mathbf{J}(\mathbf{r}, t)\rangle = -D\boldsymbol{\nabla}\langle c(\mathbf{r}, t)\rangle \tag{10.3.15}$$

where D is the diffusion coefficient. This is the simplest constitutive relation[9] that expresses our universal experience that there is an average net flow of the solute from regions of high concentration into regions of low concentration. It is a "reasonable" relationship only if there is no convective flow in the system, otherwise a convective term $c(\mathbf{r}, t)\mathbf{u}(\mathbf{r}, t)$ must be added to this equation. Combining Eq. (10.3.14) with $\sigma_c = 0$ and Eq. (10.3.15) gives the diffusion equation which is a closed equation. *Thus relaxation equations are derived by substituting constitutive relations into the conservation equations.* A systematic formalism for determining these linear constitutive relations is provided by the theory of *nonequilibrium thermodynamics.* An introduction to this subject is given in Chapter 13. Equation (10.3.15) is not entirely valid. In Section 10.6 it is shown that the diffusion flux will also be dependent on a temperature gradient. Also in a system containing three or more components a gradient in the concentration of one component induces a flux in the other components (see Chapter 13).

10 · 4 THE RAYLEIGH-BRILLOUIN SPECTRUM OF A PURE MONATOMIC FLUID

There are, as we shall see, three longitudinal and two transverse hydrodynamic modes in a one-component liquid. The transverse modes are purely diffusive shear modes and one longitudinal mode is a purely diffusive heat-diffusion mode. The remaining two longitudinal modes are propagating modes (due to sound waves). Only the longitudinal modes contribute to light scattering. These give rise, respectively, to a *triplet spectrum* consisting of a central or *Rayleigh band* at the incident light frequency ω_0 of width determined by the "thermal diffusivity" and doublets called *Brillouin doublets* shifted to frequencies $\omega_0 \pm \omega(q)$ of width $q^2\Gamma$ where Γ is the classical acoustic attenuation coefficient and $\pm\omega(q)$ $(=c_S q)$ are the Doppler shifts of the scattered light produced by the sound modes, all of which have adiabatic velocities c_S. Light scattering therefore gives useful information about the hydrodynamic parameters in the system such as the transport coefficients and sound speed.

In this section we consider the hydrodynamic modes of a monatomic liquid that arise from the conservation of mass (or particle number), momentum, and energy. Light scattering is a probe of certain of these modes and, correspondingly, the spectrum of the scattered light contains useful information about the hydrodynamic parameters of the system. Thus it behooves us in this section to show how the hydrodynamic equations are solved to give the light-scattering spectrum. We steer a middle road here, leaving a full derivation for books on fluid mechanics (e.g., see Landau and Lifshitz, 1960), but giving sufficient detail for the novice to follow the arguments.

The basic variables of fluid mechanics are the conserved densities, the number density $\rho(\mathbf{r}, t)$, the momentum density $\mathbf{g}(\mathbf{r}, t)$, and the energy density $e(\mathbf{r}, t)$.[10] The conservation of mass (or number), momentum, and energy are expressed locally by the conservation equations [see Eq. (10.3.9)],

$$\frac{\partial \rho(\mathbf{r}, t)}{\partial t} + \nabla \cdot \mathbf{J}(\mathbf{r}, t) = 0 \tag{10.4.1a}$$

$$\frac{\partial g_i(\mathbf{r}, t)}{\partial t} + \nabla_j \tau_{ij}(\mathbf{r}, t) = 0 \qquad (i = 1, 2, 3) \tag{10.4.1b}$$

$$\frac{\partial e(\mathbf{r}, t)}{\partial t} + \nabla \cdot \mathbf{J}_e(\mathbf{r}, t) = 0 \tag{10.4.1c}$$

where $\mathbf{J}(\mathbf{r}, t)$, $\tau_{ij}(\mathbf{r}, t)$ and $\mathbf{J}_e(\mathbf{r}, t)$ are respectively the fluxes of number, momentum, and energy. In the foregoing ∇_j denotes the jth component of the gradient operator, repeated indices imply summation and $\tau_{ij}(\mathbf{r}, t)$ is the jth component of the flux of the ith component of the momentum. $\tau_{ij}(\mathbf{r})$ is a second-rank symmetric tensor so that $\tau_{ij} = \tau_{ji}$. These equations apply to the microscopic "densities" as well as to locally averaged densities. They can be used to define the microscopic fluxes.

We proceed by noting that, $N(t) = \int_V d^3r\, \rho(\mathbf{r},t)$, $\mathbf{G}(t) = \int_V d^3r\mathbf{g}(\mathbf{r},t)$, and $E(t) = \int_V d^3re(\mathbf{r}, t)$ are, respectively, the number of particles, momentum, and energy of fluid in an arbitrary fixed volume V. The rates of change of these quantities are, respectively,

$$\frac{dN(t)}{dt} = \int_V d^3r \, \frac{\partial \rho(\mathbf{r}, t)}{\partial t} \tag{10.4.2a}$$

$$\frac{d\mathbf{G}(t)}{dt} = \int_V d^3r \, \frac{\partial}{\partial t} \mathbf{g}(\mathbf{r}, t) \tag{10.4.2b}$$

$$\frac{dE(t)}{dt} = \int_V d^3r \, \frac{\partial}{\partial t} e(\mathbf{r}, t) \tag{10.4.2c}$$

If $\mathbf{dS}$ is an element of the surface S bounding V, then $\mathbf{u}(\mathbf{r}, t)\cdot\mathbf{dS}$ is the volume of fluid passing through dS out of V per unit time. It follows that $-\rho(\mathbf{r}, t)\mathbf{u}(\mathbf{r}, t)\cdot d\mathbf{S}$ is the number of particles passing into V through $\mathbf{dS}$ per unit time. Also $-\mathbf{g}(\mathbf{r}, t)\mathbf{u}(\mathbf{r}, t)\cdot d\mathbf{S}$ and $-\mathbf{e}(\mathbf{r}, t)\mathbf{u}(\mathbf{r}, t)\cdot\mathbf{dS}$ are, respectively, the amounts of momentum and energy passing per unit time into V through $\mathbf{dS}$ by convection. Integrating these over the entire surface S then gives the contributions to $\dot{N}(t)$, $\dot{\mathbf{G}}(t)$, and $\dot{E}(t)$ due to convection

$$\left[\frac{dN(t)}{dt}\right]_{conv} = -\int_S \rho(\mathbf{r}, t)\,\mathbf{u}(\mathbf{r}, t) \cdot d\mathbf{S} = \frac{dN(t)}{dt} \tag{10.4.3a}$$

$$\left[\frac{dG(t)}{dt}\right]_{conv} = -\int_S \mathbf{g}(\mathbf{r}, t)\,\mathbf{u}(\mathbf{r}, t) \cdot d\mathbf{S} \tag{10.4.3b}$$

$$\left[\frac{dE(t)}{dt}\right]_{conv} = -\int_S e(\mathbf{r}, t)\,\mathbf{u} \cdot d\mathbf{S} \tag{10.4.3c}$$

The number of particles in V can change only by virtue of convection so that $dN/dt = (dN/dt)_{conv}$. Combining Eqs. (10.4.2a) and (10.4.3a) then gives

$$\int_V d^3r \, \frac{\partial \rho}{\partial t} = -\int_S \rho\mathbf{u} \cdot \mathbf{dS} \tag{10.4.4a}$$

The momentum in the volume V changes not only by convection, but also because the fluid outside V can exert a force $\mathbf{F}$ on the fluid in V. If $d\mathbf{F}$ is the force acting on an element of surface $d\mathbf{S}$, then $d\mathbf{F} = \boldsymbol{\sigma}\cdot d\mathbf{S}$ where $\boldsymbol{\sigma}$ is called the *stress tensor*. The total force acting on V is then $\mathbf{F} = \int_S \boldsymbol{\sigma} \cdot d\mathbf{S}$. Now equating $d\mathbf{G}/dt$ of Eq. (10.4.2b) to the sum of the convective contribution of Eq. (10.4.3b) and the force $\mathbf{F}$ we find

$$\int_V d^3r \, \frac{\partial}{\partial t} \mathbf{g}(\mathbf{r}, t) = \int_S [\boldsymbol{\sigma}(\mathbf{r}, t) - m\rho(\mathbf{r}, t)\,\mathbf{u}(\mathbf{r}, t)\,\mathbf{u}(\mathbf{r}, t)] \cdot \mathbf{dS} \tag{10.4.4b}$$

where we have substituted $\mathbf{g}(\mathbf{r}, t) = \mathbf{m}\rho(\mathbf{r}, t)\,\mathbf{u}(\mathbf{r}, t)$ into Eq. (10.4.3b).

The energy in volume V changes not only by convection, but also by virtue of the work done by the fluid outside V on the fluid in V and by virtue of the heat diffusion.[11] Since the force exerted on $d\mathbf{S}$ is $d\mathbf{F} = \boldsymbol{\sigma} \cdot d\mathbf{S}$, the work done on $d\mathbf{S}$ per unit time is $\mathbf{u} \cdot d\mathbf{F} = \mathbf{u} \cdot \boldsymbol{\sigma} \cdot d\mathbf{S}$, and the work done on V per unit time is $\int_S \mathbf{u} \cdot \boldsymbol{\sigma} \cdot d\mathbf{S}$. If $\mathbf{Q}$ is the diffusive flux of heat, then $-\mathbf{Q}\cdot d\mathbf{S}$ is the amount of heat diffusing into $d\mathbf{S}$ per unit time and the total heat entering V this way per unit time is $-\int_S \mathbf{Q} \cdot \mathbf{dS}$. Adding these terms to Eq. (10.4.3) we find $(d\mathbf{E}/dt)$, that is,

$$\int_V d^3r \, \frac{\partial e(\mathbf{r}, t)}{\partial t} = \int_S (\mathbf{u} \cdot \boldsymbol{\sigma} - \mathbf{Q} - e\mathbf{u}) \cdot \mathbf{dS} \tag{10.4.4c}$$

Applying Gauss's theorem to convert the surface integrals to volume integrals on the right-hand sides of Eqs. (10.4.4) gives three equations such as $\int_V d^3r\ Y = 0$. Since the volume V is arbitrary, $Y = 0$. Thus we find the local conservation laws

$$\frac{\partial \rho}{\partial t} + \nabla \cdot \rho \mathbf{u} = 0 \tag{10.4.5a}$$

$$\frac{\partial \mathbf{g}}{\partial t} + \nabla \cdot (m\rho \mathbf{u}\mathbf{u} - \boldsymbol{\sigma}) = 0 \tag{10.4.5b}$$

$$\frac{\partial e}{\partial t} + \nabla \cdot (\mathbf{Q} + e\mathbf{u} - \mathbf{u} \cdot \boldsymbol{\sigma}) = 0 \tag{10.4.5c}$$

Comparison of Eqs. (10.4.1), with Eqs. (10.4.5) gives

$$\mathbf{J}(\mathbf{r}, t) = \rho(\mathbf{r}, t)\, \mathbf{u}(\mathbf{r}, t) \tag{10.4.6a}$$

$$\tau_{ij}(\mathbf{r}, t) = m\rho(\mathbf{r}, t)\, u_i(\mathbf{r}, t)\, u_j(\mathbf{r}, t) - \sigma_{ij}(\mathbf{r}, t) \tag{10.4.6b}$$

$$\mathbf{J}_e(\mathbf{r}, t) = e(\mathbf{r}, t)\, \mathbf{u}(\mathbf{r}, t) + \mathbf{Q}(r, t) - \mathbf{u}(\mathbf{r}, t) \cdot \boldsymbol{\sigma}(\mathbf{r}, t) \tag{10.4.6c}$$

The force acting on $d\mathbf{S}$ is, as we have seen, $d\mathbf{F} = \boldsymbol{\sigma} \cdot d\mathbf{S}$. This relationship defines the stress tensor $\boldsymbol{\sigma}$. To get some feeling for the physical meaning of the stress tensor, note that the x-component of the force acting on $d\mathbf{S}$ is

$$dF_x = \sigma_{xx} dS_x + \sigma_{xy} dS_y + \sigma_{xz} dS_z$$

From this we see that the component σ_{xx} is the force per unit area exerted on a plane surface which is perpendicular to the x direction. Likewise σ_{xy} is the x component of the force per unit area exerted on a plane perpendicular to the y direction and σ_{xz} is the x component of the force per unit area exerted on a plane perpendicular to the z direction. The other six components of the stress tensor can be similarly interpreted. Because the diagonal elements σ_{xx}, σ_{yy}, σ_{zz} represent forces per unit area normal to given planes, they are called *normal components* of the stress tensor. The off-diagonal components such as σ_{xy} and σ_{xz} represent forces per unit area parallel to their planes of reference and are consequently called the *shear components* or *shear stresses*. For a fluid in overall equilibrium the shear stresses are zero and the normal stresses are equal (otherwise the fluid would move) and independent of position. Thus at equilibrium $\sigma_{xx} = \sigma_{yy} = \sigma_{zz} = -p_0$ where p_0 is the equilibrium pressure.[12] In general we define the hydrostatic pressure $p(\mathbf{r},t)$ as the normal force per unit area averaged over three mutually orthogonal planes through the point $\mathbf{r}$, that is, by[13] $p(\mathbf{r}, t) \equiv -\frac{1}{3} Tr\boldsymbol{\sigma}(\mathbf{r}, t)$ where the trace indicates the sum of the diagonal elements of the stress tensor.

The stress tensor σ_{ij} is a symmetric tensor[14] and can be subdivided into a pressure part $p\delta_{ij}$ and a viscous part $\sigma_{ij}'(\mathbf{r}, t)$; that is,

$$\sigma_{ij} = -p\delta_{ij} + \sigma_{ij}' \tag{10.4.7}$$

To proceed it is helpful to consider a "gedanken" experiment. Consider two concentric cylinders the annular region of which is filled with a fluid. Now start rotating the inner cylinder uniformly. At first the outer cylinder is stationary; nevertheless, after some time the outer cylinder begins to rotate. The inner cylinder imparts tangential momentum to the fluid. Eventually this tangential momentum diffuses to the outer cylinder, causing it to rotate. This flux of momentum is precisely the shear flux. Now this flux delivers momentum from regions of high momentum to regions of low momentum, that is, in the direction opposite to that of the velocity gradient. Moreover, the flux is experimentally found to be proportional to the magnitude of this gradient, that is, the greater the difference in rotational velocities of the inner and outer cylinders, the larger the flux. Consider the components of the velocity flow gradients $\nabla_i u_j$. These are components of a second-rank tensor. σ'_{ij} is a second-rank symmetric tensor which, we surmise, should be proportional in some sense to $\nabla_i u_j$. There are two independent symmetric irreducible second-rank tensors that can be found from $\nabla_i u_j$ in an isotropic (rotationally invariant) system.[15] These are $\nabla\cdot\mathbf{u}\delta_{ij}$; $[\nabla_i u_j + \nabla_j u_i - \frac{2}{3}\nabla\cdot\mathbf{u}\delta_{ij}]$, the first being the scalar part and the second being the traceless symmetric part of $\nabla_i u_j$. Thus we assume that σ'_{ij} is a linear combination of these two irreducible tensors so that

$$\sigma'_{ij} = \eta_s[\nabla_i u_j + \nabla_j u_i - \frac{2}{3}\nabla\cdot\mathbf{u}\delta_{ij}] + \eta_v\nabla\cdot\mathbf{u}\delta_{ij} \tag{10.4.8}$$

The two expansion coefficients η_s and η_v are called the *shear* and *bulk viscosities*, respectively. The shear viscosity is all that is required to describe our gedanken experiment. The bulk viscosity describes the viscous or dissipative part of the response to a compression. This linear constitutive relation is called the *Newtonian stress tensor*. A fluid correctly described by this form is called a *Newtonian fluid*.[16]

Combining Eqs. (10.4.7), (10.4.8), and (10.4.6b) then gives the constitutive relations for τ_{ij} as

$$\tau_{ij} = m\rho u_i u_j + p\delta_{ij} - \eta_s(\nabla_i u_j + \nabla_j u_i - \frac{2}{3}\nabla\cdot\mathbf{u}\delta_{ij}) - \eta_v\nabla\cdot\mathbf{u}\delta_{ij} \tag{10.4.9}$$

It remains to determine the energy flux $\mathbf{J}_e$. The diffusive part of the energy flux $\mathbf{Q}$ is given by *Fourier's law*

$$\mathbf{Q} = -\lambda\nabla T \tag{10.4.10}$$

According to this law, heat flows from regions of high temperature to regions of low temperature, that is, in a direction opposite to the temperature gradient. The quantity λ is called the thermal conductivity. The local energy density can be subdivided into a local kinetic energy density $\frac{1}{2}m\rho u^2$ and a local internal energy density $e'(\mathbf{r}, t)$, so that

$$e(\mathbf{r}, t) = \tfrac{1}{2}m\rho(\mathbf{r}, t)u^2(\mathbf{r}, t) + e'(\mathbf{r}, t) \tag{10.4.11}$$

Substitution of Eqs. (10.4.10) and (10.4.11) into Eq. (10.4.6c) then gives the linear constitutive relation for $\mathbf{J}_e$

$$\mathbf{J}_e = (\tfrac{1}{2}m\rho u^2 + e')\mathbf{u} + \mathbf{u}\cdot\boldsymbol{\sigma} - \lambda\nabla T \tag{10.4.12}$$

where it remains to substitute Eqs. (10.4.7) and (10.4.8) into Eq. (10.4.12).

The constitutive relations Eqs. (10.4.6a), (10.4.9), and (10.4.12), when combined with the conservation Eqs. (10.4.1), give the basic equations of fluid mechanics which are a set of five nonlinear partial differential equations involving the seven variables, ρ, $\mathbf{g}$, e, p, and T. Five equations cannot be used to determine seven quantities. The equations are closed by expressing any two variables of the set (ρ, e, p, T) in terms of the two remaining variables of the set, using the assumptions of local equilibrium and the thermodynamic equations of state.

The Onsager regression hypothesis plus the above equations of fluid mechanics can be used to determine the correlation functions required in Eq. (10.1.5). Because the density fluctuation $\rho_1 \equiv \delta\rho(\mathbf{r}, t)$, the energy fluctuation $e_1 \equiv \delta e(\mathbf{r}, t)$, the temperature fluctuation $T_1(\mathbf{r}, t) \equiv \delta T(\mathbf{r}, t)$, and the momentum fluctuations $\mathbf{g}_1 \equiv \delta\mathbf{g}(\mathbf{r}, t)$ [or velocity fluctuations $(\mathbf{u}_1 \equiv \delta\mathbf{u}(\mathbf{r}, t))$] around the equilibrium values ρ_0, e_0, T_0, $\mathbf{g}_0$ ($= 0$ or $\mathbf{u}_0 = 0$) are expected to be very small, we can use the linearized equations of fluid mechanics. These are obtained by substituting $\rho = \rho_0 + \rho_1$, $e = e_0 + e_1$, $T = T_0 + T_1$, $\mathbf{g} = \mathbf{g}_0 + \mathbf{g}_1$, $\mathbf{u} = \mathbf{u}_0 + \mathbf{u}_1$ into the equations and retaining all terms that are no higher than first order in the fluctuations. Thus, for example, in $e = (\frac{1}{2}m\rho u^2 + e')$, since $u_0 = 0$ and $u^2 = u_1{}^2$, $\frac{1}{2}m\rho u_1^2$ can be be omitted because it is at least second order in the fluctuation $\mathbf{u}_1$. To this order then $e_1 = e_1'$, and we need only consider fluctuations in the internal energy. A consistent application of these arguments then gives the *linearized equations of fluid mechanics*. Thus substitution of the linearized fluxes into the conservation laws [Eq. (10.4.5)] gives the linearized equations of fluid mechanics.

$$\frac{\partial \rho_1}{\partial t} + \rho_0 \nabla \cdot \mathbf{u}_1 = 0 \tag{10.4.13a}$$

$$m\rho_0 \frac{\partial \mathbf{u}_1}{\partial t} = -\nabla p_1 + \eta_s \nabla^2 \mathbf{u}_1 + \left(\eta_v + \frac{1}{3}\eta_s\right) \nabla(\nabla \cdot \mathbf{u}_1) \tag{10.4.13b}$$

$$\frac{\partial e_1}{\partial t} + (e_0 + p_0)\, \nabla \cdot \mathbf{u}_1 = \lambda \nabla^2 T_1 \tag{10.4.13c}$$

We note from these equations that the quantity $\psi = (\nabla \cdot \mathbf{u}_1)$ couples directly to the energy and number density.

The energy equation can be transformed somewhat. Eliminating $\psi = \nabla \cdot \mathbf{u}_1$ from Eq. (10.4.13c) by means of Eq. (10.4.13a) allows us to write Eq. (10.4.13c) as

$$\frac{\partial h_1}{\partial t} = \lambda \nabla^2 T_1 \tag{10.4.13d}$$

where

$$h_1 \equiv e_1 - \frac{(e_0 + p_0)}{\rho_0} \rho_1$$

The quantity h_1 is the fluctuation in the heat density.[17] In terms of the entropy per unit volume $ds \equiv dS/V$, the left-hand side of the equation reads $T ds$, so that $dh = T ds$ (or $h_1 = T_0 s_1$) and the energy-balance equation can be expressed in terms of the fluctuations in the entropy per unit volume.

The linearized equations of fluid mechanics are therefore

$$\frac{\partial \rho_1}{\partial t} + \rho_0 \nabla \cdot \mathbf{u}_1 = 0 \tag{10.4.14a}$$

$$m\rho_0 \frac{\partial \mathbf{u}_1}{\partial t} = -\nabla p_1 + \eta_s \nabla^2 \mathbf{u}_1 + \left(\eta_v + \frac{1}{3}\eta_s\right) \nabla(\nabla \cdot \mathbf{u}_1) \tag{10.4.14b}$$

$$T_0 \frac{\partial s_1}{\partial t} = \lambda \nabla^2 T_1 \tag{10.4.14c}$$

where the subscript $_1$ denotes a fluctuation and the subscript $_0$ denotes true equilibrium values.

There are five linear hydrodynamic equations containing the seven fluctuations $(\rho_1, u_{1x}, u_{1y}, u_{1z}, p_1, s_1, T_1)$. The local equilibrium thermodynamic equations of state can be used to eliminate two of the four scalar field quantities (p_1, s_1, T_1, ρ_1). In this chapter we chose the temperature and number density as independent variables, although we could equally well have chosen the pressure and entropy. One useful criterion for choosing a particular set is that the equilibrium fluctuations of the two variables be statistically independent. The two sets $(\rho_1 = \delta\rho, T_1 = \delta T)$ and $(p_1 = \delta p, s_1 = \delta s)$ both involve two variables that are statistically independent, that is, $\langle \delta\rho\delta T\rangle = \langle \delta p\, \delta s\rangle = 0$. This is shown in Appendix 10.C. The statistical independence of the two variables simplifies our analysis considerably. It is particularly convenient to chose the set (ρ_1, T_1) over the set (p_1, s_1) because the dielectric constant derivatives $(\partial\varepsilon/\partial\rho)_T$ and $(\partial\varepsilon/\partial T)_\rho$ are more readily obtained from experiment (other than light scattering) than are $(\partial\varepsilon/\partial S)_p$ and $(\partial\varepsilon/\partial p)_S$.

For this choice of independent variables it is necessary to eliminate the entropy, s_1, and the pressure, p_1, fluctuations in order to close the hydrodynamic equations. The assumption of local equilibrium enables us to write

$$\delta s = \left(\frac{\partial s}{\partial \rho}\right)_T \delta\rho + \left(\frac{\partial s}{\partial T}\right)_\rho \delta T; \quad \delta p = \left(\frac{\partial p}{\partial \rho}\right)_T \delta\rho + \left(\frac{\partial p}{\partial T}\right)_\rho \delta T \qquad (10.4.15)$$

where $\delta s = s_1$, $\delta p = p_1$, and $\delta T = T_1$. The derivatives can be expressed in terms of measurable quantities using well-known thermodynamic identities (cf. Landau and Lifshitz, 1960). The useful thermodynamic definitions and identities[18] (see Appendix 10.B) are

$$\chi_T \equiv \rho^{-1}\left(\frac{\partial \rho}{\partial p}\right)_T \qquad (\chi_T \text{ is the isothermal compressibility})$$

$$\chi_S \equiv \rho^{-1}\left(\frac{\partial \rho}{\partial p}\right)_S \qquad (\chi_s \text{ is the adiabatic compressibility})$$

$$\gamma \equiv c_P/c_V \equiv \chi_T/\chi_S \qquad (\text{specific heat ratio})$$

$$\left(\frac{\partial s}{\partial T}\right)_\rho = m\rho c_V/T \qquad (c_V \text{ is the specific heat at constant } V)$$

$$\left(\frac{\partial s}{\partial T}\right)_p = m\rho c_P/T \qquad (c_P \text{ is the specific heat at constant } p)$$

$$c_S^2 \equiv \left(\frac{\partial P}{\partial \rho}\right)_S = (m\rho\chi_S)^{-1} \qquad (c_S \text{ is the adiabatic sound speed})$$

$$c_T^2 \equiv \left(\frac{\partial P}{\partial \rho}\right)_T = (m\rho\chi_T)^{-1} \qquad (c_T \text{ is the isothermal sound speed})$$

$$\alpha = -\rho^{-1}\left(\frac{\partial \rho}{\partial T}\right)_P \qquad (\alpha \text{ is the thermal expansion coefficient})$$

It is shown in Appendix 10.B that [cf. Eq. (10.B.8)]

$$p_1 = mc_T^2[\rho_1 + \alpha\rho_0 T_1]; \; s_1 = -\frac{m\rho_0 c_V}{T_0}\left[\frac{(\gamma-1)}{\alpha\rho_0}\rho_1 - T_1\right] \qquad (10.4.16)$$

Taking the divergence of Eq. (10.5.25b), calling

$$\psi_1 = \nabla \cdot \mathbf{u}_1 \tag{10.4.17}$$

and substituting p_1 and s_1 from Eq. (10.4.16) into Eq. (10.4.14) leads to the linearized hydrodynamic equations

$$\frac{\partial \rho_1}{\partial t} + \rho_0 \psi_1 = 0 \tag{10.4.18a}$$

$$\frac{\partial \psi_1}{\partial t} + \frac{c_T^2}{\rho_0} \nabla^2 \rho_1 + \alpha c_T^2 \nabla^2 T_1 - D_V \nabla^2 \psi_1 = 0 \tag{10.4.18b}$$

$$\frac{\partial T_1}{\partial t} - \frac{(\gamma - 1)}{\alpha \rho_0} \frac{\partial \rho_1}{\partial t} - \gamma D_T \nabla^2 T_1 = 0 \tag{10.4.18c}$$

where D_V and D_T are defined as

$$D_V \equiv (\eta_v + \frac{4}{3} \eta_s)/m\rho_0; \quad D_T \equiv \lambda/m\rho_0 c_P \tag{10.4.19}$$

D_V is called the *longitudinal kinematic viscosity,* and D_T is called the *thermal diffusivity.* These quantities are "diffusion coefficients" for the "diffusion" of longitudinal momentum and heat, respectively.

Equation (10.4.18) consists of three linear partial differential equations in the three unknowns which are to be solved subject to the three initial conditions $\rho_1(\mathbf{r}, 0)$, $\psi_1(\mathbf{r}, 0)$, $T_1(\mathbf{r}, 0)$.

These equations are most easily solved using Fourier–Laplace analysis. Introducing the Fourier–Laplace transforms

$$\tilde{\rho}_1(\mathbf{q}, s) = \int_0^\infty dt\, e^{-st} \int d^3r\, e^{i\mathbf{q}\cdot\mathbf{r}}\, \rho_1(\mathbf{r}, t) = \int_0^\infty dt\, e^{-st}\, \rho_1(\mathbf{q}, t)$$

$$\tilde{\psi}_1(\mathbf{q}, s) = \int_0^\infty dt\, e^{-st} \int d^3r\, e^{i\mathbf{q}\cdot\mathbf{r}}\, \rho_1(\mathbf{r}, t) = \int_0^\infty dt\, e^{-st}\, \psi_1(\mathbf{q}, t)$$

$$\tilde{T}_1(\mathbf{q}, s) = \int_0^\infty dt\, e^{-st} \int d^3r\, e^{i\mathbf{q}\cdot\mathbf{r}}\, T_1(r, t) = \int_0^\infty dt\, e^{-st}\, T_1(\mathbf{q}, t)$$

The quantities $\rho_1(\mathbf{q}, t)$, $\psi_1(\mathbf{q}, t)$, and $T_1(\mathbf{q}, t)$ are the spatial Fourier transforms of the fluctuations $\rho_1(\mathbf{r}, t)$, $\psi_1(\mathbf{r}, t)$, and $T_1(\mathbf{r}, t)$ respectively. It is the correlation functions of these quantities that describes the light-scattering spectrum [see Eq. (10.1.5)].

The Fourier–Laplace transform of Eq. (10.4.16) can be expressed in matrix form as

$$\begin{pmatrix} s & \rho_0 & 0 \\ -\omega^2(q)/\gamma\rho_0 & (s + D_V q^2) & -\alpha\omega^2(q)/\gamma \\ -(\gamma - 1)s/\alpha\rho_0 & 0 & (s + \gamma D_T q^2) \end{pmatrix} \begin{pmatrix} \tilde{\rho}_1(\mathbf{q}, s) \\ \tilde{\psi}_1(\mathbf{q}, s) \\ \tilde{T}_1(\mathbf{q}, s) \end{pmatrix} =$$

$$\begin{pmatrix} 1 & 0 & 0 \\ 0 & 1 & 0 \\ -(\gamma - 1)/\alpha\rho_0 & 0 & 1 \end{pmatrix} \begin{pmatrix} \rho_1(\mathbf{q}, 0) \\ \psi_1(\mathbf{q}, 0) \\ T_1(\mathbf{q}, 0) \end{pmatrix} \tag{10.4.20a}$$

This can be written in the more concise form

$$\tilde{\mathbf{M}}(\mathbf{q}, s) \cdot \tilde{\boldsymbol{\phi}}(\mathbf{q}, s) = \mathbf{N}(\mathbf{q}) \cdot \boldsymbol{\phi}(\mathbf{q}) \tag{10.4.20b}$$

where there is a one-to-one correspondance, reading left to right, between the four matrices in Eqs. (10.5.20a) and (10.5.20b).

In these equations $\omega(q)$ is defined by

$$\omega(q) \equiv c_s q \tag{10.4.21}$$

The solution of the matrix Eq. (10.5.30b) is

$$\tilde{\boldsymbol{\phi}}(\mathbf{q}, s) = \tilde{\mathbf{M}}^{-1}(\mathbf{q}, s) \cdot \mathbf{N}(\mathbf{q}) \cdot \boldsymbol{\phi}(\mathbf{q}, t = 0) \tag{10.4.22a}$$

where $\tilde{\boldsymbol{M}}^{-1}(\mathbf{q}, s)$ is the reciprocal of $\tilde{\mathbf{M}}(\mathbf{q}, s)$. This inverse is easily found by the standard methods of matrix algebra and is

$$\tilde{\boldsymbol{M}}^{-1}(\mathbf{q}, s) = \frac{1}{\tilde{M}(s)} \begin{pmatrix} [(s + D_V q^2)(s + \gamma D_T q^2)] & [-\rho_0(s + \gamma D_T q^2)] & \left[-\dfrac{\alpha\omega^2(q)\rho_0}{\gamma}\right] \\ \left[\dfrac{\omega^2(q)}{\rho_0}(s + D_T q^2)\right] & \left[s(s + \gamma D_T q^2)\right] & \left[\dfrac{\alpha\omega^2(q)}{\gamma} s\right] \\ \left[\dfrac{(\gamma - 1)}{\alpha\rho_0} s(s + D_V q^2)\right] & \left[\dfrac{-(\gamma - 1)s}{\alpha}\right] & \left[s(s + q^2 D_V) + \dfrac{\omega^2(q)}{\gamma}\right] \end{pmatrix} \tag{10.4.23}$$

where $\tilde{M}(s)$ is the determinant of the matrix $\tilde{\mathbf{M}}(\mathbf{q}, s)$, that is,

$$\tilde{M}(s) = s^3 + (D_V q^2 + \gamma D_T q^2)\, s^2 + (\omega^2(q) + \gamma D_T q^2 D_V q^2)s + \omega^2(q) D_T q^2 \tag{10.4.24}$$

Multiplying out the matrices in Eq. (10.4.22a) then gives the solutions

$$\tilde{M}(s) \begin{bmatrix} \tilde{\rho}_1(\mathbf{q}, s) \\ \tilde{\psi}_1(\mathbf{q}, s) \\ \tilde{T}_1(\mathbf{q}, s) \end{bmatrix} = \begin{bmatrix} \left[\begin{matrix}(s + D_V q^2)(s + \gamma D_T q^2) \\ + (\gamma - 1)\,\omega(q)/\gamma\end{matrix}\right] & [-\rho_0(s + \gamma D_T q^2)] & \left[-\dfrac{\alpha\rho_0\omega^2(q)}{\gamma}\right] \\ \left[\dfrac{\omega^2(q)}{\gamma\rho_0}(s + \gamma D_T q^2)\right] & [s(s + \gamma D_T q^2)] & \left[\dfrac{\alpha\omega^2(q)s}{\gamma}\right] \\ \left[-\dfrac{(\gamma - 1)\omega^2(q)}{\alpha\rho_0\gamma}\right] & \left[-\dfrac{(\gamma - 1)s}{\alpha}\right] & \left[s(s + q^2 D_V) + \dfrac{\omega^2(q)}{\gamma}\right] \end{bmatrix} \begin{bmatrix} \rho_1(\mathbf{q},0) \\ \psi_1(\mathbf{q},0) \\ T_1(\mathbf{q},0) \end{bmatrix} \tag{10.4.25}$$

The resulting equations for $\tilde{\rho}(\mathbf{q}, s)$, $\tilde{\psi}_1(\mathbf{q}, s)$ and $\tilde{T}_1(\mathbf{q}, s)$ are complicated algebraic equations that the reader is urged to write explicitly. What we require for light scattering are the correlation functions $\langle\rho_1^*(\mathbf{q},0)\rho_1(\boldsymbol{q},t)\rangle$, $\langle\rho_1^*(\mathbf{q},0)T_1(q,t)\rangle$, and so on, or equivalently their Laplace transforms $\langle\rho_1^*(\mathbf{q},0)\tilde{\rho}_1(\mathbf{q},s)\rangle$, $\langle\rho_1^*(\mathbf{q},0)\tilde{T}_1(\mathbf{q},s)\rangle$, $\langle T_1^*(\mathbf{q},0)\tilde{T}_1(\mathbf{q},s)\rangle$. These latter functions can be found by multiplying Eq. (10.4.25) in turn by $\rho_1^*(\mathbf{q}, 0)$, $\psi_1^*(\mathbf{q}, 0)$, and $T_1^*(\mathbf{q}, 0)$ and then ensemble-averaging. The resulting equations are greatly simplified because[20] in the limit $q \to 0$: (a) $\langle\rho_1^*(\mathbf{q},0)T_1(\mathbf{q},0)\rangle = 0$, (b) $\langle\rho_1^*(\mathbf{q},0)\psi_1(\mathbf{q},0)\rangle = 0$, and (c) $\langle T_1^*(\mathbf{q},0)\,\psi_1(\mathbf{q},0)\rangle = 0$. The resulting correlation functions are

$$\frac{\langle \rho_1^*(\mathbf{q},0)\tilde{\rho}_1(\mathbf{q},s)\rangle}{\langle \rho_1^*(\mathbf{q})\rho_1(\mathbf{q})\rangle} = \frac{B(s)}{\tilde{M}(s)} \tag{10.4.26a}$$

$$\frac{\langle T_1^*(\mathbf{q},0)\,\tilde{\rho}_1(\mathbf{q},s)\rangle}{\langle T_1^*(\mathbf{q},0)T_1(\mathbf{q},0)\rangle} = \frac{C\,(s)}{\tilde{M}(s)} \tag{10.4.26b}$$

$$\frac{\langle \rho_1^*(\mathbf{q},0)\,\tilde{T}_1(\mathbf{q},s)\rangle}{\langle \rho_1^*(\mathbf{q},0)\rho_1(\mathbf{q},0)\rangle} = \frac{D(s)}{\tilde{M}(s)} \tag{10.4.26c}$$

$$\frac{\langle T_1^*(\mathbf{q},0)\,\tilde{T}_1(\mathbf{q},s)\rangle}{\langle T_1^*(\mathbf{q},0)T_1(\mathbf{q},0)\rangle} = \frac{F\,(s)}{\tilde{M}(s)} \tag{10.4.26a}$$

where

$$B(s) \equiv (s + D_V q^2)\,(s + \gamma D_T q^2) + (\gamma - 1)\frac{\omega^2(q)}{\gamma} \tag{10.4.27a}$$

$$C(s) \equiv -\,\alpha\rho_0\frac{\omega^2(q)}{\gamma} \tag{10.4.27b}$$

$$D(s) = -\left(1 - \frac{1}{\gamma}\right)\frac{\omega^2\,(q)}{\alpha\rho_0} \tag{10.4.27c}$$

$$F(s) = s(s + q^2 D_V) + \frac{\omega^2(q)}{\gamma} \tag{10.4.27d}$$

From the discussion in Section 10.1 it follows that the dominant contribution to the spectrum is from the density–density correlation function. The techniques we now use to determine this function can also be applied to the remaining correlation functions. The spectrum can be found from the Laplace transforms as follows. The required time-correlation functions are real even functions of time (see Section 11.5), so that Eq. (6.2.6) can be used; that is,

$$S_{\rho\rho}(\mathbf{q},\,\omega) = \pi^{-1}\,Re\langle \rho_1^*(\mathbf{q},\,0)\tilde{\rho}_1(\mathbf{q},\,s = i\omega)\rangle \tag{10.4.28}$$

From Eq. (10.5.33a) we see that

$$S_{\rho\rho}(\mathbf{q},\,\omega) = \pi^{-1}\,S(q)\,Re\left\{\frac{B(s = i\omega)}{\tilde{M}(s = i\omega)}\right\} \tag{10.4.29}$$

where $S(q) \equiv \langle \rho_1^*(\mathbf{q})\,\rho_1(\mathbf{q})\rangle$ is called the *structure factor* (and is discussed in Section 10.6). Substitution of the explicit forms of $B(s)$ and $\tilde{M}(s)$ from Eqs (10.4.24) and (10.4.27a) into Eq. (10.4.28) leads to the full but very complicated spectrum

$$S_{\rho\rho}(\mathbf{q},\,\omega) = \pi^{-1}\,S(q)\frac{[N_1(\omega)D_1(\omega) + N_2(\omega)\,D_2(\omega)]}{[D_1^2(\omega) + D_2^2(\omega)]} \tag{10.4.30a}$$

where

$$N_1(\omega) = -\omega^2 + \gamma D_T D_V q^4 + (1 - 1/\gamma)\,\omega^2(q) \tag{10.4.30b}$$

$$N_2(\omega) = \omega[\gamma D_T q^2 + D_V q^2] \tag{10.4.30c}$$

$$D_1(\omega) = -\omega^2[\gamma D_T q^2 + D_V q^2] + \omega^2(q)\,D_T q^2 \tag{10.4.30d}$$

$$D_2(\omega) = \omega[-\omega^2 + \omega^2(q) + \gamma D_T D_V q^4] \tag{10.4.30e}$$

From Eq. (10.4.30) it is clear that the spectrum is completely determined by the transport coefficients $(\eta_s, \eta_v, \lambda)$ and the thermodynamic properties $(\rho_0, c_S, \langle|\rho_1|^2\rangle, c_V, c_P)$. The spectrum should therefore be useful in measuring various combinations of these properties. Unfortunately Eq. (10.4.30) is so complicated that it is very difficult to interpret the important features of the spectrum. For most liquids a considerable simplification can be made by using the perturbation theory outlined in Section 6.3.

The density–density correlation function is found by Laplace inverting Eq. (10.426a). This gives

$$\frac{\langle\rho_1^*(\mathbf{q}, 0)\rho_1(\mathbf{q}, t)\rangle}{\langle\rho_1^*(\mathbf{q}, 0)\rho_1(\mathbf{q}, 0\rangle} = \sum_{i=1}^{3}\left\{\lim_{s\to si} \frac{B(s_i)(s - s_i)}{\tilde{M}(s)}\right\} \exp s_i|t| \tag{10.4.31}$$

where $s_1 = s_+$, $s_2 = s_-$, $s_3 = s_0$ are the three roots[21] of the dispersion equation

$$\tilde{M}(s) = s^2 + (D_V q^2 + \gamma D_T q^2)s^2 + (\omega^2(q) + \gamma D_T q^2 D_V q^2)s + \omega^2(q) D_T q^2 = 0 \tag{10.4.32}$$

$\tilde{M}(s)$ can be expressed in terms of its roots as

$$M(s) = (s - s_1)(s - s_2)(s - s_3) = (s - s_+)(s - s_-)(s - s_0) \tag{10.4.33}$$

There are three parameters, $D_V q^2$, $\gamma D_T q^2$, and $\omega(q)$ in Eq. (10.4.32). In most fluids $D_V q^2$ and $\gamma D_T q^2$ are very small compared with $\omega(q)$. For example, in argon at $T_0 = 235°$, $\rho_0 = 1$ g/cm^3, $c_S = 6.85 \times 10^4$ cm/sec, $D_T = 1.0 \times 10^{-3}$ cm^2/sec, and $D_V = 1.6 \times 10^{-3}$ cm^2/sec, the quantities $x \equiv \gamma D_T q^2/\omega(q)$ and $y \equiv D_V q^2/\omega(q)$ have the values $x = 1.5 \times 10^{-3}$ and $y = 2.4 \times 10^{-3}$ when $q = 10^5$ cm^{-1} (which is typical for light scattering). In the following we therefore treat $D_V q^2$ and $\gamma D_T q^2$ as small numbers compared to $\omega(q)$.

The solution of the dispersion equation Eq. (10.4.32) can be expressed as $s = s^{(0)} + s^{(1)} + s^{(2)} + \ldots$ where $s^{(n)}$ is a term of order n in any of the small quantities $q^2 D_V$ and $\gamma D_T q^2$. This allows us to write a series of perturbation equations for the zeroth, first, . . . , n^{th} order term (as in Section 6.3). These are

$$[s^{(0)}]^3 + \omega^2(q)s^{(0)} = 0 \tag{10.4.34a}$$

$$3[s^{(0)}]^2 s^{(1)} + (D_V q^2 + \gamma D_T q^2)[s^{(0)}]^2 + \omega^2(q)s^{(1)} + \omega^2(q) D_T q^2 = 0 \tag{10.4.34b}$$

The roots of the zeroth-order equation are $s_{\pm}^{(0)} = \pm i\omega(q)$ and $s_0^{(0)} = 0$. The roots of the first-order equation corresponding to these zeroth order roots are $s_{\pm}^{(1)} = -\frac{1}{2}[(\gamma - 1)D_T + D_V]q^2$ and $s_0^{(1)} = -D_T q^2$. Adding each zeroth order root to its corresponding first-order root gives the three roots of the dispersion equation which are correct to first order in $D_V q^2$ and $D_T q^2$. These are

$$\begin{cases} s_0 = -D_T q^2 \\ s_{\pm} = \pm i\omega(q) - \Gamma q^2 \end{cases} \tag{10.4.35}$$

where we have defined

$$\Gamma \equiv \tfrac{1}{2}[(\gamma - 1)D_T + D_V] = \tfrac{1}{2}\left[\frac{(\gamma - 1)\lambda}{m\rho_0 c_P} + \frac{\left(\eta_v + \frac{4}{3}\eta_s\right)}{m\rho_0}\right] \tag{10.4.36}$$

Γ is the well-known classical attenuation coefficient of sound and D_T is the thermal diffusivity [Eq. (10.4.19)]. Substitution of these roots into Eq. (10.4.31) yields[22]

$$\frac{\langle \rho_1^*(\mathbf{q},0)\rho_1(\mathbf{q},t)\rangle}{\langle \rho_1^*(\mathbf{q},0)\rho_1(\mathbf{q},0)\rangle} = \left(1 - \frac{1}{\gamma}\right) \exp -q^2 D_T|t| + \frac{1}{\gamma} \exp - q^2\Gamma|t| \cos \omega(q)|t| + \frac{1}{\gamma} b(q) \exp -q^2\Gamma|t| \sin \omega(q)|t| \tag{10.4.37}$$

where $b(q)$ is defined as

$$b(q) \equiv q\left[\frac{3\Gamma - D_V}{\gamma c_S}\right] \tag{10.4.38}$$

The corresponding spectral density is easily evaluated by Fourier transforming Eq. (10.4.37). This gives

$$\begin{aligned} S_{\rho\rho}(\mathbf{q}, \omega) = \frac{1}{\pi} V\rho^2 k_B T \chi_T \Bigg\{&\left(1 - \frac{1}{\gamma}\right)\left[\frac{D_T q^2}{\omega^2 + [D_T q^2]^2}\right] \\ &+ \frac{1}{\gamma}\left(\frac{\Gamma q^2}{[\omega - \omega(q)]^2 + [\Gamma q^2]^2} + \frac{\Gamma q^2}{[\omega + \omega(q)]^2 + [\Gamma q^2]^2}\right) \\ &+ \frac{1}{\gamma} b(q)\left(\frac{[\omega + \omega(q)]}{[\omega + \omega(q)]^2 + [\Gamma q^2]^2} - \frac{[\omega - \omega(q)]}{[\omega - \omega(q)]^2 + [\Gamma q^2]^2}\right)\Bigg\} \end{aligned} \tag{10.4.39}$$

where we have used Eq. (10.1.8).

Except for the last term, which is usually difficult to observe, the light-scattering spectrum is a sum of Lorentzian line shapes. The first term represents an unshifted line called the *Rayleigh* or *central line* which is a Lorentzian line with half-width at half maximum

$$\Delta\omega_c(q) = D_T q^2 = (\lambda/m\rho_0 c_P)q^2 \tag{10.4.40}$$

The next two terms represent a doublet, called the *Brillouin doublet*. These are two Lorentzian lines symmetrically shifted by

$$\pm\omega(q) = \pm c_S q \tag{10.4.41}$$

each of half-width at half maximum

$$\Delta\omega_B(q) = \Gamma q^2 = \frac{1}{2}\left[\frac{\left(\eta_v + \frac{4}{3}\eta_s\right)}{m\rho_0} + \frac{\lambda(\gamma - 1)}{m\rho_0 c_P}\right]q^2 \tag{10.4.42}$$

The last two terms in Eq. (10.4.39) represent a non-Lorentzian correction which shifts the apparent Brillouin peaks toward the center slightly and renders the doublets asymmetric (the whole spectrum is, nevertheless, still symmetric about $\omega = 0$) about $\pm\omega(q)$. This last term is usually very small.

The treatment here was based on the following two assumptions: (a) that the fluctuations can be described by the simple linearized hydrodynamic equations and (b) that $\gamma D_T q^2 \ll \omega(q)$; $\Gamma q^2 \ll \omega(q)$, which implies that the widths are small compared to the shifts.

Condition (a) is expected to hold in sufficiently dense fluids. In gases, if $q \times$ (mean free path) is small, (a) is expected to hold, but if $q \times$ (mean free path) is large, the hydrodynamic equations cannot be used to compute the fluctuations. It is then necessary to use the Boltzmann equation to compute the spectrum. This is discussed in Chapter 14. In the event that (a) is valid, it is still possible that for sufficiently large q, (b) will not be valid. In this case Eq. (10.4.30) must be used.

The *Rayleigh–Brillouin* spectrum arises from the inelastic interaction between a photon and the hydrodynamic modes of the fluid. The doublets can be regarded as the "Stokes" and "anti-Stokes" translational Raman spectrum of the liquid. These lines are due to the inelastic collision between a photon and the fluid in which the photon gains or loses energy to the "phonons" or sound modes in the fluid, and thus suffers a frequency shift $\pm\omega(q)$. The width of these bands gives the lifetime $(q^2\Gamma)^{-1}$ of a "classical phonon" of wave vector $\mathbf{q}$. The Rayleigh band, on the other hand, represents the scattering of the light by the entropy, or heat fluctuations, which are purely diffusive or dissipative modes of the fluid. In a fluid such as argon, the translational motions are to a good approximation classical, and the above theory is adequate. In solids or such liquids as helium and hydrogen the modes are quantized, and a full quantum-mechanical treatment is required. This is beyond the scope and intent of this book.

Measurement of the width of the central component requires high resolution studies because $D_T q^2$ is usually less than 10^7 sec^{-1}. It is often difficult to perform such measurements, but many studies confirm the predicted width.

A detailed study of the Rayleigh–Brillouin spectrum of liquid argon recently made by Fleury and Boon (1969) showed that the normalized spectrum, $S(\mathbf{q}, \omega)/S(\mathbf{q})$, is described by Eq. (10.4.30) to within experimental error. In their experiment $q \simeq 2.1 \times 10^5$ cm^{-1}, T = 85°K and $P = 592.5$ mm Hg. From Eq. (10.4.41), the sound speed is $c_S = 850 \pm 4$ m/sec; this compares very well with the low-frequency sound speed measured acoustically, $c_S = 853$ m/sec. A typical spectrum is shown in Fig. (10.4.1).

To summarize, the properties that can be measured with Brillouin scattering are

χ_T, the isothermal compressibility

γ, the heat capacity ratio c_P/c_V

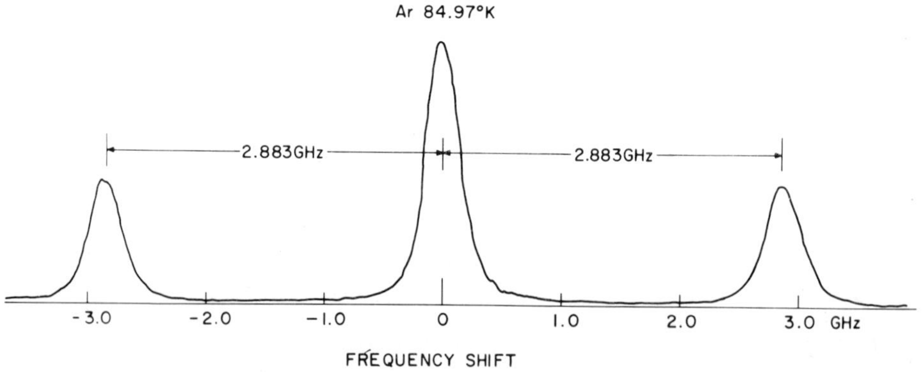

FIG. 10.4.1. Brillouin spectrum of liquid argon ($T = 84.97$°K, $\theta = 90°\ 14'$, laser wavelength = 5145 Å). (From Fleury and Boon, 1969.)

D_T, the thermal diffusivity $\lambda/m\rho_0 c_P$

Γ, the acoustic attenuation coefficient

c_S, the adiabatic sound velocity

D_V, the longitudinal kinematic viscosity $\dfrac{\left(\eta_v + \frac{4}{3}\eta_s\right)}{m\rho_0}$

If the shear viscosity is known by separate experiment it is possible to determine the bulk viscosity, a property not readily obtainable by experiment.

In the foregoing, we have ignored the temperature fluctuations entirely as a primary variable[23] (we did not ignore it as a secondary variable). This was based on the empirical observation that in simple fluids the dielectric constant is a strong function of the density and a weak function of the temperature, so that Eq. (10.1.6) is valid. Were this not the case we would require the remaining functions in Eq. (10.4.26). It is a very simple matter to compute the spectrum associated with each of these functions, which we leave as an exercise for the reader. The results are to first order in q^2D_V and q^2D_T.

$$\frac{\langle T_1^*(\mathbf{q},0)\rho_1(\mathbf{q},t)\rangle}{\langle T_1^*(\mathbf{q},0)T_1(\mathbf{q},0)\rangle} = -\frac{\alpha\rho_0}{\gamma}\left\{\exp -q^2D_T|t| - \exp - \Gamma q^2|t| \cos\omega(q)|t| + \frac{(D_T-\Gamma)q}{c_S}\exp - \Gamma q^2|t| \sin\omega(q)|t|\right\} \tag{10.4.43a}$$

$$\frac{\langle \rho_1^*(\mathbf{q},0)T_1(\mathbf{q},t)\rangle}{\langle \rho_1^*(\mathbf{q},0)\rho_1(\mathbf{q},0)\rangle} = -\frac{\left(1-\frac{1}{\gamma}\right)}{\alpha\rho_0}\left\{\exp -q^2D_T|t| - \exp - \Gamma q^2|t| \cos\omega(q)|t| + \frac{(D_T-\Gamma)q}{c_S}\exp - \Gamma q^2|t| \sin\omega(\mathrm{q})|t|\right\} \tag{10.4.43b}$$

$$\frac{\langle T_1^*(\mathbf{q},0)T_1(\mathbf{q},t)\rangle}{\langle T_1^*(\mathbf{q},0)T_1(\mathbf{q},0)\rangle} = \frac{1}{\gamma}\exp -q^2D_T|t| + \left(1-\frac{1}{\gamma}\right)\exp -q^2\Gamma|t|\cos\omega(q)|t| + \left(1-\frac{1}{\gamma}\right)\left(\frac{\Gamma-2D_T}{c_S}\right)q\exp - q^2\Gamma|t|\sin\omega(q)|t| \tag{10.4.43c}$$

where from Appendix 11.C

$$\langle|\rho_1|^2\rangle = \frac{1}{V}\rho_0^2 k_B T\chi_T \text{ and } \langle|T_1|^2\rangle = \frac{1}{V}\frac{k_B T^2}{m\rho_0 c_V} \tag{10.4.44}$$

In the event that Eq. (10.4.39) is valid, the integrated intensity [cf. Eq. (3.2.16)] is proportional to

$$I(\mathbf{q}) = \int_{-\infty}^{-\infty} d\omega S(\mathbf{q},\omega) = S(\mathbf{q}) = I_c + 2I_B \tag{10.4.45}$$

where I_C and $2I_B$ are respectively the areas under the central and two Brillouin components. In Eq. (10.4.37) the first term corresponds to the central line, and the second and third terms contribute to the Brillouin doublets. The initial values of these terms[24] are simply the integrated areas I_C, and $2I_B$,

$$I_C = \rho_0^2 k_B T \chi_T (1 - 1/\gamma);\ 2I_B = \rho_0^2 k_B T \chi_T (1/\gamma)$$

The intensity ratio, often called the *Landau–Placzek ratio,* is then

$$I = \frac{I_C}{2I_B} = (\gamma - 1) = \left(\frac{c_P}{c_V} - 1\right) \tag{10.4.46}$$

and the total integrated intensity is, as before [cf. Eq. (10.1.8)],

$$I = I_C + 2I_B = \rho_0^2 k_B T \chi_T \tag{10.4.47}$$

It is important to note that if $c_P = c_V$, the central line will not be observed. Because $c_P \simeq c_V$ in H_2O, the Rayleigh line is very hard to distinguish from the background noise.

In molecular fluids, rotational and vibrational relaxation can effect the density fluctuation. It is then necessary to supplement the equations of fluid mechanics with equations describing the molecular relaxation. We shall consider this momentarily. The whole picture developed here must be modified in the neighborhood of the critical point or near a phase transition. The long range correlations discussed in Sections 10.1 and 10.7 then affect the whole structure of the theory. See, for example, the review of Stanley, et al. (1971) and particularly references to the work of Kawasaki cited therein. Some aspects of scattering in the critical region are considered in Sec. (10.7).

10·5 THE RAYLEIGH–BRILLOUIN SPECTRUM AND INTRAMOLECULAR RELAXATION

The Brillouin spectra of molecular liquids are more complicated than the spectra of simple liquids. Molecular internal degrees of freedom generally couple to the translational motion of the molecules, thereby leading to additional relaxation mechanisms for the density fluctuations. In this section we explore a simple model of molecular liquids first proposed by Mountain (1966) in which the density fluctuations are weakly coupled to the relaxing molecular internal degrees of freedom.

The bulk and shear viscosities can be represented as integrals of time-correlation functions of microscopic fluxes[25] (see Section 11.B)

$$\begin{aligned} \eta_s(\omega) &= (Vk_BT)^{-1}\int_0^\infty dt\, e^{-i\omega t}\langle J^{xy}(0)\, J^{xy}(t)\rangle \\ \eta_v(\omega) &= (9Vk_BT)^{-1}\int_0^\infty dt\, e^{-i\omega t}\langle J^{zz}(0)\, J^{zz}(t)\rangle \end{aligned} \tag{10.5.1}$$

where J^{ab} is the $q \to 0$ limit of the momentum flux density.

Using these expression (Green–Kubo relations), Zwanzig (1965b) investigated the frequency dependences of the viscosities of a fluid composed of molecules with internal degrees of freedom which are weakly coupled to the center of mass (translational motions). He found that the bulk viscosity is

$$\eta_v(\omega) = \eta_v + \frac{(c_P - c_V)\, c_I \rho c_S^2}{(c_V - c_I)\, c_P} \int_0^\infty dt\, e^{-i\omega t} \phi(t) \tag{10.5.2}$$

where η_v is the center-of-mass part of the bulk viscosity (it is frequency-independent), c_I is the specific heat associated with the internal degrees of freedom, and the function $\phi(t)$ is defined by

$$\langle \delta E_I(0)\, \delta E_I(t) \rangle = \langle \delta E_I^2 \rangle \phi(t) \tag{10.5.3}$$

where δE_I is the fluctuation of the energy E_I in the internal degrees of freedom. Zwanzig also found that there is no corresponding dispersion (frequency-dependence) in the shear viscosity.

The dynamics of the molecular relaxation process are contained in $\phi(t)$. If the internal energy relaxes exponentially, $\phi(t) = \exp - t/\tau$, it follows from Eq. (10.5.3) that

$$\eta_v(\omega) = \eta_v + \frac{b_1}{1 + i\omega\tau} \tag{10.5.4}$$

where

$$b_1 = \left[\frac{(c_P - c_V) c_I}{(c_V - c_I) c_P}\right] \rho c_S^2 \tau = (c_\infty^2 - c_S^2)\tau \tag{10.5.5}$$

where c_∞ defined by this equation is the infinite frequency velocity of sound.[26]

The frequency-dependence of $\eta_v(\omega)$ is such that for ω small compared to the relaxation rate $1/\tau$, $\eta_v(\omega) \to \eta_v + b_1$; and for frequencies large compared with $1/\tau$, $\eta_v(\omega) \to \eta_v$. $\eta_v(\omega)$ varies rapidly in the neighborhood of $\omega = 1/\tau$. For many liquids $1/\tau$ is in the gigahertz region which is just the frequency range probed in light scattering. For these liquids, the frequency-dependence of $\eta_v(\omega)$ must somehow be incorporated into the hydrodynamics. This can be done as follows. Because $\eta_v(\omega)$ given by Eq. (10.5.4) is a Laplace transform with $s = i\omega$, a natural and very physical method for incorporating the dispersion into the transport equations is to substitute $\eta_v(s) = \eta_v(\omega = -is)$ in place of η_v into Eq. (10.4.20a). This results in the following modification: wherever $D_V = (\eta_v + \frac{4}{3}\eta_s)/\rho m$ appears in Eq. (10.4.25) replace it by $D_V(s) = D_V + D'(s)$ where $D'(s) = b_1/(1 + s\tau) m\rho_0$.

The same set of steps that led from Eq. (10.4.25) to Eq. (10.4.30a) now results in the spectrum

$$S_{\rho\rho}(\mathbf{q}, \omega) = \pi^{-1} S(\mathbf{q}) \frac{[N_1(\omega) D_1(\omega) + N_2(\omega) D_2(\omega)]}{[D_1^2(\omega)\, D_2^2(\omega)]} \tag{10.5.6}$$

where

$$N_1(\omega) \equiv -\omega^2 + \gamma D_T D_V q^4 + c_S^2 q^2 (1 - 1/\gamma) + (\gamma D_T b_1' q^4 + b_1' \omega^2 q^2 \tau)/(1 + \omega^2\tau^2)$$

$$N_2(\omega) \equiv \omega[\gamma D_T q^2 + D_V q^2 + (b_1' q^2 + \gamma D_T b_1' \tau q^4)/(1 + \omega^2\tau^2)]$$

$$D_1(\omega) \equiv -\omega^2(\gamma D_T q^2 + D_V q^2) + D_T c_s^2 q^4 + (\gamma D_T b_1' \omega^2 q^4 \tau - \omega^2 b_1' q^2)/(1 + \omega^2\tau^2)$$

$$D_2(\omega) = \omega[-\omega^2 + c_s^2 q^2 + \gamma D_T D_V q^4 + (b_1' q^2 \omega^2 \tau + \gamma D_T b_1' q^4)/(1 + \omega^2\tau^2)]$$

and where

$$b' \equiv b_1/m\rho_0$$

Gornall et al. (1966) have used the "exact" expressions given by Eq. (10.5.6) to fit their measurements in liquid CCl_4. The experimental curves (Fig. 10.5.1a) were reproduced by numerically convoluting the instrumental function for their experiment with the exact expression Eq. (10.5.6) (Fig. 10.5.1*b*). Thus the measured spectra compared favorably with the computed spectra in Fig. 10.5.1.

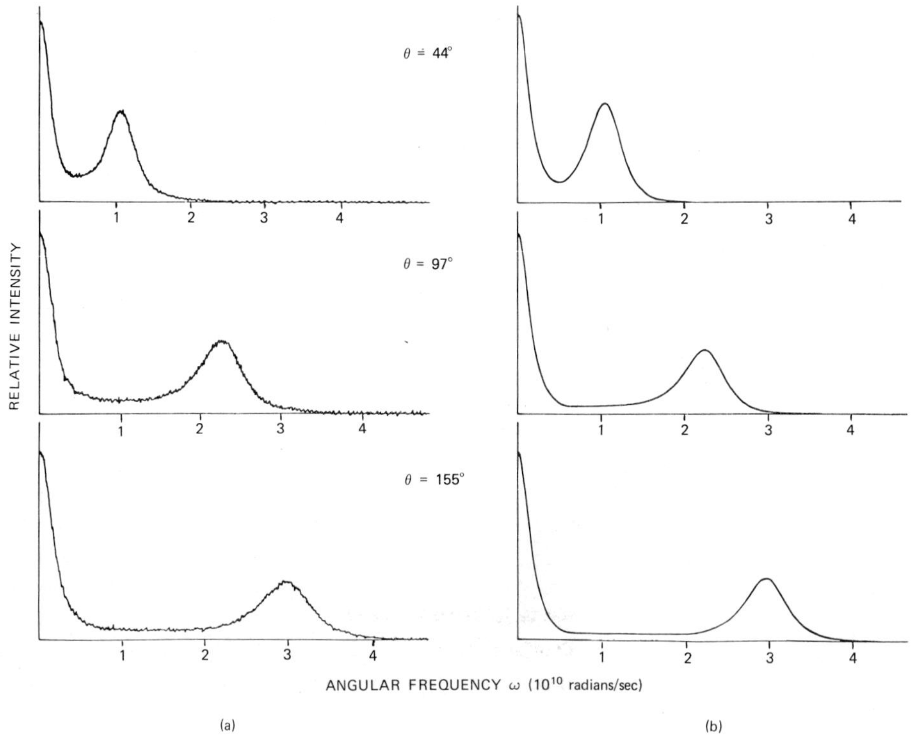

FIG. 10.5.1. Brillouin spectra of liquid CCl_4 for the scattering angles $\theta = 44$, 97, and 155° (Stokes sides only): (*a*) observed spectra; (*b*) computed spectra. (From Gornal et al., 1966.)

It is clear from Fig. 10.5.1 that the computed and observed spectra agree at all scattering angles. Several features of the spectra should be noted: (a) the prominent continuous background between the central unshifted component and doublets, (b) the background component extends symmetrically on either side of the central component to the Brillouin doublets, producing an asymmetry in the intensity on the high-frequency sides of the doublets, (c) the background component accounts for an appreciable part (approximately 20%) of the total intensity in the spectrum, and (d) the new background component is entirely due to the coupling between the translations and the internal relaxation of CCl_4 molecules. (These conclusions are corroborated by a later and more detailed study of CCl_4 by Carome, et al., 1968.)

For many molecular liquids the dispersion equation can be solved by a perturbation approximation similar to that used to derive Eq. (10.4.35). The perturbation solution gives four roots

$$s_1 = D_T q^2;\; s_2 = \frac{c_S^2}{c^2 \tau};\; s_\pm = \pm icq - \Gamma q^2 \tag{10.5.7}$$

which are given explicitly by Mountain (1966). The resulting spectrum is quite complicated. It consists of four basic contributions. The first term is due to a heat-conduction mode, which we call the *thermal mode.* The second term represents a new non-propagating mode which is entirely due to the internal degrees of freedom of the molecules and which we call the *relaxation mode.* The last terms represent the sound modes and consists of the *Brillouin doublet.*

The relaxation mode has a breadth proportional to the relaxation rate $1/\tau$. This is often quite broad compared to $q^2 D_T$ and thus gives rise to the broad background in Fig. 10.5.1.

The theoretical ratio of the intensity of the unshifted (central) components of the scattered light to the intensity of the Brillouin components is readily obtained. This ratio $I = I_C/2I_B$ is a complicated expression. It simplifies considerably in certain limits. At low sound frequencies ($cq\tau \ll 1$) this reduces to the Landau–Placzek result [cf. Eq. (10.4.46)]. At large sound frequencies ($cq\tau \gg 1$) it deviates from this ratio by a form originally derived by Rytov (1958).

The introduction of a relaxing bulk viscosity is quite general and does not require an explicit specification of the internal relaxation process. In complicated molecular fluids it is not always possible to identify the internal motions responsible for the deviations from classical behavior. In such circumstances the justification for the introduction of a relaxing variable is not obvious (see Mountain, 1968). In order to clarify the underlying mechanism Desai and Kapral (1972) focus on the description of fluids composed of small molecules where the choice of relaxing variables is easily made. For diatomic and small polyatomic molecules only the relaxation within a well-defined set of internal states needs to be taken into account in order to describe the Brillouin spectra. The fluid investigated by Gornall (1966) is an example. In liquid CCl_4 and other liquids only a few vibrational states participate, while in compressed hydrogen gas only a few rotational states need to be considered. In these situations it is clear that the variables appropriate to couple to the normal hydrodynamic variables are those variables which characterize the populations of the individual internal states.

10 · 6 BINARY MIXTURES

In a binary mixture there are six conserved variables: (a) the energy, (b) linear momentum (three components), (c) the solute concentration, and (d) the total fluid density. There is some freedom in specifying the composition of the fluid. For convenience we choose the variables used by Landau and Lifshitz (1960), and specify the composition of the fluid by giving the mass fraction of solute, that is $c = M_1/M$. The hydrodynamic state of the binary mixture is then specified by giving the local values of the mass density $m\rho$, the mass fraction c, the temperature T, and the local velocity $\mathbf{u}$.

In this section we do not present a development as detailed as that given in Section 10.4, but mention the highlights of the theory.

The first step is to write the conservation equations for[28] $m\rho$, $\mathbf{u}$, e where e is the energy density. These equations are identical to Eq. (10.4.1). There is an additional conservation equation required for c. This can be written as

$$m\frac{\partial(\rho c)}{\partial t} + \nabla \cdot \mathbf{j} = 0 \tag{10.6.1}$$

where $\mathbf{j}$ is the mass flux of the solute and $m\rho c$ is the mass density of the solute at $(\mathbf{r}, t)$. $\mathbf{j}$ consists of two parts

$$\mathbf{j} = \mathbf{i} + m\rho c\mathbf{u} \tag{10.6.2}$$

where $m\rho c\mathbf{u}$ is the mass flux of solute due to convection and $\mathbf{i}$ is the mass flux due to diffusion.

To proceed it is necessary as before to find the constitutive relations for $\boldsymbol{\tau}$, $\mathbf{Q}$ and $\mathbf{i}$. In going from a pure fluid to a binary solution, the constitutive relation for $\boldsymbol{\tau}$[cf. Eq. (10.4.9)] does not change. On the other hand the constitutive relation for $\mathbf{Q}$ [cf. Eq. (10.4.10)] does change to that [cf. Eqs. (58.11) and (58.12) of Landau and Lifshitz (1960)]

$$\mathbf{Q} = \left[K_T\left(\frac{\partial \mu}{\partial c}\right)_{p,c} - T\left(\frac{\partial \mu}{\partial T}\right)_{p,c} + \mu\right]\mathbf{i} - \lambda\nabla T \tag{10.6.3}$$

where $\mathbf{i}$ is the diffusion flux

$$\mathbf{i} = -m\rho_0\left[\nabla c + \left(\frac{K_T}{T_0}\right)\nabla T + \left(\frac{K_p}{p_0}\right)\nabla p\right] \tag{10.6.4}$$

where $\mu \equiv (\mu_1/m_1) - (\mu_2/m_2)$ is the chemical potential of the mixture [cf. Eq. (10.C.35)], K_T is the *thermal diffusion ratio,* and K_p is the barodiffusion ratio

$$K_p \equiv -\frac{\left(\dfrac{p_0}{m\rho_0^2}\right)\left(\dfrac{\partial \rho}{\partial c}\right)_{p,T}}{\left(\dfrac{\partial \mu}{\partial c}\right)_{p,T}} \tag{10.6.5}$$

Substitution of $\boldsymbol{\tau}$, $\mathbf{q}$, $\mathbf{i}$ into the conservation equations and linearization with respect to the fluctuations gives the linearized hydrodynamic equation for the binary mixture

$$\frac{\partial \rho_1}{\partial t} + \rho_0\psi_1 = 0 \tag{10.6.6a}$$

$$m\rho_0\frac{\partial \psi_1}{\partial t} = -\nabla^2 p_1 + \eta_s\nabla^2\psi_1 + \left(\eta_v + \frac{1}{3}\eta_s\right)\nabla^2\psi_1 \tag{10.6.6b}$$

$$\frac{\partial c_1}{\partial t} = D\left\{\nabla^2 c_1 + \left(\frac{K_T}{T_0}\right)\nabla^2 T_1 + \left(\frac{K_p}{p_0}\right)\nabla^2 p_1\right\} \tag{10.6.6c}$$

$$m\rho_0 c_{p,c}\frac{\partial T_1}{\partial t} = -m\rho_0 K_T\left(\frac{\partial \mu}{\partial c}\right)_{p,T}\frac{\partial c_1}{\partial t} + m\rho_0 T_0\left(\frac{\partial s}{\partial \rho}\right)_{T,c}\frac{\partial P_1}{\partial t} = \lambda\nabla^2 T_1 \tag{10.6.6d}$$

where $\psi_1 \equiv \nabla \cdot \mathbf{u}_1$ as before, and the subscripts 0 and 1 denote respectively the equilibrium value and the fluctuation in the corresponding variable.

These equations have been solved and the corresponding light scattering spectrum determined (Mountain and Deutch, 1969). Rather than repeat the details of this algebraically involved, but nonetheless routine calculation in its entirety, we consider the interesting special case of a system in which the pressure is uniform, and consequently only the concentration and temperature fluctuate. In this special case there are no sound modes, and consequently no Brillouin doublets. Nevertheless, this calculation contains several important properties common to the total solution. In the uniform pressure approximation, $p_1 = \partial p_1/\partial t = 0$ so that all pressure terms disappear from Eq. (10.6.6) and the concentration and temperature fluctuations are described by the equation [cf. Eqs. (58.14) and (58.15) of Landau and Lifshitz (1960)]

$$\begin{cases} \dfrac{\partial c_1}{\partial t} = D\left\{\nabla^2 c_1 + \dfrac{K_T}{T_0}\nabla^2 T_1\right\} \\[2ex] \dfrac{\partial T_1}{\partial t} - \dfrac{K_T}{c_p}\left(\dfrac{\partial \mu}{\partial c}\right)_{p,T}\dfrac{\partial c_1}{\partial t} = D_T \nabla^2 T_1 \end{cases} \tag{10.6.7}$$

The Fourier–Laplace transform of Eq. (10.6.7) gives

$$(s + q^2 D)\tilde{c}_1(\mathbf{q}, s) + q^2\left(\frac{K_T D}{T_0}\right)\tilde{T}_1(\mathbf{q}, s) = c_1(\mathbf{q}, 0) \tag{10.6.8}$$

and

$$-s\left(\frac{K_T}{c_p}\right)\left(\frac{\partial \mu}{\partial c}\right)_{p,T}\tilde{c}_1(\mathbf{q}, s) + (s + D_T q^2)\tilde{T}_1(\mathbf{q}, s) = T_1(\mathbf{q}, 0) - \frac{K_T}{c_p}\left(\frac{\partial \mu}{\partial c}\right)_{p,T} c_1(\mathbf{q}, 0)$$

These linear equations can be written in matrix form

$$\begin{pmatrix} [s + q^2 D] & \left[q^2\left(\dfrac{K_T D}{T_0}\right)\right] \\[2ex] \left[-s\left(\dfrac{K_T}{c_p}\right)\left(\dfrac{\partial \mu}{\partial c}\right)_{p,T}\right] & [(s + q^2 D_T)] \end{pmatrix}\begin{pmatrix} \tilde{c}_1(\mathbf{q}, s) \\[2ex] \tilde{T}_1(\mathbf{q}, s) \end{pmatrix} = \tag{10.6.9a}$$

$$\begin{pmatrix} 1 & 0 \\[2ex] -\left(\dfrac{K_T}{c_p}\right)\left(\dfrac{\partial \mu}{\partial c}\right)_{p,T} & 1 \end{pmatrix}\begin{pmatrix} c_1(\mathbf{q}) \\[2ex] T_1(\mathbf{q}) \end{pmatrix}$$

This can be written in the more concise form

$$\tilde{M}(\mathbf{q}, s) \cdot \tilde{\boldsymbol{\phi}}(\mathbf{q}, s) = \mathbf{N}(\mathbf{q}) \cdot \boldsymbol{\phi}(\mathbf{q}) \tag{10.6.9b}$$

where there is a one-to-one correspondence reading from left to right between the four matrices in Eqs. (10.6.9a, b) and the terms in Eq (10.6.9.b). The solution of the matrix equation is

$$\tilde{\boldsymbol{\phi}}(\mathbf{q}, s) = \tilde{\mathbf{M}}^{-1}(\mathbf{q}, s) \cdot \mathbf{N}(\mathbf{q}) \cdot \boldsymbol{\phi}(\mathbf{q}) \tag{10.6.9c}$$

where

$$\tilde{M}^{-1}(\mathbf{q}, s) = \frac{1}{M(s)} \begin{pmatrix} (s + q^2 D_T) & -q^2(K_T D/T_0) \\ s\left(\dfrac{K_T}{c_p}\right)\left(\dfrac{\partial \mu}{\partial c}\right)_{p,T} & (s + q^2 D) \end{pmatrix} \tag{10.6.9d}$$

where $M(s)$ is the determinant of $\tilde{\mathbf{M}}(\mathbf{q}, s)$

$$M(s) = s^2 + [\Delta q^2 + D_T q^2]s + (Dq^2)(D_T q^2) \tag{10.6.9e}$$

where

$$\Delta = D\left[1 + \left(\frac{K_T^2}{c_p T_0}\right)\left(\frac{\partial \mu}{\partial c}\right)_{p,T}\right] \tag{10.6.9f}$$

Substitution of Eq. (10.6.9d) into Eq. (10.6.9b) then gives

$$\begin{pmatrix} \tilde{c}_1(\mathbf{q}, s) \\ \tilde{T}_1(\mathbf{q}, s) \end{pmatrix} = \frac{1}{M(s)} \begin{pmatrix} \left[(s + D_T q^2) + q^2\left(\dfrac{K_T^2 D}{c_p T_0}\right)\left(\dfrac{\partial \mu}{\partial c}\right)_{p,T}\right] & \left[-q^2\left(\dfrac{K_T D}{T_0}\right)\right] \\ -\left(\dfrac{K_T D}{c_P}\right)\left(\dfrac{\partial \mu}{\partial c}\right)_{p,T} q^2 & (s + Dq^2) \end{pmatrix} \begin{pmatrix} c_1(\mathbf{q}) \\ T_1(\mathbf{q}) \end{pmatrix} \tag{10.6.10}$$

The reader is urged to write the explicit solutions for $\tilde{c}_1(\mathbf{q}, s)$ and $\tilde{T}_1(\mathbf{q}, s)$. What is required for light scattering are the correlation functions $\langle c_1^*(\mathbf{q}) c_1(\mathbf{q}, t)\rangle$, $\langle c_1^*(\mathbf{q}) T_1(\mathbf{q}, t)\rangle$, $\langle T_1^*(\mathbf{q}) c_1(\mathbf{q}, t)\rangle$, $\langle T_1^*(\mathbf{q}) T_1(\mathbf{q}, t)\rangle$ or equivalently their Laplace transforms which can be found by multiplying the solutions $\tilde{c}_1(\mathbf{q}, s)$, $\tilde{T}_1(\mathbf{q}, s)$, successively by $c_1^*(\mathbf{q}, 0)$, $T_1^*(\mathbf{q}, 0)$ and then averaging. The resulting equations are greatly simplified because $\langle c_1^*(\mathbf{q}, 0)\, T_1(\mathbf{q}, 0)\rangle = \langle T_1^*(\mathbf{q})\, c_1(\mathbf{q}, 0)\rangle = 0$. This is shown in Appendix (10.C). The resulting correlation functions are

$$\frac{\langle c_1^*(\mathbf{q}) \tilde{c}_1(\mathbf{q}, s)\rangle}{\langle |c_1(\mathbf{q})|^2\rangle} = B(s)/M(s) \tag{10.6.11a}$$

$$\frac{\langle T_1^*(\mathbf{q}) \tilde{c}_1(\mathbf{q}, s)\rangle}{\langle |T_1(\mathbf{q})|^2\rangle} = C(s)/M(s) \tag{10.6.11b}$$

$$\frac{\langle c_1^*(\mathbf{q}) \tilde{T}_1(\mathbf{q}, s)\rangle}{\langle |c_1(\mathbf{q})|^2\rangle} = D(s)/M(s) \tag{10.6.11c}$$

$$\frac{\langle T_1^*(\mathbf{q}) \tilde{T}_1(\mathbf{q}, s)\rangle}{\langle |T_1(\mathbf{q})|^2\rangle} = F(s)/M(s) \tag{10.6.11d}$$

where

$$B(s) = s + D_T q^2 + \left(\frac{K_T^2 D}{c_p T_0}\right)\left(\frac{\partial \mu}{\partial c}\right)_{p,T} q^2 \tag{10.6.11e}$$

$$C(s) = -\, q^2\left(\frac{K_T D}{T_0}\right) \tag{10.6.11f}$$

$$D(s) = -\, q^2\left(\frac{K_T D}{c_p}\right)\left(\frac{\partial \mu}{\partial c}\right)_{p,T} \tag{10.6.11g}$$

$$F(s) = s + Dq^2 \tag{10.6.11h}$$

and where $M(s)$ is given in Eq. (10.6.9e).

The time-correlation functions are found by Laplace inverting these formulas. For this we require the roots of the dispersion equation

$$M(s) = (s - s_+)(s - s_-) = 0 \tag{10.6.12}$$

These are

$$s_\pm = -\frac{1}{2}(D_T + \Delta)q^2 \pm \frac{1}{2}[(D_T + \Delta)^2 - 4D_T D]^{1/2} q^2 \tag{10.6.13}$$

The time-correlation functions are then

$$\frac{\langle c_1^*(\mathbf{q})c_1(\mathbf{q},t)\rangle}{\langle|c_1(\mathbf{q})|^2\rangle} = \frac{B(s_+)e^{s_+|t|} - B(s_-)e^{s_-|t|}}{(s_+ - s_-)} \tag{10.6.14a}$$

$$\frac{\langle T_1^*(\mathbf{q})c_1(\mathbf{q},t)\rangle}{\langle|T_1(\mathbf{q})|^2\rangle} = \frac{C(s_+)e^{s_+|t|} - C(s_-)e^{s_-|t|}}{(s_+ - s_-)} \tag{10.6.14b}$$

$$\frac{\langle c_1^*(\mathbf{q})T_1(\mathbf{q},t)\rangle}{\langle|c_1(\mathbf{q})|^2\rangle} = \frac{D(s_+)e^{s_+|t|} - D(s_-)e^{s_-|t|}}{(s_+ - s_-)} \tag{10.6.14c}$$

$$\frac{\langle T_1^*(\mathbf{q})T_1(\mathbf{q},t)\rangle}{\langle|T_1(\mathbf{q})|^2\rangle} = \frac{F(s_+)e^{s_+|t|} - F(s_-)e^{s_-|t|}}{(s_+ - s_-)}$$

From the definitions of $B(s)$, $C(s)$, $D(s)$, and $F(s)$ we note that

$$B(s_\pm) = -(s_\mp + Dq^2) \tag{10.6.14d}$$

$$C(s_\pm) = -q^2(K_T D/T_0) \tag{10.6.14e}$$

$$D(s_\pm) = -q^2\left(\frac{K_T D}{c_p}\right)\left(\frac{\partial\mu}{\partial c}\right)_{p,T} \tag{10.6.14f}$$

$$F(s_\pm) = (s_\pm + Dq^2) \tag{10.6.14g}$$

Thus the correlation functions have the correct $t \to 0$ limits; that is, the autocorrelation functions are normalized and the cross-correlation functions are initially zero.

The dielectric constant is experimentally known to be a function of (p, T, c)

$$\varepsilon = \varepsilon(p, T, c)$$

Therefore for the fixed pressure calculation

$$\delta\varepsilon(\mathbf{q}, t) = \left(\frac{\partial\varepsilon}{\partial c}\right)_{p,T} c_1(\mathbf{q}, t) + \left(\frac{\partial\varepsilon}{\partial T}\right)_{p,c} T_1(\mathbf{q}, t)$$

The intensity of the scattered light is proportional to the Fourier transform of

$$\begin{aligned}\langle\delta\varepsilon^*(\mathbf{q})\delta\varepsilon(\mathbf{q}, t)\rangle = \Big\langle &\left[\left(\frac{\partial\varepsilon}{\partial T}\right)_{p,c} T_1^*(\mathbf{q}) + \left(\frac{\partial\varepsilon}{\partial c}\right)_{p,T} c_1^*(q)\right] \\ &\times \left[\left(\frac{\partial\varepsilon}{\partial T}\right)_{p,c} T_1^*(\mathbf{q}, t) + \left(\frac{\partial\varepsilon}{\partial c}\right)_{p,T} c_1(\mathbf{q}, t)\right]\Big\rangle\end{aligned} \tag{10.6.15}$$

which is a linear combination of the correlation functions in Eqs. (10.6.14a, b). When these are substituted into Eq. (10.6.15) and the Fourier transform is explicitly taken, the spectral density of the dielectric fluctuation is

$$\pi S_{\varepsilon\varepsilon}(\mathbf{q}, \omega) = \left(\frac{\partial \varepsilon}{\partial c}\right)^2_{p,T}\left[k_B T\left[\frac{\partial c}{\partial \mu}\right]_{p,T}\right]\frac{1}{(s_+ - s_-)}\left[\frac{(s_- + Dq^2)s_+}{\omega^2 + s_+^2} - \frac{(s_+ + Dq^2)s_-}{\omega^2 + s_-^2}\right]$$
$$+ 2\left[\frac{\partial \varepsilon}{\partial c}\right]_{p,T}\left(\frac{\partial \varepsilon}{\partial T}\right)_{p,c}\left[\frac{k_B T}{c_p}\right]\left[\frac{K_T D q^2}{s_+ - s_-}\right]\left[\frac{s_+}{\omega^2 + s_+^2} - \frac{s_-}{\omega^2 + s_-^2}\right]$$
$$+ \left(\frac{\partial \varepsilon}{\partial T}\right)^2_{c,p}\left[\frac{k_B T_0^2}{c_p}\right]\frac{1}{(s_+ - s_-)}\left[\frac{(s_- + Dq^2)s_-}{\omega^2 + s_+^2} - \frac{(s_+ + Dq^2)s_-}{\omega^2 + s_-^2}\right] \qquad (10.6.16)$$

where we have replaced mean square fluctuations in Eqs. (10.6.14) by their $q \to 0$ limits

$$\langle | c_1(\mathbf{q}) |^2 \rangle \underset{q \to 0}{\rightarrow} k_B T \left(\frac{\partial c}{\partial \mu}\right)_{p,T} \qquad (10.6.17)$$
$$\langle | T_1(\mathbf{q}) |^2 \rangle \underset{q \to 0}{\rightarrow} k_B T c_p^{-1}$$

These averages have been computed in Appendix 10.C.

This spectrum consists of the superposition of two Lorentzian bands

$$\frac{s_+}{\omega^2 + s_+^2} \quad \text{and} \quad \frac{s_-}{\omega^2 + s_-^2}$$

with weights that depend on many parameters. The complexity of this spectrum arises from the coupling between particle diffusion and heat flow, which is given by the thermal diffusion ratio K_T. The widths of the two Lorentzians are (s_+) and (s_-), which both depend on D, D_T and K_T [cf. Eq. (10.6.13)]. Thus we see that this central band cannot in general be regarded as a superposition of two Lorentzians, the first arising *only* from thermal diffusivity and the second *only* from particle diffusion.

This is a matter of some urgency for us since in most biological applications it is assumed that the central (or Rayleigh) band is dependent on the diffusion coefficient and nothing else. This assumption enables the biochemist to determine diffusion coefficients (cf. Section 5.4). Let us explore this assumption further.

Under certain conditions the central line simplifies considerably. In the limit of small concentration $c \to 0$, the thermal diffusion ratio goes to zero

$$\lim_{c \to 0} K_T = 0 \qquad (10.6.18)$$

In this case the roots s_+, s_- simplify because $\Delta \to D$ [cf. Eq. (10.6.9f)]. Then

$$s_+ \cong Dq^2$$
$$s_- \cong D_T q^2 \qquad (10.6.19)$$

and the spectrum is the superposition of two Lorentzians, one due only to particle diffusion and the other only to the thermal diffusivity. The spectrum [Eq. (10.6.16)] then simplifies to

$$\pi S_{\varepsilon\varepsilon}(\mathbf{q},\omega) = \left(\frac{\partial\varepsilon}{\partial c}\right)^2_{p,T} k_B T \left(\frac{\partial c}{\partial\mu}\right)_{p,T} \left[\frac{Dq^2}{\omega^2 + [Dq^2]^2}\right] + \left(\frac{\partial\varepsilon}{\partial T}\right)^2_{c,p} \left[\frac{k_B T_0{}^2}{c_p}\right]\left[\frac{D_T q^2}{\omega^2 + [D_T q^2]^2}\right] \tag{10.6.20}$$

Thus in the limit of low concentrations the heat flow and particle diffusion are uncoupled and it is possible to determine D_T and D separately. If, as is usually true, ε is a weak function of the temperature, only the diffusion band will be important.

More important is the case when $D_T \gg D$, since this inequality is usually applicable. In liquids D is of order 10^{-5} cm^2/sec, whereas D_T is of the order 10^{-3} cm^2/sec so that the inequality is usually valid. (In dilute gases, on the other hand, the inequality is often not valid.) *In this $D_T \gg D$ limit the roots again simplify to*

$$s_+ \simeq -Dq^2 \qquad s_- \simeq -D_T q^2 \tag{10.6.21}$$

and once again the spectrum simplifies to the superposition of two Lorentzians, one due to particle diffusion and the other to thermal diffusivity. It might appear from Eq. (10.6.16) that the coefficients of these two Lorentzians will differ from those in Eq. (10.6.20). In the limit $D_T \gg D$, the spectrum actually reduces to the same form as in Eq. (10.6.20). Since $D_T \gg D$, the "diffusion component" of the spectrum appears as a very intense, sharp peak of width q^2D sitting on top of a much broader peak of width q^2D_T arising from heat flow. In this case it is possible to assign the central peak to particle diffusion. Most biological applications involve dilute macromolecular solutions in which $D \sim 10^{-6}$ cm^2/sec or smaller. In this case ($D_T \gg D$) and the sharp central component is

$$\pi S_{\varepsilon\varepsilon}(q\omega) \simeq \left(\frac{\partial\varepsilon}{\partial c}\right)^2_{p,T} k_B T \left(\frac{\partial c}{\partial\mu}\right)\left[\frac{Dq^2}{\omega^2 + [Dq^2]^2}\right] \tag{10.6.22}$$

which is similar to the spectrum derived in Section 5.4 on the basis of single-particle diffusion, with the exception that D is the mutual and not the self-diffusion coefficient[29] (see Section 13.5).

Of course pressure fluctuations occur in a real binary mixture. Our neglect of these fluctuations simplified the equations considerably. Nevertheless, the full set of equations should be analyzed. This has been done by Mountain and Deutch (1969). The full treatment of these equations gives in the limit of small q precisely the same central component that we have just calculated so that our preceding discussion is valid. The full treatment also gives Brillouin doublets. The full spectrum in the small q limit is

$$S_{\varepsilon\varepsilon}(q,\omega) = S^C_{\varepsilon\varepsilon}(q,\omega) + S^D_{\varepsilon\varepsilon}(q,\omega) \tag{10.6.23}$$

where $S^C_{\varepsilon\varepsilon}$ is the central component given in Eq. (10.6.16) and $S^D_{\varepsilon\varepsilon}(q\omega)$ is the spectrum corresponding to the Brillouin doublets

$$\pi S^D_{\varepsilon\varepsilon}(\mathbf{q},\omega) = \frac{1}{\pi}\left(\frac{\partial\varepsilon}{\partial\rho}\right)^2_{c,T}\left(\frac{k_B T_0 \rho}{\chi_S}\right)\left[\frac{\Gamma q^2}{(\omega+\omega(q))^2 + [\Gamma q^2]^2} + \frac{\Gamma q^2}{(\omega-\omega(q))^2 + [\Gamma q^2]^2}\right] \tag{10.6.24}$$

and where $\omega(q)$ is the frequency of sound of wave number q, $\omega(q) = c_S q$, where c_S is the adiabatic sound velocity, χ_S is the adiabatic compressibility, and Γ is the attenuation coefficient of the sound wave

$$\Gamma = \frac{1}{2}\left[D_T(\gamma - 1) + \frac{(\eta_v + \frac{4}{3}\eta_s)}{m\rho_0} + \frac{Dc_S^2}{m^2\rho_0^2}\left(\frac{\partial c}{\partial \mu}\right)_{p,c} B^2\right] \tag{10.6.25}$$

where γ is the specific heat ratio (c_P/c_V) and B is a complicated quantity

$$B = \left[\left(\frac{\partial p}{\partial c}\right)_{p,T} + \frac{K_T}{c_P}\left(\frac{\partial p}{\partial T}\right)_{p,c}\left(\frac{\partial \mu}{\partial c}\right)_{p,T}\right] \tag{10.6.26}$$

Γ differs from Eq. (10.4.36) by the appearance of a contribution from particle diffusion, and the thermal-diffusion ratio. We have omitted from this spectrum the small non-Lorentzian terms analogous to those in Eq. (10.4.39), which are negligible.

The spectrum associated with a binary solution consists of four contributions. The two Brillouin peaks are centered at $\pm\omega(q)$ with a width $q^2\Gamma$. The central component consists of a superposition of two Lorentzians with a height that depends on many parameters (cf. our preceding discussion).

The ratio of the central peak I_C to the Brillouin components can be evaluated[30] from Eq. (10.6.16). This gives

$$2I_B = 4\pi\left(\frac{\partial \varepsilon}{\partial p}\right)_{c,T}^2 \frac{\beta^{-1}m\rho_0}{\chi_S} \tag{10.6.27}$$

and

$$I_C = 4\pi\left\{\left(\frac{\partial \varepsilon}{\partial c}\right)_{p,T}^2 \beta^{-1}\left(\frac{\partial c}{\partial \mu}\right)_{p,T} + \left(\frac{\partial \varepsilon}{\partial T}\right)_{p,c}^2 \frac{\beta^{-1}T_0}{c_P}\right\} \tag{10.6.28}$$

and the ratio is

$$R = \frac{I_C}{2I_B} = \frac{\left\{\left(\frac{\partial \varepsilon}{\partial c}\right)_{p,T}^2\left(\frac{\partial c}{\partial \mu}\right)_{p,T} + \left(\frac{\partial \varepsilon}{\partial T}\right)_{p,c}^2\left(\frac{T_0}{c_{p,c}}\right)\right\}}{\left(\frac{\partial \varepsilon}{\partial p}\right)_{T,C}^2\left(\frac{m\rho_0}{\chi_S}\right)} \tag{10.6.29}$$

This formula was first obtained by Miller (1967) and applied to experiment byMiller and Lee (1968) to extract activity coefficients from the measured concentration dependence of $[\partial\mu/\partial c]_{p,T}$.

The full spectrum of a binary mixture has been studied recently by several investigators (e.g., see Gornall and Wang, 1972 and references cited therein). These authors show that the complete solution of the hydrodynamic equations accurately describes the observed spectrum in gaseous mixtures such as helium and xenon. This work demonstrates the importance of the coupling between heat flow and diffusion when $D_T \sim D$ as in gases. In Fig. 10.6.1 the full spectrum is analyzed in terms of the contributions S_{pp}, S_{cc}, $S_{\phi\phi}$, $S_{p,c}$ and $S_{\phi c}$ for a 50% helium–xenon mixture.

In the limit where q^{-1} is smaller than the mean free path in a gas, this hydrodynamic calculation is not valid. It is then necessary to use the kinetic theory of gases (i.e., the Boltzmann equation) to account for the observed spectrum (see Chapt. 14). Boley and Yip (1972) have performed calculations that accurately describe the experiments of Clark (1970) on mixtures of gases.

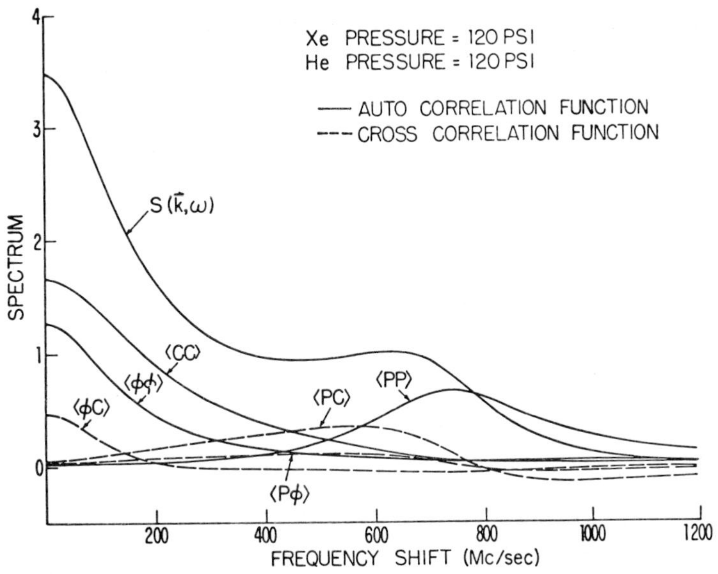

FIG. 10.6.1. Relative spectral contribution of the separate correlation functions in Eq. (10.6.15) in a 50% helium-xenon gaseous mixture. (From Gornal and Wang, 1971, Fig. 3.)

10·7 CRITICAL OPALESCENCE

The intensity of light scattered from a fluid system increases enormously, and the fluid takes on a cloudy or opalescent appearance as the gas–liquid critical point is approached. In binary solutions the same phenomenon is observed as the critical consolute point is approached. This phenomenon is called *critical opalescence*.[31] It is due to the long-range spatial correlations that exist between molecules in the vicinity of critical points. In this section we explore the underlying physical mechanism for this phenomenon in one-component fluids. The extension to binary or ternary solutions is not presented but some references are given.

For the purposes of discussing critical opalescence it is useful to define the static density–density correlation function

$$G(\mathbf{r},\mathbf{r}') \equiv \langle \delta\rho(\mathbf{r})\delta\rho(\mathbf{r}')\rangle \tag{10.7.1}$$

This function measures the correlations[32] of the fluctuations in the density at two different points of the fluid $\mathbf{r},\mathbf{r}'$ separated by $|\mathbf{r} - \mathbf{r}'|$. As $|\mathbf{r} - \mathbf{r}'| \to \infty$, the density fluctuations $\delta\rho(\mathbf{r})$ and $\delta\rho(\mathbf{r}')$ should be uncorrelated so that

$$\lim_{|\mathbf{r}-\mathbf{r}'|\to\infty} G(\mathbf{r},\mathbf{r}') = 0$$

In a spatially uniform system $G(\mathbf{r},\mathbf{r}')$ should be invariant to an arbitrary translation *a* so that $G(\mathbf{r} + \mathbf{a}, \mathbf{r}' + \mathbf{a}) = G(\mathbf{r},\mathbf{r}')$. This requires that

$$G(\mathbf{r},\mathbf{r}') = G(\mathbf{r} - \mathbf{r}') \tag{10.7.2}$$

that is $G(\mathbf{r},\mathbf{r}')$ only depends on the distance $\mathbf{R} = \mathbf{r} - \mathbf{r}'$ between the two points $\mathbf{r}$ and $\mathbf{r}'$. Because the structure factor

$$S(\mathbf{q}) = \langle |\delta\rho(\mathbf{q})|^2\rangle = \langle \sum_{ij} \exp i\mathbf{q} \cdot (\mathbf{r}_i - \mathbf{r}_j)\rangle \tag{10.7.3}$$

appearing in Eq. (10.1.7) involves $\delta\rho(\mathbf{q}) = \int_V d^3r\, e^{i\mathbf{q}\cdot\mathbf{r}}\, \delta\rho(\mathbf{r})$, it follows that

$$S(\mathbf{q}) = \int_V d^3r' \int_V d^3r\, e^{i\mathbf{q}\cdot(\mathbf{r}-\mathbf{r}')} \langle \delta\rho(\mathbf{r})\delta\rho(\mathbf{r}')\rangle \tag{10.7.4}$$

Transforming integration variables to $\mathbf{R} = \mathbf{r} - \mathbf{r}'$ and $\mathbf{r}$; substitution of Eq. (10.7.2) and integration of $\mathbf{r}'$ over V then gives

$$S(\mathbf{q}) = V \int d^3R\, e^{i\mathbf{q}\cdot\mathbf{R}} G(\mathbf{R}) \tag{10.7.5}$$

If there were no correlations between the positions of different particles, all the cross terms in Eq. (10.7.3) would disappear and $S^0(q) = \langle N\rangle$ where $\langle N\rangle$ is the average number of particles in the scattering volume V. The dependence of $S(\mathbf{q})$ on q, or better, the deviation of $S(\mathbf{q})/S^\circ(q)$ from 1 reflects the spatial correlations between different molecules.

According to Eqs. (3.2.15) and (10.1.6) the integrated intensity is proportional to the structure factor $S(\mathbf{q})$ or better to $S(\mathbf{q})/S^\circ(q)$ which from Eq. (10.7.5) is

$$\frac{S(\mathbf{q})}{S^\circ(q)} = \rho^{-1} \int_v d^3R\, e^{i\mathbf{q}\cdot\mathbf{R}} G(\mathbf{R}) \tag{10.7.6}$$

Usually for simple fluids $G(\mathbf{R})$ has a range typically of order 10 Å, which is quite small compared with the wavelength of light. The phase factor $e^{i\mathbf{q}\cdot\mathbf{R}}$, is slowly varying compared to $G(\mathbf{R})$ and may thus be replaced by unity in the integral so that

$$\frac{S(\mathbf{q})}{S^\circ(q)} = \rho^{-1} \int_V d^3R G(\mathbf{R}) \tag{10.7.7}$$

is essentially independent of q or equivalently of the scattering angle. The total scattered intensity is then expected to be independent of the scattering angle. The situation is quite different for a fluid near its gas-liquid critical point for then $G(\mathbf{R})$ is quite long-ranged (its range increasing without-limit as the critical point is approached) so that the factor $e^{i\mathbf{q}\cdot\mathbf{R}}$ cannot be ignored. The integrated intensity is consequently a strong function of the scattering angle near the critical point. Before considering the critical region, let us first consider the normal fluid.

The fluctuations in the total number of particles in the volume V can be related to $G(\mathbf{R})$. First note that $\delta N = \int_V d^3r\, \delta\rho(\mathbf{r})$. Therefore from Eqs. (10.7.1) and (10.7.3) it follows that

$$\begin{aligned}\langle \delta N^2\rangle &= \int_V d^3r \int_V d^3r' \langle \delta\rho(\mathbf{r})\delta\rho(\mathbf{r}')\rangle \\ &= \int_V d^3r \int_V d^3r' G(\mathbf{r} - \mathbf{r}')\end{aligned}$$

Transforming variables to $\mathbf{R} = \mathbf{r} - \mathbf{r}'$ and $\mathbf{r}'$ gives

$$V \int d^3R G(\mathbf{R}) = \langle \delta N^2 \rangle = \beta^{-1} V \rho^2 \chi_T \tag{10.7.8}$$

where χ_T is the isothermal compressibility. The last equality follows from the grand canonical ensemble (this is shown in Section 10.A). Equation (10.7.10) therefore becomes

$$\frac{S(\mathbf{q})}{S^\circ(q)} = \rho^{-1} \int_V d^3R G(\mathbf{R}) = \beta^{-1} \rho \chi_T \tag{10.7.9}$$

In fluids χ_T is generally a well-behaved function of the thermodynamic state. Near the critical point, however, χ_T becomes divergent (arbitrarily large). It follows that the intensity of scattered light increases very strongly as the critical point is approached. In fact there is so much scattering that the critical fluid appears cloudy or opalescent. This phenomenon, as mentioned above, is called *critical opalescence*.

Because χ_T diverges as $(\rho, T) \to (\rho_c, T_c)$, Eq. (10.7.9) predicts that the scattered intensity would diverge. From Eq. (10.7.9) we see that a large χ_T corresponds to a large value of the integral or equivalently to an increase in the range of the correlations (or the "correlation length ξ") of the density–density correlation function $G(\mathbf{R})$. Sufficiently close to T_C, the correlation length approaches the wavelength of light or q^{-1}, and the density fluctuations scatter light very strongly, thereby giving critical opalescence.

The three interrelated phenomena that are observed near the critical point are: (a) increase in the density fluctuation, (b) increase in the isothermal compressibility, and (c) increase in the correlation length of $G(\mathbf{R})$.

Because $\chi_T \to \infty$ as $(\rho, T) \to (\rho_c T_c)$, the integral in Eq. (10.7.9) must diverge, implying that the range of $G(\mathbf{R})$ diverges. Consequently in the neighborhood of the critical point the phase factor $\exp^{i\mathbf{q}\cdot\mathbf{R}}$ must be retained in Eq. (10.7.6). It is then necessary to evaluate Eq. (10.7.6).

The first important attempt to explain critical opalescence was due to Ornstein and Zernike (1914). Since this approach is revealing, we review it here. We note from Eq. (10.7.3) that terms with $i = j$ are not excluded. Thus $G(\mathbf{R})$ contains correlations between identical particles (self-correlation) separating these two sets[33] of terms ($i \neq j$ and $i = j$) in Eq. (10.7.1) gives[34]

$$G(\mathbf{r} - \mathbf{r}') = \rho\delta(\mathbf{r} - \mathbf{r}') + \rho^2 \Gamma(\mathbf{r} - \mathbf{r}') \tag{10.7.10}$$

where the first term is due to the self-correlations and the second term is due to the distinct correlations. Since G has the dimensions of ρ^2, Γ is dimensionless. We now define the *direct correlation function* $C(\mathbf{r})$ such that its spatial Fourier transform is

$$\hat{C}(\mathbf{q}) \equiv \frac{\Gamma(\mathbf{q})}{1 + \rho\Gamma(\mathbf{q})} \tag{10.7.11}$$

where $\Gamma(\mathbf{q})$ is the Fourier transform of $\Gamma(\mathbf{r})$;

From Eqs. (10.7.6), (10.7.10), and (10.7.11) it follows that

$$\frac{S(\mathbf{q})}{S^\circ(q)} = 1 + \rho\Gamma(\mathbf{q}) = [1 - \rho\hat{C}(\mathbf{q})]^{-1} \tag{10.7.12}$$

so that a determination of $\hat{C}(\mathbf{q})$ is sufficient for a discussion of critical opalescence. It is easy to show that $\hat{C}(\mathbf{q})$ does not depend strongly on temperature. First we note that as $T \to \infty$, molecular interactions become relatively unimportant so that $S(\mathbf{q})/S^\circ(q) \to 1$, and by implication from Eq. (10.7.12), $\hat{C}(\mathbf{q}) \to 0$. On the other hand, from Eq. (10.7.9) as $T \to T_c$, $\lim_{q\to 0} S(q)/S(q) \to \infty$ and $\lim_{q\to 0} C(q) \to \rho^{-1}$, which is finite. Thus $C(\mathbf{R})$ must remain short-ranged as $T \to T_c$ in marked contrast to $\Gamma(\mathbf{R})$ which, as we have seen, diverges.

If $C(\mathbf{R})$ is short-ranged, it is not unreasonable to assume that integrals of the form $\int_0^\infty d^3R\, R^l C(\mathbf{R})$ are finite. This permits $\hat{C}(\mathbf{q})$ to be expanded in powers of q. This is accomplished by expanding (in a Taylor series around $q = 0$) the phase factor $e^{i\mathbf{q}\cdot\mathbf{R}}$ in $\hat{C}(\mathbf{q})$. To second order in q, this gives[35] $\hat{C}(q) = C_0 - C_2 q^2$ where $C_0 \equiv 4\pi\int_0^\infty dR\, R^2 C(\mathbf{R})$ and $C_2 \equiv (2\pi/3)\int_0^\infty dR\, R^4 C(\mathbf{R})$. Substitution of this into Eq. (10.7.12) then gives

$$\frac{S^\circ(q)}{S(q)} = R_0^2[q_0^2 + q^2] \tag{10.7.13}$$

where

$$R_0^2 \equiv \rho C_2; \qquad q_0^2 \equiv \frac{1 - \rho C_0}{R_0^2} \tag{10.7.14}$$

Because the direct correlation function is short-ranged, the integrals C_0 and C_2 and thereby R_0 and q_0 are expected to be finite and well-behaved as $T \to T_c$. From Eqs. (10.7.13), (10.7.9), and (10.7.6) it follows that

$$\lim_{q\to 0} \frac{S^\circ(q)}{S(q)} = R_0^2 q_0^2 = \beta(\rho\chi_T)^{-1} \tag{10.7.15}$$

This gives a relationship between R_0 and q_0. R_0 is usually assumed to vary weakly with temperature and density. It follows from Eqs. (10.7.13) and (10.7.15) that

$$\frac{S(q)}{S^\circ(q)} = \beta^{-1}\,\rho\chi_T\left[\frac{1}{1 + (q/q_0)^2}\right] \tag{10.7.16}$$

This is a Lorentzian function of q or equivalently of $\sin(\theta/2)$ of width $2q_0$. This is called the *Ornstein–Zernike Approximation.* It is based on the following assumptions: (a) the direct correlation function is short-ranged, (b) $\hat{C}(\mathbf{q})$ can be expanded in a power series in q, and (c) for small values of q all terms $0(q^3)$ can be ignored. Assumption (c) restricts our attention to small q and thereby to large distances $\mathbf{R}$. Fourier inversion of Eq. (10.7.16) then gives the asymptotic form of $G(R)$ for large R as

$$G(R) \propto \frac{1}{R_0^2}\frac{\exp -q_0 R}{R} \tag{10.7.17}$$

and the range of the correlations is determined by q_0^{-1}. Thus let us define the characteristic length

$$\xi = q_0^{-1} \tag{10.7.18}$$

where ξ is called the correlation length.

It should be noted from our previous discussion that as $T \to T_c$, $\hat{C}(\mathrm{q}) \underset{q\to 0}{\to} \rho^{-1}$. From

$\hat{C}(q) = C_0 - C_2q^2$ it follows that $\lim_{T\to Tc} C_0 = \rho^{-1}$, and from the definition of $q_0{}^2(=(1-\rho C_0)/R_0)$ it follows that $\lim_{T\to Tc} q_0 = 0$. Thus as $T \to T_c$, the correlation length becomes infinite, and $G(R)$ becomes infinitely long-ranged; that is, $G(R) \propto R^{-1}$.

As the critical point is approached $q_0 \to 0$ and $\chi_T \to \infty$ so that $S(q)$ [cf. Eq. (10.2.16)] becomes more and more sharply peaked. This means that as $T \to T_c$ a larger and larger fraction of the scattered light is scattered with small values of $q(=2k_i \sin \theta/2)$ or equivalently almost all of the light is scattered into the forward direction.

In the normal fluid the intensity $S(q)$ is angle-independent, so that there is no preference for light to be scattered into any specific solid angle. The light is scattered isotropically. In the critical fluid, on the other hand, light is scattered anisotropically with a much larger fraction in the forward direction; that is, with intense scattering at small angles.

If the Ornstein–Zernike approximation is valid, a plot of $[S(q)/S°(q)]^{-1}$ [cf. Eq. (10.7.16)] against q^2 should yield a straight line with a slope $\beta/(\rho\chi_T q_0{}^2)$ [which according to Eq. (10.7.15) is practically a constant] and an intercept $\beta/(\rho\chi_T)$ that approaches zero as $T \to T_c$. Such plots for different temperatures should accordingly be a family of parallel straight lines if the Ornstein–Zernike approximation is valid. This is generally very difficult to establish because scattering at very small q is required, and moreover corrections must be made for multiple scattering. Unfortunately the data do not extend to scattering angles sufficiently small to make the test unambiguous. Nevertheless, there is indirect evidence that the Ornstein–Zernike approximation is incorrect. According to this approximation as embodied in Eq. (10.7.16), the correlation length $\xi = q_0{}^{-1}$ should diverge as $T \to T_c$ with the same critical exponent as χ_T. Experiment shows that this is not the case. Fisher has in fact derived a correction to this approximation that gives a correlation length that diverges differently from χ_T. It is beyond the scope of this book to give an extensive review of this subject. The interested reader should consult the paper by Fisher (1964) and the book by Stanley (1971).

It is stating the obvious to say that a similar treatment can be made of the critical opalescence of the scattering from binary mixtures.

APPENDIX 10.A ENSEMBLE THEORY OF FLUCTUATIONS

Let us calculate $\langle \delta N^2 \rangle$ using the grand canonical ensemble. In the grand ensemble the probability of finding N particles in the volume V is

$$p(N) = e^{\beta\mu N}\, Q_N(\beta, V)/\Xi(\beta, \mu, V) \tag{10.A.1}$$

where μ is the chemical potential of the bath, $Q_N(\beta, V)$ is the canonical partition function, and $\Xi(\beta, \mu, N) = \sum_{N=0}^{\infty} e^{\beta\mu N} Q_N(\beta, V)$ is the grand partition function. The average number of particles in V is

$$\langle N \rangle = \sum_{N=0}^{\infty} p(N)N = \sum_{N=0}^{\infty} e^{\beta\mu N} N Q_N/\Xi = \beta^{-1}\left(\frac{\partial \ln \Xi}{\partial \mu}\right)_{\beta, V} \tag{10.A.2}$$

This latter result can be written as $\Xi\langle N \rangle = \sum_{N=0}^{\infty} e^{\beta\mu N} N Q_N$. The derivative of this with respect to μ at constant β, V can be simplified by: (a) dividing through by Ξ and (b) substitution of Eq. (10.A.2). Then

$$\beta^{-1}\left(\frac{\partial \bar{N}}{\partial \mu}\right)_{\beta, V} = \sum_{N=0}^{\infty} N^2 p(N) - \langle N \rangle^2 = \langle N^2 \rangle - \langle N \rangle^2 = \langle \delta N^2 \rangle$$

where we have used $\langle N^2 \rangle = \sum_{N=0}^{\infty} N^2 e^{\beta\mu N} Q_N / \Xi$. Thus

$$\langle \delta N^2 \rangle = \beta^{-1}\left(\frac{\partial \bar{N}}{\partial \mu}\right)_{\beta, V} = V k_B T \left(\frac{\partial \bar{\rho}}{\partial \mu}\right)_{V,T} \tag{10.A.3}$$

The density $\bar{\rho}$ is only a function of intensive variables so that the derivative of $\bar{\rho}$ with respect to μ, P, T is the same whether $\bar{N}$ or V is held constant. This gives

$$\left(\frac{\partial \bar{\rho}}{\partial \mu}\right)_{V,T} = \left(\frac{\partial \bar{\rho}}{\partial \mu}\right)_{N,T} = \left(\frac{\partial \rho}{\partial P}\right)_{N,T} \left(\frac{\partial P}{\partial \mu}\right)_{N,T}$$

From the Gibbs–Duhem relation $\bar{N} d\mu = V dP - S dT$, $(\partial P/\partial \mu)_{N,T} = \bar{\rho}$, so that $(\partial \rho/\partial \mu)_{V,T} = \bar{\rho}(\partial \rho/\partial P)_{N,T} = \rho^2 \chi_T$ where the isothermal compressibility is $\chi_T = +\rho^{-1}(\partial \rho/\partial P)_{V,T}$. Combining this latter result with Eq. (10.7.24) gives Eq. (10.7.8).

APPENDIX 10.B THERMODYNAMIC IDENTITIES

A very useful method for determining thermodynamic relations is the Jacobian method (cf. Callen, 1960). Let us apply this method to show that

$$\frac{\chi_T}{\chi_S} = \gamma \equiv \frac{c_P}{c_V} \tag{10.B.1}$$

From the definition of these compressibilities we note that the left-hand side of this relation can be written[36]

$$\frac{\partial(\rho, T)}{\partial(P, T)} \cdot \frac{\partial(P, s)}{\partial(\rho, s)} = \frac{\partial(s, P)}{\partial(T, P)} \cdot \frac{\partial(T, \rho)}{\partial(s, \rho)} \tag{10.B.2}$$

Since

$$\frac{\partial(s, P)}{\partial(T, P)} = \left(\frac{\partial s}{\partial T}\right)_P = \frac{M c_P}{VT} = m\rho c_P/T$$

and

$$\frac{\partial(s, P)}{\partial(T, \rho)} = \left(\frac{\partial s}{\partial T}\right)_\rho = \frac{m\rho c_V}{T}$$

Equation (10.*B*.1) follows. The properties of the Jacobian that have been used are $\partial(x, y) = -\partial(y, x)$ and simple rearrangements of numerators and denominators. This relationship is useful in showing that the adiabatic and isothermal sound velocities are related to each other by the equation $c_S^2 = \gamma c_T^2$. These definitions and identities provide measurable values of two of the four derivatives in Eq. (10.4.15); for example,

$$\left(\frac{\partial s}{\partial T}\right)_\rho = m\rho c_V/T; \quad \left(\frac{\partial P}{\partial \rho}\right)_T = m c_T^2 = m c_S^2/\gamma \tag{10.B.3}$$

Let us now evaluate the other two derivatives. First note that

$$\left(\frac{\partial P}{\partial T}\right)_\rho = \frac{\partial(P,\rho)}{\partial(T,\rho)} = \frac{\partial(P,T)}{\partial(P,T)} \cdot \frac{\partial(P,\rho)}{\partial(T,\rho)}$$

In the last step we merely multiply by a Jacobian which is unity. Rearrangement gives

$$\left(\frac{\partial P}{\partial T}\right)_\rho = \frac{\partial(P,\rho)}{\partial(P,T)} \cdot \frac{\partial(P,T)}{\partial(T,\rho)} = -\left(\frac{\partial \rho}{\partial T}\right)_P \cdot \left(\frac{\partial P}{\partial \rho}\right)_T$$

This can be expressed in terms of c_T^2 and α, giving

$$\left(\frac{\partial P}{\partial T}\right)_\rho = m\rho\alpha c_T^2 = \frac{\alpha}{\chi_T} \tag{10.B.4}$$

This expression can be written in terms of the specific heats. Using the well-known identity (cf. Callen, p. 130), $\rho m(c_P - c_V) = (T\alpha^2/\chi_T)$ gives

$$\left(\frac{\partial P}{\partial T}\right)_\rho = \frac{\rho m(c_P - c_V)}{\alpha T} = \frac{\rho m c_V(\gamma - 1)}{\alpha T} \tag{10.B.5}$$

Now for the remaining identity we use the Maxwell relation

$$\left(\frac{\partial S}{\partial V}\right)_T = \left(\frac{\partial P}{\partial T}\right)_V \tag{10.B.6}$$

Since $(\partial S/\partial V)_T = -\rho(\partial s/\partial \rho)_T$, it follows from Eq. (10.B.6) that

$$\left(\frac{\partial s}{\partial \rho}\right)_T = -\frac{m c_V(\gamma - 1)}{\alpha T} \tag{10.B.7}$$

Combining the foregoing with Eq. (10.4.15) then gives p_1 and s_1 in terms of ρ_1 and T_1; that is, Eq. (10.4.16)

$$p_1 = m c_T^2[\rho_1 + \alpha\rho_0 T_1] \tag{10.B.8a}$$

$$s_1 = -\frac{m\rho c_V}{T_0}\left[\frac{(\gamma - 1)}{\alpha\rho_0}\rho_1 - T_1\right] \tag{10.B.8b}$$

APPENDIX 10.C THERMODYNAMIC FLUCTUATION THEORY

Consider the isolated composite system $A + B$ of Fig. (10.C.1). The thermodynamic states of the subsystems are $\mathbf{X}_A = (E_A, V_A, \mathbf{N}_A)$ and $\mathbf{X}_B = (E_B, V_B, \mathbf{N}_B)$, where E_i, V_i, and $\mathbf{N}_i$ are the energy, volume, and composition[37] of the subsystem $i = A, B$.

The entropies of A and B are given by the thermodynamic equations of state (cf. Callen, 1960) as $S_A = S_A(\mathbf{X}_A)$ and $S_B = S_B(\mathbf{X}_B)$ and the entropy of the total system is then

$$S_T = S_A(\mathbf{X}_A) + S_B(\mathbf{X}_B) \tag{10.C.1}$$

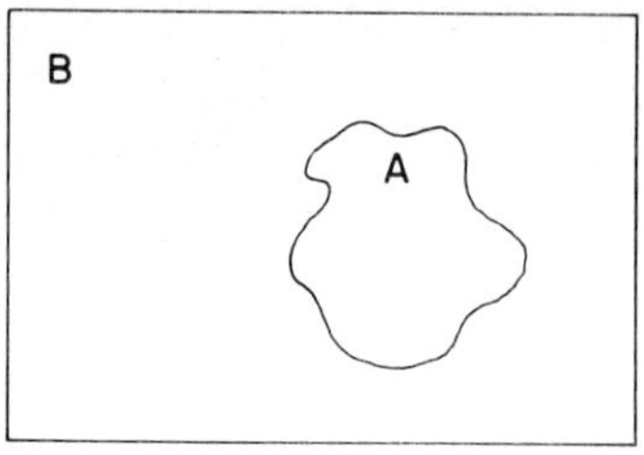

FIG. 10.C.1. Composite system T composed of two subsystems A and B that can exchange energy, volume, and number particles.

The subsystems exchange energy, volume, and particles until they come to equilibrium. The equilibrium states $\mathbf{X}_A^*$, $\mathbf{X}_B^*$ are those states which maximize the total entropy subject to the constraint $\mathbf{X}_T = \mathbf{X}_A^* + \mathbf{X}_B^*$. The equilibrium entropy is therefore

$$S_T^* = S_A(\mathbf{X}_A^*) + S_B(\mathbf{X}_B^*) = S_T(\mathbf{X}_A^*, \mathbf{X}_B^*) \tag{10.C.2}$$

If all the microstates of the isolated composite system are equally probable, the probability of observing the state $\mathbf{X}_A$, $\mathbf{X}_B$ is simply the fraction of all possible microstates consistent with this state

$$\text{prob}(\mathbf{X}_A, \mathbf{X}_B) = \frac{\Omega_A(\mathbf{X}_A)\Omega_B(\mathbf{X}_B)}{\Omega_T(\mathbf{X}_T)} \tag{10.C.3}$$

where $\Omega_A(\mathbf{X}_A)$, $\Omega_B(\mathbf{X}_B)$, and $\Omega_T(\mathbf{X}_T)$ are the number of microstates accessible to A, B, and T when they are respectively in the macroscopic states $\mathbf{X}_A$, $\mathbf{X}_B$, and $\mathbf{X}_T$. The number of states accessible to a system can be found from the Boltzmann formula $S = K \ln \Omega$. We can thus write Eq. (10.C.4) as

$$\text{prob}(\mathbf{X}_A, \mathbf{X}_B) = \exp \frac{1}{k_B} [S_A(\mathbf{X}_A) + S_B(\mathbf{X}_B) - S_T^*(\mathbf{X}_T^*)] \tag{10.C.4}$$

where $S_T^*(X_T^*)$ is the entropy of the composite system when the subsystems A and B are in equilibrium with each other. $S_A(\mathbf{X}_A)$ and $S_B(\mathbf{X}_B)$ are the entropies of the subsystems A and B when they are in the macroscopic states $\mathbf{X}_A$ and $\mathbf{X}_B$ respectively, but not necessarily in equilibrium with each other. Thus

$$\delta S_T(\mathbf{X}_A, \mathbf{X}_B) \equiv S_A(\mathbf{X}_A) + S_B(\mathbf{X}_B) - S_T^*(\mathbf{X}_T^*) \tag{10.C.5}$$

represents the entropy change involved in the transition in which the composite system goes from a state of overall equilibrium $(\mathbf{X}_A^*, \mathbf{X}_B^*)$ to a state in which A and B are each in a state of internal equilibrium $(\mathbf{X}_A, \mathbf{X}_B)$ but are not necessarily in equilibrium with each other, that is, $\mathbf{X}_A \neq \mathbf{X}_A^*$ and $\mathbf{X}_B \neq \mathbf{X}_B^*$. Thus δS_T is the entropy change corresponding to the transition $(\mathbf{X}_A^*, \mathbf{X}_B^*) \rightarrow (\mathbf{X}_A, \mathbf{X}_B)$ or equivalently it is the entropy change for the fluctuations[38] $\delta\mathbf{X}_A = \mathbf{X}_A - \mathbf{X}_A{}^*$ and $\delta\mathbf{X}_B = \mathbf{X}_B - \mathbf{X}_B{}^*$.

It follows from these considerations that the probability of a fluctuation in A is

$$p(\delta\mathbf{X}_A) = \Omega_0{}^{-1} \exp \frac{1}{k_B} \delta S_T(\mathbf{X}_A, \mathbf{X}_B) \tag{10.C.6}$$

δS_T is negative because $S(\mathbf{X}_A{}^*, \mathbf{X}_B{}^*)$ is a maximum for given $\mathbf{X}_T$. Thus the larger the entropy change $|\delta S_T|$ the more improbable the fluctuation. Ω_0 is a normalization constant. In order to use this probability we have to compute the entropy change in the total system caused by a fluctuation in the subsystem A.

The following procedure can be used to compute the probability of the fluctuation (cf. Landau and Lifshitz, 1959).

The thermodynamic equation of state of the composite system is

$$S_T = S_T(\mathbf{X}_T) = S_T(E_T, \mathbf{N}_T, V_T) \tag{10.C.7}$$

This is a monotonically increasing function of the energy E_T for fixed V_T and $\mathbf{N}_T$ as illustrated in Fig. 10.C.2. All points on the curve represent equilibrium states of the total system. By construction, point c represents a nonequilibrium state of the total system but such a state that A and B are in their own internal states of equilibrium $c = (\mathbf{X}_A, \mathbf{X}_B)$. Point b represents the equilibrium state of the total system $b = (\mathbf{X}_A'', \mathbf{X}_B'')$.

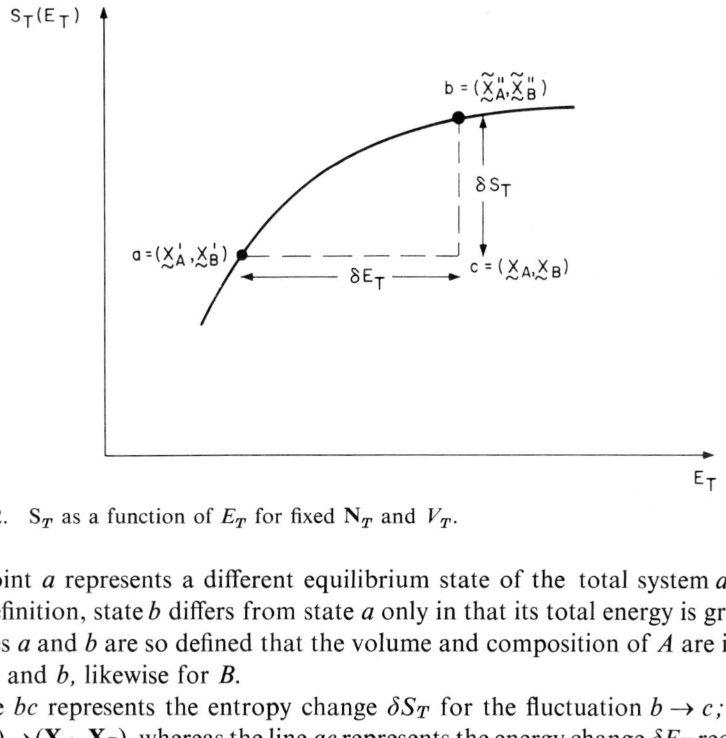

FIG. 10.C.2. S_T as a function of E_T for fixed $\mathbf{N}_T$ and V_T.

Finally point a represents a different equilibrium state of the total system $a = (\mathbf{X}_A', \mathbf{X}_B')$. By definition, state b differs from state a only in that its total energy is greater by δE_T. States a and b are so defined that the volume and composition of A are identical in states a and b, likewise for B.

The line bc represents the entropy change δS_T for the fluctuation $b \to c$; that is; $(\mathbf{X}''_A, \mathbf{X}''_B) \to (\mathbf{X}_A, \mathbf{X}_B)$, whereas the line ac represents the energy change δE_T required in the reversible transition $a \to b$ that follows the equilibrium entropy curve ab. We now make the assumption that subsystem B is much larger than A. Then S_T should change very little in going from a to b, even for a relatively large fluctuation in A and the curve ab can be approximated by a straight line of slope $(\partial S_T(E_T)/\partial E_T)_b = 1/T$. From the equation of a straight line

$$S_b = S_a + \left(\frac{\partial S_T}{\partial E_T}\right)_b (E_b - E_a) = S_a + \frac{1}{T}\,\delta E_T \tag{10.C.8}$$

Now from Fig. (10.C.2) we see that $S_c = S_a$ so that the entropy change $\delta S_T = S_b - S_c$ for the transition $c \to b$ is $\delta S_T = S_b - S_a$. When this is combined with Eq. (10.C.8) it follows that

$$\delta S_T = -\frac{1}{T}\,\delta E_T \tag{10.C.9}$$

What is required is a calculation of δE_T, the energy required to make the reversible transition from *a* to *b*.

Suppose now that a work source *C* is coupled to the subsystem *A* but not to *B* (see Fig. 10.C.3). By construction *C* can transfer energy to the composite system only by doing work on *A*. Let us now compute δE_T, which is the amount of work that *C* must do on *A* such that the composite system changes from state *a* to *c*.

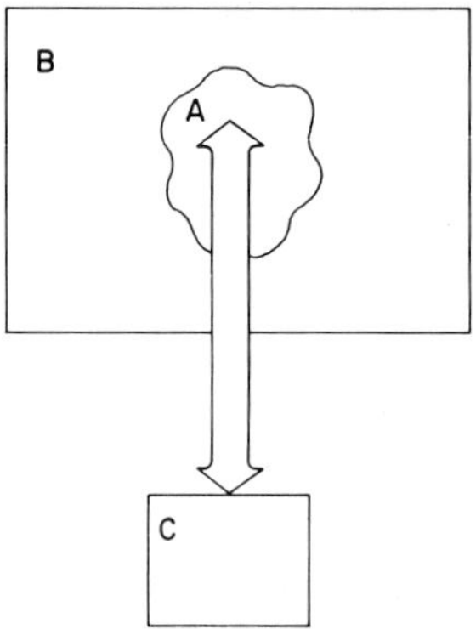

FIG. 10.C.3. Work source C can only do work on *A*. *C* cannot exchange heat or composition with *A*, neither can it couple directly to *B*.

In the transition *A* experiences a change in internal energy δE_A

$$\delta E_A = \delta E_T + P_B \delta V_B - T_B \delta S_B - \boldsymbol{\mu}_B \cdot \delta \mathbf{N}_B \tag{10.C.10}$$

where δE_T is the work done on *A* by *C*, $P_B \delta V_B$ is the work done on *A* due to *B*, $-T_B\,\delta S_B$ is the heat absorbed by *A* from *B*, and $-\boldsymbol{\mu}_B \cdot \delta \mathbf{N}_B$ is the total energy transferred to *A* from *B* by virtue of particle exchange.[39] In arriving at this formula the implicit assumption was made that a relatively large fluctuation in *A* will produce a relatively small fluctuation in *B* so that the pressure P_B, temperature T_B, and chemical potentials of *B* will be unchanged. This is tantamount to assuming that *B* is much larger than *A*. Because the volume and composition are fixed, $\delta V_B + \delta V_A = 0$, $\delta \mathbf{N}_B + \delta \mathbf{N}_A = 0$, and because the transition from *a* to *c* has been accomplished adiabatically (reversible work sources do not change the entropy of the composite system), it follows that $\delta S_A + \delta S_B = 0$. Substitution of this into Eq. (10.C.10) then leads to

$$\delta E_T = \delta E + P \delta V - T \delta S - \boldsymbol{\mu} \cdot \delta \mathbf{N} \tag{10.C.11}$$

where δE_T can be regarded as the reversible work required to produce the fluctuation δE, δV, δS, $\delta \mathbf{N}$ in *A*. The subscripts *A* and *B* have been suppressed with the understanding that P, T, μ have their equilibrium values defined by the large subsystem B.

Thus in the fluctuation $b \to c$ the entropy change is

$$\delta S_T = -\frac{1}{T}[\delta E + P\delta V - T\delta S - \boldsymbol{\mu} \cdot \delta \mathbf{N}] \tag{10.C.12}$$

where Eqs. (10.C.9) and (10.C.11) have been combined. The probability of the fluctuation follows immediately from Eqs. (10.C.6) and (10.C.12).

$$p(\delta \mathbf{X}) = \Omega_0{}^{-1} \exp -\frac{1}{k_B T}\{\delta E + P\delta V - T\delta S - \boldsymbol{\mu} \cdot \delta \mathbf{N}\} \tag{10.C.13}$$

This formula is applicable to large or small fluctuations in A; nevertheless, from its form we note that the probability rapidly decreases for large fluctuations.

The energy of the subsystem can be regarded as a function of S, V, $\mathbf{N}$, that is, $E = E(S, V, \mathbf{N})$. For small fluctuations, E can be expanded in a Taylor expansion around the equilibrium state S^*, V^*, $\mathbf{N}^*$, so that

$$\delta E = \delta^{(1)}E + \frac{1}{2}\delta^{(2)}E + \dots \tag{10.C.14}$$

where[40]

$$\delta^{(1)}E = T\delta S - P\delta V + \boldsymbol{\mu} \cdot \delta \mathbf{N} \tag{10.C.15}$$

and where

$$\delta^{(2)}E = \delta^{(1)}\delta^{(1)}E = \delta T\delta S - \delta P\delta V + \delta\boldsymbol{\mu} \cdot \delta \mathbf{N} \tag{10.C.16}$$

so that

$$\delta E = T\delta S - P\delta V + \boldsymbol{\mu} \cdot \delta \mathbf{N} + \frac{1}{2}\{\delta T\delta S - \delta P\delta V + \delta\boldsymbol{\mu} \cdot \delta \mathbf{N}\} \tag{10.C.17}$$

Equation (10.C.13) thus becomes

$$P(\delta \mathbf{X}) = \Omega_0{}^{-1} \exp -\frac{1}{2k_B T}\delta^{(2)}E \tag{10.C.18}$$

or

$$p(\delta \mathbf{X}) = \Omega_0{}^{-1} \exp \frac{-1}{2k_B T}\{\delta T\delta S - \delta P\delta V + \delta\boldsymbol{\mu} \cdot \delta \mathbf{N}\} \tag{10.C.19}$$

This is the *master formula* of fluctuation theory.

Let us now apply this formula to a one-component system. In the typical light-scattering experiment we are interested in the fluctuations that take place within the fixed volume V of the illuimnated region. In this case $\delta V = 0$ and the master formula becomes

$$p(\delta \mathbf{X}) = \Omega_0{}^{-1} \exp -\frac{1}{2k_B T}\{\delta\mu\delta N + \delta T\delta S\} \tag{10.C.20}$$

For fixed volume S and μ can be regarded as functions of N and T so that

$$\delta S = \left(\frac{\partial S}{\partial N}\right)_{T,V} \delta N + \left(\frac{\partial S}{\partial T}\right)_{N,V} \delta T \tag{10.C.21a}$$

$$\delta \mu = \left(\frac{\partial \mu}{\partial N}\right)_{T,V} \delta N + \left(\frac{\partial \mu}{\partial T}\right)_{N,V} \delta T \tag{10.C.21b}$$

From the formula for dA, $dA = pdV - SdT + \mu dN$ we find the Maxwell relation $(\partial S/\partial N)_{T,V} = -(\partial \mu/\partial T)_{N,V}$. Substitution of these results into Eq. (10.C.20) yields

$$\delta\mu\delta N + \delta T \delta S = \left\{\left(\frac{\partial \mu}{\partial N}\right)_{T,V} \delta N^2 + \left(\frac{\partial S}{\partial T}\right)_{N,V} (\delta T)^2\right\} \tag{10.C.22}$$

The derivatives can be expressed in terms of well-known properties, for instance, from the Gibbs–Duhem equation $\rho d\mu = dP - \frac{S}{V} dT$ (where ρ is the number density) it follows that $(\partial \mu/\partial P)_{T,V} = \rho^{-1}$ and

$$\left(\frac{\partial \mu}{\partial N}\right)_{T,V} = \frac{1}{V}\left(\frac{\partial \mu}{\partial \rho}\right)_{T,V} = \frac{1}{V}\left(\frac{\partial \mu}{\partial P}\right)_{T,V}\left(\frac{\partial P}{\partial \rho}\right)_{T,V} = \frac{1}{V\rho^2}\rho\left(\frac{\partial P}{\partial \rho}\right)_{T,V} \tag{10.C.23}$$

The pressure, being an intensive variable, can only be a function of intensive variables so that holding V constant in the partial derivative is of no importance. From the definition of the isothermal compressibility $\chi_T \equiv \rho^{-1}(\partial\rho/\partial P)_T$ it follows that $(\partial \mu/\partial N)_{T,V} = 1/V\rho^2\chi_T)$. Also from the equation $dE = TdS - pdV + \mu dN$, it follows that

$$\left(\frac{\partial S}{\partial T}\right)_{V,N} = \frac{C_V}{T} = \left(V\frac{m\rho c_V}{T}\right)$$

where C_V is the isochoric heat capacity of the material in the volume V and where c_V is the specific heat (cal/°gm).

Upon substitution of these quantities Eq. (10.C.20) can be written as

$$\delta\mu\delta N + \delta T\delta S = V\left\{(\rho^2\chi_T)^{-1}(\delta\rho)^2 + \left(\frac{m\rho c_V}{T}\right)(\delta T)^2\right\} \tag{10.C.24}$$

where we have replaced the fluctuation in number δN by $V\delta\rho$ where $\delta\rho$ is the fluctuation in density. Substitution into Eq. (10.C.20) then gives the joint probability of a fluctuation in density $\delta\rho$ and a fluctuation in temperature δT as

$$p(\delta\rho, \delta T) = \Omega_0^{-1} \exp - V\left\{\frac{(\delta\rho^2)}{2k_B T\rho^2\chi_T} + \frac{m\rho c_V}{2k_B T^2}(\delta T)^2\right\} \tag{10.C.25}$$

Several points should be noted about this formula.

1. p is a Gaussian distribution.
2. Density and temperature fluctuations are statistically uncorrelated

$$p(\delta\rho, \delta T) = p(\delta\rho)p(\delta T) \tag{10.C.26}$$

so that

$$\langle\delta\rho\delta T\rangle = 0, \tag{10.C.27}$$

3. Evaluation of the mean-square density and temperature fluctuations give

$$\langle \delta\rho^2 \rangle = \frac{1}{V}\,\rho^2 k_B T \chi_T$$
$$\langle \delta T^2 \rangle = \frac{1}{V}\left(\frac{k_B T^2}{m\rho c_V}\right) \tag{10.C.28}$$

4. The mean-square fluctuations of these intensive properties are inversely proportional to the volume, so that for large volumes there should be very small fluctuations.

In this simple example the volume of subsystem A is held fixed. Suppose instead that A consists of a certain mass M of material, or equivalently a certain number of molecules $N = M/m$. There will then be no fluctuations in N, but there can be fluctuations in V, S, T, and P. The master formula then becomes

$$p(\delta\mathbf{x}) = \Omega_0{}^{-1} \exp - \frac{1}{2k_B T}\,\{\delta T \delta S - \delta P \delta V\} \tag{10.C.29}$$

For fixed N, S and P can be regarded as functions of T and V

$$\delta S = \left(\frac{\partial S}{\partial T}\right)_{V,N} \delta T + \left(\frac{\partial S}{\partial V}\right)_{T,V} \delta V$$
$$\delta P = \left(\frac{\partial P}{\partial T}\right)_{V,N} T + \left(\frac{\partial P}{\partial V}\right)_{T,N} \delta V \tag{10.C.30}$$

From $dA = -PdV - SdT + \mu dN$ we find the Maxwell relation

$$\left(\frac{\partial P}{\partial T}\right)_{V,N} = \left(\frac{\partial S}{\partial V}\right)_{T,N}$$

Substitution of these equations into $\delta T \delta S - \delta P \delta V$ yields

$$\delta T \delta S - \delta P \delta V = \left(\frac{\partial S}{\partial T}\right)_{V,N} (\delta T)^2 - \left(\frac{\partial P}{\partial V}\right)_{T,N} (\delta V)^2 \tag{10.C.31}$$

As before $(\partial S/\partial T)_{V,N} = V(m\rho c_V/T)$ and since $\rho = N/V$; $-1/V\ (\partial V/\partial p)_{T,Y} = \rho^{-1}\,(\partial\rho/\partial P)_{T,N} = \chi_T$. At fixed volume $\delta\rho = -\frac{N}{V^2}(\delta V) = \frac{1}{V}\,\rho\delta V$. Substitution of these equations into Eq. (10.C.31) then gives

$$\delta T \delta S - \delta P \delta V = V\left\{\frac{m\rho c_V}{T}\,(\delta T)^2 - \frac{1}{\rho^2 \chi_T}\,(\delta\rho)^2\right\} \tag{10.C.32}$$

which when substituted into Eq. (10.C.29)

$$p(\delta\rho,\partial T) = \Omega_0{}^{-1} \exp - V\left\{\frac{(\delta\rho)^2}{2k_B T \rho^2 \chi_T} + \frac{m\rho c_V}{2k_B T^2}\,(\delta T)^2\right\} \tag{10.C.33}$$

It is important to note that this result is identical to Eq. (10.C.25), despite the fact that that equation was derived on the basis of fixed volume while this equation was derived on the basis of fixed mass or number.

It is quite generally true that the probability of fluctuations of intensive variables such as ρ and T are independent of the constraints on the extensive variables (in this case fixed volume or number). Thus we can choose those constraints for which the calculation is simplest. Needless to say, fluctuations in extensive variables are very sensitive to these constraints. For example, in the first case the volume cannot fluctuate, whereas it can in the second.

Another simple example is that of a binary mixture. Because we are interested in light scattering we should consider a fixed volume. It is easier, however, to treat the fluctuations in a binary mixture with the constraint of fixed mass

$$M = m_1 N_1 + m_2 N_2 \tag{10.C.34}$$

where m_1 and m_2 are the molecular masses of components 1 and 2, of which there are N_1 and N_2 molecules respectively at a given time. This constraint implies that fluctuations in N_1 and N_2 are related by $m_1 \delta N_1 + m_2 \delta N_2 = 0$. Then

$$\delta \boldsymbol{\mu} \cdot \delta \mathbf{N} = \delta\mu_1 \delta N_1 + \delta\mu_2 \delta N_2 = M \delta\mu \delta c \tag{10.C.35}$$

where

$$\mu \equiv \left(\frac{\mu_1}{m_1} - \frac{\mu_2}{m_2}\right); \quad c \equiv \frac{m_1 N_1}{M}$$

here $\boldsymbol{\mu}$ is an effective chemical potential per unit mass and c is the mass fraction of the system which is component 1. The master formula of fluctuation theory is then

$$p(\delta \mathbf{X}) = \Omega_0^{-1} \exp - \frac{M}{2KT} \{\delta s \delta T - \delta v \delta P + \delta \mu \delta c\} \tag{10.C.36}$$

where s and v are the entropy per unit mass and volume per unit mass of the system

$$s \equiv \frac{S}{M}; \quad v \equiv \frac{V}{M}$$

As before we must decide on a set of independent thermodynamic variables in terms of which to expand the fluctuations. For a binary mixture, three independent variables are needed. While any three of the variables will suffice for this calculation, certain choices will prove much more convenient than others. It is much easier to deal with fluctuations that are statistically independent. Thus let us try to find three variables (X_1, X_2, X_3) which are statistically independent, that is, for which

$$p(\delta \mathbf{X}) = p(\delta \mathbf{X}_1)\, p(\delta X_2)\, p(\delta X_3) \tag{10.C.37}$$

Equation (10.C.36) contains cross terms between δP and δT. It follows that δP and δT are not statistically independent; $\langle \delta P \delta T \rangle \neq 0$. It should be noted that the polynomial in δP and δT in Eq. (10.C.36) can be factorized so that

$$\delta S \delta T - \delta v \delta P + \delta \mu \delta c = \left\{ \frac{c_{P,c}}{T} (\delta \phi)^2 + \frac{\chi_{S,c}}{\rho} (\delta p)^2 + \left(\frac{\partial \mu}{\partial c}\right)_{P,T} (\delta c)^2 \right\} \tag{10.C.38}$$

where thermodynamic identities have been used and

$$\phi = T - \left[\frac{\alpha_{T,c}T}{m\rho c_{P,c}}\right]P \tag{10.C.39}$$

where

$$\delta\phi = \delta T - \frac{\alpha_{T,c}T}{m\rho c_{P,c}}\,\delta P$$

and where $c_{P,c}$ and $\chi_{S,c}$ are respectively the heat capacity at constants p and c and the adiabatic compressibility at constant c.

The set (ϕ, T, c) is statistically independent with

$$p(\delta\phi, \delta T, \delta c) = \Omega_0^{-1} \exp - \frac{M}{2k_BT}\left\{\frac{c_{P,C}}{2T}(\delta\phi)^2 + \frac{\chi_{S,c}}{\rho}(\delta P)^2 + \left(\frac{\partial\mu}{\partial c}\right)_{p,T}(\delta c)^2\right\} \tag{10.C.40}$$

It follows that

$$\langle\delta\phi^2\rangle = \frac{k_BT^2}{Mc_{P,c}}$$
$$\langle\delta p^2\rangle = \frac{\rho k_BT}{M\chi_{S,c}} \tag{10.C.41}$$
$$\langle\delta c^2\rangle = \frac{k_BT}{M}\left(\frac{\partial c}{\partial\mu}\right)_{P,T}$$

The variable ϕ is a linear combination of T and p. If one expands T as a function of the variables p, c, and s where s is the entropy per unit mass, it can be shown that

$$\delta\phi = \frac{T}{c_{P,c}}\left[\delta s + \left(\frac{\partial\mu}{\partial T}\right)_{P,c}\delta c\right] \tag{10.C.42}$$

so that, in fact, $\delta\phi$ is a linear combination of the entropy and concertration fluctuation, which are nonpropagating fluctuations, unlike δP (which is a propagating fluctuation).

There are several circumstances in which the quadratic fluctuation theory presented here breaks down. When derivatives of any of the intensive parameters with respect to the extensive parameters are very small, the corresponding fluctuations are very large. The Taylor expansion of δE in the fluctuations are then very large and the Taylor expansion of δE in the fluctuations cannot be truncated at the second-order term. For example, the mean-square density fluctuation is given by Eq. (10.C.28), where the isothermal compressibility and correspondingly $\langle\delta\rho^2\rangle$ diverges when $(\partial P/\partial V)_T \to 0$. This happens at the gas liquid critical point. Likewise at the critical consolute point of a binary mixture $\left(\frac{\partial c}{\partial\mu}\right)_{P,T}$ diverges. In order to treat these unstable regions it is necessary to repeat the kinds of arguments given in Section 10.2.

Landau and Lifshitz (1959) have introduced an extension of fluctuation theory to treat this problem. In this treatment the volume V can be regarded as being composed of small elements of volume. Let $e(\mathbf{r})$ be the energy density; that is, the energy per unit volume at $\mathbf{r}$, then the total energy in V is $E = \int_V d^3r\, e(\mathbf{r})$. The second variation in E is then $\delta^2 E = \int_V d^3r\, \delta^2 e(\mathbf{r})$, and the probability of a fluctuation that produces $\delta^2 E$ is

$$p(\delta\mathbf{X}) = \Omega_0^{-1} \exp - \frac{1}{2k_BT}\int_V d^3r\, \delta^2 e(\mathbf{r}) \tag{10.C.43}$$

where $\delta\mathbf{X}$ are fluctuations in the densities of the extensive variables. Now consider a fluctuation in the density. Such a fluctuation could lead to a heterogeneous distribution of matter in the volume V, with more particles crowded into one subregion than into another. Simply regarding $e(\mathbf{r})$ as a function of $\rho(\mathbf{r})$ is equivalent to assuming that the internal energy depends only on the average distance between the particles. Since the potential energy of the fluid in the neighborhood of a given point should be different if at a small distance away from it there are regions of higher or lower density, the energy can also depend on the gradient $\nabla\rho(\mathbf{r})$. In this case[41] $e(\mathbf{r}) = e(\rho(\mathbf{r}), \nabla\rho(\mathbf{r}))$.

The energy parameter e is a scalar quantity. In an isotropic medium, the expansion of e in the powers of ρ and $\nabla\rho$ must be rotationally invariant. To quadratic order the only form that satisfies these conditions is $\delta^2 e = a\delta\rho^2 + b[\nabla\rho]^2$. Expansion of $\delta^2 e$ in a Fourier series in the volume V and substitution into the above gives $\delta^2 E = V^{-1}\sum_q [a + bq^2]|\delta\rho(\mathbf{q})|^2$ and

$$p \propto \exp - \frac{1}{2Vk_BT}\sum_q [a + bq^2]|\delta\rho(\mathbf{q})|^2 \tag{10.C.44}$$

Thus fluctuations of different wave vectors are statistically independent. The mean-square fluctuation in $\delta\rho(\mathbf{q})$ is found by averaging over p. This is the structure factor

$$S(\mathbf{q}) = \langle|\delta\rho(\mathbf{q})|^2\rangle = \frac{Vk_BT}{(a + bq^2)} \tag{10.C.45}$$

This formula only applies at small wave vectors, otherwise higher powers in the gradients must be retained. Note that $\delta\rho(q) \underset{q\to 0}{\to} \delta N$ where $\delta\rho$ is the fluctuation in the density of particles in V. Thus $\langle|\delta\rho(\mathbf{q})|^2\rangle \underset{q\to 0}{\to} \langle\delta N^2\rangle = V\rho^2 k_BT\chi_T$. Comparison with Eq. (10.C.45) then gives $a = (\rho^2\chi_T)^{-1}$ so that $S(q)$ can be expressed as

$$\frac{S(q)}{S^\circ(q)} \equiv \beta^{-1}\rho\chi_T\left(\frac{1}{1 + (q/q_0)^2}\right) \tag{10.C.46}$$

where

$$q_0^2 = \frac{a}{b} = \frac{1}{b\rho^2\chi_T} \tag{10.C.47}$$

This is the Ornstein–Zernike approximation discussed in Section 10.7. The basic assumption in this derivation is that e can be written as a power series in ρ and $\nabla\rho$; that is, that e is analytic in these variables. The breakdown of the Ornstein–Zernike approximation indicates that this assumption is incorrect.

NOTES

1. This follows from Eqs. (3.3.3), (5.2.12), (5.2.14), and (5.2.17).
2. For example, in a nonpolar gas $(\varepsilon - 1)/(\varepsilon + 2) = (4\pi/3)\,\rho\alpha$ and $(\partial\varepsilon/\partial\rho) = 4\pi\alpha/(1 - \frac{4\pi}{3}\rho\alpha)^2$. It should be remembered that the effective polarizability is very different from the vacuum polarizability. There are many subtle questions involving local field corrections. The interested reader should consult Gelbart's review (1974) and the references cited therein.
3. $I^{\varepsilon}_{if}(\mathbf{q}) = (\mathbf{n}_i \cdot \mathbf{n}_f)^2 \left(\frac{\partial\varepsilon}{\partial\rho}\right)^2_{T_0} S(\mathbf{q})$
4. The spatial Fourier transform of the diffusion equation is $\partial \langle c(\mathbf{q}, t)\rangle/\partial t = -q^2 D \langle c(\mathbf{q}, t)\rangle$ where $\langle c(\mathbf{q}, t)\rangle = \int d^3r\, e^{i\mathbf{q}\cdot\mathbf{r}}\langle c(\mathbf{r}, t)\rangle$. The solution of this equation subject to the initial condition $\langle c(\mathbf{q})\rangle_0 = \int d^3r\, e^{i\mathbf{q}\cdot\mathbf{r}}\langle c(\mathbf{r})\rangle_0$ gives $\langle c(\mathbf{q}, t)\rangle = \langle c(q)\rangle_0 \exp -q^2 Dt$ and the inverse Fourier transform of this equation gives Eq. (10.2.1) because
$$\langle c(\mathbf{r}, t)\rangle = \left(\frac{1}{2\pi}\right)^3 \int d^3q\, e^{-i\mathbf{q}\cdot\mathbf{r}} \langle c(\mathbf{q}, t)\rangle$$
5. Eq. (10.2.2) is simply the inverse Fourier transform of Eq. (10.2.1). Note that in Fourier analysis $\mathbf{q}$ and $\mathbf{r}$ are conjugate quantities; that is, to describe properties at large values of $\mathbf{r}$ we usually require only the small q Fourier components.
6. Substitution of $\langle c(\mathbf{r}, t)\rangle = c_0 + \delta c(\mathbf{r}, t)$ into the diffusion equation followed by Fourier analysis gives Eq. (10.2.3)
7. It is often possible to express this in terms of a thermodynamic derivative, as in Section 10.C.
8. In this case $\Gamma(q)$ and $\omega(q)$ in Eq. (10.2.7) are such that
$$\lim_{q\to 0} \Gamma(q) = 0; \lim_{q\to 0} \omega(q) = 0$$
Usually hydrodynamic modes involve densities of the conserved variables (such as mass, momentum, and energy density). Sometimes, however, nonconserved variables also are hydrodynamic modes. This occurs when there is a symmetry-breaking phase transition such as the transition from paramagnetism to ferromagnetism or from an isotropic liquid to liquid crystal. Then as a consequence of the Goldstone theorem a new hydrodynamic mode must appear. This new mode is often the fluctuation in the order parameter characterizing the phase transition. For example, in a ferromagnet the fluctuations in the magnetization have a lifetime $\tau(q) \to \infty$ as $q \to 0$, and a frequency $\omega(q) = \alpha q^2$. This mode is called a spin wave. The order parameter usually does not correspond to a conserved density, nevertheless it is often a hydrodynamic mode. The number of such modes in a system is determined by the symmetry of the system. We refer the interested reader to the paper of Martin, et al. (1972).
9. If there are large concentration gradients in the system, Fick's law must be modified to include higher spatial gradients. We then obtain *Burnett equations.*
10. These densities are microscopically defined as $\hat{\rho}(\mathbf{r}, t) = \sum_{j=1}^{N} \delta(\mathbf{r} - \mathbf{r}_j)$, $\hat{\mathbf{g}}(\mathbf{r}, t) = \sum_j \mathbf{p}_j\delta(\mathbf{r} - \mathbf{r}_j)$, and $\hat{e}(\mathbf{r}, t) = \sum_j [\frac{p_j^2}{2m} + \frac{1}{2}\sum_{i\neq j} \Phi(r_{ij})]\ \delta(\mathbf{r} - \mathbf{r}_j(t))$ where the hats indicate that these are the microscopic densities and where the positions $\mathbf{r}_j$ and momenta $\mathbf{p}_j$ are taken at time t. $\Phi(r_{ij})$ is the potential energy of the pair of particles i and j with relative position $r_{ij} = |\mathbf{r}_i - \mathbf{r}_j.|$
11. Even in the absence of convection, kinetic energy can be transported into V by collisions.
12. The equilibrium pressure is the force per unit area, so that $\sigma_{xx} = \sigma_{yy} = \sigma_{zz} = -p_0$.
13. $p(\mathbf{r}, t) = -\frac{1}{3}(\sigma_{xx} + \sigma_{yy} + \sigma_{zz}) = -\frac{1}{3}Tr\boldsymbol{\sigma}$
14. Since τ_{ij} and mpu_iu_j are symmetric tensors it follows from Eq. (10.4.6b) that σ_{ij} must be a symmetric tensor.
15. The reasoning here is similar to that given in Section 7.B.
16. Most simple fluids are Newtonian. Polymer solutions and highly viscous fluids are often non-Newtonian fluids.
17. By definition the heat density is

$$dh = TdS/V.$$

From the first law of thermodynamics

$$dh = (dE + PdV)/V = [d(eV) + PdV]/V = de - \frac{(e_0 + P_0)}{\rho_0} d\rho$$

dh can also be expressed in terms of the entropy change per unit volume $dS/V = ds$. Then $ds = dS/V = dh/T$.

18. The Jacobian method for determining the identities is presented in Appendix 10.B.
19. s is the Laplace variable and should not be confused with the entropy density.
20. According to thermodynamic fluctuation theory, fluctuations in the density and temperature are statistically independent. This is demonstrated in Appendix (10. C) [cf. Eq. (10.C.27)] thus proving (a). Conditions (b) and (c) follow from the fact that

$$\rho_1(\mathbf{q}),\ T_1(\mathbf{q}),\ \psi_1(\mathbf{q}) \to -\rho_1(\mathbf{q}),\ T_1(\mathbf{q}),\ \psi_1(\mathbf{q})$$

under time reversal so that, for example,

$$\langle \rho_1{}^*(\mathbf{q}, 0)\ \psi_1(\mathbf{q}, 0)\rangle = -\ \langle \rho_1{}^*(\mathbf{q}, 0)\ \psi_1(\mathbf{q}, 0) = 0$$

Symmetry properties of this kind are discussed in Section 11.5.

21. In general the roots of a cubic consist of a pair of complex conjugate roots s_+, s_- , and a third root, s_0.
22. Substitution of these roots into Eq. (10.4.31) gives, after some tedious algebra,

$$\frac{\langle \rho_1{}^*(\mathbf{q},0)\ \rho_1(\mathbf{q},t)\rangle}{\langle |\rho_1(\mathbf{q})|^2\rangle} = \frac{1}{s_+ - s_-}\left\{\left[\frac{B(s_+)}{s_+ - s_0} - \frac{B(s_-)}{s_- - s_0}\right]\exp -q^2\Gamma|t|\ \cos\omega(q)|t|\right.$$

$$+ i\left[\frac{B(s_+)}{s_+ - s_0} + \frac{B(s_-)}{s_- - s_0}\right]\exp -q^2\Gamma|t|\ \sin\omega(q)|t|$$

$$+ \frac{B(s_0)}{(s_0 - s_+)\,(s_0 - s_-)}\exp -q^2 D_T|t|$$

The coefficients of the time-dependent functions are now expanded to first order in the small parameters $D_V q^2$ and $\gamma D_T q^2$. This gives Eq. (10.4.37).

23. The primary variables are by definition those fluctuations which contribute directly to the dielectric fluctuation. The secondary variables are those variables that are dynamically coupled to the primary variables.
24. The non-Lorentzian correction embodied in the third term in Eq. (10.5.46) does not contribute to the integrated intensity because $\sin\omega(q)t = 0$ for $t = 0$.
25. $\eta_v(\omega)$ and $\eta_s(\omega)$ are Laplace transforms of time-correlation functions with Laplace variable $s = i\omega$.
26. Because the internal degrees of freedom cannot follow the high frequency variations in pressure, the velocity of sound will be a function of frequency.
27. Including the conditions of Fig. 10.5.1.
28. $m\rho_0$ is the mass density of the solution, $m\rho = m_1\rho_1 + m_2\rho_2$ where m_i, $\rho_i (i = 1, 2)$ is the molecular mass (in grams) and the number density of component i.
29. In the limit of infinite dilution, the mutual diffusion coefficient becomes identical to the self-diffusion coefficient.
30. $\displaystyle \frac{1}{2\pi}\int_{-\infty}^{+\infty} d\omega\ \frac{2\gamma}{\omega^2+\gamma^2} = 1.$
31. For an excellent critical review of the theory of critical opalescence see Fisher (1964).
32. This function can be expressed in terms of the pair correlation function $g^{(2)}(\mathbf{r} - \mathbf{r}')$ where $\rho g^{(2)}(\mathbf{r} - \mathbf{r}')$ is the density of particles at the point $\mathbf{r}$ given that a different particle is at $\mathbf{r}'$. The relationship between G and $g^{(2)}$ is

$$G(\mathbf{r}-\mathbf{r}') = \rho^2[g^{(2)}(\mathbf{r} - \mathbf{r}') - 1] + \rho\delta(\mathbf{r} - \mathbf{r}')$$

33. From Eq. (10.7.1) and $\delta\rho(\mathbf{r}) = \sum_j \delta(\mathbf{r} - \mathbf{r}_j) -$ it follows that

$$G(\mathbf{r}, \mathbf{r}') = G(\mathbf{r} - \mathbf{r}') = \langle \sum_{i,j} \delta(\mathbf{r} - \mathbf{r}_i)\, \delta(\mathbf{r} - \mathbf{r}_j) \rangle - \rho^2$$

It is clear that $G(\mathbf{R})$ can be divided into self $i = j$ and distinct terms $i \neq j$ such that Eq. (10.7.10) follows.

34. This is equivalent to Note 32; that is,

$$\Gamma(R) = g^{(2)}(R) - 1$$

35. To second order in q, $\hat{C}(\mathbf{q}) = \int d^3R\, e^{i\mathbf{q}\cdot\mathbf{R}} C(R) = \int d^3R[1 + i\mathbf{q}\cdot\mathbf{R} - \frac{1}{2}(\mathbf{q}\cdot\mathbf{R})^2]C(R)$. Transforming to spherical polar coordinates with $\mathbf{q}$ along z, performing the angular integrations, we see that the term arising from $i\mathbf{q}\cdot\mathbf{R} = iqR\cos\theta$ is zero and C_0 and C_2 are given as above.

36. In Jacobian notation

$$\left(\frac{\partial x}{\partial y}\right)_z = \frac{\partial(x, z)}{\partial(y, z)}$$

37. The composition of subsystem i is given by the numbers of molecules of each of the n species present

$$N_i = (N_{1i}, N_{2i}, \ldots, N_{ni})$$

38. Since $\mathbf{X}_T = \mathbf{X}_A^* + \mathbf{X}_B^* = \mathbf{X}_A + \mathbf{X}_B$ it follows that $\delta\mathbf{X}_A^* + \delta\mathbf{X}_B^* = 0$ and $\delta\mathbf{X}_A$ and $\delta\mathbf{X}_B$ are not independent fluctuations.

39. $\mu_B = (\mu_{1B}, \ldots, \mu_{nB})$; $\delta\mathbf{N}_B = (\delta N_{1B} \ldots, \delta N_{nB})$

$$\mu_B \cdot \delta\mathbf{N}_B = \sum_{i=1}^{n} \mu_{iB}\, \delta N_{iB}$$

where μ_{iB} and N_{iB} are the chemical potential and number of particles of component i in the subsystem B.

40.

$$\delta^{(1)} E = \sum_{j\pm1}^{n} \left(\frac{\partial E}{\partial X_j}\right)_X \delta X_j$$

where $X_1 = S$; $X_2 = V$; $X_3 = N_1, \ldots; X_n = N_n$. Now

$$T = \left(\frac{\partial E}{\partial S}\right)_{V,N_n;p} = -\left(\frac{\partial E}{\partial V}\right)_{S,N} ; \mu = \left(\frac{\partial E}{\partial N}\right)_{S,V}$$

so that Eq. (10.C.15) follows.

41. The temperature fluctuations should also be included, but because we are interested only in the density fluctuations, and because the density and temperature fluctuations are statistically independent, we disregard the temperature fluctuations here.

REFERENCES

Berne, B. J. and Forster, D., *Ann. Rev. Phys. Chem.* **22,** 563 (1971).

Boley, C. D. and Yip, S., *Phys. Fluids* **15,** 1433 (1972).

Callen, H. B., *Thermodynamics,* Wiley, New York (1960).

Carome, E. F., Nichols, W. H., Kunsitis-Swyt, C. R., and Singal, S. P., *J. Chem. Phys.* **49,** 1013 (1968).

Clark, N. A. Ph. D. Thesis, Massachusetts Institute of Technology (1970).

DeGroot, S. R. and Mazur, P., *Non-Equilibrium Thermodynamics,* Interscience Publisters Inc.—New York, 1962.

Desai, R. C. and Kapral, R., *Phys. Rev. A***6,** 2377 (1972).

Einstein, A., *Ann. Physik,* **33,** 1275 (1910). [English Transl. in *Colloid Chemistry* (J. Alexander, ed.) Vol. I, pp. 323–339. Reinhold, New York, 1926).

Fisher, M. J., *J. Math. Phys.,* **5,** 944 (1964).

Fishman, L. and Mountain, R. D., *J. Phys. Chem.* **74,** 2178 (1970).

Fleury, R. A. and Boon, J. P., *Phys. Rev.* **186,** 244 (1969).

Forster, D., Lubensky, T. C., Martin, P. C., Swift, J., and Pershan, P., *Phys. Rev. Lett.* **26,** 1016 (1971).

Gelbart, W., *Adv. Chem. Phys.,* **26,** 1 (1974).

Gornall, W. S., Stegeman, G. I. A., Stoicheff, B. P., Stolen, R. H., and Volterra, V., *Phys. Rev. Lett.* **17,** 297 (1966).

Gornall, W. S. and Wang, C. S., *Jour. Phys. (Paris) Suppl.* No. 2–3, **33,** Cl–51 (1972).

Green, M. S., *J. Chem. Phys.* **20,** 1281 (1952).

Green, M. S., *J. Chem. Phys.* **22,** 398 (1954).

Kerker, M., *The Scattering of Light and Other Electromagnetic Radiation,* Academic New York 1969).

Landau, L. D. and Lifshitz, E. M., *Statistical Physics,* Addison-Wesley, Reading, Mass. (1959).

Landau, L. D. and Lifshitz, E. M., *Fluid Mechanics,* Addison-Wesley, Reading, Mass. (1960).

Martin, P. C. *Phys. Rev.* **1, A** 423 (1973).

Martin, P. C., Parodi, O., and Pershan, P. S., *Phys. Rev.* **6A,** 2401 (1972).

Miller, G. A., *J. Phys. Chem.* **71,** 2305 (1967).

Miller, G. A. Lee, C. S., *J. Phys. Chem.* **72,** 4644 (1968).

Mountain, R. D., *J. Res. N. B. S.* **70A,** 207 (1966).

Mountain, R. D., *J. Res. N. B. S.* **72A,** 75 (1968).

Mountain, R. D., *CRSS* **1,** 5 (1970).

Mountain, R. and Deutch, J., *J. Chem. Phys.* **50,** 1103 (1969).

Onsager, L., *Phys. Rev.* **37,** 405 (1931); **38,** 2265 (1931).

Ornstein, L. S. and Zernike, R., *Proc. Sec. Sci. K. Med. Akad. Wet.,* **17,** 793 (1914).

Pines, D. and Nozieres, P., *The Theory of Quantum Liquids,* 82–106 Benjamin, New York (1966).

Rytov, S. M. *Sov. Phys. JETP* **6,** 130, 401, 513 (1958).

Stanley, H. E., *Introduction to Phase Transitions and Critical Phenomena,* Oxford University Press (1971).

Stanley, H. E., Paul, G., and Miloservic, S., in *Physical Chemistry: an Advanced Treatise* Vol. VIIIB, Eyring, H., Henderson, D., and Jost, W. (eds.), Academic, New York (1971).

Van Hove, L., *Phys. Rev.* **95,** 249 (1954).

Zwanzig, R., *Phys. Rev.* **124,** 983 (1961).

Zwanzig, R., *Ann. Rev. Phys. Chem.* **16,** 67 (1965).

Zwanzig, R., *J. Chem. Phys.* **43,** 714 (1965b).

Zwanzig, R., in *Statistical Mechanics, New Concepts, New Problems, New Applications,* Rice, S. A., Freed, K., and Light, J. (eds.), (University of Chicago Press (1972).

CHAPTER 11

METHODS FOR DERIVING RELAXATION EQUATIONS*

11 · 1 INTRODUCTION

Until now we have discussed only elementary methods for determining correlation functions, based on ad hoc models. In this chapter a powerful formalism for computing time-correlation functions is presented. As a by-product of this formalism several useful theorems emerge which result from symmetry considerations. Moreover some of the assumptions made in Chapter 10 are shown to be valid. Throughout this chapter we treat classical systems. The methods developed here can also be applied to quantum systems. This is shown in Appendix 11.A. The formalism of this chapter is applied in Chapter 12 to the calculation of the depolarized spectrum.

11 · 2 LIOUVILLE SPACE

The state of a mechanical system of f degrees of freedom is specified by the vector $\boldsymbol{\Gamma} = (q_1, \ldots, q_f, p_1, \ldots, p_f)$, whose components are the f generalized coordinates $(q_1, \ldots, q_f)$ and f conjugate momenta $(p_1, \ldots, p_f)$. Geometrically, the state is represented by the point $\boldsymbol{\Gamma}$ in 'phase-space,' which is a $2f$ dimensional Cartesian space whose coordinate axes are labeled by $(q_1, \ldots, p_f)$ respectively. According to mechanics the state $\boldsymbol{\Gamma}$ changes with time according to the canonical equations of motion [Eq. (2.3.1)]. Equation (2.3.1) can be written in the compact vector form

$$\frac{d\boldsymbol{\Gamma}}{dt} = \{\boldsymbol{\Gamma}, H\} = iL\boldsymbol{\Gamma}; \; iL \equiv \{\ldots, H\} \tag{11.2.1}$$

where $\{\boldsymbol{\Gamma}, H\}$ is the Poisson bracket[1] of $\boldsymbol{\Gamma}$ with the Hamiltonian H and the operator L is called the *Liouvillian*. From its definition it follows that L has the explicit form

$$iL = \sum_{i=1}^{f} \left(\frac{\partial H}{\partial p_i}\frac{\partial}{\partial q_i} - \frac{\partial H}{\partial q_i}\frac{\partial}{\partial p_i}\right) \tag{11.2.2}$$

The Liouvillian is obviously a linear partial differential operator. As we shall soon see,

---

This chapter is mathematically involved and should be skipped, at least on the first reading, by those not conversant with the mathematics of quantum mechanics.

L has many of the mathematical properties of the 'Hamiltonian' operator in quantum mechanics.

The formal solution of the equations of motion [Eq. (11.2.1)] is

$$\boldsymbol{\Gamma}_t = e^{iLt}\boldsymbol{\Gamma}_0 \tag{11.2.3}$$

Here $\boldsymbol{\Gamma}_0$ and $\boldsymbol{\Gamma}_t$ are, respectively, the states of the system at times 0 and t, and the operator e^{iLt} generates the state $\boldsymbol{\Gamma}_t$ from the initial state. The operator e^{iLt} is called the *propagator*. It defines a mapping in phase space.

A mechanical property A of the system is by definition, any function of the state; that is, $A = A(\boldsymbol{\Gamma}) = A(q_1, \ldots, q_f, p_1, \ldots, p_f)$ (cf. Section 2.3). As the state $\boldsymbol{\Gamma}$ changes the mechanical properties change. The property A depends on the time only through the dependence of the state $\boldsymbol{\Gamma}$ on the time; that is, A is an implicit function of the time. It is simple to show[2] that the rate of change of A is

$$\frac{dA}{dt} = \{A, H\} = iLA \tag{11.2.4}$$

The formal solution of this equation,

$$A(\boldsymbol{\Gamma}, t) = e^{iLt}A(\boldsymbol{\Gamma}, 0) \tag{11.2.5}$$

expresses how the property A evolves in time. The time-correlation function of the property A [cf. Eq. (2.3.3)] can thus be expressed as

$$C(t) = \int d\Gamma \rho_0(\boldsymbol{\Gamma})A^*(\boldsymbol{\Gamma})\, e^{iLt}A(\boldsymbol{\Gamma}) \tag{11.2.6}$$

where $\rho_0(\boldsymbol{\Gamma})$ is the equilibrium "ensemble" distribution function discussed in Section 2.3 and the integration is over all of phase space. This formula can be written in a suggestive form. First we note that $iL\rho_0(\boldsymbol{\Gamma}) = 0$ because $\{e^{-\beta H}, H\} = 0$. Then Eq. (11.2.6) can be written as $\int d\Gamma \psi_A^*(\boldsymbol{\Gamma})e^{iLt}\psi_A(\boldsymbol{\Gamma})$ where $\psi_A(\boldsymbol{\Gamma}) \equiv \rho_0{}^{1/2}A$. The important thing to note about the function $\psi_A(\boldsymbol{\Gamma})$ is that for any property A which has a finite ensemble average, $\psi_A(\boldsymbol{\Gamma})$ will be square-integrable; that is, $\int d\Gamma |\psi_A(\boldsymbol{\Gamma})|^2 < \infty$. ψ_A can thus be thought of as a vector in the Hilbert space of functions of $\boldsymbol{\Gamma}$. This space is often called *Liouville space*. From the mathematical point of view $\psi_A(\boldsymbol{\Gamma})$ can be treated as the wave function in quantum mechanics. For example, the scalar product of two such functions $\psi_A(\boldsymbol{\Gamma})$ and $\psi_B(\boldsymbol{\Gamma})$ can be defined as $(A, B^*) \equiv \int d\Gamma \psi_B^*(\boldsymbol{\Gamma})\psi_A(\boldsymbol{\Gamma})$, but from the definition of ψ_A, ψ_B; $(A, B^*) = \int d\Gamma \rho_0(\boldsymbol{\Gamma})B^*(\boldsymbol{\Gamma})A(\boldsymbol{\Gamma})$. This is simply the ensemble average of the product of properties A and B^*. The time-correlation function [Eq. (11.2.6)] can then be written as

$$C(t) = (e^{iLt}A, A^*) = (A(t), A^*) \tag{11.2.7}$$

Becuase of the formal similarity between $C(t)$ and scalar products in "quantum mechanics" the mathematical techniques of quantum mechanics can be applied to a study of classical time-correlation functions.

Let us now formalize this. The *scalar product* of two arbitrary properties A and B is defined as

$$(B, A^*) \equiv \int d\Gamma \rho_0(\Gamma) A^*(\Gamma) B(\Gamma) \tag{11.2.8}$$

This scalar product has the following properties:

1. $(A^*, B)^* = (B^*, A)$
2. If $A = c_1 A_1 + c_2 A_2$ where A_1 and A_2 are two arbitrary functions of Γ, and c_1 and c_2 are two arbitrary constants

$$(B^*, A) = c_1(B^*, A_1) + c_2(B^*, A_2)$$
$$(A^*, B) = c_1^*(A_1^*, B) + c_2^*(A_2^*, B)$$

that is, the scalar product (B^*, A) is linear in A and antilinear in B.

3. $(A^*, A) \geqslant 0$, the equality sign appears only if A $\equiv$ 0. From condition 1 we see that the *norm* $||A|| \equiv (A^*, A)^{1/2}$ of the property A, which can be regarded as the "length" of the property A, is a real quantity. A property whose norm is unity is said to be *normalized*. Two properties, A and B, are said to be *orthogonal* if $(A^*, B) = 0$.

Thus all functions $A_i(\Gamma)$ of finite norm, together with the above definition of the scalar product define a Hilbert space. This space is called *Liouville space* because the Liouvillian generates the motion in this space.

The "Liouvillian" L is a linear Hermitian operator; that is,

$$L^+ = L \tag{11.2.9}$$

where L^+ is the Hermitian conjugate[3] of L. The proof of this follows the same lines as the proof in quantum mechanics.[4]

By analogy with quantum mechanics, $C(t)$ [c.f. Eq. (11.2.7)] can be regarded as an "expectation value" of e^{iLt} in the "state" $\psi_A = \rho_0^{1/2}A$. Let us recall that in quantum mechanics the Hamiltonian operator $\hat{H}$ is Hermitian and the operator exp $[i\hat{H}t/\hbar]$ is unitary. Correspondingly here the "Liouvillian" L is Hermitian and the propagator

$$\hat{G}(t) \equiv e^{iLt} \tag{11.2.10}$$

is unitary[5]. [The Hermitian conjugate of $G(t)$ is $\hat{G}^+(t) = e^{-iL^+t} = e^{-iLt}$. Since L is Hermitian $L^+ = L$ and the last equality follows. Now e^{-iLt} is the inverse of e^{iLt} so that $G^+(t)G(t) = G(t)G^+(t) = 1$, thus proving the unitarity of $G(t)$.]

11 · 3 PROJECTION OPERATORS AND RELAXATION EQUATIONS

In previous chapters phenomenological relaxation equations were used together with the Onsager regression hypothesis to compute time correlation functions. In this section we present a microscopic derivation of "generalized relaxation equations" (Zwanzig, 1961; Berne, Mori, 1965 and 1971). These equations can be used to compute time-correlation functions under circumstances where the usual phenomenological equations do not apply.

In this section it is shown that arbitrary dynamical properties in complicated systems can be described by equations which are analogous to the Langevin equation of Brownian motion theory [cf. Section (5.9)]. For example, the arbitrary property A is described by the equation

$$\frac{dA}{dt} = i\Omega A(t) - \int_0^t d\tau K(\tau)A(t-\tau) + F(t) \tag{11.3.1}$$

where Ω, $K(t)$, and $F(t)$ are well-defined functions, called the *frequency, memory function,* and *random force* respectively. Even more generally, a set of properties $\{A_1, \ldots, A_M\}$ is described by the matrix equation

$$\frac{dA_\mu}{dt} = \sum_\nu \left(i\Omega_{\mu\nu}A_\nu(t) - \int_0^t d\tau K_{\mu\nu}(\tau)A_\nu(t-\tau)\right) + F_\mu(t) \tag{11.3.2}$$

where the matrix elements are well-defined functions. These equations are of considerable use in light scattering since time-correlation functions and their corresponding spectra can be computed from them.

Having expressed time-correlation functions in the language of Hilbert space, we can give a geometrical interpretation of these functions. Let A be a vector in Liouville space representing the initial value $A(0)$ then $A(t) = e^{iLt}A$ is another vector representing the property at time t. The operator e^{iLt} is unitary; it preserves the norm of A. We can regard the time evolution of A, therefore, as a simple rotation in Liouville space. This is illustrated in Fig. 11.3.1a.

The scalar product of $A(0)$ with $A(t)$ is indicated in Fig. 11.3.1a, where it is clear that the time-correlation function of A can be regarded as a "projection" of $A(t)$ onto $A(0)$. The particular projection of interest here is the one onto the initial property A. An operator which *projects* an arbitrary vector onto A is

$$P \equiv (\ldots, A^*)(A, A^*)^{-1}A \tag{11.3.3}$$

When P acts on an arbitrary vector B it gives $PB = (B, A^*)(A, A^*)^{-1}A$ (see Fig. 11.3.1b).

The operator P has the following properties: (a) $PA = A$, (b) $P^2 = P$, and (c) $(g^*, Pf)^* = (f^*, Pg)$. Any operator which has the property (b) is said to be *idempotent.* Property (c) implies that P is Hermitian. Properties (b) and (c) are necessary and sufficient conditions for an operator to be a *projection operator.*

The operator

$$Q \equiv 1 - P \tag{11.3.4}$$

also satisfies (b) and (c) and is therefore a projection operator. Moreover, Q has the property that when it acts on arbitrary vector B, $(QB, A^*) = (B, A^*) - (B, A^*)(A, A^*)^{-1} \times (A, A^*) = 0$. In other words, *$QB$ is orthogonal to A* so that Q is a projector onto a subspace orthogonal to A. Moreover, $QA = 0$. It is obvious from these definition that the sum of P and Q is the unit operator

$$P + Q = 1 \tag{11.3.5}$$

These properties of Q are illustrated in Fig. 11.3.1c, and we summarize them as follows: (a) $QA = 0$, (b) $Q^2 = Q$, and (c) $(QB, A^*) = 0$

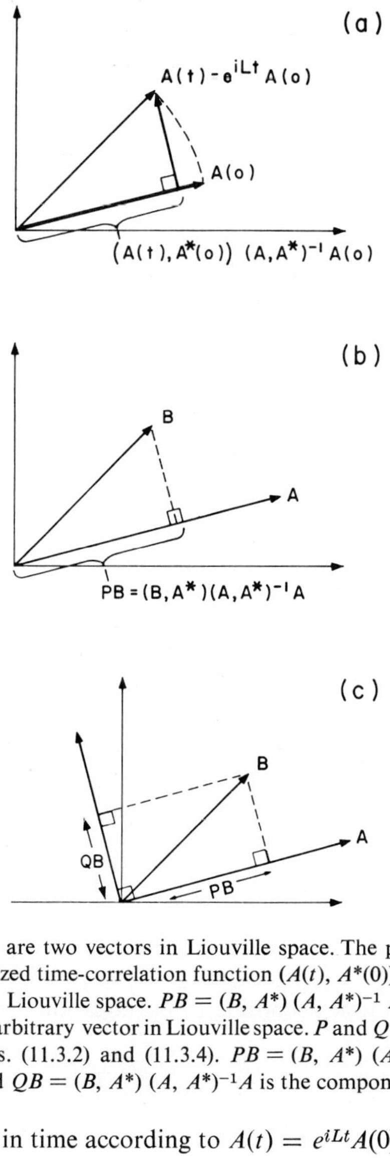

FIG. 11.3.1(*a*). $A(t)$ and $A(0)$ are two vectors in Liouville space. The projection of $A(t)$ onto $A(0)$ is the normalized time-correlation function $(A(t), A^*(0)) (A, A^*)^{-1}$. (*b*) A and B are two vectors in Liouville space. $PB = (B, A^*) (A, A^*)^{-1} A$ is the projection of B onto A. (*c*) B is an arbitrary vector in Liouville space. P and Q are the projection operators defined in Eqs. (11.3.2) and (11.3.4). $PB = (B, A^*) (A, A^*) \; A$ is the projection of B on A and $QB = (B, A^*) (A, A^*)^{-1}A$ is the component of B orthogonal to A.

The function A evolves in time according to $A(t) = e^{iLt}A(0)$ so that its time derivative is

$$\frac{d}{dt} A(t) = e^{iLt} iLA(0) \tag{11.3.6}$$

where $A(t) \equiv A(\boldsymbol{\Gamma}_t)$ and $A(0) \equiv A(\boldsymbol{\Gamma}_0)$. The propagator e^{iLt} satisfies the following identity[6]

$$e^{iLt} = e^{iO_1Lt} + \int_0^t d\tau \, e^{iL(t-\tau)} \, iO_2L \, e^{iO_1L\tau} \tag{11.3.7}$$

where O_1 and O_2 are any operators which have the property that $O_1 + O_2 = 1$. The operators P and Q satisfy this condition [cf. Eq. (11.3.5)] so that taking $O_1 = Q$, $O_2 = P$ in Eq. (11.3.7) gives

$$e^{iLt} = e^{iQLt} + \int_0^t d\tau\, e^{iL(t-\tau)} iPL\, e^{iQL\tau} \tag{11.3.8}$$

The foregoing provides sufficient background for the derivation of Eq. (11.3.1). We first note that the identity $(P + Q = 1)$ can be substituted into Eq. (11.3.6)

$$\frac{dA(t)}{dt} = e^{iLt}(P + Q)\, iLA \tag{11.3.9}$$

without changing anything. From the definition of the projector P we also note that

$$e^{iLt}PiLA = e^{iLt}A(iLA, A^*)\,(A, A^*)^{-1} = i\Omega A(t)$$

where we define the frequency Ω

$$\Omega \equiv (LA, A^*)\,(A, A^*)^{-1} \tag{11.3.10}$$

Equation (11.3.9) thus becomes

$$\frac{dA(t)}{dt} = i\Omega A(t) + e^{iLt}QiLA \tag{11.3.11}$$

Substitution of the identity Eq. (11.3.8) into Eq. (11.3.11) then gives

$$\frac{dA(t)}{dt} = i\Omega A(t) + e^{iQLt}QiLA + \int_0^t d\tau\, e^{iL(t-\tau)} iPLe^{iQL\tau}QiLA \tag{11.3.12}$$

For simplicity we define the quantity G as

$$G \equiv QiLA \tag{11.3.13}$$

This property evolves in time according to the equation

$$G(\tau) = e^{iL\tau}G \tag{11.3.14}$$

Note, however, that one of the propagators that appear in Eq. (11.3.12) is not $e^{iL\tau}$ but $e^{iQL\tau}$. We thus define the quantity

$$F(\tau) \equiv e^{iQL\tau}G \tag{11.3.15}$$

This quantity is called the *random force*. It should not be confused with $G(\tau)$. In general, $G(\tau) \neq F(\tau)$ except at $\tau = 0$ where $F(0) = G(0)$. The quantity $F(\tau)$ can also be written[7]

$$F(\tau) = Qe^{iQL\tau}G = QF(\tau) \tag{11.3.16}$$

This means that $F(\tau)$ *is a vector orthogonal to* A, or

$$(F(\tau), A^*) = 0 \tag{11.3.17}$$

Thus there is no correlation between $A(0)$ and the random force $F(\tau)$. This is a very important formal conclusion. *It is precisely this lack of correlation between $F(\tau)$ and $A(0)$ that is at the foundation of the Onsager regression hypothesis.*

The last integral in Eq. (11.3.12) involves the term $iPLF(\tau)$. From the preceding remarks and definitions it follows that this term can be expressed as

$$iPLF(\tau) = iPLQF(\tau) = (iLQF(\tau), A^*)(A, A^*)^{-1}A.$$

Because Q and L are both Hermitian operators it follows that $(iLQF(\tau), A^*) = -(F(\tau), (QiLA)^*) = -(F(\tau), F^*(0))$. Consequently

$$iPLF(\tau) = -\,(F(\tau),\ F^*(0))\ (A,\ A^*)^{-1}A \tag{11.3.18}$$

We now define the *memory function* $K(\tau)$ as

$$K(\tau) \equiv (F(\tau), F(0))^*(A, A^*)^{-1} \tag{11.3.19}$$

Combining Eqs. (11.3.12), (11.3.18), and (11.3.19) we obtain the generalized Langevin equation

$$\frac{dA(t)}{dt} = i\Omega A(t) - \int_0^t d\tau\ K(\tau)\ A(t-\tau) + F(t) \tag{11.3.20}$$

where Ω, $K(\tau)$, and $F(t)$ are defined in Eqs. (11.3.10), (11.3.19), and (11.3.15) respectively.

From Eq. (11.3.19) we note that the memory function is proportional to the autocorrelation function of the random force. This is called the *second fluctuation–dissipation theorem* (Kubo, 1966).

Generally what is wanted is an equation for the time correlation function[8] $C(t) = (A(t), A^*(0))$. Taking the scalar product of Eq. (11.3.20) with A^* and using Eq. (11.3.17) therefore gives

$$\frac{dC(t)}{dt} = i\Omega C(t) - \int_0^t d\tau\ K(\tau)C(t-\tau) \tag{11.3.21}$$

This is called the *memory-function equation.* In this equation the random force appears implicitly[9] in $K(\tau)$.

The foregoing discussion was restricted to a consideration of a single variable, A. Many circumstances arise in which we will be interested in the time evolution of many coupled variables. For example, in hydrodynamics the mass density, momentum density, and energy density are coupled. It is possible to extend the previous analysis to the case of many variables $\{A_1, \ldots, A_M\}$. These properties can be represented by the column matrix

$$\mathbf{A} = \begin{pmatrix} A_1 \\ \vdots \\ A_M \end{pmatrix} \tag{11.3.22}$$

For convenience these properties are chosen such that their equilibrium values are zero,

$\langle \mathbf{A} \rangle = 0$. Moreover we demand that the set be linearly independent; that is, that none of the A_i are a linear combination of the others.

Let us now define the *correlation matrix*

$$\mathbf{C}(t) = (\mathbf{A}(t), \mathbf{A}^{+}(0)) \tag{11.3.23}$$

where $\mathbf{A}(t)$ is the column matrix $e^{iLt}\mathbf{A}$ and $\mathbf{A}^{+}$ is the Hermitian conjugate of $\mathbf{A}$, that is, the row matrix $\mathbf{A}^{+} = (A_1^*, \ldots, A_M^*)$. $\mathbf{C}(t)$ is an $M \times M$ matrix whose ij^{th} element is the correlation function $C_{ij}(t) = \langle A_i(t)A_j^*(0)\rangle = (A_i(t), A_j^*)$. The initial value of the correlation matrix $\mathbf{C}(0)$ will be denoted $\beta^{-1}\boldsymbol{\chi}$ where $\beta = (k_B T)^{-1}$ and $\boldsymbol{\chi}$ is called the static *susceptibility matrix,*[10]

$$\mathbf{C}(0) = (\mathbf{A}, \mathbf{A}^{+}) \equiv \beta^{-1}\boldsymbol{\chi} \tag{11.3.24}$$

Since $(A_i, A_j^*)^* = (A_j, A_i^*)$, the matrix $\boldsymbol{\chi}$ is a Hermitian matrix, so that $\boldsymbol{\chi}^{+} = \boldsymbol{\chi}$.

The properties $\{A_1, \ldots, A_M\}$ define a subspace of Liouville space. This subspace is the set of all vectors that can be expressed as linear combinations of $\{A_1, \ldots, A_M\}$. Let us now determine the projection operator P that projects a vector onto this M-dimensional subspace. The projection operator must satisfy the conditions $P\mathbf{A} = \mathbf{A}$ and $P^2 = P$. Note that

$$P \equiv (\ldots, \mathbf{A}^{+}) \cdot (\mathbf{A}, \mathbf{A}^{+})^{-1} \cdot \mathbf{A} = \beta(\ldots, \mathbf{A}^{+}) \cdot \boldsymbol{\chi}^{-1} \cdot \mathbf{A} \tag{11.3.25}$$

satisfies these requirements. As before, the operator $Q \equiv 1 - P$ is a projector onto the subspace orthogonal to $\{A_1, \ldots, A_M\}$. Following the same arguments used to derive the generalized Langevin equation for the single variable yields a generalized Langevin equation for the column vector $\mathbf{A}$

$$\frac{d\mathbf{A}(t)}{dt} = i\boldsymbol{\Omega} \cdot \mathbf{A}(t) - \int_0^t d\tau \mathbf{K}(\tau) \cdot \mathbf{A}(t - \tau) + \mathbf{F}(t) \tag{11.3.26}$$

where $\boldsymbol{\Omega}$ is a matrix of *frequencies,*

$$\boldsymbol{\Omega} \equiv (L\mathbf{A}, \mathbf{A}^{+}) \cdot (\mathbf{A} \cdot \mathbf{A}^{+})^{-1} = \beta(L\mathbf{A}, \mathbf{A}^{+}) \cdot \boldsymbol{\chi}^{-1} \tag{11.3.27}$$

$\mathbf{K}(\tau)$ is a matrix of *memory functions,*

$$\mathbf{K}(\tau) \equiv (\mathbf{F}(\tau), \mathbf{F}^{+}(0)) \cdot (\mathbf{A}, \mathbf{A}^{+})^{-1} = \beta(\mathbf{F}(\tau), \mathbf{F}(0)) \cdot \boldsymbol{\chi}^{-1}, \tag{11.3.28}$$

$\mathbf{F}(\tau)$ is the random force, $\mathbf{F}(\tau) = e^{iQL\tau}\mathbf{G}$, *and* $\mathbf{G}$ is $iQL\mathbf{A}$. Equation (11.3.28) is the multidimensional *second-fluctuation dissipation theorem.* Again we note that $F(\tau)$ is orthogonal to $\mathbf{A}$, so that

$$(\mathbf{F}(\tau), \mathbf{A}^{+}) = 0 \tag{11.3.29}$$

Taking the scalar product of Eq. (11.3.26) with $\mathbf{A}^{+}$ gives the equation for the correlation function matrix

$$\frac{d\mathbf{C}(t)}{dt} = i\boldsymbol{\Omega} \cdot \mathbf{C}(t) - \int_0^t d\tau \mathbf{K}(\tau) \cdot \mathbf{C}(t - \tau) \tag{11.3.30}$$

In terms of their components Eqs. (11.3.26) and (11.3.30) become

$$\frac{dA_\nu(t)}{dt} = \sum_\mu \{i\Omega_{\nu\mu}A_\mu(t) - \int_0^t d\tau K_{\nu\mu}(\tau)\, A_\mu(t-\tau)\} + F_\nu(t) \tag{11.3.31}$$

$$\frac{dC_{\nu\mu}(t)}{dt} = \sum_\lambda \left(i\Omega_{\nu\lambda}C_{\lambda\mu}(t) - \int_0^t d\tau K_{\nu\lambda}(\tau)C_{\lambda\mu}(t-\tau)\right) \tag{11.3.32}$$

where the components of the frequency matrix $\Omega_{\nu\lambda}$ and the memory matrix $K_{\nu\lambda}(\tau)$ are

$$\Omega_{\nu\lambda} = \beta \sum_\kappa (LA_\nu, A_\kappa^*)\chi_{\kappa\lambda}^{-1} \tag{11.3.33}$$

$$K_{\nu\lambda}(\tau) = \beta \sum_\kappa (F_\nu(\tau), F_\kappa^*(0))\chi_{\kappa\lambda}^{-1} \tag{11.3.34}$$

where $\chi_{\kappa\lambda}^{-1}$ is the $\kappa\lambda$th element of the inverse of the susceptibility matrix. The random force $\mathbf{F}(t)$ is of course

$$F_\mu(t) = e^{iQLt}QiLA_\mu = e^{iQLt}G_\mu = QF_\mu(t) \tag{11.3.35}$$

In Eqs. (11.3.26) and (11.3.30) the dot denotes matrix multiplication. If $\mathbf{K}(t)$ and $\mathbf{F}(t)$ are known, these equations represent a set of closed equations from which the time evolution of the properties $\{A_1, \ldots, A_M\}$ can be computed. *Equations (11.3.26) and (11.3.30) are an exact consequence of the equations of motion.*

11 · 4 SLOW AND FAST VARIABLES

The *generalized Langevin equation* and the *memory function equation* simplify considerably when the set $\{A_1, \ldots, A_M\}$ relaxes much more slowly than all other properties. If *all* such slowly relaxing variables are included in the set $\{A_1, \ldots, A_M\}$, the set is called a "*good set of variables.*" At the outset it is important to note that there are no rules by which a "good set of variables" can be chosen. Generally this is a matter of one's intuition. It is, however, the crucial step in the application of the Zwanzig–Mori formalism to specific problems. There are several possible reasons for a given set of variables to be regarded as "slow" with respect to all other variables.

In Section 10.3 it was shown that the Fourier component $\delta A(\mathbf{q}, t)$ of the fluctuation of a conserved density has a lifetime $\tau(q)$ such that $\tau(q) \to \infty$ as $q \to 0$; that is, $\delta A(\mathbf{q}, t)$ varies slowly for small q. Thus we expect that the small ($q \to 0$) wave number Fourier components of the "densities" of all the conserved properties form a "good set" of variables. For example, in an isotropic monatomic fluid we surmise that a "good set" consists of the low q Fourier components of the mass, linear momentum, and energy densities.

Another good example of a separation of time scales is Brownian motion. Because the Brownian particle is much more massive than the solvent particles, it moves much more slowly. Thus the position and velocity of the Brownian particle should constitute a "good set" of variables.

Highly anisotropic molecules reorient slowly in dense fluids and liquid crystals. In these fluids the "conserved densities" do not by themselves constitute a "good set." It is necessary to include "densities of orientational properties." This is made more specific later.

Let us assume that we can list the independent variables whose decay is slow; that is, those whose relaxation times τ_r satisfy

$$\tau_r \gg \tau_c$$

where τ_c typifies the longest relaxation times for all other variables. We choose these slowly relaxing variables as our set **A**, and thus the subspace of slowly relaxing variables is the subspace spanned by $\{A_1, \ldots, A_M\}$. Then the projection operator P projects onto a subspace of Liouville space containing *all* "slowly" relaxing properties of the system and the projector Q projects onto the orthogonal complement of this "*slow*" *subspace* which by construction contains all of the rapidly decaying properties. Since the "random force" is always in this "*fast*" *subspace,*[11] it fluctuates rapidly, and its time-correlation function, (i.e., memory function) should decay on the time scale τ_c. Thus there will be a large separation in the time scales and **A** will decay much more slowly than **F**. Hence for times $t \gg \tau_c$ it is permissible to treat the memory function as a very rapidly decaying function so that

$$\mathbf{K}(t) = 2\boldsymbol{\Gamma}\delta(t) \tag{11.4.1}$$

This is called the *Markov* approximation. Substitution of this into Eqs. (11.3.26) and (11.3.30) then yields the equations

$$\frac{d\mathbf{A}(t)}{dt} = i\boldsymbol{\Omega} \cdot \mathbf{A}(t) - \boldsymbol{\Gamma} \cdot \mathbf{A}(t) + \mathbf{F}(t) \tag{11.4.2}$$

and

$$\frac{d\mathbf{C}(t)}{dt} = i\boldsymbol{\Omega} \cdot \mathbf{C}(t) - \boldsymbol{\Gamma} \cdot \mathbf{C}(t) \tag{11.4.3}$$

where $\boldsymbol{\Gamma}$ is called the *relaxation matrix.*

Integration of Eq. (11.4.1) then gives the *relaxation matrix* $\boldsymbol{\Gamma}$ in terms of the *memory matrix* $\mathbf{K}(t)$ as

$$\boldsymbol{\Gamma} = \int_0^\infty d\tau \mathbf{K}(\tau) = \beta \int_0^\infty d\tau (\mathbf{F}(\tau), \mathbf{F}^+(0)) \cdot \boldsymbol{\chi}^{-1} \tag{11.4.4}$$

where the second equality follows from Eq. (11.3.28). The relaxation matrix $\boldsymbol{\Gamma}$ can be regarded as the product of two parts: a *kinetic coefficient* Λ and an inverse susceptibility $\boldsymbol{\chi}^{-1}$; that is,

$$\boldsymbol{\Gamma} = \boldsymbol{\Lambda} \cdot \boldsymbol{\chi}^{-1} \tag{11.4.5}$$

where

$$\boldsymbol{\Lambda} \equiv \beta \int_0^\infty d\tau (\mathbf{F}(\tau), \mathbf{F}^+(0)) \tag{11.4.6}$$

The kinetic coefficients can be expressed in terms of ordinary time-correlation functions. Such relations are called Green–Kubo relations (see Section 11.B).

One consequence of Eq. (11.4.5) follows immediately. In the neighborhood of certain points of instability such as the gas–liquid critical point or order disorder phase transitions, the susceptibilities corresponding to the fluctuations in the order parameters become very large. Thus if Λ does not increase as rapidly as χ, the corresponding relaxation rates Γ will become small. This phenomenon is called "critical slowing" of the fluctuations. There has been much recent work on this phenomena (Swinney, 1974).

11·5 SYMMETRY PROPERTIES OF THE RELAXATION EQUATIONS

The set of variables in Eqs. (11.4.2) and (11.4.3) must include *all* of the slowly relaxing variables. When the Hamiltonian has certain symmetry properties, the set of Eqs. (11.3.26) and (11.4.2) can be separated into groups of uncoupled equations. Since, in general, we do not know how to compute the time-correlation functions $\langle \mathbf{F}(\tau), \mathbf{F}^{+}(0)\rangle$, the elements of $\boldsymbol{\Gamma}$ should be regarded as quantities to be determined from a comparison between theory and experiment. However, symmetry can be used to relate the off-diagonal elements of $\boldsymbol{\Gamma}$ to each other and thereby to reduce the number of independent quantities.

In this section we show how the *symmetry* of the Hamiltonian can be used to simplify the relaxation equations. We also present several important theorems involving time-correlation functions and memory functions. We begin by discussing time reversal symmetry.

Time Reversal Symmetry

Suppose the properties $\{A_\mu\}$ in the set $\mathbf{A}$ transform like

$$A_\mu \to \gamma_\mu A_\mu \tag{11.5.1}$$

where $\gamma_\mu = \pm 1$ under the transformation of phase space $(q, p) \to (q, -p)$. This transformation merely inverts all of the momenta p leaving the positions q unchanged. Equation (11.5.1) implies that the set $\mathbf{A}$ consists of properties that are either odd $\gamma_\mu = -1$, or even, $\gamma_\mu = +1$, functions of the momenta.

Since the Hamiltonian H of a conservative system is a quadratic function of the momenta, H is invariant to this transformation. The equilibrium distribution function $\rho_0(\boldsymbol{\Gamma})$ is a functional of H so that it is invariant to this transformation. The Liouvillian, on the other hand, contains terms such as

$$\frac{\partial H}{\partial q}\frac{\partial}{\partial p}; \quad \text{and} \quad \frac{\partial H}{\partial p}\frac{\partial}{\partial q}$$

[cf. Eq. (11.2.2)] and therefore changes sign under this transformation. In summary, under the transformation $(q, p) \to (q, -p)$; $H \to H$, $iL \to -iL$, and $\rho_0(\boldsymbol{\Gamma}) \to \rho_0(\boldsymbol{\Gamma})$.

This transformation[12] is called the *time-reversal transformation.* A property that transforms like Eq. (11.5.1) is said to have definite *time-reversal symmetry* and γ_μ is called the *signature* of A_μ under time reversal. Let us now investigate the consequences of this kind of symmetry. We proceed by proving a certain set of theorems. These theorems only apply to the set **A** if all A_μ in the set have definite time-reversal symmetry, which will be the case in all the applications.

Theorem 1

The scalar product $\langle A_\mu A_\nu^* \rangle = (A_\mu, A_\nu^*)$ *vanishes if* A_ν and A_μ *have different time-reversal symmetries; that, is, if* $\gamma_\mu \neq \gamma_\nu$.

Proof First we note that

$$(A_\mu, A_\nu^*) = \int d\boldsymbol{\Gamma} \rho_0(\boldsymbol{\Gamma}) A_\mu(\boldsymbol{\Gamma}) A_\nu^*(\boldsymbol{\Gamma})$$

Under the transformation $(q, p) \to (q, -p)$ the volume $d\boldsymbol{\Gamma}$ and the distribution function $\rho_0(\boldsymbol{\Gamma})$ are unchanged, whereas $A_\mu \to \gamma_\mu A_\mu$ and $A_\nu^* \to \gamma_\nu A_\nu^*$ by hypothesis. Thus

$$(A_\mu, A_\gamma^*) = \gamma_\mu \gamma_\nu (A_\mu, A_\nu^*) \tag{11.5.2}$$

This last equation shows that if A_μ and A_ν have different time-reversal symmetries, $\gamma_\mu \gamma_\nu = -1$ and $(A_\mu, A_\nu^*) = 0$, thus proving the theorem. The elements of the susceptibility matrix $\chi_{\mu\nu}$ also satisfy Eq. (11.5.2) [cf. Eq. (11.3.24)]

$$\chi_{\mu\nu} = \gamma_\mu \gamma_\nu \chi_{\mu\nu} \tag{11.5.3}$$

so that $\chi_{\mu\nu}$ *is different from zero only if* A_μ *and* A_ν *have the same time-reversal symmetry.*

The set of variables **A** can be divided into two subsets $\mathbf{A}_E$ and $\mathbf{A}_0$ such that the subset $\mathbf{A}_E$ contains all of the *even* properties and the subset $\mathbf{A}_0$ contains all of the *odd* properties; that is, $\mathbf{A}_E \to \mathbf{A}_E$ and $\mathbf{A}_0 \to -\mathbf{A}_0$ under time reversal. The column matrix **A** can then be written as

$$\mathbf{A} = \begin{bmatrix} \mathbf{A}_E \\ \mathbf{A}_0 \end{bmatrix} \tag{11.5.4}$$

By Theorem 1 the scalar product between any element in $\mathbf{A}_E$ and any element in $\mathbf{A}_0$ is zero. It follows that

$$(\mathbf{A}, \mathbf{A}^+) = \begin{bmatrix} (\mathbf{A}_E, \mathbf{A}_E^+) & \mathbf{0} \\ \mathbf{0} & (\mathbf{A}_0, \mathbf{A}_0^+) \end{bmatrix} \tag{11.5.5}$$

that is, $(\mathbf{A}, \mathbf{A}^+)$ reduces to block-diagonal form. The susceptibility matrix likewise reduces to block-diagonal form

$$\boldsymbol{\chi} = \begin{bmatrix} \boldsymbol{\chi}_{EE} & 0 \\ 0 & \boldsymbol{\chi}_{00} \end{bmatrix} \tag{11.5.6}$$

From this it follows that the inverse of the susceptibility matrix χ^{-1} is also block diagonal

$$\boldsymbol{\chi}^{-1} = \begin{bmatrix} \boldsymbol{\chi}_{EE}^{-1} & 0 \\ 0 & \boldsymbol{\chi}_{00}^{-1} \end{bmatrix} \tag{11.5.7}$$

so that $(\boldsymbol{\chi}^{-1})_{\kappa\lambda}$ is zero if A_κ and A_λ have different time reversal symmetries. We conclude from Theorem 1 that the susceptibility matrix and its inverse couple properties of the same time-reversal symmetry.

Theorem 2

The frequency $\Omega_{\mu\nu}$ vanishes unless the properties A_ν and A_μ have different time-reversal symmetries.

Proof. According to Eq. (11.3.33), the frequency $\Omega_{\mu\nu}$ is given by

$$\Omega_{\mu\nu} = \beta \sum_\kappa (LA_\mu, A_\kappa)\, \chi_{\kappa\nu}^{-1}$$

First we note that under time reversal

$$(LA_\mu, A_\kappa^*) = -\,\gamma_\mu\gamma_\kappa(LA_\mu, A_\kappa{}^*) \tag{11.5.8}$$

where the $-$ sign follows from the oddness of L under time reversal. Thus (LA_μ, A_κ^*) couples properties of different time-reversal symmetry.[14] From Theorem 1 we note that $\chi_{\kappa\nu}^{-1}$ transforms to $\gamma_\kappa\gamma_\nu\chi_{\kappa\nu}^{-1}$ under time reversal. Combining these two results shows that $\Omega_{\mu\nu}$ transforms like

$$\Omega_{\mu\nu} = -\,\gamma_\mu\gamma_\kappa\gamma_\kappa\gamma_\nu\Omega_{\mu\nu} = -\gamma_\mu\gamma_\nu\Omega_{\mu\nu} \tag{11.5.9}$$

where the last equality follows from $\gamma_\kappa^2 = (\pm 1)^2 = +1$. Clearly if $\gamma_\mu = \gamma_\nu$, $\Omega_{\mu\nu} = -\,\Omega_{\mu\nu} = 0$. *Thus the frequency matrix* $\boldsymbol{\Omega}$ *couples properties of different time-reversal symmetry.* $\boldsymbol{\Omega}$ can be written as [cf. Eq. (11.5.4)]

$$\boldsymbol{\Omega} = \begin{bmatrix} 0 & \boldsymbol{\Omega}_{E0} \\ \boldsymbol{\Omega}_{0E} & 0 \end{bmatrix} \tag{11.5.10}$$

Theorem 3

The elements of the correlation function matrix $\mathbf{C}\,(t)$ *have the following transformation properties under time reversal*

$$C_{\mu\nu}(t) = \gamma_\nu\gamma_\mu C_{\mu\nu}(-t) = \gamma_\mu\gamma_\nu C_{\nu\mu}^*(t) \tag{11.5.11}$$

Proof. First we note from Eq. (11.3.24) that

$$C_{\mu\nu}(t) = (e^{iLt}A_\mu, A_\nu{}^*)$$

Since the Liouvillian L transforms to $-L$ under time reversal, the correlation function $C_{\mu\nu}(t)$ transforms like

$$C_{\mu\nu}(t) = \gamma_\mu\gamma_\nu(e^{-iLt}A_\mu, A_\nu^*) \tag{11.5.12}$$

The scalar product on the right-hand side of this equation is simply $C_{\mu\nu}(-t)$, thus proving the first equality of the theorem $C_{\mu\nu}(t) = \gamma_\mu\gamma_\nu C_{\mu\nu}(-t)$. As we have already observed, L is an Hermitian operator and e^{iLt} is a unitary operator. It follows from the unitarity of e^{iLt} that $(e^{-iLt}A_\mu, A_\nu^*) = (A_\mu, (e^{iLt}A_\nu)^*)$. This latter result is the complex conjugate of $(e^{iLt}A_\nu, A_\mu^*) = C_{\nu\mu}(t)$ so that $(e^{-iLt}A_\mu, A_\nu^*) = C_{\nu\mu}^*(t)$. Substitution of this result into Eq. (11.5.12) proves the second equality of the theorem.

Theorem 3 contains much important information about time-correlation functions. First we note that if A_μ and A_ν have the same time-reversal symmetry

$$C_{\mu\nu}(t) = C_{\mu\nu}(-t) = C_{\nu\mu}^*(t)$$

The first equality shows that $C_{\mu\nu}(t)$ is an *even function* of the time. Now we note that if A_μ and A_ν have different time-reversal symmetry

$$C_{\mu\nu}(t) = -C_{\mu\nu}(-t) = -C_{\nu\mu}^*(t)$$

The first equality shows that $C_{\mu\nu}(t)$ is an *odd function* of the time.

Applying Theorem 3 to the autocorrelation functions $C_{\mu\mu}(t)$

$$C_{\mu\mu}(t) = C_{\mu\mu}(-t) = C_{\mu\mu}^*(t)$$

shows that the *autocorrelation functions* $C_{\mu\mu}(t)$ *are real even functions of the time.*[15]

The observant reader should have noted that Theorem 1 is a special case of Theorem 3 for $t = 0$.

Theorem 4

The elements of the random-force matrix $(\mathbf{F}(t), \mathbf{F}^+(0))$ *have the following properties under time reversal*

$$(F_\mu(t), F_\nu^*) = \gamma_\mu\gamma_\nu(F_\mu(-t), F_\nu^*) = \gamma_\mu\gamma_\nu(F_\nu(t), F_\mu^*)^* \tag{11.5.13}$$

Proof. First we note that the projection operator

$$P = \beta \sum_{ij} (\ .\ .\ .\ , A_i^*)\, \chi_{ij}^{-1} A_j$$

transforms to $\gamma_i\gamma_i\gamma_j\gamma_j P$. Since $\gamma_i^2 = \gamma_j^2 = 1$, P and $Q = 1 - P$ are even under time reversal. Next we note that since Q and L are Hermitian operators, QLQ is Hermitian $[(QLQ)^+ = Q^+L^+Q^+ = QLQ]$ and e^{iQLQt} is a unitary operator. Now[16]

$$(F_\mu(t), F_\nu^*) \equiv ((e^{iQLQt}QiLA_\mu), (QiLa_\nu)^*)$$

Under time reversal this random-force correlation functions transforms to

$$\begin{aligned}(F_\mu(t), F_\nu^*) &= \gamma_\mu\gamma_\nu(e^{-iQLQt}QiLA_\mu, (QiLA_\nu)^*)\\ &= \gamma_\mu\gamma_\nu(F_\mu(-t), F_\nu^*)\end{aligned} \tag{11.5.14}$$

This proves the first equality. The unitarity of e^{iQLQt} implies that the first equality in Eq. (11.5.14) can be expressed as

$$(F_\mu(-t), F_\nu^*) = \gamma_\mu\gamma_\nu(QiLA_\mu, (e^{iQLQt}QiLA_\nu)^*)$$
$$= \gamma_\mu\gamma_\nu(F_\mu, F_\nu^*(t))$$
$$= \gamma_\mu\gamma_\nu(F_\nu(t), F_\mu^*)^*$$

thus proving the second equality of Theorem 4. Thus *the random force matrix has the same time-reversal symmetry as the time-correlation function matrix.* It follows that $(F_\mu(t), F_\mu^*)$ is a real even function of the time.

Theorem 5

The kinetic coefficients have the symmetry properties under time reversal

$$\Lambda_{\mu\nu} = \gamma_\mu\gamma_\nu\Lambda_{\nu\mu}^* \tag{11.5.15}$$

Proof. Combining the definition of $\Lambda_{\mu\nu}$ in Eq. (11.4.6) with Theorem 4 gives[17] Theorem 5.

Equation (11.5.15) forms the basis of the Onsager reciprocal relations. These symmetry relations can be expressed in matrix form as

$$\Lambda = \begin{bmatrix} \Lambda_{EE} & \Lambda_{EO} \\ \Lambda_{OE} & \Lambda_{OO} \end{bmatrix} \tag{11.5.16}$$

where $\Lambda_{EE}^+ = \Lambda_{EE}$, $\Lambda_{EO}^+ = -\Lambda_{OE}$, $\Lambda_{OO}^+ = \Lambda_{OO}$.

Equation (11.4.5) can be expressed as $\boldsymbol{\Gamma} \cdot \boldsymbol{\chi} = \Lambda$. If the set **A** is so chosen that $\boldsymbol{\chi}$ is diagonal,[18] Eq. (11.4.5) simplifies to $\Gamma_{\nu\mu}\chi_{\mu\mu} = \Lambda_{\nu\mu}$. Likewise, $\Gamma_{\mu\nu}^* \cdot \chi_{\nu\nu}^* = \Lambda_{\mu\nu}^* = \gamma_\mu \gamma_\nu\Lambda_{\nu\mu}$. Since $\chi_{\nu\nu}^* = \beta^{-1}(A_\nu, A_\nu^*)^* = \chi_{\nu\nu}$ it follows that $\Gamma_{\mu\nu}^*\chi_{\nu\nu} = \gamma_\mu\gamma_\nu\Gamma_{\nu\mu}\chi_{\mu\mu}$ or

$$\Gamma_{\nu\mu} = \gamma_\nu\gamma_\mu \frac{\chi_{\nu\nu}}{\chi_{\mu\mu}} \Gamma_{\mu\nu}^* \tag{11.5.17}$$

*From this we see that the diagonal elements of the damping matrix $\boldsymbol{\Gamma}$ are real if the set **A** is chosen such that $\boldsymbol{\chi}$ is diagonal. This is always possible.* Note that $\boldsymbol{\chi}$ is diagonal if and only if

$$(A_\mu, A_\nu^*) = \delta_{\mu\nu}(A_\mu, A_\mu^*) \tag{11.5.18}$$

In this eventuality the set **A** is called an *orthogonal set* of properties. It is sometimes useful to use orthogonal sets. For example, using orthogonal sets, it is readily proved that the relaxation equations are stable, and the resulting time-correlation matrix decays to zero

A word of caution is necessary. In the presence of an external field the time-reversal symmetry of the Hamiltonian may be removed. For example, in a magnetic field **B** the spin–dipole interaction with the field is of the form$(-\mathbf{J} \cdot \mathbf{B})$, where **J** is an angular momentum and **B** is the magnetic field. **J** has odd time-reversal symmetry (like $\mathbf{r} \times \mathbf{p}$) so that if $\mathbf{B} \neq 0$ none of the above theorems hold unless $\mathbf{B} \rightarrow -\mathbf{B}$ is also imposed. Thus for example Theorem 3 would become

$$C_{\mu\nu}(t, \mathbf{B}) = \gamma_\nu\gamma_\mu C_{\mu\nu}(-t, \mathbf{B}) = \gamma_\nu\gamma_\mu C_{\nu\mu}^*(t, -\mathbf{B})$$

Parity

Suppose the properties $\{A_\mu\}$ in the set $\boldsymbol{A}$ transform like

$$A_\mu \to \varepsilon_\mu A_\mu \tag{11.5.19}$$

under the inversion of phase space $(q, p) \to (-q, -p)$ or equivalently $\boldsymbol{\Gamma} \to -\boldsymbol{\Gamma}$ where $\varepsilon_\mu = \pm 1$. This transformation merely inverts all positions and momenta. Equation (11.5.19) implies that the set $\mathbf{A}$ consists of properties that are either odd, $\varepsilon_\mu = -1$, or even, $\varepsilon_\mu = +1$ functions of $\boldsymbol{\Gamma}$.

A molecular assembly in which the molecules have centers of inversion and on which no symmetry breaking external forces act is described by an Hamiltonian which is invariant to inversion $\boldsymbol{\Gamma} \to -\boldsymbol{\Gamma}$. Likewise the distribution function, being a functional of H, is invariant to inversion. The Liouvillian contains terms like

$$\frac{\partial \mathrm{H}}{\partial q}\frac{\partial}{\partial p}; \text{ and } \frac{\partial H}{\partial p}\frac{\partial}{\partial q}$$

which are also invariant to inversion. To summarize, under the parity transformation $\boldsymbol{\Gamma} \to -\boldsymbol{\Gamma}, H \to H, \rho_0(\boldsymbol{\Gamma}) \to \rho_0(\boldsymbol{\Gamma})$, and $iL \to iL$.

A property that transforms like Eq. (11.5.19) is said to have definite *parity* and ε_μ *is called the parity of the property*. Let us now investigate the consequences of this kind of symmetry. Again we proceed by proving a certain set of theorems. These theorems apply to the set $\mathbf{A}$ only if all the A_μ have definite parity.

Theorem 6

The time-correlation function $C_{\mu\nu}(t) = (A_\mu(t), A_\nu^*)$ vanishes if A_μ and A_ν have different parities.

Proof. First we note that

$$(A_\mu(t), A_\nu^*) = \int d\boldsymbol{\Gamma} \rho_0(\boldsymbol{\Gamma}) A_\nu^*(\boldsymbol{\Gamma}) e^{iLt} A_\mu(\boldsymbol{\Gamma})$$

Under the transformation $\boldsymbol{\Gamma} \to -\boldsymbol{\Gamma}$ the volume element $d\boldsymbol{\Gamma}$, $\rho_0(\boldsymbol{\Gamma})$ and e^{iLt} are unchanged, whereas $A_\mu \to \varepsilon_\mu A_\mu$ and $A_\nu \to \varepsilon_\nu A_\nu^*$ by hypothesis. Thus

$$(A_\mu(t), A_\nu^*) = \varepsilon_\mu \varepsilon_\nu (A_\mu(t), A_\nu(0))$$

or

$$C_{\mu\nu}(t) = \varepsilon_\mu \varepsilon_\nu C_{\mu\nu}(t) \tag{11.5.20}$$

This last equation shows that if $\varepsilon_\mu \neq \varepsilon_\nu$, then $C_{\mu\nu}(t) = 0$, thus proving the theorem. This means that *in a system such that the Hamiltonian has inversion symmetry, properties of different parity are totally uncorrelated for all time.* The initial value of Eq. (11.5.20) gives

$$\chi_{\mu\nu} = \varepsilon_\mu \varepsilon_\nu \chi_{\mu\nu} \tag{11.5.21}$$

so that $\chi_{\mu\nu} = 0$ *if* A_μ and A_ν *are properties of different parity.*

Theorem 7

The frequencies $\Omega_{\mu\nu}$ vanish if A_μ and A_ν have different parities in a system where H has even parity.

Proof. First note that under inversion

$$(LA_\mu, A_\kappa^*) = \varepsilon_\mu \varepsilon_\kappa (LA_\mu, A_\kappa^*) \tag{11.5.22}$$

since both L and the average denoted by $(\ldots\ldots, \ldots)$ are invariant to inversion. Thus since $\chi_{\kappa\nu}^{-1} = \varepsilon_\kappa \varepsilon_\nu \chi_{\kappa\nu}$ and $\varepsilon_\kappa{}^2 = 1$ it follows that

$$\Omega_{\mu\nu} = \beta \sum_\kappa (LA_\mu, A_\kappa^*)\, \chi_{\kappa\nu}^{-1} = \varepsilon_\mu \varepsilon_\nu \Omega_{\mu\nu} \tag{11.5.23}$$

This last equation shows that if $\varepsilon_\mu \neq \varepsilon_\nu$, $\Omega_{\mu\nu} = 0$ thus proving the theorem.

Theorem 8

The elements of the random-force matrix $(\mathbf{F}(t), \mathbf{F}^{+}(0))$ *have the following property due to inversion symmetry*

$$(F_\mu(t), F_\nu^*) = \varepsilon_\mu \varepsilon_\nu (F_\mu(t), F_\nu^{+}(0)) \tag{11.5.24}$$

so that the random forces corresponding to properties of different parity $(\varepsilon_\mu \neq \varepsilon_\nu)$ *are totally uncorrelated for all times.*

Proof. First we note that

$$P = \beta \sum_{ij} (\,.\,.\,.\,, A_i^*)\, \chi_{ij}^{-1} A_j$$

transforms to $\varepsilon_i \varepsilon_i \varepsilon_j \varepsilon_j P = P$ under inversion so that both P and Q have even parity. Thus $QLQ \rightarrow QLQ$ under inversion and

$$\begin{aligned}(F_\mu(t), F_\nu{}^*) &= (e^{iQLQt} QiLA_\mu, QiLA_\nu) \\ &= \varepsilon_\mu \varepsilon_\nu (F_\mu(t), F_\nu{}^*)\end{aligned}$$

Thus for properties such that $\varepsilon_\mu \neq \varepsilon_\nu$, $(F_\mu(t), F_\nu{}^*(0)) = 0$. It follows from this that the kinetic coefficients [cf. Eq. (11.4.5)] also satisfy

$$\Lambda_{\mu\nu} = \varepsilon_\mu \varepsilon_\nu \Lambda_{\mu\nu} \tag{11.5.25}$$

so that *if* $\varepsilon_\mu \neq \varepsilon_\nu$, $\Lambda_{\mu\nu} = 0$.

The chief consequence of this theorem is that the memory function $K_{\mu\nu}(t)$ defined by Eq. (.1.3.35) has the property, by virtue of inversion symmetry, that

$$K_{\mu\nu}(t) = \varepsilon_\mu \varepsilon_\nu K_{\mu\nu}(t) \tag{11.5.26}$$

so that if $\varepsilon_\mu \neq \varepsilon_\nu$, $K_{\mu\nu}(t) = 0$.

It can be concluded that *the subset of properties in* **A** *that have* $\varepsilon = +1$ *are totally uncorrelated with the subset of properties which have* $\varepsilon = -1$.

A word of caution is required. If a system contains molecules that do not have an inversion center, that is, if the system contains optically active molecules, the Hamiltonian does not have even parity and none of these theorems apply. Furthermore even if the molecules are optically inactive; if a symmetry breaking field is turned on the Hamiltonian may lose its inversion symmetry. Notable in this regard is the electric dipole interaction $-\mathbf{P} \cdot \mathbf{E}$ where **P** is the electric polarization of the system which has odd parity and **E** is the external electric field. In the presence of **E**, the Hamiltonian $H = H_0 - \mathbf{P} \cdot \mathbf{E}$, loses its parity and properties of different parity can be coupled. In irreversible thermodynamics such symmetry-breaking fields are responsible for interesting new couplings. For example, the imposition of a temperature gradient *might* have the effect of inducing an electrical polarization.

Reflection Symmetry

Suppose that the properties $\{A_\mu\}$ in the set A transform like

$$A_\mu \rightarrow \alpha_\mu A_\mu \tag{11.5.27}$$

where $\alpha_\mu = \pm 1$ under the transformation of phase space $(x_j, y_j, z_j, p_{jx}, p_{jy}, p_{jz}) \rightarrow (x_j, -y_j, z_j, p_{jx}, -p_{jy}, p_{jz})$ for all particles j. This transformation merely *reflects* all the positions and momenta through the (x, z) plane. Equation (11.5.27) implies that the set **A** consists of properties that are either even, $\alpha_\mu = +1$, or odd, $\alpha_\mu = -1$ with respect to reflection through the (x, z) plane.

Let us consider systems in which the Hamiltonian is invariant to this reflection. Then $\rho_0(\boldsymbol{\Gamma})$ is invariant to reflection. The Liouvillian contains terms such as

$$\frac{\partial H}{\partial y}\frac{\partial}{\partial p_y}; \text{ and } \frac{\partial H}{\partial p_y}\frac{\partial}{\partial y}$$

and is consequently invariant to reflection. Properties such as those given in Eq. (11.5.27) are said to have *definite reflection symmetry* and α_μ is called the signature of A_μ under reflection through the (x, z) plane. The following theorems apply to such properties in a system with reflection symmetry, that is, in systems in which H is invariant to reflection ($\alpha = +1$). The reader familiar with the treatment of time-reversal symmetry and parity should have no trouble proving the following theorems.

Theorem 9

The time-correlation function $C_{\mu\nu}(t)$ *vanishes if* A_μ *and* A_ν *have different reflection symmetries; that is,*

$$C_{\mu\nu}(t) = \alpha_\mu \alpha_\nu C_{\mu\nu}(t) \tag{11.5.28}$$

According to this theorem, in a system with reflection symmetry properties which transform differently under reflection are uncorrelated for all time. The initial value of Eq. (11.5.28) then gives

$$\chi_{\mu\nu} = \alpha_\mu \alpha_\nu \chi_{\mu\nu} \tag{11.5.29}$$

so that $\chi_{\mu\nu} = 0$ *if* A_μ *and* A_ν *have different reflection symmetries.*

Theorem 10

In a system with reflection symmetry

$$\Omega_{\mu\nu} = \alpha_\mu \alpha_\nu \Omega_{\mu\nu} \tag{11.5.30}$$

and the frequencies $\Omega_{\mu\nu} = 0$ *if* A_μ *and* A_ν *transform differently under reflection.*

Theorem 11

In a system with reflection symmetry

$$(F_\mu(t), F_\nu^*) = \alpha_\mu \alpha_\nu (F_\mu(t), F_\nu^*) \tag{11.5.31}$$

so that the random forces corresponding to properties A_μ, A_ν *with different reflection symmetries are totally uncorrelated.*

This last theorem leads to the same conclusions for the memory functions and for the kinetic coefficients.

It can be concluded that *in a system with definite reflection symmetry, the subset of properties with* $\alpha = +1$ *are totally uncorrelated with the subset of properties with* $\alpha = -1$. This symmetry can also be broken with external fields. We shall find this particular symmetry very powerful.

Any other plane of symmetry with respect to which the properties **A** have definite reflection symmetry also gives identical results.

It should be clear from the foregoing that other symmetries can be used to simplify the relaxation equations in an analogous manner. For example, rotational symmetry simplifies the analysis greatly. Unfortunately, the full exploration of rotational symmetry would divert us too long from our goal of understanding light scattering. It is also important to note that particular combinations of symmetries might also help to simplify the equations. In Chapter 12 we introduce, when needed, these other symmetry arguments. Before doing this we prove some additional useful theorems involving symmetry with regard to specific functions.

The Densities of Conserved Variables

Let us investigate the symmetry properties of quantities such as

$$A_\mu(\mathbf{q}) = \sum_{j=1}^{N} a_\mu(\mathbf{r}_j, \mathbf{p}_j) \exp i\mathbf{q} \cdot \mathbf{r}_j$$

Here $\mathbf{r}_j$ is the position vector of particle j with Cartesian components (x_j, y_j, z_j) and $\mathbf{p}_j$ is the momentum of particle j with Cartesian components (p_{jx}, p_{jy}, p_{jz}). $a_\mu(\mathbf{r}_j, \mathbf{p}_j)$ is a property of the jth particle which depends only on its position and momentum. $A_\mu(\mathbf{q})$ can be regarded as the spatial Fourier transform of the density $\sum_j a_\mu^j \delta(\mathbf{r} - \mathbf{r}_j)$. Let us choose a coordinate system such that $\mathbf{q}$ is parallel to the z axis. Then

$$A_\mu(\mathbf{q}) = \sum_{j=1}^{N} a_\mu^j \, e^{iqz_j} \tag{11.5.32}$$

The index μ labels the specific density of interest. A typical set of variables might be

$$\mathbf{A}(\mathbf{q}) = \begin{pmatrix} \rho(\mathbf{q}) \\ g_z(\mathbf{q}) \\ g_x(\mathbf{q}) \\ g_y(\mathbf{q}) \end{pmatrix} \tag{11.5.33}$$

where $\rho(\mathbf{q}) = \sum_j \exp iqz_j$, $g_\mu(\mathbf{q}) = \sum_j p_{\mu j} \exp iqz_j$ (where μ runs through x, y, z) are the respective Fourier transforms of the number density and the momentum densities. Before considering the simplifications caused by the symmetries already introduced, let us consider a new symmetry and how it relates to this choice of variables.

Translational Symmetry

First consider the coordinate transformation $(\mathbf{r}_j, \mathbf{p}_j) \to (\mathbf{r}_j + \mathbf{a}, \mathbf{p}_j)$ for $j = 1, \ldots N$, where $\mathbf{a}$ is an arbitrary vector. This corresponds to a shift or translation in the origin of the coordinate system by the arbitrary vector $\mathbf{a}$. Now in general the intermolecular potential V is a function of the relative positions, $(\mathbf{r}_i - \mathbf{r}_j)$; thus under the above transformation $V_{ij}(\mathbf{r}_i - \mathbf{r}_j) \to V_{ij}(\mathbf{r}_i - \mathbf{r}_j)$. This implies that in the absence of external forces, both the Hamiltonian H and its corresponding Liouvillian are invariant to this transformation. We say that *H and L are translationally invariant.*

The properties in Eq. (11.5.32) transform like

$$A_\mu(\mathbf{q}) \to e^{i\mathbf{q}\cdot\mathbf{a}} A_\mu(\mathbf{q})$$
$$A_\nu(\mathbf{q}') \to e^{i\mathbf{q}'\cdot\mathbf{a}} A_\nu(\mathbf{q}')$$

Moreover, in an homogeneous system, $\rho_0(\boldsymbol{\Gamma})$ and $d\boldsymbol{\Gamma}$ are invariant under this transformation. The correlation function therefore transforms like

$$(A_\mu(\mathbf{q}, t), A_\nu^*(\mathbf{q}')) = e^{i(\mathbf{q}-\mathbf{q}')\cdot\mathbf{a}} (A_\mu(\mathbf{q}, t), A_\nu^*(\mathbf{q}'))$$

This relationship must be satisfied for arbitrary translations $\mathbf{a}$ of the coordinate system. This implies that if $\mathbf{q} \neq \mathbf{q}'$ the correlation function is identically zero. We thus surmise that $\langle A_\mu(\mathbf{q}, t) A_\nu^*(\mathbf{q}', 0)\rangle$ *is zero unless* $\mathbf{q} = \mathbf{q}'$.

Theorem 12

Translational symmetry implies that for homogeneous systems

$$(A_\mu(\mathbf{q}, t), A_\nu^*(\mathbf{q}')) = (A_\mu(\mathbf{q}, t), A_\nu^*(\mathbf{q}))\, \delta_{\mathbf{q},\mathbf{q}'} \tag{11.5.34}$$

This means that different wave-vector components of "fluctuations" are uncorrelated. In solids the theorem is still valid if we restrict attention to vectors in the same Brillouin zone.

Now let us investigate the implications of reflection symmetry. We note that all of the properties in Eq. (11.5.33) have reflection symmetry with respect to reflections through either the x–z or the y–z planes, but not with respect to reflections through the x–y plane, for then $z_j \to -z_j$ and $\exp iqz_j \to \exp -iqz_j$. Thus

$$\mathbf{A}(\mathbf{q}) \xrightarrow[\text{reflection through } x\text{–}z \text{ plane}]{} \begin{pmatrix} \rho(\mathbf{q}) \\ g_z(\mathbf{q}) \\ g_x(\mathbf{q}) \\ -g_y(\mathbf{q}) \end{pmatrix} \tag{11.5.35a}$$

$$\mathbf{A}(\mathbf{q}) \xrightarrow[\text{reflection through } y\text{–}z \text{ plane}]{} \begin{pmatrix} \rho(\mathbf{q}) \\ g_z(\mathbf{q}) \\ -g_x(\mathbf{q}) \\ g_y(\mathbf{q}) \end{pmatrix} \tag{11.5.35b}$$

From Theorems 9–11 it follows that g_x and g_y are each uncorrelated with all other variables in the set **A**. Thus reflection symmetry yields the result

$$(\mathbf{A}(\mathbf{q}, t), \mathbf{A}^{+}(\mathbf{q})) = \begin{pmatrix} (\rho(\mathbf{q}, t), \rho^*(\mathbf{q}, 0)) & (\rho(\mathbf{q}, t), g_z^*(\mathbf{q})) & 0 & 0 \\ (g_z(\mathbf{q}, t), \rho^*(\mathbf{q})) & (g_z(\mathbf{q}, t), g_z^*(\mathbf{q})) & 0 & 0 \\ 0 & 0 & (g_x(\mathbf{q}, t), g_x^*(\mathbf{q})) & 0 \\ 0 & 0 & 0 & (g_y(\mathbf{q}, t), g_y^*(\mathbf{q})) \end{pmatrix} \tag{11.5.36}$$

Because of reflection symmetry, the set **A** separates into three uncoupled subsets: $(\rho(\mathbf{q}), g_z(\mathbf{q}))$, $g_x(\mathbf{q})$, and $g_y(\mathbf{q})$. g_z represents the particle flux along $\mathbf{q}$. This flux is called the *longitudinal* momentum. $g_x(\mathbf{q})$ and $g_y(\mathbf{q})$ represent the particle flux perpendicular or transverse to $\mathbf{q}$. These fluxes are called the *transverse momentum*. The longitudinal and transverse fluxes in an isotropic system of optically inactive molecules are therefore uncoupled. Here we have derived a result which is well known in hydrodynamics.

In a system of optically active molecules or in the presence of a symmetry-breaking external field, the longitudinal and transverse fields are coupled and the resulting relaxation equations should contain cross effects.

The implications of inversion symmetry are a bit more subtle. First we note that under inversion, properties such as Eq. (11.5.33) transform like

$$A_\mu(\mathbf{q}) \to \varepsilon_\mu A_\mu(-\mathbf{q})$$

where ε_μ is the signature of a_μ^j under inversion. The time-correlation function $C_{\mu\nu}(\mathbf{q}, t)$ then transforms under inversion to $(A_\mu(\mathbf{q}, t), A_\nu^*(\mathbf{q})) = \varepsilon_\mu\varepsilon_\nu(A_\mu(-\mathbf{q}, t), A_\nu^*(-\mathbf{q}))$. That is,

$$C_{\mu\nu}(\mathbf{q}, t) = \varepsilon_\mu\varepsilon_\nu C_{\mu\nu}(-\mathbf{q}, t) \tag{11.5.37a}$$

It follows from the initial value of $C_{\mu\nu}(\mathbf{q}, t)$ that

$$\chi_{\mu\nu}(\mathbf{q}) = \varepsilon_\mu\varepsilon_\nu\chi_{\mu\nu}(-\mathbf{q}) \tag{11.5.37b}$$

Likewise for the frequencies and memory functions

$$\Omega_{\mu\nu}(\mathbf{q}) = \varepsilon_\mu\varepsilon_\nu\Omega_{\mu\nu}(-\mathbf{q}) \tag{11.5.37c}$$

and

$$K_{\mu\nu}(\mathbf{q}, t) = \varepsilon_\mu \varepsilon_\nu K_{\mu\nu}(-\mathbf{q}, t) \tag{11.5.37d}$$

It is clear from the foregoing that $C_{\mu\nu}$, $\chi_{\mu\nu}$, $\Omega_{\mu\nu}$, and $K_{\mu\nu}$ are odd functions of $\mathbf{q}$ if a_μ^j and a_ν^j have different parities, and are even functions of $\mathbf{q}$ if a_μ and a_ν have the same parities. Inversion symmetry allows us to predict how these important quantities behave at small q.

In Appendix 11.B we prove other useful theorems concerning time-correlation functions.

11·6 RELAXATION OF A SINGLE CONSERVED VARIABLE

To illustrate the general applicability of the relaxation equations of Section 11.4 let us study the simple case of a single conserved variable $A(\mathbf{q}, t)$ which has the form given by Eq. (11.5.32). The property a_j of the jth molecule is presumed to have definite time-reversal symmetry and parity.

First we note that according to Theorem 2 of Section 11.5

$$\Omega = \beta(LA(\mathbf{q}), A^*(\mathbf{q}))\,(A(\mathbf{q}), A\,(\mathbf{q})^*)^{-1} = 0 \tag{11.6.1}$$

Second we note from Eq. (11.5.37b)

$$\chi(\mathbf{q}) = \beta^{-1}(A\,(\mathbf{q}), A^*(\mathbf{q})) = \chi(-\mathbf{q})$$

Thus $\chi(\mathbf{q})$ is an even function of $\mathbf{q}$. This follows from considerations of inversion symmetry [cf. (Eq. (11.5.37b)]. The chief consequence of this is that

$$\lim_{q\to 0} \chi(\mathbf{q}) = \chi^0 \neq 0 \tag{11.6.2}$$

These considerations also show [cf. Eq. (11.5.37d)] that the random-force autocorrelation function and thereby $\Gamma(\mathbf{q})$ are also even functions of $\mathbf{q}$; that is,

$$\Gamma(\mathbf{q}) = \Gamma(-\mathbf{q})$$

Because $\Gamma(\mathbf{q})$ is even in $\mathbf{q}$ a Taylor expansion of $\Gamma(\mathbf{q})$ around $q = 0$ should then have the form

$$\Gamma(\mathbf{q}) = \Gamma^{(0)} + q^2\Gamma^{(2)} + \ldots \tag{11.6.3}$$

where $\Gamma^{(2n)}$ is the coefficient of q^{2n}. It is easy to see that $\Gamma^{(0)} = 0$ and that the first nonzero term is $q^2\Gamma^{(2)}$. In the following we perform this expansion explicitly.

First we note that because $A(\mathbf{q}, t)$ is a conserved variable it satisfies the conservation equation [Eq. (10.3.9)]

$$\dot{A}(\mathbf{q}) = iLA(\mathbf{q}) = iqJ(\mathbf{q}) \tag{11.6.4}$$

where $J(\mathbf{q})$ is the flux of A. Equation (11.6.4) defines $J(\mathbf{q})$ as

$$J(\mathbf{q}) \equiv \frac{1}{iq} iLA(\mathbf{q}) = \frac{1}{iq} \sum_j [\dot{a}_j + iqa_j\dot{z}_j] \exp iqz_j$$

The right-hand side of this equation can be expanded in powers of q. To order $O(q^2)$ this gives

$$J(\mathbf{q}) = \sum_j [\dot{a}_j z_j + a_j\dot{z}_j] + O(q^2) \equiv J^{(0)}(\mathbf{q}) + O(q^2) \tag{11.6.5}$$

where we have used the fact that[19] $\sum_j \dot{a}_j = 0$. $J^0(\mathbf{q})$ is the zeroth order term—it is independent of q.

The random force is [cf. Eq. (11.3.15)]

$$F(t) = e^{iQLt}i(1 - P)LA = iqe^{iQLt}J(\mathbf{q})$$

where we have substituted Eq. (11.6.4) and used the fact that $PiLA = 0$ [cf. Eq. (11. 6.1)]. It follows that

$$(F(t), F^*(0)) = q^2\langle J^*(\mathbf{q})\, e^{iQL\tau}J(\mathbf{q})\rangle \tag{11.6.6a}$$

Our goal is to expand this function to lowest order in q. This can be accomplished by using Eq. (11.3.8) in the slightly rearranged form

$$e^{iQLt} = e^{iLt} - \int_0^t d\tau\, e^{iL(t-\tau)}\, iPLe^{iQL\tau} \tag{11.6.6b}$$

Now for an arbitrary property B,
$PiLB = (iLB, A^*)(A, A^*)^{-1}A = (B, (iLA)^*)\,(A, A^*)^{-1}A = iq(B, J^*(\mathbf{q}))\,(A, A^*)^{-1}A$. The second equality follows from the Hermitian property of L and the third equality follows from $iLA = iqJ(\mathbf{q})$. Thus applying e^{iQLt} in the form given by Eq. (11.3.8) to an arbitrary vector B gives to lowest order in q, e^{iLt}, that is,

$$e^{iQLt} \underset{q\to 0}{\longrightarrow} e^{iLt} \tag{11.6.7}$$

Thus to lowest order in q.

$$F(t) = iqe^{iLt}J^{(0)}(\mathbf{q}) = iqJ^0(t) \tag{11.6.8}$$

where $J^0(\mathbf{q})$ is given by Eq. (11.6.5).

Substitution of Eq. (11.6.8) gives to lowest order in q

$$(F(t), F^*(0)) = q^2(J^{(0)}(t), J^{(0)*}(0)) = q^2\langle J^{(0)}(t)J^{(0)*}(0)\rangle \tag{11.6.9}$$

The important thing to note about this result is that the complicated propagator e^{iQLt} has been replaced by the simpler propagator e^{iLt}.

The memory function thus becomes to lowest order in q

$$K(t) = q^2\beta\langle J^{(0)*}(0)\, J^{(0)}(t)\rangle\chi_0^{-1} \tag{11.6.10}$$

where χ_0 is the limit of the susceptibility $\chi(\mathbf{q})$ as $q \to 0$, [cf. Eq. (11.6.2)]; that is,

$$\chi^0 \equiv \lim_{q\to 0+} \beta\langle|A(\mathbf{q})|^2\rangle \tag{11.6.11}$$

χ_0 can be determined by the methods given in Section (10.3) and Appendix 10. C.

Substitution of Eq. (11.6.10) into Eq. (11.3.52) gives the damping coefficient[20] $\Gamma(\mathbf{q})$ to lowest order in q

$$\Gamma(\mathbf{q}) = q^2\Gamma^{(2)} = q^2\Lambda\chi_0^{-1} \tag{11.6.12}$$

where the "kinetic coefficient" Λ is found by combining Eq. (11.6.9) and Eq. (11.4.4)

$$\Lambda = \lim_{\eta\to 0+} \beta \int_0^\infty dt\, e^{-\eta t} \langle J^{(0)}(0)J^{(0)}(t)\rangle \tag{11.6.13}$$

The limit has been inserted to insure the convergence of this integral. This is an example of a *Green–Kubo* relation that relates the transport coefficient Λ of a property A to a time integral of the ordinary time-correlation function of its corresponding flux $J^{(0)}$ (e.g., see Zwanzig, 1965, Berne and Forster, 1971).

The net result of these considerations is that the property $A(\mathbf{q})$ satisfies the relaxation equation

$$\frac{\partial A(\mathbf{q}, t)}{\partial t} = -q^2\Gamma^{(2)}A(\mathbf{q}, t) + F(\mathbf{q}, t) \tag{11.6.14}$$

where $F(\mathbf{q}, t)$ is the random force. The autocorrelation function of $A(\mathbf{q}, t)$ is then to lowest order in q

$$\langle A^*(\mathbf{q}, 0)A(\mathbf{q}, t)\rangle = \beta^{-1}\chi^0 \exp -q^2\Gamma^{(2)}|t| \tag{11.6.15}$$

and $A(\mathbf{q}, t)$ is a purely diffusive hydrodynamic mode. $\Gamma^{(2)}$ is the diffusion coefficient of this mode.

We have succeeded in showing that the phenomenological equation for A naturally arises from microscopic consideration in the small q limit. In the process we have obtained a formal microscopic definition of the transport coefficient Λ. It should be noted that in this derivation we did not postulate a linear constitutive relation between J and A.

The foregoing can be applied to the transverse momentum $g_x(\mathbf{q})$ which is uncoupled from the other hydrodynamic modes because of reflection symmetry [see Eq. (11.5.36)]. Then taking

$$A(\mathbf{q}) \equiv g_x(\mathbf{q}) = \sum_{j=1}^{N} p_{jx} \exp iqz_j \tag{11.6.16}$$

other quantities are

$$\chi_0 = \beta\langle|\sum_{j=1}^{N} p_{jx} \exp iqz_j|^2\rangle = Nm \tag{11.6.17}$$

$$J^0(\mathbf{q}) \equiv J_{xz} \equiv \sum_j [F_{xj}z_j + \frac{1}{m} p_{xj}p_{zj}] \tag{11.6.18}$$

where Eq. (11.6.17) follows from the statistical independence of the momenta, and the equipartition theorem. F_{xj} in Eq. (11.6.18) is the xth component of the force acting on particle j. The damping coefficient is thus [cf. Eq. (11.6.12)]

$$\Gamma^{(2)} = D_s = (NmkT)^{-1} \lim_{\varepsilon \to 0} \int_0^\infty dte - \varepsilon|t| \langle J_{xz}(0) J_{xz}(t) \rangle \tag{11.6.19}$$

where the symbol D_s is defined as $\Gamma^{(2)}$.

The correlation function is [see Eq. (11.6.15)]

$$\langle g_x^*(\mathbf{q}, 0) g_x(\mathbf{q}, t) \rangle = (Nmk_B T)^{-1} \exp - q^2 D_s |t|$$

A hydrodynamic calculation of this same correlation function gives

$$\langle g_x^*(\mathbf{q}, 0) g_x(\mathbf{q}, t) \rangle \underset{hydro}{\rightarrow} \exp - q^2 \frac{\eta_s}{m\rho_0} |t|$$

Thus the diffusion coefficient of transverse momentum D_s is related to the shear viscosity

$$D_s = \frac{\eta_s}{m\rho_0}$$

and Eq. (11.6.19) relates the transport coefficient η_s to a time-correlation function. This is a Green–Kubo relation for the shear viscosity.

The foregoing can also be applied to the simplified theory of particle diffusion. Guided by the hydrodynamic theory of diffusion in a binary mixture (cf. Section 10.6) we take $A(\mathbf{q}) = \delta c(\mathbf{q})$ where $\delta c(\mathbf{q})$ is the Fourier component of the fluctuations in the solute mass fraction at $(\mathbf{r}, t)$, that is, $c(\mathbf{r}, t) = m_1(\mathbf{r}, t)/m(\mathbf{r}, t)$ where m_1, m_2 and m are respectively the mass densities of solute, solvent, and solution. Since $m_1 = \bar{m}_1 + \delta m_1$, $m_2 = \bar{m}_2 + \delta m_2$ it follows that $\delta c = (\bar{m})^{-2}[\bar{m}_2 \delta m_1 - \bar{m}_1 \delta m_2]$; therefore, we take

$$A(\mathbf{q}) = \delta c(\mathbf{q}) = (\bar{m})^{-2}[\bar{m}_2 \delta m_1(\mathbf{q}) - \bar{m}_1 \delta m_2(\mathbf{q})] \tag{11.6.20}$$

where the microscopic forms of $m_1(\mathbf{q})$ and $m(\mathbf{q})$ are $\delta m_1(\mathbf{q}) \equiv M_1 \sum_{j \in S_1} \exp iqz_j$ and $\delta m_2(\mathbf{q}) = M_2 \sum_{j \in S_2} \exp iqz_j$ where M_1 and M_2 are the molecular masses of molecules of type 1 and 2 and S_1 and S_2 designate the sets of these molecules. Because mass is conserved $\delta c(\mathbf{q}, t)$ must satisfy a continuity equation so that $iL\delta c(\mathbf{q}, t) = iqJ(\mathbf{q}, t)$. This defines the flux as $J(\mathbf{q}) = \bar{m}^{-2} [\bar{m}_2 \delta m_1(\mathbf{q}) - \bar{m}_1 \delta m_2(\mathbf{q})](iq)^{-1}$. From the explicit form of δm_1 and δm_2 it follows that $\delta m_1 = iqg_1^z$ and $\delta m_2 = iqg_2^z$ where g_1^z and g_2^z are respectively the z components of the momentum densities of components 1 and 2; that is $g_1^z \equiv \sum_{j \in S_1} M_1 \dot{z}_j \exp iqz_j$ and $g_2^z = \sum_{j \in S_2} M_2 \dot{z}_j \exp iqz_j$. Substitution of these results into J gives

$$J(\mathbf{q}) = \frac{1}{\bar{m}^2} [\bar{m}_2 g_1^z(\mathbf{q}) - \bar{m}_1 g_2^z(\mathbf{q})] \tag{11.6.21}$$

which becomes in the small q limit

$$J(\mathbf{q}) \underset{q \to 0}{\rightarrow} J^{(0)} = \frac{1}{\bar{m}^2} [\bar{m}_2 \sum_{j \in S_1} p_j^z - \bar{m}_1 \sum_{j \in S_1} p_j^z] \tag{11.6.22}$$

where p_j^z is the zth component of the momentum of particle j. From Eq. (10.6.21) it follows that

$$\langle|\delta c(\mathbf{q})|^2\rangle \underset{q\to 0}{\to} \frac{k_BT}{mV}\left(\frac{\partial c}{\partial \mu}\right)_{T,P} \tag{11.6.23}$$

Combining all of these results it follows that in the Markovian limit

$$\frac{\partial}{\partial t}\,\delta c(\mathbf{q}, t) = -\Gamma(\mathbf{q})\delta c(\mathbf{q}, t) + F(t) \tag{11.6.24}$$

where

$$\Gamma(\mathbf{q}) = q^2 D \tag{11.6.25}$$

with the diffusion coefficient

$$D = \lim_{\eta\to 0+} mV\left(\frac{\partial \mu}{\partial c}\right)_{T,P} \beta\int_0^\infty dt\, e^{-\eta t}\langle J^{(0)}(0)\, J^{(0)}(t)\rangle \tag{11.6.26}$$

In the limit of infinite dilution this becomes

$$D = \lim_{\eta\to 0+}\int_0^\infty dt\, e^{-\eta t}\langle v_z(0)v_z(t)\rangle \tag{11.6.27}$$

where $v_z(t)$ is the zth component of the velocity of a typical solute particle [cf. Eq. (5.9.14)].

This theory of diffusion in a binary solution is incomplete. We have left out other slow variables [cf. Section 10.6] and have thereby not used a good set. Nevertheless this discussion was useful because it acted as a vehicle for drawing together our ideas.

APPENDIX 11.A PROJECTION OPERATORS IN QUANTUM STATISTICAL MECHANICS

In quantum mechanics observables are represented by linear Hermitian operators $\hat{A}$, which change in time according to the equation of motion

$$\frac{d\hat{A}}{dt} = \frac{1}{i\hbar}[\hat{A}, \hat{H}] \equiv iL\hat{A} \tag{11.A.1}$$

where $[\hat{A}, \hat{H}]$ is the commutator and $\hat{H}$ is the Hamiltonian operator. The operator L defined by Eq. (11.A.1) is the quantum mechanical *Liouvillian.* The formal operator solution of this equation is

$$\hat{A}(t) = e^{iLt}\,\hat{A} = \exp\left(\frac{i\hat{H}t}{\hbar}\right)\hat{A}\exp\left(-\frac{i\hat{H}t}{\hbar}\right)$$

The quantum mechanical autocorrelation function is then

$$\langle A^+(0)A(t)\rangle = Tr\hat{\rho}_0\hat{A}^+(0)\,\hat{A}(t) = Tr\,\hat{\rho}_0\hat{A}^+\,e^{iLt}\hat{A} \tag{11.A.2}$$

where $\hat{\rho}_0$ is the equilibrium density matrix (e.g., in the canonical ensemble $\hat{\rho}_0 = Q^{-1} e^{-\beta\hat{H}}$). The order of the operators in the correlation function is important. In fact $\langle A^+(0)$

$A(t)\rangle$ and $\langle A(t)A^+(0)\rangle$ are not equal, but are complex conjugates of one another. Although a projection-operator formalism for these "one sided" correlation functions can be developed, it is more convenient for the purposes of comparing with classical correlation functions to deal with the symmetrized function

$$\langle A^+(0)A(t)\rangle_s \equiv \langle \tfrac{1}{2}\{\hat{A}^+(0)\hat{A}(t) + \hat{A}(t)\hat{A}^+(0)\}\rangle \tag{11.A.3}$$

These are real quantities that go in the classical limit $\hbar \to 0$ to the classically defined correlation function. It is then possible to define the scalar product of two Hermitian operators as

$$(A, B^+) = Tr\hat{\rho}_0 \frac{1}{2}\{AB^+ + B^+A\}$$

and to define the projection operator onto A as

$$\hat{P} = (\,.\,.\,.\,.\,, A^+)\,(A, A^+)^{-1}\hat{A} \tag{11.A.4}$$

Then the whole formalism of this chapter can be applied to evaluate $\langle A^+(0)A(t)\rangle_s$.

Another correlation function that often appears in quantum statistical mechanics is the Kubo transformed correlation function (cf. Zwanzig, 1965). This function can be related to $\langle A(0)A(t)\rangle_s$ so that it is unnecessary to define new scalar products and projection operators, although the Kubo transform itself can be fit into this context (cf. Mori, 1965).

APPENDIX 11.B AN EXPRESSION FOR THE RELAXATION RATE IN TERMS OF ORDINARY TIME-CORRELATION FUNCTIONS

According to our previous discussion, $\mathbf{F}(t)$ and $\mathbf{G}(t)$ [(cf. Eqs. (11.3.33)] have different dynamical behavior. $\mathbf{K}(t)$ is proportional to the autocorrelation function of $\mathbf{F}(t)$. Let us define the time-correlation function of $\mathbf{G}(t)$;

$$\mathbf{M}(t) = (\mathbf{G}(t), \mathbf{G}^+(0)) \cdot (\mathbf{A}, \mathbf{A}^*)^{-1} \tag{11.B.1}$$

This is an ordinary time-correlation function. There is an important relationship between $\mathbf{M}(t)$ and $\mathbf{K}(t)$ that is best stated in terms of their respective Laplace transforms $\tilde{\mathbf{M}}(s)$ and $\tilde{\mathbf{K}}(s)$ as

$$\tilde{\mathbf{K}}(s) = (\mathbf{I} - (s\mathbf{I} - i\Omega)^{-1} \cdot \tilde{\mathbf{M}}(s))^{-1} \cdot \tilde{\mathbf{M}}(s) \tag{11.B.2}$$

where $\boldsymbol{I}$ is the unit matrix. The derivation of this relation would require too much space. The reader should consult the references.

Now why is this relationship important? To answer this question let us return to the case of slow variables. Then Laplace transformation to Eq. (11.4.1) yields $\tilde{\mathbf{K}}(s) = \boldsymbol{\Gamma}$. Solving Eq. (11.B.2) for $\tilde{\mathbf{M}}(s)$ subject to $\tilde{\mathbf{K}}(s) = \boldsymbol{\Gamma}$ gives

$$\tilde{M}(s) = \boldsymbol{\Gamma} - \boldsymbol{\Gamma} \cdot (s\mathbf{I} - i\Omega + \boldsymbol{\Gamma})^{-1} \cdot \boldsymbol{\Gamma} \tag{11.B.3}$$

Laplace inversion of this result gives

$$\mathbf{M}(t) = 2\boldsymbol{\Gamma}\delta(t) + \boldsymbol{\Gamma} \cdot e^{(i\Omega - \Gamma)t} \cdot \boldsymbol{\Gamma} \tag{11.B.4}$$

or in the single variable case

$$M(t) = 2\Gamma\delta(t) - \Gamma^2 e^{(i\Omega-\Gamma)\,t} \tag{11.B.5}$$

This is shown in Fig. 11.B.1. $M(t)$ includes the slow process with a negative long time tail and its time integral

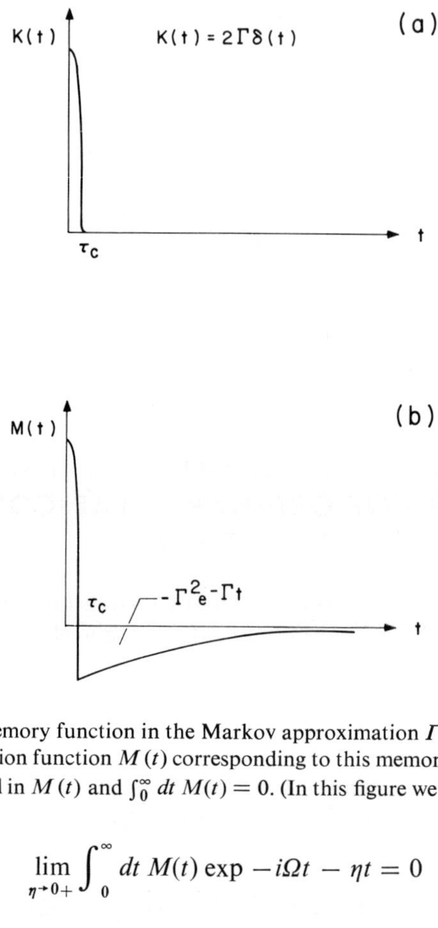

FIG. 11.B.1(*a*). The memory function in the Markov approximation $\Gamma = \int_0^\infty dtK(t) > 0$. (*b*) The correlation function $M(t)$ corresponding to this memory function. There is a long-time tail in $M(t)$ and $\int_0^\infty dt\, M(t) = 0$. (In this figure we take $\Omega = 0$.)

$$\lim_{\eta \to 0+} \int_0^\infty dt\, M(t) \exp -i\Omega t - \eta t = 0 \tag{11.B.6}$$

vanishes due to the cancelation of the negative long tail by the positive fast decay. This is in strong contrast to the memory function $K(t)$ in which the slow process does not appear.

In the case of $\Gamma\tau_c \ll 1$, the fast process $K(t)$ and the slow process $C(t)$ are sufficiently well separated that the relaxation rate Γ can be obtained from Eq. (11.B.5)

$$\Gamma \simeq \int_0^\tau dt\, M(t) \tag{11.B.7}$$

where τ is a time satisfying the inequality $\tau_c \ll \tau \ll \Gamma^{-1}$. In practice the integral is evaluated as a function of τ. The integral goes through a plateau region. Only in this plateau region does Eq. (11.B.7) have a meaning.

The same arguments can be applied to the matrix $\mathbf{M}(t)$ so that the damping matrix is

$$\boldsymbol{\Gamma} \simeq \int_0^\tau dt\, \mathbf{M}(t) \tag{11.B.8}$$

APPENDIX 11.C ADDITIONAL THEOREMS CONCERNING TIME-CORRELATION FUNCTIONS AND MEMORY FUNCTIONS

As we have already seen, the Liouvillian L is an Hermitian operator and the propagator e^{iLt} is unitary. Likewise since Q is Hermitian, QLQ is Hermitian $(QLQ)^+ = Q^+L^+Q^+ = (QLQ)$. It follows that e^{iQLQt} is a unitary operator. These properties allow us to prove the following theorems.

Theorem C.1

The correlation function $\langle A_\nu^*(t)A_\mu(t + \tau)\rangle$ is a stationary random process; that is, it is independent of time t.

$$\langle A_\nu^*(t)A_\mu(t + \tau)\rangle = \langle A_\nu^*(0)A_\mu(\tau)\rangle \tag{11.C.1}$$

Proof: First note that

$$A(t + \tau) = e^{iL(t+\tau)}\, A_\mu;\ A_\nu^*(t) = (e^{iLt}A_\nu)^*$$

so that

$$\begin{aligned}\langle A_\nu^*(t)A_\mu(t + \tau)\rangle &= ((e^{iLt}A_\nu)^*, e^{iL(t+\tau)}A_\mu)\\ &= (A_\nu^*, e^{-iLt}\, e^{iL(t+\tau)}A_\mu) = (A_\nu^*, e^{iL\tau}A_\mu) \quad \text{Q.E.D.}\end{aligned}$$

The second equality follows from the unitarity of e^{iLt}.

As a special case of this theorem, note that for $\tau = 0$

$$\langle A_\nu^*(t)A_\mu(t)\rangle = \langle A_\nu^*(0)A_\mu(0)\rangle \tag{11.C.2}$$

Stationarity follows from the fact that the time-correlation functions are defined as averages over equilibrium (stationary) ensembles. In such ensembles, it should not, and by Theorem C.1 does not, matter what time is chosen as the initial time. Time correlation functions in stationary ensembles are invariant to a shift in the origin of time.

Theorem C.2

Time-correlation functions satisfy the inequality.

$$0 \leq |\langle A_\mu(t)A_\nu^*(0)\rangle| \leq (\langle|A_\mu|^2\rangle\langle|A_\nu|^2\rangle)^{1/2} \tag{11.C.3}$$

Proof: Given the definition of the scalar product, it is a simple matter to prove Schwartz's inequality,[21] that is,

$$0 \leq |(B, A^*)| \leq \{(A, A^*)(B, B^*)\}^{1/2}$$

Now if we let $A = A_\nu$ and $B = e^{iLt}A_\mu$, this inequality becomes

$$0 \leq |(e^{iLt}A_\mu, A_\nu^*)| \leq \{(A_\nu A_\nu^*)((e^{iLt}A_\mu)^*, (e^{iLt}A_\mu))\}^{1/2} \tag{11.C.4}$$

From the unitarity of e^{iLt} the last bracket on the right is simply (A_μ^*, A_μ). Equation (11.C.4) is thus equivalent to Eq. (11.C.3), hence proving the theorem. Specializing Theorem 3 to autocorrelation functions gives

$$0 \leq |\langle A_\mu^*(0)A_\mu(t)\rangle| \leq \langle |A_\mu|^2\rangle \tag{11.C.5}$$

Since the autocorrelation function is real it follows that it is bounded from above by $\langle |A_\mu|^2\rangle$ and from below by $-\langle |A_\mu|^2\rangle$.

The same theorem can be proved for the memory functions. This gives

$$0 \leq |\langle F_\mu(t), F_\nu^*(0)\rangle| \leq (\langle |G_\mu|^2\rangle\langle |G_\nu|^2\rangle)^{1/2} \tag{11.C.6}$$

This follows from the unitarity of e^{iQLQt}. This theorem shows that time-correlation functions are bounded.

We conclude by showing that the spectrum of $\langle A^*(0)A(t)\rangle$ is always positive

$$I(\omega) = \frac{1}{2\pi}\int_{-\infty}^{+\infty} dt\, e^{-i\omega t}\langle A^*(0)A(t)\rangle \geq 0 \tag{11.C.7}$$

This property is important since it implies that the light-scattering spectrum ($A = E_s$) is always positive.

A simple but nonrigorous version of this proof is the following. Since the Liouville operator is Hermitian its eigenvalues λ are real and its eigenfunctions $\phi_\lambda(\boldsymbol{\Gamma})$ are orthogonal where

$$L\phi_\lambda(\boldsymbol{\Gamma}) = \lambda\phi_\lambda(\boldsymbol{\Gamma}) \tag{11.C.8}$$

The property $A(\boldsymbol{\Gamma})$ can be expanded in the eigenfunctions $\{\phi_\lambda\}$ so that

$$A(\boldsymbol{\Gamma}) = \sum_\lambda (\phi_\lambda^*, A)\,\phi_\lambda(\boldsymbol{\Gamma}) \tag{11.C.9}$$

where (ϕ_λ^*, A) is a scalar product defined above. The spectrum of L may be part discrete and part continuous so that the sum $\sum_\lambda$ really consists of a sum over the discrete part and an integral over the continuous part. From Eqs. (11.C.8) and (11.C.9) it follows that

$$e^{iLt}A = \sum_\lambda (\phi_\lambda^*, A)\phi_\lambda(\boldsymbol{\Gamma})\, e^{i\lambda t}$$

From the orthogonality of $\{\phi_\lambda\}$ the correlation function is clearly

$$\langle A^*(0)\, A(t)\rangle = (A^*, e^{iLt}A) = \sum_{\lambda} |(\phi_\lambda{}^*, A)|^2\, e^{i\lambda t} \qquad \text{(11.C.10)}$$

where λ is a real quantity. Equation (11.C.10) shows that the correlation function might be quasiperiodic if L only has a discrete spectrum, but will be nonperiodic if L has a partly continuous spectrum. Substitution of Eq. (11.C.10) into Eq. (11.C.7) then yields[22]

$$I(\omega) = \sum_{\lambda} |(\phi_\lambda{}^*, A)|^2\, \delta(\omega - \lambda)$$

Since $|(\phi_\lambda{}^*, A)|^2 > 0$ and $\delta(\omega - \lambda) > 0$, the spectral density $I(\omega)$ is positive, thus proving the theorem. There are more rigorous proofs of the positivity (e.g., see Feller, 1966), but this is the most straightforward for our purposes.

NOTES

1. The Poisson Bracket { , } is defined such that for any two functions of Γ, F, and G,

$$\{F, G\} = \sum_{i=1}^{f} \left(\frac{\partial F}{\partial q_i}\frac{\partial G}{\partial p_i} - \frac{\partial F}{\partial p_i}\frac{\partial G}{\partial q_i}\right)$$

Writing Eq. (11.2.1) in component form gives, for example, $\dot{q}_j = \{q_j, H\}$. Now use the definition of the Poisson bracket

$$\{q_j, H\} = \sum_{i=1}^{f} \left(\frac{\partial q_j}{\partial q_i}\frac{\partial H}{\partial p_i} - \frac{\partial q_j}{\partial p_i}\frac{\partial H}{\partial q_i}\right)$$

Since the q's and $p'q$ are independent variables, $\partial q_j/\partial q_i = \delta ij$ and $\partial q_j/\partial p_j = 0$, From this it follows that $\dot{q}_j = \partial H/\partial p_j$, therefore proving our assertion.

2. Note that

$$\frac{dA}{dt} = \sum_{i=1}^{f} \left(\frac{\partial A}{\partial q_i}\dot{q}_i + \frac{\partial A}{\partial p_i}\dot{p}_i\right) = \sum_{i=1}^{f} \left(\frac{\partial A}{\partial q_i}\frac{\partial H}{\partial p_i} - \frac{\partial A}{\partial p_i}\frac{\partial H}{\partial q_i}\right)$$

where the last equality follows from a substitution of Eq. (2.2.3). This right-hand side is equivalent to Eq. (11.2.4).

3. L^+ is defined by $(g^*, Lf)^* = (f^*, L^+g)$ for all f, g in Liouville space.
4. L is a linear differential operator. It is a simple matter to show by an integration by parts that for any pair of functions g, f in Liouville space $(g^*, Lf)^* = (f^*, Lg)$. This proves that L is Hermitian.
5. An operator G is unitary if $G^+G = GG^+ = 1$. A unitary operator does not change the norm of a vector, that is, $((Gf)^*,(Gf)) = (f^*, f)$. This follows from the definition of unitarity.
6. The interested reader will note that this identity follows from the operator identity $A^{-1} - B^{-1} = A^{-1}(B - A)B^{-1}$. If we choose $A = (s - iL)$; $B = (s - iO_1L)$ then

$$\frac{1}{s - iL} = \frac{1}{s - iO_1L} + \frac{1}{s - iL}\, i(1 - O_1)L\, \frac{1}{s - iO_1L}$$

Now because $O_1 + O_2 = 1$, $(1 - O_1) = O_2$ and Laplace inversion yields Eq. (11.3.7).

7. Equation (11.3.16) follows from the identity $e^{iQL\tau}Q = Qe^{iQL\tau}Q$. To verify this identity we expand the right-hand side in a Taylor series

$$Qe^{iQL\tau}Q = Q(1 + iQL\tau + \ldots)\, Q$$
$$= (1 + iQL\tau + \ldots.)\, Q$$

where the second inequality follows from $Q^2 = Q$. The term in brackets is $e^{iQL\tau}$, thus proving the identity. Thus

$$F(\tau) = e^{iQL\tau}QiLA = Qe^{iQL\tau}QiLA = QF(\tau).$$

By exactly the same reasoning it follows that

$$e^{iQL\tau}Q = e^{iQL\tau}Q$$

8. By definition $(A(t), A^*(0)) = \langle A^*(0)A(t)\rangle$.
9. This equation was derived without appealing to the Onsager regression hypothesis.
10. This equation defines the susceptibility matrix.
11. See the definition of the random force, Eq. (11.3.35), where the presence of Q on the left insures that **F** lies entirely in the fast subspace.
    It should be noted that if one or more slow variables are omitted from the set defined by P, $F(\tau)$ will contain slow variables, and Eq. (11.4.1) will be invalid.
12. Under this transformation $iL \to -iL$ so that $e^{iLt} \to e^{-iLt}$ and the transformation is equivalent in some sense to replacing t by $-t$.
13. Since the γ's can be either $+1$ or -1 if $\gamma_\mu \neq \gamma_\nu$, then $\gamma_\mu\gamma_\nu = -1$.
14. Thus if $\gamma_\mu = \gamma_\kappa$, $\gamma_\mu\gamma_\kappa = +1$ and $(LA_\mu, A_\kappa^*) = 0$ whereas if $\gamma_\mu \neq \gamma_\kappa$, $\gamma_\mu\gamma_\kappa = -1$ and $(LA_\mu\ A_\kappa)$ can be nonzero.
15. In the matrix form, $\mathbf{C}^+{}_{EE}(t) = \mathbf{C}_{EE}(t)$, $\mathbf{C}^+{}_{OO}(t) = \mathbf{C}_{OO}(t)$, and $\mathbf{C}^+{}_{EO}(t) = -\mathbf{C}_{EO}(t)$.
16. See Note 7.
17. It is possible to define frequency-dependent kinetic coefficients as

$$\Lambda(\omega) \equiv \beta \lim_{\eta\to 0+} \int_0^\infty dt(\mathbf{F}(\tau), \mathbf{F}^+(0)) \exp -i\omega t - \eta|t|$$

Again using Theorem 4 gives

$$\Lambda_{\mu\nu}(\omega) = \gamma_\mu\gamma_\nu\Lambda^*{}_{\nu\mu}(-\omega)$$

18. This is very easy to do. One uses a Schmidt orthogonalization technique.
19. Because $A(\mathbf{q} \to 0)$ is conserved, $dA/dt(\mathbf{q} \to 0) = 0$, implying that $\sum_j \dot{a}_j = 0$.
20. This shows that $\Gamma^{(0)} = 0$ in Eq. (11.6.3).
21. Any comprehensive book on quantum mechanics gives the proof of this inequality.
22. Here we use the integral representation of the delta function.

REFERENCES

Berne, B. J. *Physical Chemistry*, Vol. VIII B, Eyring, H., Henderson, D., and Jost, W. (eds), Academic, New York (1971).

Berne, B. J. and Forster, D., *Ann. Rev. Phys. Chem.* **22,** 563 (1971).

Feller, W., *An Introduction to Probability Theory and its Applications,* Wiley, New York (1966).

Kubo, R., *Rep. Prog. Phys.* **29,** 255 (1966).

Mori, H., *Prog. Theor. Phys. (Kyoto)* **33,** 423 (1965); **34,** 399 (1965).

Stanley, H., Paul, G., and Milosivic, S. *Physical Chemistry,* Vol. VIIIB, Eyring, H., Henderson, D., and Jost, W. (eds.), Academic, New York (1971). This review contains an extensive list of references. See, for example, references to the work of Fixman, Kananoff, Swift, and Kawasaki.

Swinney, H. L. in *Photon Correlation and Light Scattering Spectroscopy,* Cummins, H. Z. and Pike, E. R. (eds.), Plenum, New York (1974).

Zwanzig, R. in *Boulder Lectures in Theoretical Physics,* Vol. III, p. 106, Wiley-Interscience, New York (1961).

Zwanzig, R., *Ann. Rev. Phys. Chem.,* **16,** 67 (1965).

CHAPTER 12

COOPERATIVE EFFECTS IN DEPOLARIZED LIGHT SCATTERING

12·1 INTRODUCTION

The light scattered from a fluid containing optically isotropic molecules was considered in Chapter 10, where it was shown that the Navier–Stokes equations of fluid mechanics correctly describe the observed Rayleigh–Brillouin spectrum in the low-frequency regime from 0 to about 10 cm^{-1}. Many new features arise in the spectrum of light scattered from fluids containing optically anisotropic molecules. First and foremost is the appearance of depolarized components in the spectrum. In Chapter 7 we have had occasion to discuss the depolarized spectrum in the particularly simple case when the rotational motions of different molecules are uncorrelated. In dense fluids, this is certainly not valid. The question then arises as to how the collective motions of the fluid can be described when rotations are considered. Unfortunately we do not have a simple set of "hydrodynamic equations" which describe this situation. Otherwise we would follow precisely the same program outlined in Chapter 10 in connection with the Navier–Stokes equations. In this section we use the results of Chapter 11 for deriving the "hydrodynamic equations" which include rotations. The depolarized spectrum is then calculated and compared with experiment.

By way of introduction let us note that the depolarized spectrum $I_{VH}(\omega)$ calculated in Section 7.5 for independent rotors consists of a superposition of Lorentzian bands all centered at zero frequency. In the simplest case of symmetric top rotors the spectrum consists of a single band with a width $[q^2D + 6\Theta]$ which depends only on the translational self-diffusion coefficient D and on the rotational diffusion coefficient Θ. This should be compared and contrasted with the depolarized spectrum $I_{VH}(\omega)$ of certain pure liquids (e.g., aniline, nitrobenzene, quinoline, hexafluorobenzene) shown schematically in Fig. 12.1.1. The spectrum appears to be split. This entirely novel fea-

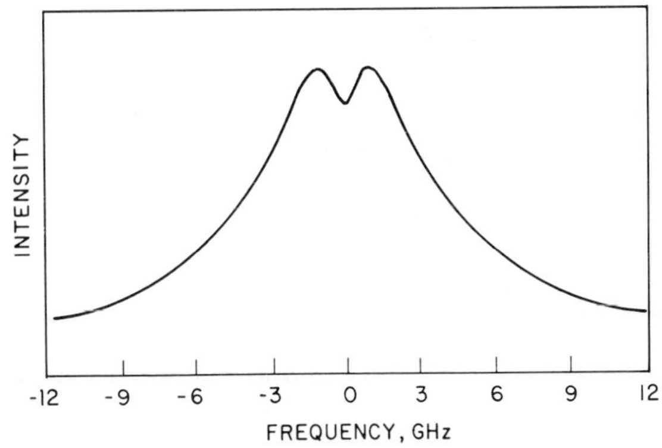

FIG. 12.1.1. A schematic drawing of the splitting in the depolarized spectrum I_{VH} observed in many nonassociated liquids.

ture was first extensively studied several years ago by Stegeman and Stoicheff (1968). It is the object of this section to present a simple theory of this spectrum from a molecular point of view. In Section 12.3 we present a comparison between the single particle and collective reorientation times as measured respectively in NMR and depolarized Rayleigh scattering.

12·2 KINETIC EQUATIONS FOR ORIENTATIONAL RELAXATION IN DEPOLARIZED SCATTERING

For simplicity we restrict our attention to the VH component of the scattering in fluids containing symmetric top molecules. Moreover scattering geometry I of Section 3.4 is used throughout this section. From Eqs. (3.2.17) and (3.4.1) we note that the polarizability fluctuation $\delta\alpha_{VH}(\mathbf{q}, t)$ that gives rise to the VH scattering involves only the xy and yz components of the polarizability fluctuation tensor $\delta\alpha_{\alpha\beta}(\mathbf{q}, t)$, that is,

$$\delta\alpha_{VH}(\mathbf{q}, t) = \delta\alpha_{yx}(\mathbf{q}, t) \sin\frac{\theta}{2} - \delta\alpha_{yz}(\mathbf{q}, t) \cos\frac{\theta}{2} \tag{12.2.1}$$

First we consider the symmetry properties of $\delta\alpha_{\alpha\beta}(\mathbf{q}, t)$. As we have shown in Appendix 7.B, the polarizability of a symmetric top is given by Eq. (7.B.1). Combining this with Eq. (3.3.4) gives

$$\delta\alpha_{\alpha\beta}(\mathbf{q}, t) = \alpha \sum_{j=1}^{N} \exp iqz_j + \beta \sum_{j=1}^{N} \left[u_\alpha^j u_\beta^j - \frac{1}{3}\delta_{\alpha\beta}\right] \exp iqz_j \tag{12.2.2}$$

where we have taken $\mathbf{q}$ in the z direction, where α and β are defined in Eq. (7.B.1) and where the unit vector $\mathbf{u}^j$ specifies the orientation of the principal axis of the $\mathbf{j}^{th}$ molecule. $\delta\alpha_{\alpha\beta}(\mathbf{q})$ is clearly a *symmetric* tensor; that is, $\delta\alpha_{\alpha\beta}(\mathbf{q}, t) = \delta\alpha_{\alpha\beta}(\mathbf{q}, t)$. Furthermore because $\alpha_{\alpha\beta} \propto u_\alpha^j u_\beta^j - \frac{1}{3}\delta_{\alpha\beta}$ it follows that $\alpha_{\alpha\beta}^j$ and consequently $\delta\alpha_{\alpha\beta}(\mathbf{q}, t)$ *have definite symmetry with respect to reflections in xz and yz planes.* For example, $\delta\alpha_{xy}$ and $\delta\alpha_{yz}$ both reverse sign on reflection through the xz plane (that is, for $u_y^j \to -u_y^j$ for all j). Thus $\delta\alpha_{VH}(\mathbf{q}, t)$, which depends only on $\delta\alpha_{xy}$ and $\delta\alpha_{yz}$ reverses sign on reflection, so that according to Theorem 9 of Section 11.5, $\delta\alpha_{VH}(\mathbf{q}, t)$ *can couple only to properties with odd reflection symmetry in the xz and yz planes.*

It should be noted that a reflection through the xz plane followed by a reflection through the yz plane (or vice versa) reverses the sign of $\delta\alpha_{yz}(\mathbf{q}, t)$ but leaves $\delta\alpha_{xy}(\mathbf{q}, t)$ unchanged. As a consequence of this, the correlation functions $\langle\delta\alpha_{xy}^*(\mathbf{q}, 0)\, \delta\alpha_{yz}(\mathbf{q}, t)$ and $\langle\delta\alpha^*_{yz}(\mathbf{q}, 0)\delta\alpha_{xy}(\mathbf{q}, t)\rangle$ are zero. Substitution of Eq. (12.2.1) into Eq. (3.3.13) then gives, aside from the multiplicative constant

$$I_{VH}(\mathbf{q}, \omega) = \frac{1}{2\pi}\int_{-\infty}^{+\infty} dt\, e^{-i\omega t} \left\{\langle\delta\alpha^*_{yx}(\mathbf{q}, 0)\, \delta\alpha_{yx}(\mathbf{q}, t)\rangle \sin^2\frac{\theta}{2} \right.$$
$$\left. + \langle\delta\alpha^*_{yz}(\mathbf{q}, 0)\, \delta\alpha_{yz}(\mathbf{q}, t)\rangle \cos^2\frac{\theta}{2}\right\} \tag{12.2.3}$$

It follows from Eq. (12.2.2) that $\delta\alpha_{\alpha\beta}(\mathbf{q}, t)$ transforms to $e^{iq\Delta z}\,\delta\alpha_{\alpha\beta}(\mathbf{q}, t)$ under the arbitrary translation Δz along the z direction, that $\delta\alpha_{\alpha\beta}(\mathbf{q}, t)$ has even time reversal symmetry, and that $\delta\alpha_{\alpha\beta}(\mathbf{q}, t)$ transforms to $\delta\alpha_{\alpha\beta}(-\mathbf{q}, t)$ under inversion symmetry.

Having discussed the symmetry properties of the primary variables $\delta\alpha_{\alpha\beta}(\mathbf{q}, t)$, let us now pass to the derivation of the relaxation equations that describe evolution of $\delta\alpha_{\alpha\beta}(\mathbf{q}, t)$.

In applying the Zwanzig–Mori formalism we follow the following prescription.

STEP 1. Determine the "primary" variables. In the case of light scattering these are the elements of the polarizability tensor. For the part of the spectrum not involving very large frequency shifts, we are only interested in the parts of the polarizability fluctuations that relax slowly. In this eventuality $\delta\alpha_{\alpha\beta}(\mathbf{q})$ lies wholly in the slow subspace of Liouville space, and can either be taken as one of the "slow variables" in the formalism or as a linear combination of the other slow variables of the same symmetry. In the following we choose $\delta\alpha_{\alpha\beta}(\mathbf{q}, t)$ as the primary variables and do not express it as a linear combination of other variables. The correlation functions of the primary variables are then calculated using the formalism developed in Chapter 11.

STEP 2. Find the symmetry properties of the Hamiltonian and of the slowly relaxing variables. Useful symmetry operations are reflection of the positions and momenta of all molecules through a plane, translation of the entire fluid, and permutation of identical molecules.

STEP 3. Using physical intuition or independent information, find all the independent dynamical variables having long relaxation times and the same symmetry properties as the variables of interest. For convenience, choose them to be orthogonal (i.e., so that the $\boldsymbol{\chi}$ matrix is diagonal) and to be even or odd under time reversal.

STEP 4. Calculate the $\boldsymbol{\chi}$ and $\boldsymbol{\Omega}$ matrices. This involves the calculation of equilibrium averages.

STEP 5. Determine the relationships between the off-diagonal elements of the $\boldsymbol{\Gamma}$ matrix. Detailed microscopic expressions for the nonzero elements can be obtained from Eq. (11.4.4).

STEP 6. Substitute the results of 4 and 5 into Eq. (11.4.2) and thus obtain kinetic equations for the correlation functions of interest.

Aside from the assumption of classical mechanics, the only assumptions that must be made to derive the kinetic equations are the number and choice of variables with long relaxation times. This choice is usually made from physical considerations, and once made no additional assumptions about the nature of the fluid state or of relaxation processes in that state are necessary to obtain the form of these kinetic equations. This technique, is very powerful and also quite general.

In the following discussion, only the derivation of the $\delta\alpha_{yz}(\mathbf{q}, t)$ correlation function is given in detail. The results for the $\delta\alpha_{xy}$ correlation function are obtained in a similar way and are therefore merely summarized.

As noted above, some of the relevant symmetry properties of $\delta\alpha_{yz}(\mathbf{q}, t)$ are *reflection* in the xz and yz planes and *translation* by any amount Δz along the z axis. The signatures of $\delta\alpha_{yz}(\mathbf{q}, t)$ under these operations are respectively -1, $+1$ and $e^{iq\Delta z}$. The only slowly relaxing variables that come immediately to mind are the spatial Fourier components of the conserved densities that appear in hydrodynamics; that is, the number density, $\rho(\mathbf{q}, t)$, the three components of the velocity (or equivalently the momentum), density $v_x(\mathbf{q}, t)$, $v_y(\mathbf{q}, t)$, or $v_z(\mathbf{q}, t)$, and the energy density $e(\mathbf{q}, t)$. Of these variables, the only one with the same symmetry properties as $\alpha_{yz}(\mathbf{q})$ is

$$v_y(\mathbf{q}) = \sum_{j=1}^{N} v_y^j \exp iqz_j \tag{12.2.4}$$

That is, the signatures of $v_y(\mathbf{q})$ under reflections in the xz and yz planes and translation by Δz along the z axis are respectively -1, $+1$, and $e^{iq\Delta z}$. It follows from Theorem 9 of Section 11.5 that the only "hydrodynamic" variables that couple to $\alpha_{yz}(\mathbf{q})$ are $v_y(\mathbf{q})$. Thus the "simplest" relaxation theory imaginable is the one in which $v_y(\mathbf{q})$ and $\alpha_{yz}(\mathbf{q})$ are the two variables A_1 and A_2 in the Zwanzig–Mori Formalism

$$\begin{Bmatrix} A_1(\mathbf{q}) \equiv v_y(\mathbf{q}) \\ A_2(\mathbf{q}) \equiv \delta\alpha_{yz}(\mathbf{q}) \end{Bmatrix} \tag{12.2.5}$$

The next step is the evaluation of the susceptibility matrix $\boldsymbol{\chi}$ and the frequency matrix [cf. Eqs. (11.3.25) and (11.3.34)]. First we note that $A_1(\mathbf{q})$ and $A_2(\mathbf{q})$ have opposite time-reversal symmetries (-1 and $+1$ respectively), so that according to Theorem 1 of Section 11.5 $(A_1, A_2^*) = (A_2, A_1^*) = 0$ and the susceptibility matrix $\chi(\mathbf{q})$ is consequently diagonal. In addition, $A_1(\mathbf{q})$ and $A_2(\mathbf{q})$ transform under inversion as $A_1(+\mathbf{q}) \to -A_1(-\mathbf{q})$ and $A_2(\mathbf{q}) \to A_2(-\mathbf{q})$, so that (A_1, A_1^*) and (A_2, A_2^*) are even functions of q. We therefore surmise from symmetry considerations that

$$\boldsymbol{\chi}(\mathbf{q}) = \begin{pmatrix} \chi_{11}(\mathbf{q}) & 0 \\ 0 & \chi_{22}(\mathbf{q}) \end{pmatrix} \tag{12.2.6}$$

where $\chi_{11}(\mathbf{q})$ and $\chi_{22}(\mathbf{q})$ are even functions of $\mathbf{q}$. The elements have the form

$$\chi_{11}(\mathbf{q}) = \beta\langle |v_y(\mathbf{q})|^2\rangle = \frac{N}{m}$$

$$\chi_{22}(\mathbf{q}) = \beta\langle |\alpha_{yz}(\mathbf{q})|^2\rangle \tag{12.2.7}$$

We note that $\chi_{11}(\mathbf{q})$ is strictly independent of $\mathbf{q}$ whereas $\chi_{22}(\mathbf{q})$, which can be expressed in terms of the angular dependent pair-correlation function, can depend on $\mathbf{q}$ albeit as an even function of $\mathbf{q}$. In the absence of long-range orientational order[2], $\chi_{22}(\mathbf{q})$ can be regarded as a constant independent of $\mathbf{q}$ (for the small values of q that are probed in light scattering).

The matrix

$$\boldsymbol{\theta}(\mathbf{q}) = (L\mathbf{A}(\mathbf{q}), \mathbf{A}^+(\mathbf{q})) \tag{12.2.8}$$

that appears in $\boldsymbol{\Omega}$ also simplifies. Since $A_1(\mathbf{q})$ and $A_2(\mathbf{q})$ have definite time-reversal symmetry, an application of Theorem 2 of Section 11.5 shows that the diagonal elements $\theta_{11}(\mathbf{q}) = \theta_{22}(\mathbf{q}) = 0$. The off-diagonal elements are also zero. This follows from an explicit consideration of the matrix element $(iLv_y(\mathbf{q}), \alpha_{yz}^*(\mathbf{q}))$. First we note that $iLv_y(\mathbf{q}) = \sum_{j=1}^{N}$

$\left[\frac{1}{m} F_{yj} + iq\, v_z^j v_y^j\right] \exp iqz_j$ where F_{yj} is the y component of the force acting on particle j; that is, $F_{yj} = -\partial\Phi/\partial y_j$ where Φ is the potential energy function of the system. The matrix element can thus be written as

$$\theta_{12}(\mathbf{q}) = -i\left\{-\left(\sum_j \exp iqz_j \frac{\partial}{\partial y_j} \Phi, \alpha_{yz}^*(\mathbf{q})\right) + iq\left(\sum_j v_y^j v_z^j \exp iqz_j, \alpha_{yz}^*(\mathbf{q})\right)\right\} \qquad (12.2.9)$$

The second term on the right-hand side is zero because $\alpha_{yz}^*(\mathbf{q})$ is independent of the velocity and $\langle v_y^j v_z^j\rangle = 0$. The first term on the right can be simplified further. First we note that $e^{-\beta H}(\partial\Phi/\partial y_j) \exp iqz_j$ can be written as $-\beta^{-1}(\partial e^{-\beta H}/\partial y_j)) \exp -iqz_j$. Substituting this result into the first term of the right-hand side of Eq. (12.2.9), then integrating by parts, we obtain

$$\theta_{12}(\mathbf{q}) = +i(\beta^{-1})\left(\sum_j \exp iqz_j, \frac{\partial}{\partial y_j} \alpha_{yz}^*(\mathbf{q})\right)$$

Because $\alpha_{yz}(\mathbf{q}) = \beta \sum_l u_y^l u_z^l \exp iqz_l$ is independent of y_j, $\dfrac{\partial \alpha_{yz}^*(\mathbf{q})}{\partial y_j} = 0$, so that

$$\theta_{12}(\mathbf{q}) = \mathbf{0} = \theta_{21}(\mathbf{q}) \qquad (12.2.10)$$

We have just shown that for the particular set of variables given by Eq. (12.2.5), the matrix $\boldsymbol{\theta}(\mathbf{q})$ is a null matrix and correspondingly the frequency matrix $\boldsymbol{\Omega}(\mathbf{q})$ is a null matrix; that is,

$$\boldsymbol{\Omega}(\mathbf{q}) = \begin{pmatrix} 0 & 0 \\ 0 & 0 \end{pmatrix} \qquad (12.2.11)$$

Let us now consider the damping matrix $\boldsymbol{\Gamma}$. Because A_1 and A_2 have different time-reversal symmetry and the random forces $F_1 = QiLA_1$ and $F_2 = QiLA_2$ also have different time-reversal symmetries, it follows from Eq. (11.5.15) that the kinetic coefficients have the symmetry relation

$$\Lambda_{12}(\mathbf{q}) = -\Lambda_{21}^*(\mathbf{q}) \qquad (12.2.12)$$

In addition, since the variables transform under inversion as $A_1(\mathbf{q}) \to -A_1(-\mathbf{q})$ and $A_2(-\mathbf{q}) \to +A_2(\mathbf{q})$, it follows that the corresponding random forces transform under inversion as $F_1(\mathbf{q}) \to -F_1(-\mathbf{q})$ and $F_2(\mathbf{q}) \to F_2(-\mathbf{q})$ with the important consequence that $\Lambda_{11}(\mathbf{q}) = \Lambda_{11}(-\mathbf{q})$, $\Lambda_{12}(\mathbf{q}) = -\Lambda_{12}(-\mathbf{q})$ and $\Lambda_{22}(\mathbf{q}) = \Lambda_{22}(-\mathbf{q})$.
The matrix of kinetic coefficients consists of diagonal elements that are even functions of $\mathbf{q}$ *and off-diagonal elements that are odd functions of* $\mathbf{q}$.

An investigation of the explicit form of $iLA_1 = \sum_j [a_y^j + iqv_y^j v_z^j] \exp iqz_j$ (where a_y^j is the y component of the acceleration of particle j) shows that as $q \to 0$ only the second term[3] contributes so that as $q \to 0$, iLA_1 and consequently F_1 are first order in $\mathbf{q}$; that is, $F_1 = 0(\mathbf{q})$. This is not true of iLA_2 or F_2. From these considerations it follows that $\Lambda_{11}(\mathbf{q})$ depends quadratically on q in the small q limit and consequently vanishes as $q \to 0$, whereas $\Lambda_{22}(\mathbf{q})$ does not vanish as $q \to 0$. This merely reflects the fact that as $q \to 0$, A_1 is a component of the total linear momentum which is a conserved variable, whereas A_2 is not a conserved variable.

We have already observed that $\Lambda_{12}(\mathbf{q})$ is an odd function of $\mathbf{q}$. Thus for fluids whose intermolecular potential is short-ranged, $\Lambda_{12}(\mathbf{q})$ is at least linear in $\mathbf{q}$. It is convenient to define

$$\Lambda_{12}(\mathbf{q}) \equiv q\,\Lambda'_{12}(\mathbf{q}) \tag{12.2.13}$$

where $\Lambda'_{12}(q)$ is an even function of $\mathbf{q}$ (which in the limit of small q, is a constant independent of $\mathbf{q}$. In the ensuing paragraphs we will always factor the dominant q-dependence at low q out of the matrix elements and *will use the convention that the number of primes written on a matrix element denotes the number of* q*s factored from it.*

We therefore surmise from time-reversal symmetry and the transformation properties of A_1 and A_2 under inversion that the matrix of kinetic coefficients can be written as

$$\Lambda(\mathbf{q}) = \begin{pmatrix} q^2\Lambda''_{11}(q) & q\Lambda'_{12}(q) \\ -q\Lambda'^*_{12}(q) & \Lambda_{22}(q) \end{pmatrix} \tag{12.2.14}$$

where we have used Eq. (11.7.12) in taking $\Lambda_{21}(\mathbf{q}) = -\Lambda^*_{12}(\mathbf{q}) = -q\Lambda'^*_{12}(q)$. The primed elements are all even functions of q, which are constant as $q \to 0$. Substitution of Eqs. (12.2.6) and (12.2.14) into Eq. (11.4.5) gives the damping matrix

$$\boldsymbol{\Gamma}(\mathbf{q}) = \begin{pmatrix} q^2\Lambda''_{11}(\mathbf{q})\chi^{-1}_{11}(\mathbf{q}) & q\Lambda'_{12}(\mathbf{q})\chi^{-1}_{22}(\mathbf{q}) \\ -q\Lambda'^*_{12}(\mathbf{q})\chi^{-1}_{11}(\mathbf{q}) & \Lambda_{22}(\mathbf{q})\chi^{-1}_{22}(\mathbf{q}) \end{pmatrix} \tag{12.2.15a}$$

$\boldsymbol{\Gamma}(\mathbf{q})$ can be expressed as

$$\boldsymbol{\Gamma}(\mathbf{q}) = \begin{pmatrix} q^2\Gamma''_{11}(\mathbf{q}) & q\Gamma'_{12}(\mathbf{q}) \\ q\Gamma'_{21}(\mathbf{q}) & \Gamma_{22}(\mathbf{q}) \end{pmatrix} \tag{12.2.15b}$$

with the obvious definitions of the primed matrix elements. *It is important to remember that the primed matrix elements all approach constants in the limit of small q.*

Substituting these expressions for $\boldsymbol{\Omega}$ and $\boldsymbol{\Gamma}$ into Eq. (11.4.2), we obtain the relaxation equations

$$\begin{aligned} \frac{dA_1}{dt} &= -q^2\Gamma''_{11}(\mathbf{q})\,A_1 - q\Gamma'_{12}(\mathbf{q})\,A_2 + F_1 \\ \frac{dA_2}{dt} &= -q^2\Gamma_{22}(\mathbf{q})\,A_2 - q\Gamma'_{21}(\mathbf{q})\,A_1 + F_2 \end{aligned} \tag{12.2.15c}$$

Solving these linear equations by the method of Laplace transforms and alternately finding the scalar products of the solutions with $A_1(\mathbf{q}, 0)$ and $A_2(\mathbf{q}, 0)$ gives

$$\begin{aligned} \langle A_1^*(\mathbf{q}, 0)\,\tilde{A}_1(\mathbf{q}, s)\rangle &= \beta^{-1}\,\chi_{11}(\mathbf{q})\left[\frac{(s + \Gamma_{22}(\mathbf{q}))}{s^2 + \Gamma_s(\mathbf{q})s + \omega_s^2(\mathbf{q})}\right] \\ \langle A_2^*(\mathbf{q}, 0)\,\tilde{A}_2(\mathbf{q}, s)\rangle &= \beta^{-1}\chi_{22}(\mathbf{q})\left[\frac{(s + q^2\Gamma''_{11}(\mathbf{q}))}{s^2 + \Gamma_s(\mathbf{q})s + \omega_s^2(\mathbf{q})}\right] \end{aligned} \tag{12.2.16}$$

where s is the Laplace variable and we have defined the quantities

$$\Gamma_s(\mathbf{q}) \equiv \Gamma_{22}(\mathbf{q}) + q^2\Gamma''_{11}(\mathbf{q})$$
$$\omega_s^2(\mathbf{q}) \equiv q^2[\Gamma''_{11}(\mathbf{q})\Gamma_{22}(\mathbf{q}) - \Gamma'_{12}(\mathbf{q})\Gamma'_{21}(\mathbf{q})] \tag{12.2.17}$$

Substitution of the explicit forms of Γ'_{12} and Γ'_{21} from Eq. (12.2.15a) gives

$$\begin{aligned}\omega_s^2(\mathbf{q}) &= q^2\{[\Lambda''_{11}(\mathbf{q})\,\Lambda_{22}(\mathbf{q}) + |\Lambda'_{12}(\mathbf{q})|^2]/\chi_{11}\chi_{22}\} \\ &= q^2\alpha^2(\mathbf{q})\end{aligned} \tag{12.2.18}$$

where $\alpha(\mathbf{q})$ is defined in the last equation.

It may be seen from Eq. (12.2.3) that the depolarized spectrum $I_{VH}(\mathbf{q}, \omega)$ contains

$$S_{yz}(\mathbf{q}, \omega) = \frac{1}{2\pi}\int_{-\infty}^{\infty} dt\, e^{-i\omega t}\langle \delta\alpha_{yz}^*(\mathbf{q}, 0)\, \delta\alpha_{yz}(\mathbf{q}, t)\rangle \tag{12.2.19}$$

The quantity $A_2(\mathbf{q}, t) = \delta\alpha_{yz}(\mathbf{q})$ is even under time reversal so that by virtue of Eq. 11. 5.1), $\langle A_2^*(\mathbf{q}, 0)A_2(\mathbf{q}, t)\rangle$ is an even function of time. We may therefore write [cf. Eq. (6.2.6)]

$$\begin{aligned}S_{yz}(\mathbf{q}, \omega) &= \pi^{-1} Re \int_0^{\infty} dt\, e^{-i\omega t}\langle A_2^*(\mathbf{q}, 0)\, A_2(\mathbf{q}, t)\rangle \\ &= \pi^{-1} Re\,\langle A_2^*(\mathbf{q}, 0)\, \tilde{A}_2(\mathbf{q}, s = i\omega)\rangle\end{aligned} \tag{12.2.20}$$

Substitution of eq. (12.2.16) then gives the spectral density

$$S_{yz}(\mathbf{q}, \omega) = \pi^{-1}\langle|\alpha_{yz}(\mathbf{q})|^2\rangle\left\{\frac{\omega^2\Gamma_s(\mathbf{q}) + q^2\Gamma''_{11}(\mathbf{q})[\omega_s^2(\mathbf{q}) - \omega^2]}{[\omega_s^2(\mathbf{q}) - \omega^2]^2 + \omega^2\Gamma_s^2(\mathbf{q})}\right\} \tag{12.2.21}$$

It remains to calculate [cf. Eq. (12.2.3)]

$$S_{xy}(\mathbf{q}, \omega) = \frac{1}{2\pi}\int_{-\infty}^{-\infty} dt\, e^{-i\omega t}\langle \delta\alpha_{xy}(\mathbf{q}, 0)\, \delta\alpha_{xy}(\mathbf{q}, t)\rangle \tag{12.2.22}$$

Again because $\delta\alpha_{yx}(\mathbf{q})$ is even under time reversal

$$S_{xy}(\mathbf{q}, \omega) = \pi^{-1} Re\langle \delta\alpha_{xy}^*(\mathbf{q}, 0)\, \delta\tilde{\alpha}_{xy}(\mathbf{q}, s = i\omega)\rangle \tag{12.2.23}$$

Now $\delta\alpha_{xy}(\mathbf{q})$ is odd with respect to reflections through the xz and yz planes, whereas none of the other components of the polarizability tensor and none of the other hydrodynamic variables $\rho(\mathbf{q})$, $v_x(\mathbf{q})$, $v_y(\mathbf{q})$, $v_z(\mathbf{q})$, and the energy density have this symmetry. It follows from Theorems 9–11 of Section 11.5 that $\delta\alpha_{xy}(\mathbf{q})$ is not coupled to these other variables, and is consequently the only slow variable with this symmetry, so that the relaxation equation describing this variable is

$$\frac{d\alpha_{xy}}{dt} = -\Gamma_{22}^{xy}(\mathbf{q})\alpha_{xy} + F_{xy} \tag{12.2.24}$$

which gives

$$S_{xy}(\mathbf{q}, \omega) = \pi^{-1}\langle|\alpha_{xy}(\mathbf{q})|^2\rangle\left\{\frac{\Gamma_{22}^{xy}(\mathbf{q})}{\omega^2 + [\Gamma_{22}^{xy}(\mathbf{q})]^2}\right\} \tag{12.2.25}$$

Note that $\Gamma_{22}^{xy}(\mathbf{q})$ is not the same quantity as $\Gamma_{22}(\mathbf{q})$. In the limit $q \to 0$ we expect that $\Gamma_{22}^{xy}(0) \to \Gamma_{22}(0)$. This can be shown by explicit consideration of microscopic quantities if no long-range forces are present. Because $\omega_s(\mathbf{q})$, $\Gamma_s(\mathbf{q})$, and $\Gamma_{22}^{xy}(\mathbf{q})$ are functions of q, the spectral densities $S_{xy}(\mathbf{q}, \omega)$ and $S_{yz}(\mathbf{q}, \omega)$ are complicated functions of q. These functions simplify considerably for small q, because $\omega_s^2(\mathbf{q})$, $\Gamma_s(\mathbf{q})$, and $\Gamma_{22}^{xy}(\mathbf{q})$ can be replaced by their leading terms in q. In Eqs. (12.2.14) through (12.2.25) and the rest of this chapter we use the convention that when the q argument is omitted, the limiting value at $q = 0$ is understood. Thus, for example, $\Gamma_{22}^{xy}(0)$, $\Gamma_{22}(0)$, and $\Gamma_{11}''(0)$ are written as Γ_{22}^{xy}, Γ_{22}, and Γ_{11}''. Thus for small q we write Eqs. (12.2.21) and (12.2.25) as

$$S_{yz}(\mathbf{q}, \omega) = \pi^{-1}\langle|\alpha_{yz}|^2\rangle \left\{\frac{\omega^2[\Gamma_{22} + q^2\Gamma_{11}''] + q^2\Gamma_{11}''[q^2\alpha^2 - \omega^2]}{[q^2\alpha^2 - \omega^2]^2 + \omega^2\,[\Gamma_{22} + q^2\Gamma_{11}'']^2}\right\} \quad (12.2.26)$$

$$S_{xy}(\mathbf{q}, \omega) = \pi^{-1}\langle|\alpha_{xy}|^2\rangle \left\{\frac{\Gamma_{22}^{xy}}{\omega^2 + [\Gamma_{22}^{xy}]^2}\right\} \quad (12.2.27)$$

where α^2 is the $q \to 0$ limit of $\alpha^2(\mathbf{q})$ defined in Eq. (12.2.18). It should be noted that in the $q \to 0$ limit $\langle|\alpha_{yz}|^2\rangle = \langle|\alpha_{xy}|^2\rangle$, and $\Gamma_{22}^{xy} = \Gamma_{22}$.

The constant α^2 can be related to the shear viscosity η_s and to Γ_{22} in the following way. The autocorrelation function of $A_1 = v_y(q)$ given by Eq. (12.2.16) should reduce in the $q \to 0$ limit to the well-known hydrodynamic result [see Eq. (11.6.19)] for the transverse velocity, $\langle v_y^*(\mathbf{q}, 0)\, \tilde{v}_y(\mathbf{q}, s)\rangle = \langle|v_y|^2\rangle[q^2\nu_s/(s+q^2\nu_s)]$ where ν_s is the kinematic shear viscosity $\nu_s = \eta_s/m\rho_0$. Thus for small q and $s = 0$, $\langle v_y^*(q)\tilde{v}_y(\mathbf{q},\ s = 0)\rangle = \langle|v_y|^2\rangle(q^2\nu_s)^{-1}$. Equation (12.2.61) reduces to this limit, as it must, if[4]

$$\alpha = [\nu_s\Gamma_{22}]^{1/2} = [\eta_s\Gamma_{22}/m\rho_0]^{1/2} \quad (12.2.28)$$

It will prove convenient to define a quantity R such that

$$R \equiv [1 - \Gamma_{11}''/\nu_s] \text{ or } \Gamma_{11}'' = \nu_s(1 - R) \quad (12.2.29)$$

Substitution of Eqs. (12.2.28) and (12.2.29) into Eqs. (12.2.26) and (12.2.27), together with the foregoing identities $\langle|\alpha_{yx}|^2\rangle = \langle|\alpha_{yz}|^2\rangle$ and $\Gamma_{22}^{xy} = \Gamma_{22}$ gives, for the depolarized spectrum [Eq. (12.2.3)]

$$I_{VH}(\mathbf{q}, \omega) = \pi^{-1}\langle|\alpha_{yz}|^2\rangle \left\{\frac{\Gamma_{22}[\omega^2 + (\nu_s q^2)^2(1 - R)] \cos^2 \dfrac{\theta}{2}}{[\omega^2 - q^2\nu_s\Gamma_{22}]^2 + \omega^2[\Gamma_{22} + q^2\nu_s(1 - R)]^2} + \frac{\Gamma_{22}}{\omega^2 + [\Gamma_{22}]^2} \sin^2 \frac{\theta}{2}\right\} \quad (12.2.30)$$

In the limit $q^2\nu_s/\Gamma_{22} \ll 1$, Eq. (12.2.30) reduces to the sum of two Lorentzians

$$I_{VH}(\mathbf{q}, \omega) = \pi^{-1}\langle|\alpha_{yz}|^2\rangle \left\{\frac{\Gamma_{22}}{\omega^2 + \Gamma_{22}^2} - R\left[\frac{\nu_s q^2}{\Gamma_{22}}\right]\frac{q^2\nu_s}{\omega^2 + [q^2\nu_s]^2} \cos^2 \frac{\theta}{2}\right\} \quad (12.2.31)$$

This equation is derived by solving the dispersion equation $s^2 + \Gamma_s(\mathbf{q})s + \omega^2{}_s(\mathbf{q}) = 0$ of Eq. (12.2.16) by perturbation theory to order q^2, and then evaluating the Fourier transform (cf. Section (10.4).

It is clear from Eq. (12.2.31) that for $q^2\nu_s/\Gamma_{22} \ll 1$, the first Lorentzian is broad compared to the second Lorentzian. The corresponding *spectrum therefore consists of a* *broad band of width Γ_{22} subtracted from which is a narrow Lorentzian that leads to a dip or splitting of half-width* $(q^2\nu_s)^{1/2}$. The depth of the dip at $\omega = 0$ is $R\cos^2\theta/2$. It follows that the amplitude of the dip is directly proportional to the constant R whereas the half-width of the splitting is proportional is $q\sqrt{\nu_s}$. Equation (12.2.30) can also lead to a dip, but the analysis is somewhat less transparent. When $R = 0$ both Eqs. (12.2.30) and (12.2.31) reduce to one Lorentzian with half-width Γ_{22}. Thus $I_{VH}(\mathbf{q}, \omega)$ exhibits no doublet when $R = 0$.

The parameter R is so important in this theory of light scattering that it behooves us to interpret its meaning. From the definition of R in Eq. (12.2.29) it follows that $R = (\nu_s - \Gamma''_{11})/\nu_s$. From Eqs. (12.2.28) and (12.2.18) it is clear that $\nu_s = \Gamma''_{11} - |\Lambda'_{12}|^2/\chi_{11}\chi_{12}\Gamma_{22}$. Combining these last two results one obtains

$$R = \frac{|\Lambda'_{12}|^2}{\nu_s\Gamma_{22}\chi_{11}\chi_{22}} \tag{12.2.32}$$

Since $F_1 = QiLv_y(\mathbf{q}) \underset{q\to 0}{=} iq\sigma_{yz}$ where $\sigma_{yz} \equiv \sum_j [F_{yj} + mv_y^j v_z^j]$ is a component of the stress tensor and $F_2 = QiL\alpha_{yz} = \dot{\alpha}_{yz}$ it follows that

$$R = \frac{\beta^2}{\nu_s\Gamma_{22}\chi_{11}\chi_{22}} \left| \int_0^\infty dt \, \langle \sigma_{yz}^*(0) \, \dot{\alpha}_{yz}(t) \rangle_Q \right|^2 \tag{12.2.33}$$

$\dot{\alpha}_{yz}$ is the time rate of change of α_{yz}, which in the $q \to 0$ limit has the explicit form $\propto \sum_l [\dot{u}_y^l u_z^l + u_y^l \dot{u}_z^l]$. Since $\dot{\mathbf{u}}_l = (\boldsymbol{\omega}_l \times \mathbf{u}_l)$, where $\boldsymbol{\omega}_l$ is the angular velocity, we see that the value of R *is determined in part by the dynamic coupling between the shear stress and the angular momentum of the rotors.* Unfortunately there do not exist any explicit calculations of R at this time.

In recent years, experimental investigation of the depolarized Rayleigh scattering of several liquids composed of optically anisotropic molecules has confirmed the existence of a doublet-symmetric about zero frequency change and with a splitting of approximately 0.5 GHz (see Fig. 12.1.1). The existence of this doublet had been predicted on the basis of a hydrodynamic theory several years previously by Leontovich (1941). This theory assumes that local strains set up by a transverse shear wave are relieved by collective reorientation of individual molecules. Later, Rytov (1957) formulated a more general hydrodynamic theory for viscoelastic fluids that reduces to the Leontovich theory in the appropriate limit. The theories of Rytov and Leontovitch are different from the present two-variable theory, in that the *primary variable* is the stress tensor and not the polarizability.

Stegeman (1968) and Stegeman and Stoicheff (1968) have made an extensive experimental study of the doublet in the depolarized light-scattering spectrum for several liquids and claim reasonable agreement with the Rytov theory with the exception that an essential requirement for the validity of the theory does not hold—namely that the shear viscosity η_s, shear modulus μ_s, and shear relaxation time τ must satisfy the relation

$$\frac{\mu_s\tau}{\eta_s} = 1. \tag{12.2.34}$$

Stegeman and Stoicheff found that this ratio was less than unity for all liquids that they studied.

Since the experiments of Stegeman and Stoicheff (1968) several authors have formmulated theories of the doublet in order to resolve this discrepancy and to place the somewhat hazy assumptions of the Leontovich–Rytov theories on a firmer statistical mechanical foundation. Most of these recent theories are based on the methods presented in Chapter 11. This work is reviewed in the recent review article by Fleury and Boon (1973).

Only limited comparisons of these microscopic theories with experiment have been made (cf. Enright, et al., 1972; Andersen and Pecora, 1971). However all these comparisons of theory to experiment are (in terms of the two-variable theory presented here) for liquids at temperatures for which $q^2\nu_s/\Gamma_{22} \ll 1$. In this limit, the general spectral equation reduces to a simple two-Lorentzian form [cf. Eq. (12.2.31)]. In order to test Eq. (12.2.30) in the region where

$$\frac{q^2\nu_s}{\Gamma_{22}} \approx 1 \tag{12.2.35}$$

and to obtain values of the parameters appearing in the theory, Alms, et al. (1973a) have measured the depolarized light-scattering spectrum of anisaldehyde (4-methoxy benzaldehyde) in the temperature range 6–79 °C.

Anisaldehyde was chosen for this study since Γ_{22} is sufficiently small so that values of $(q^2\nu_s/\Gamma_{22}) \approx 1$ could be achieved by performing measurements at appropriate temperatures. Pecora et al. have fit these spectra to the parameters. Γ_{22}, R, the coupling parameter that corresponds to the depth of the dip, and $q^2\nu_s$.

The depolarized light-scattering spectrum of anisaldehyde for $\theta = 90°$ at temperatures of 6–79 °C is shown in Fig. 12.2.1. These spectra were fit to Eq. (12.2.30) with a nonlinear, least-square fitting program. The fitted values of Γ_{22}, R, and $q^2\nu_s$ are listed in Table 12.2.1. The rms errors for these computer fits ranged from. 3 to 1.0%. Standard deviations for Γ_{22} were 2–5%, for $q^2\nu_s$, 3–5%, and for R, 1–3%. Only at the highest

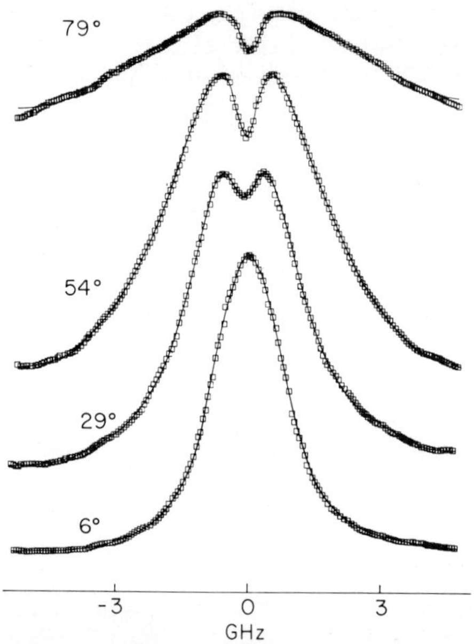

FIG. 12.2.1. Measured depolarized spectrum, I_{VH} (q), of anisaldehyde for $\theta = 90°$ at temperatures of 6, 29, 54 and 79°C. (From Alms et al., 1973, Fig. 2.)

TABLE 12.2.1

Fitted Parameters

T (°C)	Γ_{22}[a]	R	$q^2\eta_s/m\rho_0(\exp)$[a]	rms error %
6.2	0.67	0.43	0.97	0.5
12.2	0.80	0.42	0.77	0.6
16.3	0.86	0.46	0.69	0.6
21.6	1.04	0.45	0.57	0.6
28.8	1.24	0.44	0.49	0.5
32.6	1.38	0.45	0.43	0.4
38.3	1.75	0.45	0.37	0.7
44.7	2.02	0.42	0.33	0.7
53.8	2.28	0.44	0.27	0.3
62.5	2.91	0.42	0.22	0.5
70.6	3.40	0.41	0.22	0.5
79.0	3.93	0.41	0.22	1.0

a. Γ_{22} and $\frac{q^2\eta_s}{m\rho_0}$ are given in GHz.

temperature, $T = 79\,°\mathrm{C}$ were the error limits outside these ranges. At this temperature the standard deviation in $q^2\nu_s$ rose to 7% and the standard deviation in R rose to 7.5%.

The value of $q^2\nu_s$ calculated from the viscosity and density of anisaldehyde at the various temperatures is compared to the value derived from the light-scattering spectra in Table 12.2.2. The accuracy of the fitted parameters Γ_{22}, R, and $q^2\nu_s$ was tested by

TABLE 12.2.2[a]

Comparison of Values of $\frac{q^2\eta_s}{m\rho_0}$ from Light-Scattering, Viscosity, and Density Measurements

T(°C)	η_s	$m\rho_0$	$\frac{q^2\eta_s}{m\rho_0}$ (direct)[c]	$\frac{q^2\eta_s}{m\rho_0}$ (light scattering)
6.2	8.30	1.136	0.99	0.97
12.2	6.60	1.131	0.79	0.77
16.3	5.60	1.127	0.67	0.69
21.6	4.70	1.123	0.57	0.57
28.8	3.95	1.117	0.48	0.49
32.6	3.47	1.113	0.42	0.43
38.3	2.98	1.108	0.36	0.37
44.7	2.62	1.102	0.32	0.33
53.8	2.10	1.094	0.26	0.27
62.5	1.77	1.086	0.22	0.22
70.6	1.50	1.079	0.19	0.22
79.0	1.35	1.072	0.17	0.22

a. $\frac{q^2\eta_s}{m\rho_0}$ is in GHz, η_s in cP, $m\rho_0$ in g/ml.
b. Data from Perjus (1896) and Jaeger (1917).
c. q calculated using 1.60 as the value of the refractive index of anisaldehyde at 4880 Å.

generating theoretical spectra corresponding to the experimental spectra and then using the fitting program to reproduce the values of Γ_{22}, R, and $q^2\nu_s$. The generated theoretical spectra were then numerically convoluted with the digitized instrumental spectrum by computer and then fit for the values of Γ_{22}, R, and $q^2\nu_s$(see Table 12.2.3). In this way the effect of convolution of the actual light-scattering spectrum with the instrumental function was determined. The reported values of Γ_{22}, R, and $q^2\nu_s$ are corrected for the effect of convolution.

TABLE 12.2.3[a]

Effect of Convolution

$\frac{q^2\eta_s}{\rho_0\Gamma_{22}}$	Γ_{22} (GHz)	Γ_{22} con (GHz)	R	R_{con}	$\frac{q^2\eta_s}{m\rho_0}$ (GHz)	$\frac{q^2\eta_s}{m\rho_0}$ con (GHz)
0.065	3.80	3.77	0.40	0.36	0.247	0.295
0.23	1.90	1.90	0.40	0.38	0.437	0.481
0.80	0.95	0.96	0.45	0.44	0.760	0.803

a. The instrumental half-width at half-height = 0.089 GHz. (Finesse = 60.)
b. "con" denotes the value of the parameters after convolution with the instrumental lineshape.

Two criteria are useful in determining the agreement between theory and experiment for this work: (a) the "goodness of fit" of Eq. (12.2.30) to the data, and (b) the agreement between the value of $q^2\nu_s$ derived from the spectra and that determined by the more usual methods of measuring viscosity and density. As shown in Tables 12.2.1 and 12.2.2, both these criteria are met very well. Alms et al. also fit the data to the two Lorentzians form of Eq. (12.2.31). For all spectra the fit is significantly worse than the fit to Eq. (12.2.30). The fit to two Lorentzians does improve at high temperatures where the condition $q^2\nu_s/\Gamma_{22} \ll 1$ is approached, but is never as good as the fit to the general expression [Eq. (12.2.30)].

The only disagreement between Eq. (12.2.30) and experiment occurs for $q^2\nu_s$ at the highest temperature considered (79 °C). At this temperature the theoretical fit to the spectrum also has the highest rms error with the largest standard deviation for both $q^2\nu_s$ and R. This relatively poor fit is probably due to the decreased signal to noise ratio resulting from the fact that the total anisaldehyde scattering decreases with increasing temperature.

The value of R derived from the spectra is independent of temperature. This result is in agreement with the results of Stegeman and Stoicheff for a variety of liquids. Thus, for anisaldehyde and the liquids studied by Stegeman and Stoicheff, R, the fraction of the viscosity due to dynamic coupling between shear stress and the angular momentum, is essentially independent of T. The value of Γ_{22} at 22.5° interpolated from the data is 1.07 ± .04 and is in excellent agreement with Stegeman and Stoicheff's value of 1.05 ± .05 at the same temperature.

We conclude that the "two-variable" light-scattering theory describes the low-frequency depolarized light-scattering spectra of anisaldehyde in the region where $(q^2\nu_s/\Gamma_{22}) \approx 1$ very accurately. Although R has a rigorous statistical mechanical definition and an interesting physical interpretation, it has been used simply as a parameter to fit the experimental data. So far, no calculations of R from the basic definition [Eq. (12.2.33)] have been made for particular fluids. Given the complexity of Eq. (12.2.33) it appears unlikely that any calculations of this type will be forthcoming in the near future. The most that can be expected are empirical correlations of R to some simple molecular parameters (e.g., the molecular geometrical anistropy). As discussed above, R is not very temperature-dependent, therefore temperature studies are not likely to yield useful information. Perhaps solution studies of the concentration-dependence of R would aid in finding such correlations. For instance, recent observations have shown that the doublet for nitrobenzene persists in solution in certain solvents down to 50% concentration (Lucas and Jackson, 1971).

12 · 3 COMPARISON BETWEEN SINGLE-PARTICLE AND COLLECTIVE REORIENTATION TIMES

In Section 12.2 we found that when $R \simeq 0$, the depolarized spectrum is a simple Lorentzian

$$I_{VH}(\mathbf{q}, \omega) = \pi^{-1}\langle|\alpha_{yz}|^2\rangle \left\{\frac{\Gamma_{22}(\mathbf{q})}{\omega^2 + \Gamma_{22}^2(\mathbf{q})}\right\} \tag{12.3.1}$$

where Γ_{22} is the relaxation rate of the variable $\alpha_{yz}(\mathbf{q})$ given by Eq. (12.2.2). From Eq. (12.2.2) it follows that $\alpha_{yz}(\mathbf{q})$ varies in time by virtue of two processes: (a) tumbling of the molecules [through $\mathbf{u}_j(t)$] and (b) translational motions of the molecules [through $\exp iqz_j(t)$]. In most liquids the orientations relax on a much faster time scale than the translational phase factors $\exp iqz_j(t)$. Moreover, the range of static correlations is short compared to q^{-1}. From this it follows that when $R = 0$ the translational phase factors can be omitted entirely from consideration, and from the Mori formalism

$$\Gamma_T \equiv \Gamma_{22} = \int_0^\infty dt\, \langle\dot{\alpha}_{yz}^{T*}\, e^{iQLt}\, \dot{\alpha}_{yz}^T\rangle\langle|\alpha_{yz}^T|^2\rangle^{-1} \tag{12.3.2}$$

where $\alpha_{yz}^T = \sum_j \alpha_{yz}^j = (\alpha_\parallel - \alpha_\perp) \sum_j u_y^j u_z^j$, and where $P = 1 - Q$ is the projector onto α_{yz}^T. Because $\dot{\mathbf{u}}_j = (\boldsymbol{\omega}_j \times \mathbf{u}_j)$, it is clear that Γ_{22} depends on angular momentum decay times—among other things. Γ_T can be regarded as the collective tumbling rate of the N symmetric top molecules in the system. Substitution of $\dot{\alpha}_{yz} = (\alpha_\parallel - \alpha_\perp) \sum_j iL(u_y^j u_z^j)$ into Eq. (12.3.2) shows that the numerator and denominator of Γ_T consist of N terms that involve the same particle index and $N(N - 1)$ terms that involve different particle indices. The former terms we call *single-particle terms,* and the latter terms we call *pair terms.* Thus the light-scattering reorientational relaxation rate is determined not only by the relaxation of "single molecules" but also by the relaxation of "pairs of molecules."

Light scattering is not the only method for determining reorientational relaxation rates. Infrared absorption and Raman band profiles, NMR spin lattice relaxation times, and depolarization of fluorescence are among some of the methods used to study orientational relaxation (see Chapter 15). These methods, in marked contrast to light scattering, determine the *single molecule* relaxation rates[5]

$$\Gamma_S \equiv \Gamma_{22}^{(S)} = \int_0^\infty dt \langle \dot{\alpha}_{yz}^{(1)} \exp iQ_sLt \, \dot{\alpha}_{yz}^{(1)} \rangle \langle |\alpha_{yz}^{(1)}|^2 \rangle^{-1} \tag{12.3.3}$$

where $P_s = 1 - Q_s$ is the projector onto the single particle polarizability $\alpha_{yz}^{(1)} = (\alpha_\parallel - \alpha_\perp) u_y^1 u_z^1$. In the rotational diffusion approximation Γ_s is related to the rotational diffusion coefficient Θ by

$$\Gamma_S = 6\Theta \tag{12.3.4}$$

On intuitive grounds we expect Γ_T for a solution in which the solvent is optically isotropic and the solute is anisotropic to reduce to Γ_S in the limit of infinite dilution. This can be checked by performing light-scattering and NMR measurements on each of a series of solutions of decreasing concentration. The intuitive notion is born out by such experiments, as we shall see.

It should be noted that Γ_T and Γ_S differ in the following respects: (a) projection operators are different, (b) static correlation functions $\langle |\alpha_{yz}^T|^2 \rangle$ and $\langle |\alpha_{yz}^{(1)}|^2 \rangle$ are different, and (c) Γ_S involves only single-particle terms, whereas Γ_T involves pair terms as well. It is of considerable interest to find a relationship between Γ_T and Γ_S. This can be done using the Mori formalism, as we now show.

The two variables of interest are $\alpha_{yz}^{(1)}$ and α_{yz}^T where $\alpha_{yz}^{(1)} = (\alpha_\parallel - \alpha_\perp) u_y^1 u_z^1$ and $\alpha_{yz}^T = (\alpha_\parallel - \alpha_\perp) \sum_j u_y^j u_z^j$. Actually it is more convenient to choose the two variables as

$$\begin{cases} A_1 \equiv \alpha_{yz}^{(1)} \\ A_2 \equiv \alpha_{yz}^T - (\alpha_{yz}^T, \alpha_{yz}^{(1)}) (\alpha_{yz}^{(1)}, \alpha_{yz}^{(1)})^{-1} \alpha_{yz}^{(1)} \end{cases} \tag{12.3.5}$$

We have suppressed the complex conjugate star in these equations because $\alpha_{yz}^{(1)}$ and α_{yz}^T are real quantities. It should be noted that for this choice, A_1 and A_2 are orthogonal quantities; that is, $(A_1, A_2) = 0$, which is the reason for our choice of these variables. It is a simple matter to apply the Mori formalism to these variables. Before pursuing this, we define the quantities

$$f \equiv (\alpha_{yz}^{(1)}, \alpha_{yz}^{(2)}) / \langle |\alpha_{yz}^{(1)}|^2 \rangle \tag{12.3.6a}$$

$$g = \int_0^\infty dt \langle \dot{\alpha}_{yz}^{(1)} e^{iQLt} \dot{\alpha}_{yz}^{(2)} \rangle \Big/ \int_0^\infty dt \langle \dot{\alpha}_{yz}^{(1)} e^{iQLt} \dot{\alpha}_{yz}^{(1)} \rangle \tag{12.3.6b}$$

where $P = 1 - Q$ is the projection operator onto the subspace $\{A_1, A_2\}$. f measures the static correlation of any two distinct particles relative to $\langle |\alpha_{yz}^{(1)}|^2 \rangle$, and g measures the dynamic correlation between any two distinct particles relative to the single-particle dynamic correlation.

The scalar product (A_2, A_2) involves the coefficient f as follows:

$$(A_2, A_2) = (\alpha_{yz}^T, \alpha_{yz}^T) - (\alpha_{yz}^T, \alpha_{yz}^{(1)})^2 (\alpha_{yz}^{(1)}, \alpha_{yz}^{(1)})^{-1} \tag{12.3.7}$$

Now

$$\begin{aligned} (\alpha_{yz}^{(1)}, \alpha_{yz}^T) &= (\alpha_{yz}^{(1)}, \alpha_{yz}^{(1)}) + (N-1)(\alpha_{yz}^{(1)}, \alpha_{yz}^{(2)}) \\ &= \langle |\alpha_{yz}^{(1)}|^2 \rangle [1 + Nf] \end{aligned} \tag{12.3.8}$$

and

$$(\alpha_{yz}^T, \alpha_{yz}^T) = N\langle|\alpha_{yz}^{(1)}|^2\rangle + N(N-1)(\alpha_{yz}^{(1)}, \alpha_{yz}^{(2)})$$
$$= N\langle|\alpha_{yz}^{(1)}|^2\rangle[1 + Nf] \tag{12.3.9}$$

From this it follows that

$$A_2 = \alpha_{yz}^T - (1 + Nf)\alpha_{yz}^{(1)} \tag{12.3.10}$$

so that

$$(A_2, A_2) = \langle|\alpha_{yz}^{(1)}|^2\rangle\{N(1 + Nf) - (1 + Nf)^2\} \tag{12.3.11}$$

where we have taken $N - 1 \simeq N$ because $N \gg 1$. Also, because intermolecular forces are short-ranged, it is reasonable to assume[6] that Nf is of order one or equivalently, that f is of order $1/N$. It follows from this that $N(1 + Nf) \gg (1 + Nf)^2$, because $N \gg 1$ so that to an excellent approximation

$$(A_2, A_2) = \langle|\alpha_{yz}^{(1)}|^2\rangle N(1 + Nf) \tag{12.3.12}$$

Combining these results gives the matrix

$$\boldsymbol{\chi} = \beta(\mathbf{A}, \mathbf{A}^+) = \beta\begin{pmatrix} \langle|\alpha_{yz}^{(1)}|^2\rangle & 0 \\ 0 & \langle|\alpha_{yz}^{(1)}|^2\rangle N(1 + Nf) \end{pmatrix} \tag{12.3.13}$$

Now let us look at the random forces and correspondingly the matrix of kinetic coefficients. Let us first denote

$$\lambda_{11} \equiv \int_0^\infty dt\,\langle\dot{\alpha}_{yz}^{(1)}\, e^{iQLt}\,\dot{\alpha}_{yz}^{(1)}\rangle \equiv \frac{\Lambda_{11}}{\beta} \tag{12.3.14a}$$

$$\lambda_{12} \equiv \int_0^\infty dt\,\langle\dot{\alpha}_{yz}^{(1)}\, e^{iQLt}\dot{\alpha}_{yz}^{(2)}\rangle \tag{12.3.14b}$$

Because $\alpha_{yz}^{(1)}$ is real and has even time reversal symmetry, $\lambda_{12} = \lambda_{21}$. From Eq. (12.3.6b) it follows that

$$g = \lambda_{12}/\lambda_{11} \tag{12.3.15}$$

The kinetic coefficients, Λ_{11}, Λ_{12}, Λ_{21}, and Λ_{22}, can be expressed in terms of λ_{11} and λ_{12}. First we note that

$$\beta\int_0^\infty dt\,\langle\dot{\alpha}_{yz}^{(1)}\, e^{iQLt}\,\dot{\alpha}_{yz}^T\rangle = \beta[\lambda_{11} + N\lambda_{12}] = \beta\lambda_{11}(1 + Ng) \tag{12.3.16}$$

and

$$\beta\int_0^\infty dt\,\langle\dot{\alpha}_{yz}^T\, e^{iQLt}\,\dot{\alpha}_{yz}^T\rangle = \beta[N\lambda_{11} + N(N-1)\lambda_{12}]$$
$$= \beta N\lambda_{11}(1 + Ng) \tag{12.3.17}$$

where again we have used Eq. (12.3.15). It follows that

$$\Lambda_{11} = \beta\lambda_{11}$$

$$\Lambda_{12} = \beta\int_0^\infty dt\,\langle\dot{A}_1\,e^{iQLt}\dot{A}_2\rangle = \beta N\lambda_{11}(g-f) = \Lambda_{21}$$

$$\Lambda_{22} = \beta\int_0^\infty dt\,\langle\dot{A}_2\,e^{iQLt}\dot{A}_2\rangle = \beta[N\lambda_{11}(1+Ng) - (1+Nf)^2\lambda_{11}] \qquad (12.3.18)$$

Again we assume that for short-ranged forces Ng is of order one or equivalently g is of order $1/N$. This allows us to neglect $(1 + Nf)^2$ compared with $N(1 + Ng)$ so that

$$\Lambda_{22} = \beta\lambda_{11}\,N(1+Ng) \qquad (12.3.19)$$

The matrix of kinetic coefficients is to highest order in N

$$\Lambda = \beta\begin{pmatrix}\lambda_{11} & \lambda_{11}N(g-f)\\ \lambda_{11}N(g-f) & \lambda_{11}N(1+Ng)\end{pmatrix} \qquad (12.3.20)$$

The damping matrix can now be evaluated

$$\boldsymbol{\Gamma} = \Lambda\cdot\boldsymbol{\chi}^{-1} = 6\Theta\begin{pmatrix}1 & \dfrac{(g-f)}{(1+Ng)}\\ N(g-f) & \dfrac{(1+Ng)}{(1+Nf)}\end{pmatrix} \qquad (12.3.21)$$

where we define

$$6\,\Theta \equiv \frac{\lambda_{11}}{\langle|\alpha_{yz}^{(1)}|^2\rangle} \qquad (12.3.22)$$

The relaxation equations are thus

$$\begin{cases}\dfrac{dA_1}{dt} = -6\Theta A_1 - 6\Theta\dfrac{(g-f)}{(1+Nf)}A_2 + F_1 & (12.3.23a)\\[2ex] \dfrac{dA_2}{dt} = -6\Theta N(g-f)\,A_1 - 6\Theta\dfrac{(1+Ng)}{(1+Nf)}A_2 + F_2 & (12.3.23b)\end{cases}$$

Solving these equations for the Laplace transforms of the correlation functions gives

$$\langle A_1(0)\tilde{A}_1(s)\rangle = \langle|\alpha_{yz}^{(1)}|^2\rangle\frac{\left[s + 6\Theta\left[\dfrac{1+Ng}{1+Nf}\right]\right]}{\Delta(s)} \qquad (12.3.24a)$$

$$\langle A_2(0)\tilde{A}_2(s)\rangle = \langle|\alpha_{yz}^{(1)}|^2\rangle N(1+Nf)\frac{[s+6\Theta]}{\Delta(s)} \qquad (12.3.24b)$$

where

$$\Delta(s) = [s+6\Theta]\left[s + 6\Theta\left[\frac{1+Ng}{1+Nf}\right]\right] - (6\Theta)^2\left[\frac{N(g-f)^2}{1+Nf}\right] \qquad (12.3.24c)$$

The second term on the right-hand side of $\Delta(s)$ is of order $1/N$ and should therefore be negligible compared with the first term, so that

$$\Delta(s) \cong (s + 6\Theta)\left(s + 6\Theta\left[\frac{1 + Ng}{1 + Nf}\right]\right) \tag{12.3.25}$$

By substituting Eq. (12.3.25) into Eqs. (12.3.24a) and (12.3.24b) and inverse Laplace transforming the resulting equation we obtain

$$\begin{cases} \langle A_1(0)A_1(t)\rangle = \langle|\alpha_{yz}^{(1)}|^2\rangle\, e^{-6\Theta t} & (12.3.26a) \\ \langle A_2(0)A_2(t)\rangle = N(1 + Nf)\langle|\alpha_{yz}^{(1)}|^2\rangle \exp - 6\Theta\left[\dfrac{1 + Ng}{1 + Nf}\right]t & (12.3.26b) \end{cases}$$

Equation (12.3.26a) is the single-particle correlation function and Θ can therefore be regarded as the rotational diffusion coefficient. The light-scattering correlation function is $\langle\alpha_{yz}^T(0)\,\alpha_{yz}^T(t)\rangle$. Since $\alpha_{yz}^T = A_2 + (1 + Nf)A_1$ [cf. Eq. (12.3.10)] we find to highest order in N that

$$\begin{aligned} \langle\alpha_{yz}^T(0)\,\alpha_{yz}^T(t)\rangle &= \langle A_2(0)A_2(t)\rangle \\ &= \langle|\alpha_{yz}^T|^2\rangle \exp - 6\Theta\left[\frac{1 + Ng}{1 + Nf}\right]t \end{aligned} \tag{12.3.27}$$

Comparing Eqs. (12.3.27) and (12.3.26a) it is found that[7]

$$\Gamma_T = \Gamma_S\frac{[1 + Ng]}{[1 + Nf]} \tag{12.3.28}$$

or equivalently in terms of correlation times $\Gamma_T = (\tau_T)^{-1}$ and $\Gamma_S = (\tau_S)^{-1}$

$$\tau_T = \frac{[1 + Nf]}{[1 + Ng]}\tau_S \tag{12.3.29}$$

f is a static correlation function. For a symmetric top molecule it is easy to show that

$$Nf = \left\langle\sum_{i\neq j}\frac{P_2(\mathbf{u}_i \cdot \mathbf{u}_j)}{(N - 1)}\right\rangle \tag{12.3.30}$$

so that f can be positive or negative. If the molecules surrounding a given molecule tend to be oriented parallel to it f is positive, whereas if they tend to be oriented perpendicular to it f is negative. Thus if $g = 0$ we expect

$\tau_T < \tau_S$ for $f < 0$ (perpendicular orientations)

$\tau_T > \tau_S$ for $f > 0$ (parallel orientations)

Of course there is no a priori reason for $g = 0$. Equation (12.3.27) provides a useful interpretation of light-scattering experiments. In principle $(1 + Nf)$ can be determined from integrated intensities.[8] τ_T can be determined from the spectral width; τ_S can be determined from an NMR experiment or other method, and Eq. (12.3.29) can then be used to find g.

These ideas are illustrated by the depolarized light scattering and C–13 NMR spin relaxation experiments of Alms et al. (1973b) on chloroform and chloroform solutions.

Chloroform is a symmetric top molecule so that light scattering measures the reorientation time of the molecular symmetry axis. The C-13 experiments measure the reorientation time of the carbon—hydrogen bond axis, which in this case coincides with the molecular symmetry axis. Thus in the absence of pair correlations between molecules the experiments should give identical reorientation times. Any difference in the results of the two experiments is attributed to the effect of pair correlations, that is, the factors f and g in Eq. (12.3.29).

The light-scattering experiments were performed as a function of concentration at a constant viscosity of .56 centipoise and a constant temperature of 22°C. The solvents used (mixtures of carbon tetrachloride and isopentane) contributed very little to the depolarized scattering at high chloroform concentrations. These solvents, however, contributed approximately 10% to the intensity of the 40% chloroform solution and approximately 20% to the intensity of the 20% chloroform solution. The contribution of the solvent was subtracted for these chloroform solutions. The measured light-scattering reorientation times for the chloroform solutions are shown in Fig. 12.3.1.

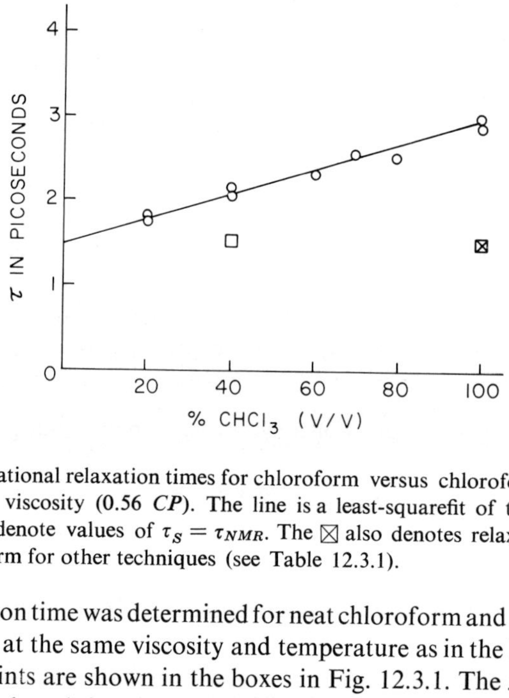

FIG. 12.3.1. Reorientational relaxation times for chloroform versus chloroform concentration at constant viscosity (0.56 *CP*). The line is a least-squarefit of the τ_T data (○). The squares denote values of $\tau_S = \tau_{NMR}$. The ⊠ also denotes relaxation times for neat chloroform for other techniques (see Table 12.3.1).

The NMR relaxation time was determined for neat chloroform and a 40% (by volume) chloroform solution at the same viscosity and temperature as in the light-scattering experiments. These points are shown in the boxes in Fig. 12.3.1. The x in Fig. 12.3.1 indicates that other authors have also determined the single-particle relaxation time for neat chloroform. These values are shown in Table 12.3.1. Note that the values for the single-particle relaxation times for neat chloroform as determined by C-13 nuclear relaxation, Raman band widths, and nuclear quadrupole relaxation are in excellent agreement. The light-scattering time is, however, about twice as long as that measured by these experiments.

From Fig. 12.3.1 we see that the $CHCl_3$ single-particle relaxation time is constant at constant solution viscosity. The light-scattering reorientation time extrapolated to infinite dilution is in excellent agreement with this single-particle time. Thus the concentration dependence of τ_T at higher concentrations is due to the increasing importance

TABLE 12.3.1

Relaxation Times for Chloroform Determined by Various Techniques

	τ_{LS}^{9} (psec)	τ_{RAM} (psec)	τ_{NMR} (psec)	τ_{NQR} (psec)
$CHCl_3$ (neat)	2.95[a]	1.5[b]	1.5,[a] 1.4[c]	1.7[d]
$CHCl_3$ (inf. dilution)	1.5[a]		1.5[a]	

a. Alms et al. (1973b).
b. Bartoli and Litovitz (1972).
c. Farrar et al. (1972).
d. Huntress (1969).

of pair correlations at higher chloroform concentrations. At infinite dilution light scattering yields the single-particle time.

From absolute total intensity light scattering (Malmberg and Lippincott, 1965) measurements the value of the static orientational pair-correlation term fN for neat chloroform is found to be $1.0 \pm .01$. Using this value of fN, the concentration-dependence of τ_T, and the single-particle reorientation time τ_S in Eq. (12.3.29), the value of the dynamic correlation parameter gN is found to be 0.0 ± 0.1 for neat chloroform. This value of gN is also consistent with the value of τ_T for the whole range of solutions studied.

Thus we conclude that the tendency of chloroform molecules to orient with their symmetry axes parallel causes the total (light-scattering) correlation function to relax more slowly than the single-particle correlation function at large chloroform concentrations.

It would indeed be interesting to have measurements of this kind on a wide variety of liquids. If it can be determined that g is usually negligible, as it is in the case for chloroform, then the single-particle and total relaxation times are related simply by the static orientational correlation factor f.

NOTES

1. This form ignores local field or collision-induced contributions and should be regarded as approximately valid.
2. As that found in a solid or a liquid crystal.
3. As $q \to 0$, $\sum_j a_y^j \exp iqz_j \to \sum_j a_y^j$. This sum is zero because in an isolated system the total force is zero.
4. From Eq. (12.2.16) it is clear that $\langle A_1^* (\mathbf{q}, 0)\tilde{A}(\mathbf{q}, s = 0)\rangle = \beta\chi_{11}^{-1}(\mathbf{q}) \dfrac{\Gamma_{22}(\mathbf{q})}{\omega_s^2(\mathbf{q})}$ which in the small q limit is $\langle|v_y|^2\rangle\Gamma_{22}(q^2\alpha^2)^{-1}$. Equating this to $\langle|v_y|^2\rangle(q^2 v_s)^{-1}$ gives Eq. (12.2.28).
5. The absorption and NMR relaxation rates are not precisely given by Eq. (12.3.3) but involve other equivalent quantities.
6. Only near liquid to liquid–crystal phase transitions should this argument be inapplicable.
7. Another method for deriving this result is given by Keyes and Kivelson (1972).
8. From Eq. (12.3.26b) note that $\langle\alpha_{yz}^T(o)\,\alpha_{yz}^T(o)\rangle = N(1 + Nf)\,\langle|\alpha_{yz}^{(1)}|^2\rangle$.
9. τ_{LS} is the relaxation time in a light-scattering experiment and is identical to τ_T in the text.

REFERENCES

Alms, G. R., Bauer, D. R., Brauman, J. I. and Pecora, R., *J. Chem. Phys.*, **59**, 5304 (1973a); **59**, 5310 (1973b).

Andersen, H. C. and Pecora, R., *J. Chem. Phys.* **54**, 2584 (1971).

Bartoli, F. J. and Litovitz, T. A., *J. Chem. Phys.* **56**, 413 (1972).

Enright, G. D., Stegeman, G. I. A., and Stoicheff, B. P., *J. Phys.* (*Paris*) **33**, Cl-207 (1972).

Farrar, T. C., Druck, S. J., Shoup, R. R., and Becker, E. D., *J. Am. Chem. Soc.* **94**, 699 (1972).

Fleury, P· and Boon, J. P. *Adv. Chem. Phys* **24**, 1 (1973)

Huntress, W. T., Jr., *J. Phys. Chem.* **73**, 103 (1969).

Jaeger, F. M., *Z. Anorg. Chem.* **101**, 1 (1917).

Fleury, P. and Boon, J. P. Adv. Chem. Phys **24**, 1 (1973)

Keyes, R. and Kivelson, D., *J. Chem. Phys.* **56**, 1057 (1972).

Leontovich, M. A., *Izv. Acad. Mauk SSR,* Ser. Fiz. **5**, (1941). [*Bull. Acad. Sci. U S S R, Phys. Ser.* **4**, 499 (1941)].

Lucas, H. C. and Jackson, D. A., *Mol. Phys.* **20**, 801 (1971).

Malmberg, M. S. and Lippincott, E. R., *J. Colloid Interface Sci.* **27**, 591 (1965).

Perjus, W. H., *J. Chem. Soc.* (*Lond*). **69**, 1025 (1896).

Rytov, S. M., *Zh. Eksp. Reor. Fiz.* **33**, 166, 514, 671 (1957). [*Sov. Phys. JETP* **6**, 130, 401, 513 (1958)].

Stegeman, G. I. A., Ph. D. Thesis, University of Toronto (1969).

Stegeman, G. I. A. and Stoicheff, B. P., *Phys. Rev. Lett.* **21**, 202 (1968).

CHAPTER 13

NONEQUILIBRIUM THERMODYNAMICS, DIFFUSION, AND ELECTROPHORESIS

13 · 1 INTRODUCTION

In this chapter the formalism of nonequilibrium thermodynamics, is reviewed. This formalism is then applied to the theory of isothermal diffusion and electrophoresis. It is shown that this theory is important in determining the relations between the transport coefficients measured by light scattering and those measured by classical macroscopic techniques. Since much of this material is covered in other chapters, this chapter is very brief. Our presentation closely follows that of Katchalsky and Curran (1965). Other books that can be consulted are those of DeGroot and Mazur (1962) and Prigogine (1955).

13 · 2 THE EQUATION OF ENTROPY BALANCE

The local entropy density, $s(\mathbf{r}, t)$, was used frequently in Chapter 10. The total entropy S of the fluid contained in a region of volume V is then

$$S = \int_V d^3r s(\mathbf{r}, t) \tag{13.2.1}$$

According to the second law of thermodynamics the entropy of a system is not a conserved quantity in an irreversible process. Applying Eq. (10.3.8) to Eq. (13.2.1) then gives

$$\frac{\partial s(\mathbf{r}, t)}{\partial t} + \boldsymbol{\nabla} \cdot \mathbf{J}_s = \sigma(\mathbf{r}, t) \tag{13.2.2}$$

This is the local equation for the entropy density. It is of fundamental importance in what follows. The quantity σ, the local *entropy production,* is the entropy produced irreversibly per unit time per unit volume and is analogous to the property σ_A defined prior to Eq. (10.3.4). The quantity $\mathbf{J}_s$ is the *entropy flux* and $\boldsymbol{\nabla} \cdot \mathbf{J}_s$ represents the rate of change of the local entropy due to an inflow of entropy from neighboring regions of the fluid.

The overall change of entropy in volume V can be expressed as

$$\frac{dS}{dt} = \frac{d_eS}{dt} + \frac{d_iS}{dt} \tag{13.2.3}$$

where d_eS/dt is the rate of change due to entropy exchange with the surrounding fluid and d_iS/dt is the rate of change due to the internal entropy production. These quantities are clearly

$$\frac{d_eS}{dt} = -\int_V d^3r \nabla \cdot \mathbf{J}_S = -\int_V d\mathbf{S} \cdot \mathbf{J}_s \tag{13.2.4}$$

$$\frac{d_iS}{dt} = \int_V d^3r\, \sigma(\mathbf{r}, t) \tag{13.2.5}$$

In an irreversible process the total entropy production as well as the local entropy production must be positive ($\sigma \geqslant 0$).

Thus we see that according to Eq. (13.2.2) the entropy change of a local fluid element is due to the exchange of entropy with neighboring fluid elements and the internal production of entropy in an irreversible process.

13 · 3 CALCULATION OF THE ENTROPY PRODUCTION

The aim of this section is to express σ in terms of "flows" and "thermodynamic forces." Our starting point is the Gibbs equation, according to which changes in the thermodynamic properties of a system are related by

$$TdS = dE + PdV - \sum_{i=1}^{n} \mu_i dn_i \tag{13.3.1}$$

where S, E, P, V, μ_i, and n_i are the *total* entropy, internal energy, pressure, volume, chemical potential of species i, and the number of moles of species i respectively. The integrated form of Eq. (13.3.1) is

$$TS = E + PV - \sum_{i=1}^{n} \mu_i n_i \tag{13.3.2}$$

Dividing this latter equation by V gives an expression for the local densities s, e, and c_i corresponding to S, E, and n_i respectively. This is

$$Ts = e + P - \sum_i \mu_i c_i \tag{13.3.3}$$

where

$$s \equiv \frac{S}{V}; \qquad e = \frac{E}{V}; \qquad c_i = \frac{n_i}{V}$$

Likewise substituting sV for S, eV for E, and c_iV for n_i in Eq. (13.3.1) gives

$$Td\,(Vs) = d\,(Ve) + PdV - \sum_i \mu_i d(Vc_i)$$

which upon rearrangement is

$$V\,(Tds - de + \sum_i \mu_i dc_i) = dV(-Ts + e + p - \sum_i \mu_i c_i) \tag{13.3.4}$$

Now according to Eq. (13.3.3) the right-hand side of Eq. (13.3.4) is zero and we obtain

$$Tds = de - \sum_{i=1}^{n} \mu_i dc_i \tag{13.3.5}$$

This equation relates the change in entropy density to the changes in the internal energy density and the molar concentration. This equation can be alternatively expressed as

$$Tds = dq - \sum_{i=1} \mu_i dc_i \tag{13.3.6}$$

where dq is the increment of "pure heat" per unit volume[1].

One of the important postulates of irreversible thermodynamics is the postulate of local equilibrium discussed in Chapter 10. Accordingly, the local rate of change of the entropy density is

$$T\frac{\partial s}{\partial t} = \frac{\partial q}{\partial t} - \sum_{i=1}^{n} \mu_i \frac{\partial c_i}{\partial t} \tag{13.3.7}$$

The local equations of change (see Section 10.3)

$$\frac{\partial s}{\partial t} + \nabla \cdot \mathbf{J}_s = \sigma \tag{13.3.8a}$$

$$\frac{\partial q}{\partial t} + \nabla \cdot \mathbf{J}_q = 0 \tag{13.3.8b}$$

$$\frac{\partial c_i}{\partial t} + \nabla \cdot \mathbf{J}_i = \bar{\phi}_i = \nu_i J_r \tag{13.3.8c}$$

can be substituted in Eq. (13.3.7) yielding

$$-\nabla \cdot \mathbf{J}_s + \sigma = -\frac{1}{T}\nabla \cdot \mathbf{J}_q - \sum_{i=1}^{n} \frac{\mu_i}{T}(-\nabla \cdot \mathbf{J}_i + \bar{\phi}_i) \tag{13.3.9}$$

where $\bar{\phi}_i$ is the local production of component i per unit volume due to chemical reaction. This can be expressed in terms of the stoichiometric coefficient ν_i and the time derivatives of the extent of reaction. Eq(13.3.9) can be rearranged algebraically[2] to give for σ

$$\sigma = \mathbf{J}_q \cdot \nabla\left(+\frac{1}{T}\right) + \sum_{i=1}^{n} \mathbf{J}_i \cdot \nabla\left(-\frac{\mu_i}{T}\right) + J_r\frac{A}{T} \tag{13.3.10}$$

where

$$A = -\sum_i \nu_i \mu_i$$

is the *affinity* of the chemical reaction.[3]

Equation (11.3.10) shows that σ can be expressed as

$$\sigma = \sum_{\alpha=1}^{m} J_\alpha X_\alpha \tag{13.3.11}$$

where the set J_α are "flows" and the set X_α are "conjugate thermodynamic forces." The choice of the flows determines the conjugate forces. For example, choosing J_q as the heat flow determines $\nabla(1/T)$ as the conjugate force, while choosing the matter flow $\mathbf{J}_i$ determines $\nabla(-\mu_i/T)$ as the conjugate force. The forces in Eq. (13.3.11) are not necessarily linearly independent.

In electrochemical systems these results must be modified somewhat because electrical work has been excluded from our preceeding equations. The total charge carried by a mole of species i is $z_i F$ where z_i is the valence and F is the Faraday (96,500 Coulombs/mole). If the electrical potential at a point is ψ, the electrical work required to increase the number of moles n_i of species i at that point by dn_i is $z_i F \psi dn_i$, and the Gibbs equation (13.3.1) can be written

$$TdS = dE + PdV - \sum_{i=1}^{n} (\mu_i + z_i F \psi)\, dn_i \tag{13.3.12}$$

where we now have in addition to PV work, the electrical work $z_i F \psi dn_i$. Equation (13.3.12) can also be written as

$$TdS = dE + PdV - \sum_i^n \tilde{\mu}_i dn_i \tag{13.3.13}$$

where

$$\tilde{\mu}_i \equiv \mu_i + z_i F \psi \tag{13.3.14}$$

is a quantity called the *electrochemical potential.* Following the same procedure as before we obtain the entropy production in the charged system

$$\sigma = \mathbf{J}_q \cdot \nabla\left(\frac{1}{T}\right) + \sum_{i=1}^{n} \mathbf{J}_i \cdot \nabla\left(-\frac{\tilde{\mu}_i}{T}\right) + \frac{J_r \tilde{A}}{T} \tag{13.3.15a}$$

where

$$\tilde{A} = -\sum \nu_i \tilde{\mu}_i \tag{13.3.15b}$$

so that the chemical potential is to be replaced everywhere by the electrochemical potential.

13 · 4 THE PHENOMENOLOGICAL EQUATIONS

Onsager (1931) used a set of equations that expresses in an explicit manner the linear dependence of the thermodynamic flows on the thermodynamic forces. These equations, known as the *phenomenological equations,* can be expressed as

$$J_\alpha = \sum_{\beta=1}^{n} L_{\alpha\beta} X_\beta \qquad (\alpha, \beta = 1, \ldots, n) \tag{13.4.1}$$

where J_α is the αth flow in the system and X_α is its conjugate force. Where we have already noted that $J_\alpha X_\alpha$ is one contribution to the entropy production σ [cf. Eq. (13.3.11)]. Note that the αth flow can be coupled to the βth force if the "coupling coefficient" $L_{\alpha\beta} \neq 0$. Onsager suggested this "linear law" only for systems sufficiently "close" to equilibrium, where the thermodynamic forces are small. Such linear laws are well known in physics—for example, Ohm's law, Fourier's law, and Fick's law.

Equation (13.4.1) is a matrix equation and can be easily inverted to give $X_\alpha = \sum_\beta R_{\alpha\beta} J_\beta$ where $R_{\alpha\beta}$ are coupling coefficients that give the forces as linear function of the fluxes ($\mathbf{R} = \mathbf{L}^{-1}$). The coefficients $L_{\alpha\beta} = \left(\frac{J_\alpha}{X_\beta}\right)_{X=0} = 0$ is a flow per unit force and has the dimensions of a generalized mobility, whereas $R_{\alpha\beta} = \left(\frac{X_\alpha}{J_\beta}\right)_{J=0}$ is a force per unit flow and has the dimension of a generalized friction (or resistance).

The overall symmetry of the system can be used to show that some coefficients in the L or $\mathbf{R}$ matrices are zero. If, for example, the force X_β is a vector quantity but the flow J_α is a scalar flow, the coefficient $L_{\alpha\beta}$ must be a vector quantity. This is, however, impossible in an isotropic homogeneous system in the absence of external forces. Thus a scalar force cannot induce a vector flow and $L_{\alpha\beta} = 0$. An example is that of a mixture in which there are chemical reactions. According to the above, the chemical affinity, a scalar force, cannot induce a flow of matter $\mathbf{J}_i$ in any particular direction; thus *simultaneous* diffusion and chemical reaction cannot be coupled.

A general statement of this argument is that in an isotropic system flows and forces of different tensorial orders are not coupled. This is known as the *Curie principle*. Systems that are anisotropic often have some elements of symmetry which reduce the number of nonzero coefficients from the maximum of n^2. To prove these relations one must apply the arguments of Chapter 11 involving parity, reflection symmetries, rotational symmetries, and time-reversal symmetries.

Onsager (1931) in his celebrated theorem on the reciprocal relations, was able to show that, as long as the forces and flows appearing in Eq. (13.4.1) are obtained in such a way that Eq. (13.3.11) is valid, and the forces are linearly independent, the phenomenological coefficients $L_{\alpha\beta}$ satisfy the relation

$$L_{\alpha\beta} = L_{\beta\alpha} \tag{13.4.2}$$

that is, the matrix $\mathbf{L}$ is a symmetric matrix. Thus the maximum number of independent phenomenological coefficients is reduced from n^2 to $n(n + 1)/2$. Hence it is important to evaluate the entropy production in choosing the fluxes and forces. As we showed in Chapter 11, Onsager's theorem is based on time-reversal symmetry, that is, on microscopic reversibility.

Substitution of Eq. (13.4.1) into Eq. (13.3.11) gives the entropy production

$$\sigma = \sum_{\alpha\beta} X_\alpha L_{\alpha\beta} X_\beta = \mathbf{X}^+ \cdot \mathbf{L} \cdot \mathbf{X} \tag{13.4.3}$$

where $\mathbf{X}$ is the column vector

$$\mathbf{X} = \begin{pmatrix} X_1 \\ \vdots \\ X_n \end{pmatrix}$$

$\mathbf{X}^+$ is its Hermitian conjugate, and $\mathbf{L}$ is the matrix **L**. For any choice of the magnitudes of the forces **X**, that is, for arbitrary **X**, σ must be positive or zero; that is,

$$\mathbf{X}^+ \cdot \mathbf{L} \cdot \mathbf{X} \geq 0 \tag{13.4.4}$$

This means that **L** must be positive-semidefinite. A matrix **L** is a positive-semidefinite matrix if and only if **L** is Hermitian and

$$L_{\alpha\alpha} \geq 0 \tag{13.4.5a}$$

and

$$\det \mathbf{L} = |\mathbf{L}| \geq 0 \tag{13.4.5b}$$

If **L** is symmetric—and according to Onsager it is—it follows from Eq. (13.4.5b) that

$$L_{\alpha\alpha}L_{\beta\beta} \geq L_{\alpha\beta}^2 \tag{13.4.5c}$$

This gives a bound on the cross coefficients.

It should be noted that Eq. (13.4.2) implies that

$$\left(\frac{J_\alpha}{X_\beta}\right)_{X_\lambda=0} = \left(\frac{J_\beta}{X_\alpha}\right)_{X_\lambda=0}$$

that is, the flux J_α per unit force X_β is identical to the flux J_β per unit force X_α.

13 · 5 ISOTHERMAL DIFFUSION OF UNCHARGED MOLECULES IN A TWO-COMPONENT SYSTEM

In an isothermal isobaric system in which no chemical reactions can occur, no free charges are present, and upon which no external forces act, Eq. (13.3.10) reduces to

$$T\sigma = \sum_{i=1}^{n} \mathbf{J}_i \cdot \nabla(-\mu_i) \tag{13.5.1}$$

where we have introduced the constancy of the temperature T. Now the forces appearing in Eq. (13.5.1) are not independent. According to the Gibbs–Duhem equation

$$SdT - VdP + \sum_{i=1}^{n} n_i d\mu_i = 0 \tag{13.5.2}$$

In an isothermal–isobaric system $dT = 0$, $dP = 0$, and

$$\sum_{i=1}^{n} c_i d\mu_i = 0 \tag{13.5.3}$$

where we have divided Eq. (13.5.2) by V. Thus locally Eq. (13.5.3) is satisfied and the gradients of the chemical potentials are related by

$$\sum_{i=1}^{n} c_i \nabla \mu_i = 0 \tag{13.5.4}$$

This shows that not all the gradients of the chemical potentials are independent. It is convenient to consider the component present in excess as *solvent* (labeled ω) and to express its chemical potential gradient $\nabla \mu_\omega$ in terms of the chemical potential gradients of the remaining components (solutes). Then

$$\nabla(-\mu_\omega) = -\frac{1}{c_\omega} \sum_{i=1}^{n-1} c_i \nabla(-\mu_i) \tag{13.5.5}$$

Combining this with Eq. (13.5.1) gives the entropy production

$$T\sigma = \sum_{i=1}^{n-1} \left(\mathbf{J}_i - \frac{c_i}{c_\omega} \mathbf{J}_\omega \right) \cdot \nabla(-\mu_i) = \sum_{i=1}^{n-1} \mathbf{J}_i{}^d \cdot \nabla(-\mu_i) \tag{13.5.6}$$

Thus for an n component solution there are $n - 1$ independent forces $\nabla(-\mu_i)$ corresponding to the fluxes

$$\mathbf{J}_i{}^d = \mathbf{J}_i - \frac{c_i}{c_\omega} \mathbf{J}_\omega \tag{13.5.7}$$

The physical significance of these fluxes becomes apparent when we substitute $\mathbf{J}_i = c_i \mathbf{V}_i$ and $\mathbf{J}_\omega = c_\omega \mathbf{V}_\omega$ where $\mathbf{V}_i$ and $\mathbf{V}_\omega$ are the local velocities of component i and solvent ω.

$$\mathbf{J}_i{}^d = c_i(\mathbf{V}_i - \mathbf{V}_\omega)$$

Consequently $\mathbf{J}_i^d$ is determined by the velocities of the solutes relative to that of the solvent. In any experiment (including light scattering), it is the absolute fluxes $\mathbf{J}_i$, that is, the fluxes of the solutes relative to some laboratory-fixed coordinate frame that are measured and not the relative flux $\mathbf{J}_i{}^d$. Theory, however, deals with the relative flux $\mathbf{J}_i{}^d$. It will thus be necessary after the calculation to transform $\mathbf{J}_i{}^d$ back to the laboratory-fixed flux. We shall deal with this later.

Let us first consider a *binary solution.* Then $n - 1 = 1$ and Eq. (13.5.6) becomes

$$T\sigma = \mathbf{J}_s{}^d \cdot \nabla(-\mu_s) \tag{13.5.8}$$

where s denotes the solute. Thus the flux is $J_s{}^d$ and the conjugate force is $\nabla(-\mu_s)$. Accordingly we write [cf. Eq. (13.4.1)]

$$\mathbf{J}_s{}^d = \mathrm{L}_s{}^d \nabla(-\mu_s) \tag{13.5.9}$$

where $L_s{}^d$ is the single phenomenological coefficient characterizing diffusion in a two-component system.

The chemical potential of the solute μ_s is a function of T, P, and c_s. Thus

$$d\mu_s = \left(\frac{\partial \mu_s}{\partial T}\right)_{P,c_s} dT + \left(\frac{\partial \mu_s}{\partial P}\right)_{T,c_s} dP + \left(\frac{\partial \mu_s}{\partial c_s}\right)_{P,T} dc_s$$

Thus in an isobaric, isothermal system ($dP = dT = 0$)

$$\nabla \mu_s = \left(\frac{\partial \mu_s}{\partial c_s}\right)_{T,P} \nabla c_s$$

Substitution of this into Eq. (13.5.9) gives

$$\mathbf{J}_s{}^d = -L_s{}^d(\partial \mu_s/\partial c_s)_{T,P} \nabla c_s = -D^d \nabla c_s \tag{13.5.10a}$$

where the diffusion coefficient is

$$D^d = L_s{}^d \left(\frac{\partial \mu_s}{\partial c_s}\right)_{T,P} \tag{13.5.10b}$$

This is the diffusion coefficient in the relative coordinate system.[4]

The next step is to compute $\mathbf{J}_s$; that is, to compute the flux in the laboratory-fixed coordinate system from $\mathbf{J}_s{}^d$. We recall that $\mathbf{J}_i$ is the number of moles of i passing through a unit area per second. Returning now to the n-component solution, if $\bar{V}_i$ is the partial molar volume of i, which is usually independent of $\mathbf{r}$, t, it follows that $\mathbf{J}_i\bar{V}_i$ is the flow of volume due to the flow of i and

$$\mathbf{J}_v = \sum_{i=1}^{n} \mathbf{J}_i \bar{V}_i = 0 \tag{13.5.11a}$$

is the total flow of volume. In diffusion $\mathbf{J}_v = 0$ hence the last equality. Solving this for $\mathbf{J}_\omega$ the solvent flux gives

$$\mathbf{J}_\omega = -\frac{1}{\bar{V}_\omega} \sum_{i=1}^{n-1} \bar{V}_i \mathbf{J}_i = c_\omega \sum_{i=1}^{n-1} \frac{\phi_i}{\phi_\omega c_i} \mathbf{J}_i \tag{13.5.11b}$$

where $\phi_i \equiv c_i \bar{V}_i$ is the volume fraction of the ith component.

Now for the binary solution

$$\mathbf{J}_\omega = -\frac{c_\omega \phi_s}{c_s \phi_\omega} \mathbf{J}_s \tag{13.5.12}$$

From Eqs. (13.5.7) and (13.5.12), it follows that

$$\mathbf{J}_s{}^d = \mathbf{J}_s - \frac{c_s}{c_\omega} \mathbf{J}_\omega = \left(1 + \frac{\phi_s}{\phi_\omega}\right) \mathbf{J}_s \tag{13.5.13}$$

Since $\phi_s + \phi_\omega = 1$ it follows that

$$\mathbf{J}_s = \phi_\omega \mathbf{J}_s{}^d = -\phi_\omega D^d \nabla c_s = -D \nabla c_s \tag{13.5.14}$$

Thus in the laboratory-fixed coordinate system the diffusion coefficient is

$$D = \phi_\omega L_s{}^d \left(\frac{\partial \mu_s}{\partial c_s}\right)_{T,P} = L_s \left(\frac{\partial \mu_s}{\partial c_s}\right)_{T,P} \tag{13.5.15}$$

where $L_s = L_s{}^d \phi_\omega$. In sufficiently dilute solution the volume fraction of solvent ϕ_ω is essentially unity so that to a very good approximation $D^d \cong D$.

Combining Eqs. (13.3.8c) (in the nonreactive case) and (13.5.14) then gives the diffusion equation

$$\frac{\partial c_s}{\partial t} = D\nabla^2 c_s \tag{13.5.16}$$

Substituting $c_s = c_s{}^0 + \delta c_s$ gives a diffusion equation for the fluctuation, which when solved yields

$$\langle \delta c_s^*(\mathbf{q}, 0)\, \delta c_s(\mathbf{q}, t)\rangle = \langle |\delta c_s(\mathbf{q}, 0)|^2\rangle \exp -q^2 Dt \tag{13.5.17}$$

Light-scattering experiments on binary solutions therefore determine D, not D^d, and not k_BT/ζ. In infinitely dilute ideal solutions D should reduce to the self-diffusion coefficient. In Section 11.6 this is demonstrated.

The friction coefficient f_s of a solute molecule can be defined in the following way. The thermodynamic force per mole $\mathbf{X}_s$ acting on the solute induces a flow of the solute which is, in the relative coordinate system,

$$\mathbf{J}_s{}^d = c_s(\mathbf{V}_s - \mathbf{V}_\omega) = L_s{}^d \mathbf{X}_s \tag{13.5.18a}$$

where $\mathbf{X}_s = \nabla(-\boldsymbol{\mu}_s)$. In the steady state the thermodynamic force per molecule of solvent $\mathbf{X}_s/N_0$ is balanced by the frictional force $f_s(\mathbf{V}_s - \mathbf{V}_\omega)$ so that

$$\mathbf{X}_s = N_0 f_s(\mathbf{V}_s - \mathbf{V}_\omega) \tag{13.5.18b}$$

where N_0 is Avogardro's number and f_s is by definition the *friction coefficient*. Combining Eqs. (13.5.18a) and (13.5.18b) gives

$$L_s{}^d = \frac{c_s}{(N_0 f_s)} \tag{13.5.18c}$$

Substitution of this into Eq. (13.5.15) then gives the diffusion coefficient in terms of the friction coefficient

$$D = \frac{\phi_\omega c_s}{N_0 f_s}\left(\frac{\partial \mu_s}{\partial c_s}\right)_{T,P} \tag{13.5.19}$$

It should be noted that $\phi_\omega = (1 - \phi_s)$, c_s, f_s, and $(\partial\mu_s/\partial c_s)$ are quantities that depend on the concentration c_s. In an ideal solution, that is, a solution at infinite dilution ($\phi_\omega \sim 1$)

$$\mu_s \approx \mu_s{}^0 + RT\, ln\, c_s \tag{13.5.20a}$$

and

$$\left(\frac{\partial \mu_s}{\partial c_s}\right) \approx \frac{RT}{c_s} \tag{13.5.20b}$$

Combining this with Eq. (13.5.19) gives

$$D^0 = \frac{k_B T}{\zeta_s} \tag{13.5.21}$$

where ζ_s is the friction coefficient at infinite dilution

$$\zeta_s = \lim_{c_s \to 0} f_s \qquad (13.5.22)$$

Eq. (13.5.21) is the Einstein relation.

Solutions of macromolecules are often sufficiently dilute that Eq. (13.5.21) applies. Moreover for large molecules ζ_s can be computed from hydrodynamics. For a sphere with stick boundary conditions $\zeta_s = 6\pi\eta a_s$. Thus in dilute solutions D^0 and thereby a_s, the particle radius, can be determined (see Chapters 5 and 8). Since D^0 depends on the temperature and the solvent, it is important to report the data in a standardized manner. Usually the measurements are performed at room temperature and are extrapolated to inifinite dilution. Thus for example the notation $D^0_{20,\omega}$ denotes the diffusion coefficient of the solute at 20°C in the solvent H_2O extrapolated to infinite dilution. For nonideal solutions

$$\mu_s = \mu_s{}^0 + RT \ln (y_s c_s) \qquad (13.5.23a)$$

where y_s is the molar activity coefficient of the solute. Clearly $y_s \to 1$ as $c_s \to 0$ and $\mu_s{}^0$ is the chemical potential of a hypothetical one-molar ideal solution. Then

$$\left(\frac{\partial \mu_s}{\partial c_s}\right) = \frac{RT}{c_s}\left[1 + c_s\left(\frac{\partial \ln y_s}{\partial c_s}\right)\right] \qquad (13.5.23b)$$

and this quantity has a complicated dependence on the solute concentration.

Now it can be shown (Tanford, 1961) that

$$B_2 = \lim_{c_s \to 0}\left(\frac{\partial \ln y_s}{\partial c_s}\right)_{T,P} \qquad (13.5.24a)$$

is the second virial coefficient of the osmotic pressure. In a nonelectrolyte solution, B_2 is equal to the molar-excluded voulme of the solute, which for rigid spherical particles is (Tanford, 1961)

$$B_2 = 8\bar{V}_s \qquad (13.5.24b)$$

For arbitrary geometry we write

$$B_2 = K_t \bar{V}_s \qquad (13.5.24c)$$

Thus to first order in c_s Eq. (13.5.24b) can be expressed as

$$\left(\frac{\partial \mu_s}{\partial c_s}\right) = \frac{RT}{c_s}[1 + K_t \phi_s] \qquad (13.5.25)$$

where we recall $\phi_s = \bar{V}_s c_s$ is the volume fraction of the solute.

The friction constant f_s is also a function of the concentration, and at low concentrations can also be expanded in powers of c_s. To first order in c_s we find

$$\frac{(1 - \phi_s)}{f_s} = \frac{1}{f_s{}^0}[1 - K_f \phi_s] \qquad (13.5.26)$$

where K_f is an empirical constant that can be obtained in principle from sedimentation experiments in which f_s is determined.

Combining Eqs. (13.5.19), (13.5.25), (13.5.26), and $\phi_\omega = 1 - \phi_s$, we find to first order in c_s (or equivalently ϕ_s) that for rigid spherical molecules ($K_t = 8$)

$$D = \frac{k_B T}{f_s^0}[1 + (8 - K_f)\phi_s] + 0(c_s^2) \tag{13.5.27}$$

There have been several theoretical treatments directly applicable to a determination of K_f, or spherical molecules. The predicted values of K_f are

$$K_f = \begin{cases} 6.55 \text{ (Batchelor, 1972)} \\ 6.86 \text{ (Burgers, 1941, 1942)} \\ 7.2 \text{ (Pyun and Fixman, 1964)} \end{cases} \tag{13.5.28}$$

Thus since $K_f \sim 7$ we expect that the first-order term (in c_s) will be very small, and D should show a much smaller concentration-dependence than the sedimentation coefficient or equivalently f_s. In fact D might either increase or decrease with c_s, depending on which theory is applied.

It is important to note that Eqs. (13.5.24b) and (13.5.27) do not apply to electrolyte solutions because $(\partial\mu_s/\partial c_s)$ for an electrolyte depends not only on the volume of the particle but also on its charge. We discuss this important case in Section 13.6.

In general, to first order in c_s

$$D = \frac{k_B T}{f_s^0}[1 + K_D\phi_s] + 0(c_s^2) \tag{13.5.29}$$

There have been several experiments interpreted in terms of this "virial expansion." Herbert and Carlson (1971, 1972) found for the muscle protein myosin that $K_D = 0$. Pusey, et al. (1972) found that in solutions of R-17 viruses K_D could be positive, negative, or zero depending on NaCl concentration (see Fig. 13.5.1). Raj and Flygare (1974) have systematically studied the dependence of the diffusion coefficient of BSA on pH, BSA concentration, and ionic strength. Their data for pH = 3.1 are shown in Fig. 13.5.2. It should be noted that the slope and therefore K_D increases rapidly as the ionic strength is lowered. This undoubtedly arises from the fact that at low ionic strength, neighboring BSA molecules are not "screened" from each other, whereas at high ionic strength they are. In this latter case the macroions should behave like uncharged rigid spheres, in which case $K_D \sim 0$. From their detailed analysis, Raj and Flygare were able to conclude that BSA when highly charged expands due to electrostatic repulsions between its charges. Needless to say, in interpreting the data in these experiments it is important to realize that $\bar{V}_s$ is the partial molar volume of the solute in the solution under study and not in any other solution.

Newman et al. (1974) have recently studied the concentration-dependence of the diffusion coefficient and the sedimentation coefficient in a highly monodisperse solution of the single-stranded circular DNA from the *fd Bacteriophage*. Their results are shown in Fig. 13.5.3. From these data it is possible to determine the coefficients in the expression

$$D = D^\circ[1 + (K_t - K_f)\phi_s] \tag{13.5.30}$$

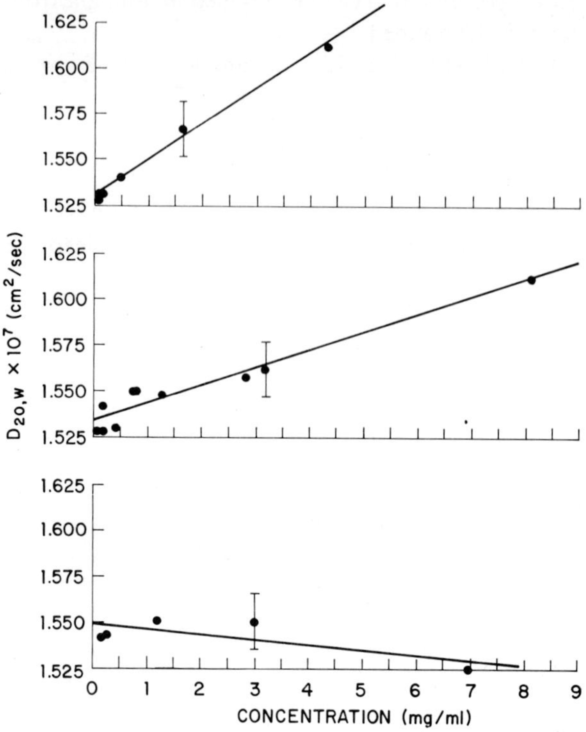

FIG. 13.5.1. Diffusion coefficient $D_{20,\omega}$ of R-17 virus as a function of virus concentration in 1, 0.15, and 0.015M NaCl. (From Pusey, et al., 1972.)

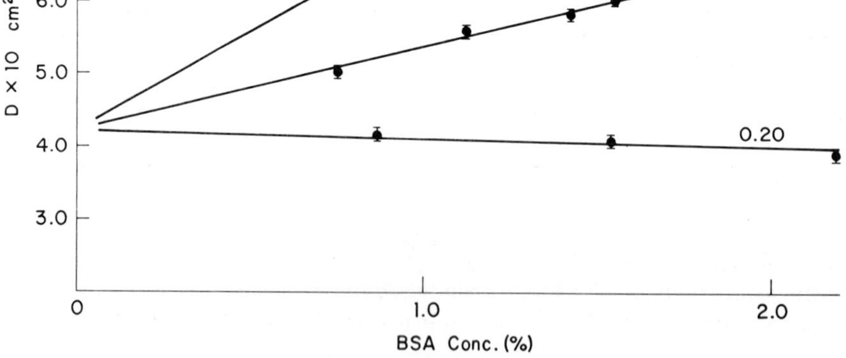

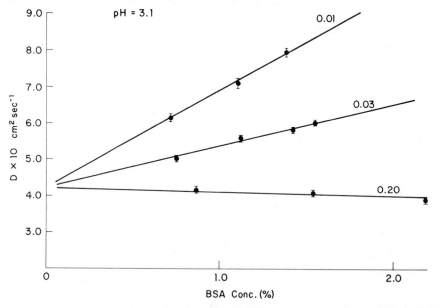

FIG. 13.5.2. Dependence of D on BSA concentration at pH 3.1 and at ionic strengths (= 0.01, 0.03, 0.20. (From Raj and Flygare, 1974.)

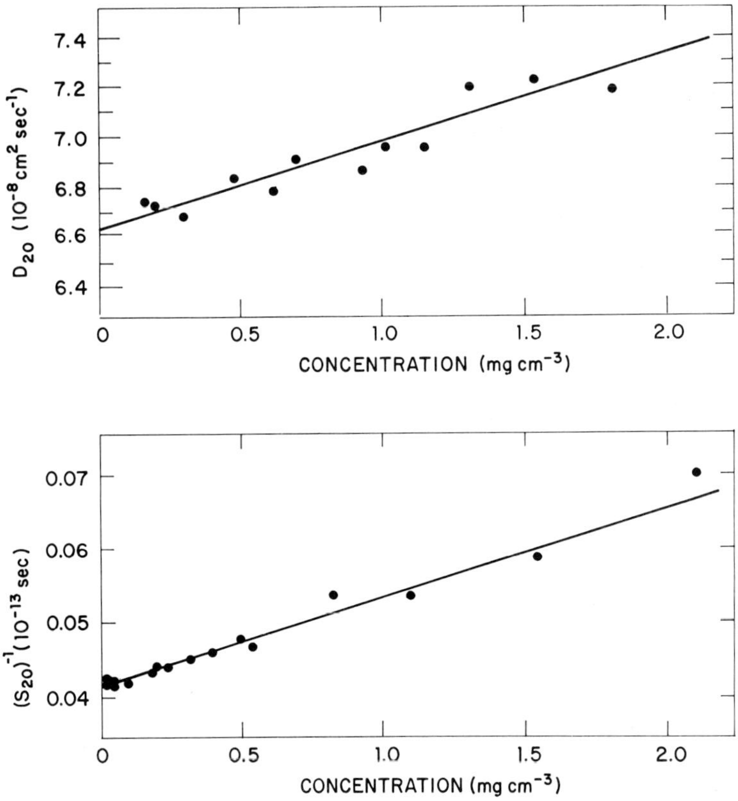

FIG. 13.5.3. The diffusion coefficient D_{20} and the sedimentation coefficient S_{20} at 20°C for a solution of fd DNA as a function of concentration. (From Newman and Swinney, 1974.)

These results are given in Table 13.5.1.

TABLE 13.5.1[a]

Dimensionless Virial Coefficients (from Newman et. al (1974))

Virial	Hard sphere model	Experiment
K_s (sedimentation)	6.55[b]	6.7 ± 0.8
K_t (thermodynamic)	8	7.9 ± 2.6[c]
		7.6 ± 3.9[d]
K_t-K_s (diffusion)	1.45[e]	1.2 ± 0.4
	1.3–1.9[f]	

a. Errors represent 95% confidence limits.
b. Batchelor (1972).
c. Deduced from measured values of K_s and K_t-K_s.
d. From equilibrium sedimentation (Berkowitz and Day, 1974).
e. From Batchelor's value of K_s.
f. Altenberger and Deutch (1973) (for the range of angles studied in these experiments.)

The above experiments involve systems that are, strictly speaking, multicomponent electrolyte solutions. The theory presented in this section is thus not entirely appropriate to the analysis of these results. Unfortunately, as we show in subsequent sections, the appropriate expressions are sufficiently complicated that ingenuity must be exercised in the analysis of the data. Nevertheless we expect that, at the high ionic strengths at which some of the above experiments were performed, the macroions are effectively screened by the counterions and salts, and that the above theory is appropriate.

In conclusion we would like to mention the article by Altenberger and Dutch (1973) which discusses the contribution of the hydrodynamic interaction (Oseen interaction) between solute particles to the concentration dependence of the diffusion coefficient. This treatment yields $K_D = 1$ for small spheres if the ϕ_ω is introduced into the treatment.

13 · 6 ISOTHERMAL DIFFUSION IN AN UNCHARGED MULTICOMPONENT SYSTEM

In an n-component system Eq. (13.5.4) becomes

$$T\sigma = \sum_{i=1}^{m} \mathbf{J}_i{}^d \cdot \nabla(-\mu_i) \tag{13.6.1}$$

where $m = n - 1$ and the force corresponding to the flux $\mathbf{J}_i{}^d$ is $\nabla(-\mu_i)$. Thus according to Onsager [Eq. (13.4.1)] we can write

$$\mathbf{J}_i{}^d = -\sum_{j=1}^{m} L_{ij}^d \nabla\mu_j \tag{13.6.2}$$

where $L_{ij}^d = L_{ji}^d$. The chemical potentials can again be eliminated in favor of the concentrations so that

$$\nabla\mu_j = \sum_{k=1}^{m} \left[\frac{\partial\mu_j}{\partial c_k}\right] \nabla c_k \tag{13.6.3}$$

Combining Eqs. (13.6.2) and (13.6.3) gives

$$\mathbf{J}_i{}^d = -\sum_k D_{ik}^d \nabla c_k \tag{13.6.4a}$$

where the diffusion coefficients are

$$D_{ik}^d = \sum_j L_{ij}^d \left(\frac{\partial\mu_j}{\partial c_k}\right) \tag{13.6.4b}$$

This equation shows that a gradient in concentration c_k can induce a flow of component i. D_{ij}^d is called a cross-diffusion coefficient. Equation (13.6.4a) can be expressed in matrix form

$$\mathbf{J}^d = -\mathbf{D}^d \cdot \nabla\mathbf{c} \tag{13.6.5}$$

where $\mathbf{D}^d$ is the $m \times m$ matrix of diffusion coefficients whose elements are given by

Eq. (13.6.4b), **c** is a column vector of the m concentrations, and $\mathbf{J}^d$ is a column vector of the fluxes $\mathbf{J}_1^d \ldots \mathbf{J}_m^d$.

To complete the picture we must find $\mathbf{J}_i$ from $\mathbf{J}_i^d$. Combining Eqs. (13.5.7) and (13.5.11b) gives

$$\mathbf{J}_i^d = \mathbf{J}_i - \frac{c_i}{c_\omega}\mathbf{J}_\omega = \mathbf{J}_i + \sum_{j=1}^{m} \frac{c_i\phi_j}{c_j\phi_\omega}\mathbf{J}_j \tag{13.6.6}$$

This gives $\mathbf{J}_i^d$ as a linear combination of the fluxes $\mathbf{J}_j$. This can also be expressed as

$$\mathbf{J}_i^d = \sum_j T_{ij}J_j \tag{13.6.7a}$$

where

$$T_{ij} = (1 + \phi_i/\phi_\omega)\,\delta_{ij} + (1 - \delta_{ij})\frac{c_i\phi_j}{c_i\phi_\omega} \tag{13.6.7b}$$

or in matrix form

$$\mathbf{J}^d = \mathbf{T}\cdot\mathbf{J} \tag{13.6.7c}$$

where **T** is the matrix in Eq. (13.6.7b). Thus multiplying Eq. (13.6.7c) from the left by the inverse of **T** gives the desired **J**.

$$\mathbf{J} = \mathbf{T}^{-1}\cdot\mathbf{J}^d \tag{13.6.8}$$

Substituting Eq. (13.6.5) into Eq. (13.6.8) then gives the desired laboratory-fixed fluxes

$$\mathbf{J} = -(\mathbf{T}^{-1}\cdot\mathbf{D}^d)\cdot\nabla\mathbf{c} = -\,\mathbf{D}\cdot\nabla\mathbf{c} \tag{13.6.9}$$

where the laboratory-fixed diffusion coefficient is

$$\mathbf{D} = (\mathbf{T}^{-1}\cdot\mathbf{D}^d) \tag{13.6.10a}$$

or

$$D_{ik} = \sum_j(\mathbf{T}^{-1})_{ij}D_{jk}^d \tag{13.6.10b}$$

Equation (13.6.9) in component form is thus

$$\mathbf{J}_i = -\sum_j^m D_{ij}\nabla c_j \qquad i = 1, \ldots, n-1 \tag{13.6.11}$$

Substitution of Eq. (13.6.11) into Eq. (13.6.8c) then gives the coupled-diffusion equations

$$\frac{\partial c_i}{\partial t} = \sum_{j=1}^{m} D_{ij}\nabla^2 c_j \qquad i = 1, \ldots, m \tag{13.6.12}$$

These equations can be solved by the Fourier–Laplace transform techniques introduced in Chapter 6. Instead of doing this we shall specialize to the case of a three-component solution. Then from Eq. (13.6.7b) we find

$$\mathbf{T} = \begin{pmatrix} \left(1 + \dfrac{\phi_1}{\phi_\omega}\right) & \dfrac{c_1\phi_2}{c_2\phi_\omega} \\ \dfrac{c_2\phi_1}{c_1\phi_\omega} & \left(1 + \dfrac{\phi_2}{\phi_\omega}\right) \end{pmatrix} \tag{13.6.13a}$$

$$\mathbf{T}^{-1} = \begin{pmatrix} (1 - \phi_1) & \dfrac{-c_1\phi_2}{c_2} \\ -\dfrac{c_2\phi_1}{c_1} & (1 - \phi_2) \end{pmatrix} \tag{13.6.13b}$$

and

$$\det \mathbf{T} = \frac{1}{\phi_\omega} \tag{13.6.13c}$$

Thus from Eq. (13.6.10a) we find the diffusion coefficients that appear in Eq. (13.6.12)

$$\mathbf{D} = \begin{pmatrix} \left[(1 - \phi_1) D_{11}{}^d - \dfrac{c_1\phi_2}{c_2} D_{21}{}^d\right] & \left[(1 - \phi_1)D_{12}{}^d - \dfrac{c_1\phi_2}{c_2} D_{22}{}^d\right] \\ \left[-\dfrac{c_2\phi_1}{c_1} D_{12}{}^d + (1 - \phi_2)D_{21}{}^d\right] & \left[-\dfrac{c_2\phi_1}{c_1} D_{12}{}^d + (1 - \phi_2)D_{22}{}^d\right] \end{pmatrix} \tag{13.6.13d}$$

Hence the diffusion coefficients are in general quite complicated. In dilute solutions $\phi_1 \approx \phi_2 \approx 0$ and $\mathbf{D} \approx \mathbf{D}^d$.

It is important to recognize that these equations have been derived for an isothermal–isobaric system. In the event that there are temperature or pressure gradients, the equations are more complicated. In this case there is a coupling between the concentrations and other hydrodynamic modes. We have investigated this coupling for binary solutions in Section 10.6. The formalism of nonequilibrium thermodynamics enables this to be done systematically for any number of components.

13 · 7 ELECTROLYTE SOLUTIONS

It is possible to describe solutions of electrolytes in much the same way that we considered nonelectrolytes in the preceding sections.

An electrolyte solution in thermodynamic equilibrium is locally in an overall state of electroneutrality; that is, there are equal numbers of opposite charges in the neighborhood of any point in the fluid. This means that

$$\sum_i z_i F c_i{}^0 = 0 \tag{13.7.1}$$

where the superscript zero denotes the equilibrium concentration. It is well to remember that fluctuations may occur that produce local deviations from electroneutrality, but such fluctuations are strongly hindered by the strong electrostatic restoring forces, that is, the forces of attraction between oppositely charged ions.

In the following we consider the strong electrolyte $A_{\nu A}B_{\nu B}$ which fully dissociates in solution into ions of valence z_A and z_B

$$A_{\nu_A}B_{\nu_B} \rightleftarrows \nu_A A + \nu_B B \tag{13.7.2}$$

so that $c_A = \nu_A c_s$ and $c_B = \nu_B c_s$ where c_s is the salt concentration. The condition of electroneutrality then gives at equilibrium

$$z_A c_A{}^0 + z_B c_B{}^0 = (\nu_A z_A + \nu_B z_B) c_s = 0 \tag{13.7.3a}$$

In the general case of a salt dissociating into many ionic species

$$\sum_i z_i \nu_i = 0 \tag{13.7.3b}$$

It is clear from Eqs. (13.3.13) and (13.3.15a) that the results of Section 13.5 apply where μ_i is replaced by the electrochemical potential $\tilde{\mu}_i$ of Eq. (13.3.14). It follows that

$$\mathbf{J}_i{}^d = -\sum_j L_{ij}^d \nabla \tilde{\mu}_j \qquad i = 1, 2 \tag{13.7.4a}$$

where the reciprocal relations are

$$L_{ij}^d = L_{ji}^d \tag{13.7.4b}$$

and where the subscripts 1 and 2 refer respectively to the ions of types A and B. There are thus two ion fluxes $\mathbf{J}_1{}^d$ and $\mathbf{J}_2{}^d$. The solution can be regarded as a ternary mixture consisting of the uncharged solvent (which is usually H_2O) and ions of types A and B.

The ion fluxes in the laboratory-fixed coordinate system are evaluated using Eq. (13.6.8) and Eq. (13.7.4), which give for an isothermal, isobaric, nonchemically reacting system

$$\mathbf{J}_i = -\sum_k L_{ik} \nabla \tilde{\mu}_k \tag{13.7.5}$$

where

$$L_{ik} = \sum_j (\mathbf{T}^{-1})_{ij} L_{jk}{}^d \tag{13.7.6a}$$

For the ternary system the matrix $\mathbf{L}$ can be found using Eq. (13.6.13b)

$$\mathbf{L} = \begin{pmatrix} (1-\phi_1) & -\dfrac{c_1\phi_2}{c_2} \\ -\dfrac{c_2\phi_1}{c_1} & (1-\phi_2) \end{pmatrix} \begin{pmatrix} L_{11}^d & L_{12}^d \\ L_{21}^d & L_{22}^d \end{pmatrix}$$

This gives

$$\mathbf{L} = \begin{pmatrix} \left[(1-\phi_1) L_{11}^d - \dfrac{c_1\phi_2}{c_2} L_{21}^d\right] & \left[(1-\phi_1) L_{12}^d - \dfrac{c_1\phi_2}{c_2} L_{22}^d\right] \\ \left[-\dfrac{c_2\phi_1}{c_1} L_{11}^d + (1-\phi_2) L_{21}^d\right] & \left[(1-\phi_2) L_{22}^d - \dfrac{c_2\phi_1}{c_1} L_{12}^d\right] \end{pmatrix} \tag{13.7.6b}$$

It should be noted that

$$L_{ij} \neq L_{ji} \tag{13.7.6c}$$

that is, the reciprocal relations only apply to the phenomenological coefficients defined with respect to the relative fluxes.

Substitution of Eq. (13.3.14) together with the electric field

$$\mathbf{E} = -\nabla\psi \tag{13.7.7}$$

into Eq. (13.7.5) gives the absolute fluxes as

$$\mathbf{J}_i = -\sum_k L_{ik}\nabla\mu_k + \sum_k (L_{ik}z_kF)\mathbf{E} \tag{13.7.8}$$

Before using Eq. (13.7.8) to compute the correlation functions of concentration fluctuations it is useful to review its more standard applications.

The electrical conductance of electrolyte solutions is measured under isothermal, isobaric conditions with uniform concentration throughout the cell, in which case $\nabla\mu_k = 0$ and Eq. (13.7.8) becomes

$$\mathbf{J}_i = (\sum_k L_{ik}z_kF)\mathbf{E} \tag{13.7.9a}$$

It is important to remember that $\mathbf{J}_i$ is the molar flux of species i.

The electric current due to all the ionic species is consequently

$$\mathbf{I} = \sum_i z_iF\mathbf{J}_i = z_1F\mathbf{J}_1 + z_2F\mathbf{J}_2 \tag{13.7.9b}$$

where z_iF is the charge carried by a mole of species i and $z_iF\mathbf{J}_i$ is the flux of charge. Substitution of Eq. (13.7.9a) into Eq. (13.7.9b) gives Ohm's law

$$\mathbf{I} = \kappa\mathbf{E} \tag{13.7.10a}$$

where κ is the electrical conductance of the solution.

$$\kappa = [\sum_{ik} z_iL_{ik}z_k]F^2 \tag{13.7.10b}$$

From Eq. (13.7.9b) we note that the fractions of the current carried by species 1 and 2 are respectively

$$t_1 = \frac{z_1 FJ_1}{I} = \frac{\sum_k z_1L_{1k}z_k}{\sum_{ik} z_iL_{ik}z_k} \tag{13.7.11a}$$

$$t_2 = \frac{z_2FJ_2}{I} = \frac{\sum_k z_2L_{2k}z_k}{\sum_{ik} z_iL_{ik}z_k} \tag{13.7.11b}$$

where the second equalities follow from Eqs. (13.7.9a) and (13.7.10b). Obviously

$$t_1 + t_2 = 1 \tag{13.7.11c}$$

The quantities t_1 and t_2 are called *transference numbers*.

It should be noted that $\mathbf{J}_i = c_i^0 \mathbf{V}_i$ so that the steady-state velocity of species i follows from Eq. (13.7.9a)

$$\mathbf{V}_i = \left[\sum_k \frac{L_{ik} z_k}{c_i^0}\right] F\mathbf{E} \tag{13.7.12a}$$

Comparison with Eq. (5.8.13) then gives the *mobility* u_i for species i

$$u_i = \left[\frac{\sum_k L_{ik} z_k F}{c_i^0}\right] \tag{13.7.12b}$$

in terms of the phenomenological coefficients.

It should be noted that the current in the relative frame I^d is

$$\mathbf{I}^d = \sum_i z_i F \mathbf{J}_i^d = \mathbf{I} \tag{13.7.13}$$

Substituting Eqs. (13.5.7) and (13.7.3a) gives the second equality. Thus we conclude that the electrical current is the same in the relative and laboratory-fixed coordinate systems. Substituting Eq. (13.7.4a) into Eq. (13.7.13) with $\nabla \mu_j = 0$ and comparing the result with Eq. (10.7.10) leads to the conclusion that

$$\kappa = \sum_{ik} z_i L_{ik} z_k F^2 = \sum_{ik} z_i L_{ik}^d z_k F^2 = \kappa^d \tag{13.7.14}$$

so that the conductivity κ^d calculated in the relative coordinate system is identical to that in the space-fixed system.

Transference numbers t_1^d and t_2^d ($t_1^d + t_2^d = 1$) in the relative coordinate system are $t_i^d \equiv z_i F J_i^d / I$. Substitution of Eq. (13.5.7) then gives

$$t_1^d = t_1 - \frac{z_1 c_1}{c_\omega} F \left(\frac{J_\omega}{I}\right) = t_1 \tag{13.7.15a}$$

$$t_2^d = t_2 - \frac{z_2 c_2}{c_\omega} F \left(\frac{J_\omega}{I}\right) = t_2 \tag{13.7.15b}$$

where the last equality follows from the fact that in a conductance experiment there is no net transport of solvent and $\mathbf{J}_\omega = 0$. Thus

$$t_i^d = \frac{\sum_k z_i L_{ik}^d z_k}{\sum_{i,k} z_i L_{ik}^d z_k} = t_i \tag{13.7.15c}$$

In a diffusion experiment, there is no electric field and consequently no electric current; that is,

$$\mathbf{I} = \sum_i z_i F \mathbf{J}_i = 0$$

or (13.7.16)

$$\sum_i z_i \mathbf{J}_i = 0$$

Upon substituting the fluxes from Eq. (13.7.8) into Eq. (13.7.13) with $\mathbf{E} = 0$, it follows that

$$\sum_{ik} z_i L_{ik} \nabla \mu_k = 0 \tag{13.7.17a}$$

This gives $\nabla\mu_1$ in terms of $\nabla\mu_2$.

$$\nabla\mu_1 = -\left[\frac{z_1 L_{12} + z_2 L_{22}}{z_1 L_{11} + z_2 L_{21}}\right]\nabla\mu_2 \tag{13.7.17b}$$

The chemical potential of the neutral salt is

$$\mu_s = \nu_1\mu_1 + \nu_2\mu_2$$

so that

$$\nabla\mu_s = \nu_1\nabla\mu_1 + \nu_2\nabla\mu_2 \tag{13.7.18}$$

Eliminating $\nabla\mu_2$ from this equation by using Eq. (13.7.17b) gives $\nabla\mu_1$ in terms of $\nabla\mu_s$. Substitution of the resulting expressions into Eq. (13.7.8) (with $\mathbf{E} = 0$) then gives

$$\mathbf{J}_1 = \frac{z_1 z_2}{\nu_2}\left(\frac{L_{11}L_{22} - L_{12}L_{21}}{z_1^2 L_{11} + z_1 z_2(L_{12} + L_{21}) + z_2^2 L_{22}}\right)\nabla\mu_s \tag{13.7.19a}$$

$$\mathbf{J}_2 = \frac{z_1 z_2}{\nu_1}\left(\frac{L_{11}L_{22} - L_{12}L_{21}}{z_1^2 L_{11} + z_1 z_2(L_{12} + L_{21}) + z_2^2 L_{22}}\right)\nabla\mu_s \tag{13.7.19b}$$

The flow of the neutral salt is clearly

$$\mathbf{J}_s = \frac{1}{\nu_1}\mathbf{J}_1 = \frac{1}{\nu_2}\mathbf{J}_2 = \frac{z_1 z_2}{\nu_1\nu_2}\left(\frac{L_{11}L_{22} - L_{12}L_{21}}{z_1^2 L_{11} + z_1 z_2(L_{12} + L_{21}) + z_2^2 L_{22}}\right)\nabla\mu_s \tag{13.7.19c}$$

But in a diffusion experiment

$$\mathbf{J}_s = -D\nabla c_s \tag{13.7.19d}$$

so that

$$D = -\frac{z_1 z_2}{\nu_1\nu_2}\left(\frac{L_{11}L_{22} - L_{12}L_{21}}{z_1^2 L_{11} + z_1 z_2(L_{12} + L_{21}) + z_2^2 L_{22}}\right)\left(\frac{\partial\mu_s}{\partial c_s}\right) \tag{13.7.20}$$

It is of interest to examine the relationship between the transport coefficients in the relative and laboratory-fixed coordinate systems. According to Eqs. (13.7.14) and (13. 7.15c), $\kappa = \kappa^d$ and $t_i = t_i{}^d$ are identical in both coordinate systems, and are thus determined directly by L_{ik}^d. On the other hand D as given by Eq. (13.7.20) has not yet been explicitly given in terms of the L_{ik}^d. This is simply done as follows. First note from Eqs. (13.7.6a) and (13.7.6b) that

$$\begin{aligned} \det \mathbf{L} = L_{11}L_{22} - L_{12}L_{21} &= (\det \mathbf{T}^{-1})\cdot(\det \mathbf{L}^d) \\ &= \phi_\omega[L_{11}^d L_{22}^d - L_{12}^{d2}] \end{aligned} \tag{13.7.21}$$

where in the last equation the reciprocal relation Eq. (13.7.4b) and Eq. (13.6.13c) has been used. Substituting Eqs. (13.7.19) and (13.7.21) into Eq. (13.7.20) gives

$$D = \phi_\omega D^d \tag{13.7.22a}$$

where

$$D^d = -\left(\frac{z_1 z_2}{\nu_1 \nu_2}\right)\left(\frac{L_{11}^d L_{22}^d - L_{12}^{d2}}{\sum_{ik} z_i L_{ik}^d z_k}\right)\left(\frac{\partial \mu_s}{\partial c_s}\right) \tag{13.7.22b}$$

is the diffusion coefficient in the relative coordinate system. Now Eqs. (13.7.14), (13.7.15c), and (13.7.22) relate three independent transport coefficients to three independent phenomenological coefficients L_{11}^d, L_{22}^d, and L_{12}^d. Solving for the L_{ij}^d in terms of κ, D, t_1, t_2 gives

$$\mathbf{L}^d = \begin{pmatrix} \left[\dfrac{\nu_1^2 D}{\phi_\omega\left(\dfrac{\partial \mu_s}{\partial c_s}\right)} + \kappa\left(\dfrac{t_1}{z_1 F}\right)^2\right] & \left[\dfrac{\nu_1 \nu_2 D}{\phi_\omega\left(\dfrac{\partial \mu_s}{\partial c_s}\right)} + \dfrac{\kappa t_1 t_2}{z_1 z_2 F^2}\right] \\ \left[\dfrac{\nu_1 \nu_2 D}{\phi_\omega\left(\dfrac{\partial \mu_s}{\partial c_s}\right)} + \dfrac{\kappa t_1 t_2}{z_1 z_2 F^2}\right] & \left[\dfrac{\nu_2^2 D}{\phi_\omega\left(\dfrac{\partial \mu_s}{\partial c_s}\right)} + \kappa\left(\dfrac{t_2}{z_2 F}\right)^2\right] \end{pmatrix} \tag{13.7.23}$$

Substitution of these elements into Eq. (13.7.6b) then gives the laboratory-fixed phenomenological coefficients

$$L_{11} = (1-\phi_1)\left(\frac{\nu_1^2 D}{\phi_\omega\left(\dfrac{\partial \mu_s}{\partial c_s}\right)} + \kappa\left(\frac{t_1}{z_1 F}\right)^2\right) - \frac{c_1\phi_2}{c_2}\left(\frac{\nu_1\nu_2 D}{\phi_\omega\left(\dfrac{\partial \mu_s}{\partial c_s}\right)} + \frac{\kappa t_1 t_2}{z_1 z_2 F^2}\right) \tag{13.7.24a}$$

$$L_{12} = (1-\phi_1)\left(\frac{\nu_1\nu_2 D}{\phi_\omega\left(\dfrac{\partial \mu_s}{\partial c_s}\right)} + \frac{\kappa t_1 t_2}{z_1 z_2 F^2}\right) - \frac{c_1\phi_2}{c_2}\left(\frac{\nu_2^2 D}{\phi_\omega\left(\dfrac{\partial \mu_s}{\partial c_s}\right)} + \kappa\left(\frac{t_2}{z_2 F}\right)^2\right) \tag{13.7.24b}$$

$$L_{21} = (1-\phi_2)\left(\frac{\nu_1\nu_2 D}{\phi_\omega\left(\dfrac{\partial \mu_s}{\partial c_s}\right)} + \frac{\kappa t_1 t_2}{z_1 z_2 F^2}\right) - \frac{c_2\phi_1}{c_2}\left(\frac{\nu_1^2 D}{\phi_\omega\left(\dfrac{\partial \mu_s}{\partial c_s}\right)} + \frac{\kappa t_1 t_2}{z_1 z_2 F^2}\right) \tag{13.7.24c}$$

$$L_{22} = (1-\phi_2)\left(\frac{\nu_2^2 D}{\phi_\omega\left(\dfrac{\partial \mu_s}{\partial c_s}\right)} + \kappa\left(\frac{t_2}{z_2 F}\right)^2\right) - \frac{c_2\phi_1}{c_1}\left(\frac{\nu_1\nu_2 D}{\phi_\omega\left(\dfrac{\partial \mu_s}{\partial c_s}\right)} + \frac{\kappa t_1 t_2}{z_1 z_2 F^2}\right) \tag{13.7.24d}$$

In the limit of infinitely dilute solutions $\phi_1 \simeq \phi_2 = 0$ and $\phi_\omega = 1$. In this case $\mathbf{L} = \mathbf{L}^d$. Measurements of κ, t_1, t_2, and D have been used to study the concentration-dependence of L_{ij}. It is usually found that at low concentrations L_{11} and L_{22} are linear functions of c_s whereas L_{12} is a higher order function of c_s which rapidly goes to zero as $c_s \to 0$ (see Harned and Owen, 1950).

In the limit of infinite dilution the mobility defined by Eq. (13.7.12c) will be independent of the concentration

$$u_1{}^0 = \lim_{c_s \to 0} \sum_k \left[\frac{L_{ik} z_k}{c_s}\right] F \simeq \lim_{c_s \to 0} \frac{L_{ii}}{c_s} z_i \tag{13.7.25}$$

where we have used the above discussion to eliminate $L_{12}/c_s \to 0$ and the superscript 0 indicates infinite dilution. Likewise in this limit the solution is ideal so that

$$\mu_s = \mu_s{}^0 + \nu_1 RT \ln \nu_1 c_s + \nu_2 RT \ln \nu_2 c_s$$

and

$$\left(\frac{\partial \mu_s}{\partial c_s}\right) = \frac{RT}{c_s}(\nu_1 + \nu_2)$$

Substitution of this into D^0, the infinite dilution limit of Eq. (13.7.20) gives after some rearrangement

$$D^0 = -\frac{(\nu_1 + \nu_2)\, RT}{\nu_1 z_1 F} \frac{u_1^0 u_2^0}{(u_1^0 - u_2^0)} \tag{13.7.26}$$

Now if the mobilities are given by the Brownian motion limits

$$u_1^0 = \frac{z_1 F D_1}{RT} \qquad u_2^0 = \frac{z_2 F D_2}{RT} \tag{13.7.27}$$

where D_1 and D_2 are the Stokes–Einstein diffusion coefficients of 1 and 2 respectively, it follows that

$$D^0 = \left[\frac{(\nu_1 + \nu_2)\, D_1^0 D_2^0}{\nu_2 D_1^0 + \nu_1 D_2^0}\right] \tag{13.7.28}$$

For a 1–1 electrolyte

$$D^0 = \frac{2 D_1^0 D_2^0}{D_1^0 + D_2^0} \tag{13.7.29}$$

If one of the ions, say 1, is a macromolecule that is highly charged such that $\nu_2 D_1^0 \gg \nu_1 D_2^0$.

$$D^0 \to \frac{(\nu_1 + \nu_2)}{\nu_2} D_2^0 \simeq D_2^0 \tag{13.7.30}$$

and it should be possible to observe much larger diffusion coefficients for polymers than would be expected on the basis of their radii.

These infinite dilution results are similar to those derived in Chapter 9. However we are now able to calculate the light-scattering spectra in more general circumstances.

13 · 8 ELECTROPHORETIC FLUCTUATION THEORY

In this section the foregoing analysis is applied to the analysis of concentration fluctuations (Freidhof and Berne, 1975). This section covers essentially the same ground as Section 9.2, but from the point of view of nonequilibrium thermodynamics. There are several different conclusions.

Our starting point is Eqs. (13.7.8) and (13.3.8c) (with no reaction). Expanding $\nabla \mu_k$ in terms of the molar concentration gives $\nabla \mu_k = \sum_l (\partial \mu_k / \partial c_l) \nabla c_l$. When this is substituted into Eq. (13.7.8) the fluxes become

$$\mathbf{J}_i = -\sum_l D_{il}\nabla c_l + u_i c_i \mathbf{E} \qquad i = 1, 2 \tag{13.8.1a}$$

where

$$D_{il} \equiv \sum_k L_{ik}\left(\frac{\partial \mu_k}{\partial c_l}\right) \tag{13.8.1b}$$

and

$$u_i \equiv \frac{1}{c_i}\sum_k L_{ik} z_k F = \frac{\kappa t_i}{(z_i F c_i)} \tag{13.8.1c}$$

where the last equation follows from Eqs. (13.7.10b) and (13.7.11a). These quantities depend on the concentration, and as we have seen before only in the limit of infinite dilution is u_i independent of the concentration.

Substitution of Eq. (13.8.1) into Eq. (13.3.8c) gives the generalized diffusion equation of electrophoresis

$$\frac{\partial c_i}{\partial t} = \sum_l \nabla \cdot D_{il}\nabla c_l - \nabla \cdot (u_i c_i \mathbf{E}) \qquad i = 1, 2 \tag{13.8.2}$$

This equation—the primary equation of this section—is treated along the lines proposed in Section 9.3.

The field $\mathbf{E}$ in Eq. (13.8.2) consists of a homogeneous external part $\mathbf{E}_0$ and an internal part $\mathbf{E}_l$, that is,

$$\mathbf{E} = \mathbf{E}_0 + \mathbf{E}_l \tag{13.8.3a}$$

where $\mathbf{E}_l$ satisfies the Poisson equation

$$\nabla \cdot \mathbf{E}_l = \frac{4\pi}{\varepsilon}\sum_{i=1}^{2} z_i F c_i \tag{13.8.3b}$$

As before, we express

$$c_i = c_i^0 + \delta c_i(\mathbf{r}, t) \tag{13.8.4}$$

Substitution of this into Eq. (13.8.3b) and using Eq. (13.7.3a) it follows that $\mathbf{E}_l$ is a linear functional of δc_i. Substitution of Eqs. (13.8.3a) and (13.8.4) into Eq. (13.8.2) and taking note of the linear dependence of $\mathbf{E}_l$ on δc_i we find to linear order in δc_i that Eq. (13.8.2) becomes

$$\frac{\partial \delta c_i}{\partial t} = \sum_l\left(D_{il}\nabla^2\delta c_l - \left[\frac{\partial(u_i c_i)}{\partial c_l}\right]_0 \mathbf{E}_0 \cdot \nabla\delta c_l\right) - u_i^0 c_i^0\, \nabla \cdot \mathbf{E}_l \tag{13.8.5}$$

where[5]

$$\nabla(u_i c_i) = \sum_l \left[\frac{\partial(u_i c_i)}{\partial c_l}\right]_0 \nabla\delta c_l \tag{13.8.6}$$

has been substituted.

The spatial Fourier transform of Eq. (13.8.5) is

$$\frac{\partial \delta c_i(\mathbf{q}, t)}{\partial t} = -\sum_l \left[q^2 D_{il} - i\omega_{il}(\mathbf{q})\right]\delta c_l(\mathbf{q}, t) - u_i^0 c_i^0 (\nabla \cdot \mathbf{E})_q \tag{13.8.7a}$$

where the frequencies $\omega_{il}(\mathbf{q})$ are by definition

$$\omega_{il}(\mathbf{q}) \equiv \frac{\partial(u_i c_i)}{\partial c_l} (\mathbf{q} \cdot \mathbf{E}_0) \tag{13.8.7b}$$

Substitution of the Fourier transform of Eq. (13.8.3b) into Eq. (13.8.7a) then gives

$$\frac{\partial \delta c_i(\mathbf{q}, t)}{\partial t} = - \sum_l [q^2 D_{il} + \lambda_{il} - i\omega_{il}(\mathbf{q})]\, \delta c_l(\mathbf{q}, t) \tag{13.8.8a}$$

where

$$\lambda_{il} \equiv \frac{4\pi}{\varepsilon} (z_l F)\, (c_i{}^0 u_i{}^0) = \frac{4\pi}{\varepsilon} F^2 (\sum_k L_{ik} z_k z_l) \tag{13.8.8b}$$

Comparison of Eqs. (13.8.8b), (13.7.10b), and (10.7.11) gives

$$\lambda_{il} = \left(\frac{4\pi}{\varepsilon}\right) \kappa (z_i{}^{-1} t_i z_l) \tag{13.8.8c}$$

Equation (13.8.8a) can be expressed as a matrix equation

$$\frac{\partial \boldsymbol{\delta} \mathbf{c}(\mathbf{q}, t)}{\partial t} = - \mathbf{M}(\mathbf{q}) \cdot \boldsymbol{\delta} \mathbf{c}(\mathbf{q}, t) \tag{13.8.9a}$$

where

$$\boldsymbol{\delta} \mathbf{c}(\mathbf{q}, t) = \begin{pmatrix} \delta c_1(\mathbf{q}, t) \\ \delta c_2(\mathbf{q}, t) \\ \vdots \\ \delta c_m(\mathbf{q}, t) \end{pmatrix} \tag{13.8.9b}$$

and $\mathbf{M}$ is an $m \times m$ matrix

$$\mathbf{M}_{il}(\mathbf{q}) = q^2 D_{il} + \lambda_{il} - i\omega_{il}(\mathbf{q}) \tag{13.8.9c}$$

Solution of the Laplace transformation of Eq. (10.8.10) for $\delta \mathbf{c}(\mathbf{q}, s)$ gives

$$\delta \tilde{\mathbf{c}}(\mathbf{q}, s) = (s\mathbf{I} + \mathbf{M}(\mathbf{q}))^{-1} \cdot \delta \mathbf{c}(\mathbf{q}, 0) \tag{13.8.10a}$$

Now we specialize to the case of two ionic species. Then

$$(s\mathbf{I} + \mathbf{M}(\mathbf{q})) = \begin{pmatrix} s + q^2 D_{11} + \lambda_{11} - i\omega_{11} & q^2 D_{12} + \lambda_{12} - i\omega_{12} \\ q^2 D_{21} + \lambda_{21} - i\omega_{21} & s + q^2 D_{22} + \lambda_{22} - i\omega_{22} \end{pmatrix} \tag{13.8.10b}$$

where

$$\begin{aligned} \Delta(s) = \det(s\mathbf{I} + \mathbf{M}(\mathbf{q})) = [s + q^2 D_{11} + \lambda_{11} - i\omega_{11}]\,[s + q^2 D_{22} + \lambda_{22} - i\omega_{22}] \\ - [q^2 D_{21} + \lambda_{21} - i\omega_{21}]\,[q^2 D_{12} + \lambda_{12} - i\omega_{12}] \end{aligned} \tag{13.8.10c}$$

and

$$(s\mathbf{I} + \mathbf{M}(\mathbf{q}))^{-1} = \frac{1}{\Delta(s)}\begin{pmatrix} [(s + q^2D_{22} + \lambda_{22} - i\omega_{22}] & [-(q^2D_{12} + \lambda_{12} - i\omega_{12})] \\ [-(q^2D_{21} + \lambda_{21} - i\omega_{21})] & [(s + q^2D_{11} + \lambda_{11} - i\omega_{11}] \end{pmatrix} \tag{13.8.10d}$$

As before, (Section 9.3) we require the roots of the dispersion equation

$$\Delta(s) = 0$$

or

$$\begin{aligned} &s^2 + [q^2D_{11} + q^2D_{22} + \lambda_{11} + \lambda_{22} - i\omega_{11} - i\omega_{22}]\, s \\ &+ q^4[D_{11}D_{22} - D_{12}D_{21}] + [\lambda_{11}\lambda_{22} - \lambda_{12}\lambda_{21}] - [\omega_{11}\omega_{22} - \omega_{12}\omega_{21}] \\ &+ q^2[\lambda_{22}D_{11} + \lambda_{11}D_{22} - \lambda_{12}D_{21} - \lambda_{21}D_{12}] \\ &- iq^2[\omega_{22}D_{11} + \omega_{11}D_{22} - \omega_{12}D_{21} - \omega_{21}D_{12}] \\ &- i[\lambda_{11}\omega_{22} + \lambda_{22}\omega_{11} - \lambda_{12}\omega_{21} - \lambda_{21}\omega_{12}] = 0 \end{aligned} \tag{13.8.11}$$

A few of the terms in this sum are zero. Moreover the term $\omega_{11}\omega_{22} - \omega_{12}\omega_{21}$ is second order in $\mathbf{E}_0$ and can therefore be neglected.

Actually we require the roots to second order in q and to first order in the applied electric field $\mathbf{E}_0$. Eq. (13.8.11) simplifies considerably because the matrix $\boldsymbol{\lambda}$ defined by Eq. (13.8.8c) is

$$\boldsymbol{\lambda} = \left(\frac{4\pi}{\varepsilon}\kappa\right)\begin{pmatrix} t_1 & \dfrac{1}{z_1}t_1z_2 \\ \dfrac{1}{z_2}t_2z_1 & t_2 \end{pmatrix} \tag{13.8.12a}$$

and has the determinant

$$\det \boldsymbol{\lambda} = \lambda_{11}\lambda_{22} - \lambda_{12}\lambda_{21} = \left(\frac{4\pi}{\varepsilon}\kappa\right)^2 [t_1t_2 - t_2t_1] = 0 \tag{13.8.12b}$$

Thus the fourth term in Eq. (13.8.11) vanishes, and to second order in q and first order in $\mathbf{E}_0$, Eq. (13.8.11) becomes

$$\begin{aligned} &s^2 + [q^2D_{11} + q^2D_{22} + \lambda_{11} + \lambda_{22} - i\omega_{11} - i\omega_{22}] \\ &+ q^2[\lambda_{22}D_{11} + \lambda_{11}D_{22} - \lambda_{12}D_{21} - \lambda_{21}D_{12}] \\ &- i[\lambda_{22}\omega_{11} + \lambda_{11}\omega_{22} - \lambda_{12}\omega_{21} - \lambda_{21}\omega_{12}] = 0 \end{aligned} \tag{13.8.13}$$

Remembering that λ_{ij} is zero order in q, ω_{ij} is first order in q, and q^2D_{ij} is second order in q, we can apply the perturbation theory used in preceding sections (e.g., Section 9.3) to obtain the roots to order q^2. There are

$$s_{\pm} = \begin{cases} -\Gamma_f - q^2D_f + i\omega_f(\mathbf{q}) & \text{(fast)} \\ -q^2D_s + i\omega_s(\mathbf{q}) & \text{(slow)} \end{cases} \tag{13.8.14a}$$

where

$$D_s = \left[\frac{\lambda_{11}D_{22} + \lambda_{22}D_{11} - \lambda_{12}D_{21} - \lambda_{21}D_{12}}{\lambda_{11} + \lambda_{22}}\right] \tag{13.8.14b}$$

$$\omega_s(\mathbf{q}) \equiv \left[\frac{\lambda_{11}\omega_{22} + \lambda_{22}\omega_{11} - \lambda_{12}\omega_{21} - \lambda_{21}\omega_{12}}{\lambda_{11} + \lambda_{22}}\right] \tag{13.8.14c}$$

$$D_f \equiv D_{11} + D_{22} - D_s \tag{13.8.14d}$$

$$\omega_f(\mathbf{q}) = \omega_{11}(\mathbf{q}) + \omega_{22}(\mathbf{q}) - \omega_s(\mathbf{q}) \tag{13.8.14e}$$

$$\Gamma_f = \lambda_{11} + \lambda_{22} = \frac{4\pi}{\varepsilon}\kappa \tag{13.8.14f}$$

where the last equality follows from Eq. (13.8.8). Thus the "ionic relaxation rate" Γ_f is proportional to the conductance of the solution κ.

It is important to calculate $\omega_s(\mathbf{q})$ and D_s explicitly. Substitution of Eqs. (13.8.8c), (13.8.7b), and (13.8.1c) for λ_{il}, $\omega_{il}(\mathbf{q})$, and u_i respectively into Eq. (13.8.14c) gives, after some trivial algebraic manipulation and using Eq. (13.7.3a) and,

$$\frac{\partial}{\partial c_s} = \frac{\partial c_1}{\partial c_s}\frac{\partial}{\partial c_1} + \frac{\partial c_2}{\partial c_s}\frac{\partial}{\partial c_2} = \nu_1\frac{\partial}{\partial c_1} + \nu_2\frac{\partial}{\partial c_2}$$

$$\omega_s(\mathbf{q}) = \frac{1}{2}(\mathbf{q}\cdot\mathbf{E}_0)\,\kappa\left[\frac{1}{\nu_1 z_1 F}\left(\frac{\partial t_1}{\partial c_s}\right) + \frac{1}{\nu_2 z_2 F}\left(\frac{\partial t_2}{\partial c_s}\right)\right] \tag{13.8.15a}$$

Likewise we find that

$$D_s = D \tag{13.8.15b}$$

where D is the diffusion coefficient, Eq. (13.7.20) determined by a macroscopic diffusion experiment. Thus D_s measured in a light-scattering experiment is simply the diffusion coefficient measured in a macroscopic experiment, as we expect.

Equation (13.8.10a) can be inverted to give the concentration-fluctuation time correlation function, $F_{ij}(\mathbf{q}, t) \equiv \langle\delta c_i^*(\mathbf{q}, 0)\delta c_j(\mathbf{q}, t)\rangle$. We find for the small q limit that F_{ij} is a superposition of two exponential decays

$$\exp -[\Gamma_f + q^2 D_f - i\omega_f(\mathbf{q})]|t| \text{ and } \exp -[q^2 D_s - i\omega_s(\mathbf{q})]|t|$$

These decay on fast and slow time scales, respectively. The fast decay is determined by the conductance of the solution through the quantity $\Gamma_f = (4\pi/\varepsilon)\kappa$, whereas the slow decay occurs in the absence of an applied field on the time scale $(q^2 D_s)^{-1}$. In the presence of a field the electrophoretic shift $\omega_s(\mathbf{q})$ is given by Eq. (13.8.15a). Pursuing the analysis given at the end of Section 13.7 it is easy to show in the limit of infinite dilution that,

$$\lim_{c_s \to 0} \omega_s(\mathbf{q}) = 0$$

that is, no Doppler shift should be observed at very low concentration. However at finite concentrations $\omega_s(\mathbf{q})$ should be observable. This should be compared with the simple theory given in Chapter 9 where it was shown that for $q \ll q_0$, ω_s is always zero.

It is not very difficult to extend the analysis given here to ternary solutions. This is not presented because of its algebraic complexity. It should also be noted that we have assumed in this fluctuation theory that the pressure and temperature are uniform and constant in time. It is clear that this is not the case. It is also clear that the formalism is sufficiently tractable that these fluctuations could have been included. The result would have been a set of coupled hydrodynamic and diffusion equations much like those in Section 10.6 with the important difference that there are three components here, there is an applied field E_0, and there is a local field to be incorporated. The resulting spectrum should display Doppler shifts due to the applied field (electrophoresis) as well as shifts due to the sound modes. There would also be a coupling between diffusion and thermal conduction.

It should also be noted that this section represents a novel application of irreversible thermodynamics to systems with long-range forces. The local field has been dealt with self-consistently. In the macroscopic theory of Section 13.7 local electroneutrality was imposed through Eq. (13.7.10), whereas in the fluctuation theory there is no constraint of electroneutrality. However because we applied Eq. (13.8.3b) we see that deviations from local electroneutrality decay on the time scale Γ_f^{-1}. This is the "ionic relaxation time." In Section 9.4 only an approximate theory was presented.

This theory has not yet been applied to the study of macroions in solution by light scattering. One difficulty in its application is that the theory applies only to systems of the form

$$A_{\nu_A}B_{\nu_B} \rightleftarrows \nu_A A + \nu_B B$$

Thus only two types of ions may be present. It is difficult in practice to achieve this condition for macroions although it is expected that experiments of this type will be performed within the next several years. The extension of the theoretical approach given above to solutions containing ions of many types is algebraically very complex, but suitable approximations made in the general equations should yield useful results for the study of these complicated and important systems.

NOTES

1. See Chapter 7 Katchalsky and Curran (1965).
2. This is found-by substituting into Eq. (13.3.9) $\frac{1}{T}\,\nabla\cdot\mathbf{J}_q = \nabla\cdot\left(\frac{\mathbf{J}_q}{T}\right) - \mathbf{J}_q\cdot\nabla\left(\frac{1}{T}\right)$ and

   $\frac{\mu_i}{T}\nabla\cdot\mathbf{J}_i = \nabla\cdot\left(\frac{\mu_i\mathbf{J}_i}{T}\right) - \mathbf{J}_i\cdot\nabla\left(\frac{\mu_i}{T}\right)$ This gives

   $$\sigma = -\sum_i \frac{\nu_i\mu_i}{T}J_r + \nabla\cdot J_s - \nabla\cdot\left[(\mathbf{J}_q - \sum_i \mu_i\mathbf{J}_i)\frac{1}{T}\right] + \mathbf{J}_q\cdot\nabla\left(\frac{1}{T}\right) + \sum_i \mathbf{J}_i\cdot\nabla\left(\frac{\mu_i}{T}\right)$$

   Both sides must be scalar quantities. This means that $\mathbf{J}_s = \dfrac{\mathbf{J}_q - \sum_i \mu_i\mathbf{J}_i}{T}$
3. We are considering only a single chemical reaction here. If there are many reactions in which i participates we must sum over all of these reactions.
4. This simplifies considerably for an ideal solution where $\mu_s = \mu_s^0 + RT\ln c_s$ so that $(\partial\mu_s/\partial c_s) = RT/c_s$

   $$D^d = RT\,L_s/c_s$$

   This means that $L_s = c_s D^d/RT = c_s u_s$ where u_s is the solute mobility. Thus we see that L_s depends on concentration rather directly. It should also be noted that this expression has the form given in Eq. (11.4.5) with $(\partial c_s/\partial\mu_s)$ being the susceptibility.

5. This follows from the fact that the mobility u_i is generally a function of the concentrations of the ions. The diffusion coefficients are also functions of the concentrations but to first order in δc_i, $\nabla \cdot D_{il} \nabla \delta c_l = D_{il} \nabla^2 \delta c_l$.

REFERENCES

Altenberger, A. R. and Deutch, J. M., *J. Chem. Phys.* **59,** 894 (1973).

Berkowitz, S. A. and Day, L. A., Biochem. **13,** 4825 (1974).

Batchelor, G. K., *J. Fluid Mech.* **52,** 243 (1972).

Burgers, J. M., *Proc. Acad. Sci. Amst.* **44,** 1045, 1177 (1941); **45,** 9, 126 (1942).

Carlson, F. D. and Herbert, T. J., *J. Phys.* (*Paris*) **33-C1,** 157 (1972).

Cummins, H. Z. and Pike, E. R. (eds.) *Photon Correlation and Light Beating Spectroscopy*, Plenum, New York (1974). See particularly the articles by Cummins and Pusey.

De Groot, S. R. and Mazur, P., *Non-Equilibrium Thermodynamics*, North-Holland, Amsterdam (1962).

Freidhof, L. and Berne, B. J., Biopol., in press (1975)

Harned, H. S. and Owen, B. B., *Physical Chemistry of Electrolyte Solutions,* Reinhold, New York (1950).

Herbert, T. J. and Carlson, F. D., *Biopolymers,* **10,** 2231 (1971).

Katchalsky, A. and Curran, P. F., *Non-Equilibrium Thermodynamics*, Harvard University Press (1965).

Newman, J., Swinney, H. L., Berkowitz, S. A. and Day, L. A., Biochem. **13,** 4832 (1974).

Onsager, L., *Phys. Rev.* **37,** 405; **38,** 2265 (1931).

Prigogine, I., *Introduction to Thermodynamics of Irreversible Processes*, Wiley-Interscience, New York (1955).

Pusey, P. N., Schaefer, D. W., Koppel, D. E., Camerini-Otero, R. D., and Franklin, R. M., *J. Phys.* **33-C1,** 163 (1972).

Pyun, C. W. and Fixman, M., *J. Chem. Phys.* **41,** 937 (1964).

Raj, T. and Flygare, W. H., *Biochem.* **13,** 3336 (1974).

Tanford, C., *Physical Chemistry of Macromolecules*, Wiley, New York (1961).

CHAPTER 14

COLLISION-INDUCED LIGHT SCATTERING AND LIGHT SCATTERING BY GASES

14 · 1 INTRODUCTION

In this book we have often asserted that because atoms are optically isotropic, there should be no depolarized light scattering; that is, $I_{VH} = 0$. Nevertheless, it has been demonstrated recently (see the pioneering work of McTague and Birnbaum, 1968 and the review by Fleury and Boon, 1974) that there is a substantial depolarized component from atomic fluids. The observed depolarized spectrum from monatomic fluids has an exponential frequency-dependence. Fleury et al. (1971) have studied $I_{VH}(\omega)$ over a wide range of densities and temperatures and have found that $I_{VH}(\omega)$ can be fitted to a two-branch exponential spectrum

$$\begin{aligned} I_{VH}(\omega) &= I_0 \exp -\left[\frac{\omega}{\Delta_1}\right] \qquad && \omega \leq \omega_0 \\ &= I_0 A(\omega_0) \exp -\left[\frac{\omega}{\Delta_2}\right] \qquad && \omega \geq \omega_0 \end{aligned} \tag{14.1.1}$$

where $A(\omega_0) = \exp[(\Delta_2^{-1} - \Delta_1^{-1})\omega_0]$ and where Δ_1, Δ_2, and ω_0 are parameters that depend on the thermodynamic state. Depolarized spectra for argon at 300°K for different densities ranging from 200 to 905 amagats(1) are plotted in Fig. 14.1.1. The parameters Δ_1 and Δ_2 fitted to these spectra are plotted in Fig. 14.1.2 against the density in amagats.[1] It was found that ω_0 does not change substantially.

Although no dynamical theory has yet been developed that correctly gives the observed spectrum for all densities, the spectrum at low densities is fairly well understood, and there has been some progress in understanding the higher densities. As we shall see, a treatment of this problem involves a reconsideration of the manner in which radiation interacts with matter, as well as a prescription for calculating collisional dynamics. Our aim in this chapter is to present a brief description of this developing subject. Clearly, these effects occur not only in atomic fluids, but also molecular fluids where they often cannot be separated from the usual depolarized component arising from the permanent molecular optical anisotropy. Thus to determine rotational dynamics by depolarized light scattering (cf. Chapter 7) it behooves the experimentalist to find a method for subtracting the component due to the collision-induced optical anisotropy. Unfortunately there exists no unique prescription for this subtraction at present. Much work in this area is expected in the future.

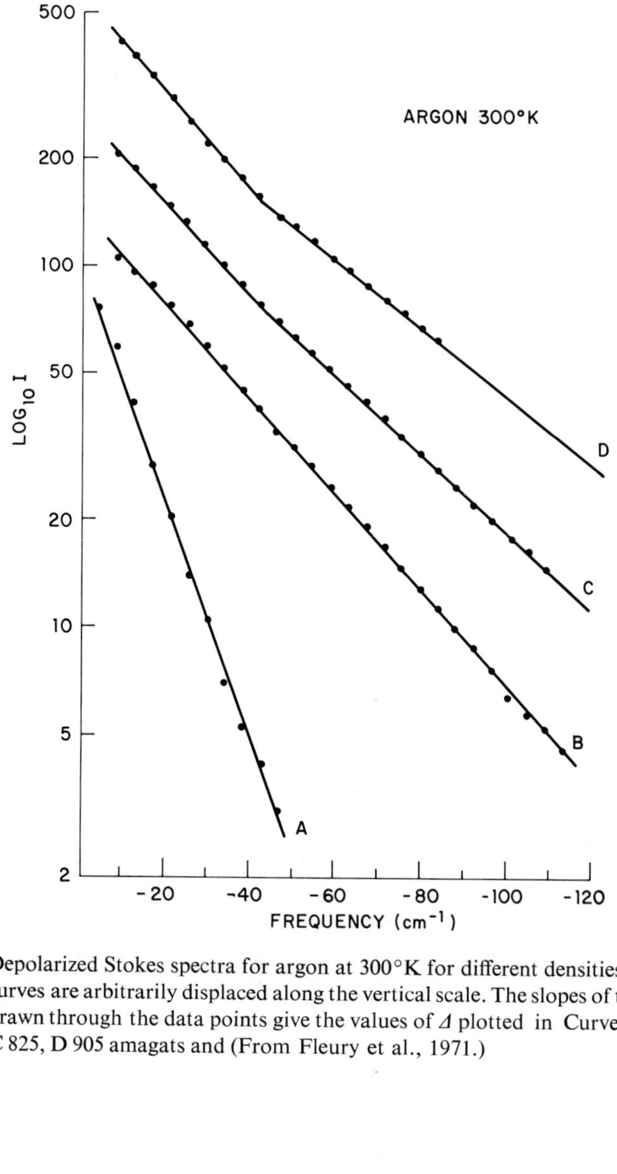

FIG. 14.1.1. Depolarized Stokes spectra for argon at 300°K for different densities. For display the curves are arbitrarily displaced along the vertical scale. The slopes of the straight lines drawn through the data points give the values of Δ plotted in Curves A 200, B 750, C 825, D 905 amagats and (From Fleury et al., 1971.)

14 · 2 A SIMPLE COLLISIONAL MODEL

Let us consider what happens during an atomic collision between atoms i and j. At large internuclear distances ($r_{ij} = |\mathbf{r}_i - \mathbf{r}_j| \rightarrow \infty$) the atoms do not interact and the electrons of each atom should be spherically distributed about their respective nuclei. As the atoms approach each other, their electron clouds should begin to interact and distort each other. The electrons will then be distributed with axial symmetry about the internuclear vector ($\mathbf{r}_{ij} = \mathbf{r}_i - \mathbf{r}_j$). This axially symmetric charge distribution should give rise to an axially symmetric polarizability tensor, $\boldsymbol{\alpha}(\mathbf{r}_{ij})$ and thereby to a depolarized component in the scattered light. The collisional pair can therefore be regarded as a "quasilinear" molecule, at least for the duration τ_c of the collision.

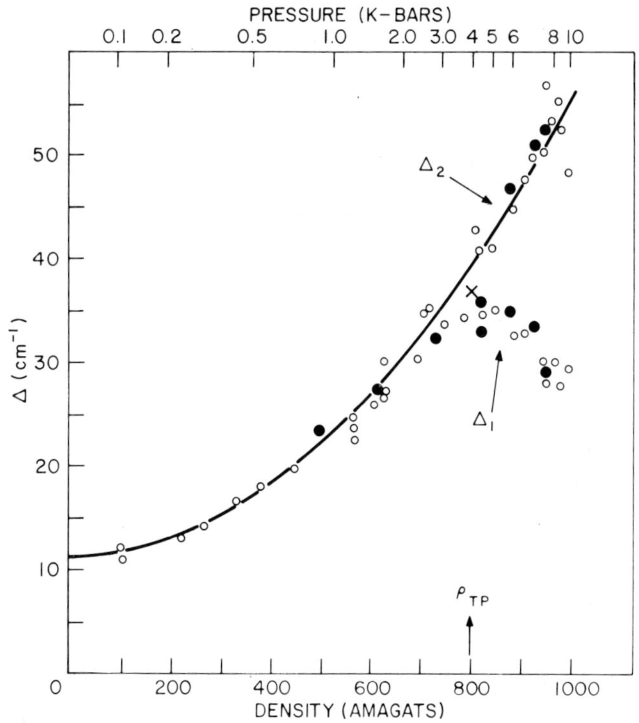

FIG. 14.1.2. Density and temperature-dependence of the spectral exponents in argon. Δ_1 and Δ_2 respresent the slopes of the low (0–50 cm^{-1}) and high (> 50 cm^{-1}) frequency regions of the spectra, respectively. Open circles are data at 300°K; filled circles are data at 180°K scaled by $(k_B T)^{1/2}$; X is the liquid at 90 °K scaled by $(k_B T^{1/2})$ The P scale gives the pressure at 300 K corresponding to the density ρ indicated along the bottom scale. The solid line is given by $\Delta = [1 + \rho/\rho_0)^2]\Delta_0$, as described in the reference. (From Fleury, et al., 1971.)

We should be able to apply the symmetry arguments of Section 7.B to guess the form of the polarizability tensor. Thus from Eq. (7.B.1) we write

$$\boldsymbol{\alpha}(\mathbf{r}_{ij}) = \alpha(r_{ij})\,\mathbf{I} + \beta(r_{ij})\left[\hat{\mathbf{r}}_{ij}\hat{\mathbf{r}}_{ij} - \frac{1}{3}\,\mathbf{I}\right] \tag{14.2.1a}$$

where $\hat{\mathbf{r}}_{ij}$ is a unit vector along $\mathbf{r}_{ij}$, $\alpha(r_{ij})$ is the scalar part and $\beta(r_{ij})$ is the anisotropic part of the polarizability tensor; that is,

$$\alpha(r_{ij}) = \frac{1}{3}\,[\alpha_{\parallel}(r_{ij}) + 2\alpha_{\perp}(r_{ij})] \tag{14.2.1b}$$

$$\beta(r_{ij}) = \alpha_{\parallel}(r_{ij}) - \alpha_{\perp}(r_{ij}) \tag{14.2.1c}$$

where $\alpha_{\parallel}(r_{ij})$ and $\alpha_{\perp}(r_{ij})$ are respectively the components of the full polarizability tensor parallel and perpendicular to the vector $\mathbf{r}_{ij}$.

Unfortunately the precise dependence of $\alpha(r_{ij})$ and $\beta(r_{ij})$ on distance r_{ij} is not directly accessible to experimental measurement and must therefore be calculated. There has been some progress recently in computing these quantities (see Heller et al., 1975). It is clear from the preceding that

$$\lim_{r_{ij}\to\infty} \alpha(r_{ij}) = \alpha_0 + \alpha_0 = 2\alpha_0 \tag{14.2.2}$$

$$\lim_{r_{ij}\to\infty} \beta(r_{ij}) = 0 \tag{14.2.3}$$

where Eq. (14.2.2) merely reflects the fact that at infinite separations the scalar polarizability should reduce to the sum of the noninteracting atom polarizabilities. It is therefore useful to write

$$\alpha(r_{ij}) = 2\alpha_0 + \delta\alpha(r_{ij}) \tag{14.2.3c}$$

where

$$\delta\alpha(r_{ij}) \equiv \alpha(r_{ij}) - 2\alpha_0 \tag{14.2.3d}$$

is by definition the collision-induced scalar polarizability. Eq. (14.2.2) reflects the fact that at infinite separations the atoms behave independently as optically isotropic entities so that the optical anisotropy β is zero. $\beta(r_{ij})$ is thus totally collision-induced.

A quantum-mechanical calculation of $\delta\alpha(r_{ij})$ and $\beta(r_{ij})$ for large r_{ij} ($r_{ij} \to \infty$) can be carried out via perturbation theory. This calculation is the analog of the calculation of London-dispersion forces. The important asymptotic results are

$$\beta(r_{ij}) \to 6\,\frac{\alpha_0{}^2}{r_{ij}^3} \tag{14.2.4a}$$

$$\delta\alpha(r_{ij}) \to \frac{4\alpha_0{}^3}{r_{ij}^6} \tag{14.2.4b}$$

and we see that $\delta\alpha(r_{ij})$ is much shorter ranged than $\beta(r_{ij})$. This is the polarizability anisotropy that arises from a point-dipole induced dipole mechanism which we denote DID. Clearly this asymptotic result does not apply to small separations where it is expected that the electron clouds of the colliding atoms should strongly overlap. Overlap effects should clearly overwhelm the DID mechanism.

Although these collision-induced effects will contribute to both the polarized and depolarized spectra, it is only in the depolarized spectrum that they contribute against a background of "zero." That is, in the absence of collision-induced effects we should observe no depolarized spectrum in dilute atomic gases. In the remainder of this chapter, only the depolarized spectrum $I_{VH}(\omega)$ is considered.

It is reasonable (see McTague and Birnbaum, 1971) to regard each pair of atoms (ij) in the fluid as a quasilinear diatomic molecule with an axis $\hat{\mathbf{r}}_{ij}$ defined by the interatomic vector $\mathbf{r}_{ij}$. Thus to each of the $N(N-1)$ pairs (ij) of atoms we ascribe the polarizability tensor given by Eq. (14.2.1a) and a center-of-mass distance $\mathbf{R}_{ij} = (\mathbf{r}_i + \mathbf{r}_j)/2$. It then follows from Section 7.B that the depolarized spectrum is the Fourier transform of the time-correlation function

$$I_{VH}(t) = \langle \sum_{i\neq j} \sum_{l\neq m} \beta(r_{lm}(0))\, \beta(r_{ij}(t))\, P_2(\hat{\mathbf{r}}_{lm}(0) \cdot \hat{\mathbf{r}}_{ij}(t))\rangle \tag{14.2.5}$$

At the risk of stating the obvious, it is important to note that the collision-induced polarizability in a many-atom system should contain contributions from two, three, four . . . , body interactions, and cannot be ascribed entirely to the two-body polarizability as in Eq. (14.2.5). In dilute gases, where the two-body contributions dominate, Eq. (14.2.5) should suffice, but in dense systems, there should be additional terms. Nevertheless, there is some evidence from computer simulations that Eq. (14.2.5) is adequate for the determination of the details of the line shape of $I_{VH}(\omega)$.

In a dilute gas where the molecular dynamics is dominated by independent binary collisions Eq. (14.2.5) reduces to

$$I_{VH}(t) = \langle \beta(r_{ij}(0)) \, \beta(r_{ij}(t)) \, P_2(\hat{\mathbf{r}}_{ij}(0) \cdot \hat{\mathbf{r}}_{ij}(t)) \rangle \tag{14.2.6}$$

where we have omitted a density-dependent multiplicative constant. This correlation function depends on a radial part $\beta(r_{ij}(0))\,\beta(r_{ij}(t))$ which varies as the internuclear distance varies and an orientational part $P_2(\hat{\mathbf{r}}_{ij}(0) \cdot \hat{\mathbf{r}}_{ij}(t))$ that varies as $\hat{\mathbf{r}}_{ij}(t)$ reorients]. It is clear that $I_{VH}(t)$ should decay on the time scale

$$\tau_0 = r_0 \left(\frac{3k_B T}{\mu} \right)^{-1/2} \tag{14.2.7}$$

where r_0 characterizes the "range" of $\beta(r)$ and $(3k_B T/\mu)^{+1/2}$ is the rms relative velocity of the particle (μ is the reduced mass); τ_0 is effectively the "duration" of the collision induced anisotropy $\beta(r)$. Thus on intuitive grounds, the line width of $I_{VH}(\omega)$ in a dilute gas is expected to be proportional to the inverse duration of a collision which varies with temperature as $T^{1/2}$ and is independent of the density. This behavior is observed in dilute gases.

In order to compute the correlation function it is necessary to know the precise dependence of $\beta(r)$ on r and the precise intermolecular potential. It is then always possible to calculate by computer the collision trajectories and thereby the quantity in the brackets in [Eq. (14.2.6)] for a large sample of impact parameters and relative velocities. The results are then weighted with the Maxwell distribution of relative velocities and averaged over the sample of trajectories. Lallemand (1970) has computed Eq. (14.2.5) in this way for the DID model of $\beta(r_{ij})$ [Eq. (14.2.4a)] and for three different potentials: (a) hard sphere, (b) Lennard–Jones (12–6), and (c) the modified Buckingham potential. In Fig. 14.2.1 experiments on gaseous CH_4 and theory (using the Lennard–Jones potential) are compared. It would thus seem that for dilute gases, the simple independent binary collision model in the DID approximation successfully accounts for the data. These calculations not only give the correct depolarized spectrum, they also give the correct integrated depolarized intensity. The as yet unaswered question immediately arises as to why the DID approximation works so well. Perhaps the intermolecular potential is so strongly repulsive at internuclear distances for which the overlap effects contribute to $\beta(r)$ that only the improbable very high energy collisions sample these distances.

It is possible to show for many collision models and even for the simple model of linear trajectories that $I_{VH}(\omega)$ will have an exponential line shape (to within a factor that is some power of the frequency); that is, $I_{VH}(\omega) \sim I_0 e^{-\omega/\Delta}$ with $\Delta \propto \frac{1}{\tau_0}$. Thus the observed exponential wing seems to be a consequence of the binary collision approximation.

The experiments of McTague et al. (1969) show that $I_{VH}(\omega)$ of liquid argon is very different from that of its vapor at the same temperature and pressure. Clearly then Δ cannot be a density-independent quantity as would be expected from the simple binary

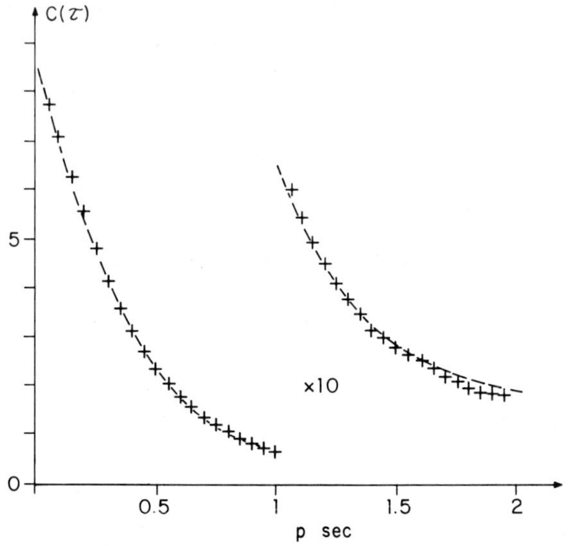

FIG. 14.2.1. Velocity-averaged correlation function. Dashed line is computed for a Lennard–Jones potential; crosses are calculated from the zx spectrum scattered by CH_4 at 70 amagats and 20°. Notice a change of scale by a factor 10 for $t \geq 1$ psec. (From Lallemand, 1970.)

collision picture. In order to separate the temperature and density-dependence, Fleury et al. (1971) have investigated the spectra of argon and neon along isotherms as a function of density. The results of these measurements are shown in Figs. 14.1.1 and 14.1.2. Unfortunately there is as yet no simple theory capable of explaining these effects.

Equation (14.2.5), and not Eq. (14.2.6), is expected to apply to dense systems. This can be expressed as

$$I_{VH}(t) = \frac{3}{2} \langle \boldsymbol{\beta}(0) : \boldsymbol{\beta}(t) \rangle \tag{14.2.8a}$$

where

$$\boldsymbol{\beta}(t) \equiv \sum_i \left\{ \sum_{j \neq i} \beta(r_{ij}(t)) \left[\hat{\mathbf{r}}_{ij}(t)\hat{\mathbf{r}}_{ij}(t) - \frac{1}{3}\mathbf{I} \right] \right\} = \sum_i \boldsymbol{\beta}_i(t) \tag{14.2.8b}$$

The term in $\{\cdot\cdot\}$ can be regarded as the net anisotropy to be associated with atom i. This is denoted $\boldsymbol{\beta}_i$. It is clear that if atom i is surrounded by a spherically symmetric distribution of neighbors, $\boldsymbol{\beta}_i = 0$. Thus $I_{VH}(t)$ can be regarded as being caused by density fluctuations that break the spherical symmetry of the cage surrounding any atom. Equation (14.2.8b) consists of terms such as $\langle \boldsymbol{\beta}_i(0):\boldsymbol{\beta}_i(t) \rangle$ and $\langle \boldsymbol{\beta}_i(0):\boldsymbol{\beta}_j(t) \rangle$ $(i \neq j)$. The first kind of term involves two- and three-body correlations, whereas the second involves two-, three-, and four-body correlations. Equations (14.2.5) and (14.2.8) consequently involve all correlated two-body and higher order collisions, in marked contrast with the simple binary collision picture. To evaluate the full correlation function analytically for dense systems is as yet impossible. In order to understand the underlying dynamic contributions, Berne, et al. (1973) have calculated Eq. (14.2.5) for two thermodynamic states of liquid argon by molecular dynamics. Comparing this with similar calculations of the radial function

$$C_2(t) \equiv \langle \sum_{i \neq j} \beta(r_{ij}(0)) \sum_{l \neq m} \beta(r_{lm}(t)) \rangle \tag{14.2.9a}$$

and angular function

$$C_3(t) \equiv \langle \sum_{l \neq m} \sum_{i \neq j} P_2(\hat{\mathbf{r}}_{ij}(0) \cdot \hat{\mathbf{r}}_{lm}(t)) \rangle \tag{14.2.9b}$$

they found that for the several models of $\beta(r)$ considered $I_{VH}(t)$ and $C_3(t)$ decay to zero on the same time scale whereas $C_2(t)$ stays constant over this time scale. Moreover they show that the isolated binary collision model does not agree with the full calculation. Different dynamical processes contribute to Eqs. (14.2.5) and (14.2.6) in a dense system.

In some sense one can argue that in a dense fluid, it is not the duration of a single collision that matters, but the duration of an anisotropy fluctuation. Molecules are continuously within the "range" of their neighbors. Thus the two time scales may reflect anisotropy fluctuations of the shell of nearest neighbors on the one hand and the next-nearest neighbors on the other. There are a variety of explanations. Until there is a successful model or a more detailed study of molecular dynamics we do not think it is profitable to review the long list of "physical" explanations here. The interested reader should consult the extensive reviews by Gelbart (1974) and Fleury and Boon (1974).

Alder et al. (1973) have computed Eq. (14.2.5) for hard spheres and the Lennard–Jones model over a wide range of densities and find that for $\beta(r)$ given by Eq. (14.2.4a) the resulting spectra as a function of density resemble those measured by Fleury et al. (1971). However they find in another study (Alder, et al., 1973) that they can not account for the measured density-dependence of the integrated intensity.

Much remains to be done in this area. The role of the many-body polarizability has yet to be explored, a simple dynamical model has yet to be presented, and a microscopic justification of Eq. (14.2.5) has yet to be developed. Moreover the important practical question concerning the collision-induced scattering from molecular liquids has yet to be answered. Only after this effect is assessed can it be subtracted from the depolarized scattering in such a manner that the remaining spectrum gives information about molecular tumbling.

The only microscopic theory that has been carried through explicitly is the DID theory. When a linearly polarized electric field is applied to a monatomic fluid, the electric field experienced by atom i is not the applied field, but a local field. This local field differs from the applied field in several respects. It does not propagate with the phase velocity given by the vacuum wave number k_i, but rather with the wave number nk_i, where n is the refractive index of the fluid. In addition its amplitude differs from the incident field amplitude by a factor usually given by $(n^2 + 2)/3$, the Lorentz–Lorentz factor. There remains a more subtle difference. Consider an atom at the point $\mathbf{r}_i$ of the fluid. The applied field polarizes atom j; that is, induces a dipole moment (and higher order multipole moments) on that atom, which therefore gives rise to an electric field at $\mathbf{r}_i$ in addition to the applied field at $\mathbf{r}_i$. If only the induced dipole is included it is possible to show that the resulting spectrum is well approximated by the DID approximation. Thus collision-induced scattering can be regarded as arising from local field effects. The microscopic theory, however, has not been sufficiently extended to include the overlap effects.

For further information about this interesting topic we refer the reader to the extensive review by Gelbart (1974).

14 · 3 THE KINETIC THEORY OF GASES

An interesting application of light scattering has been to the study of the kinetic theory of gases. In Section 10.4 $S(\mathbf{q}, \omega)$ was calculated using the Navier–Stokes equations of hydrodynamics. Implicit in the application of the equations of fluid dynamics is the assumption that the "mean free path" λ_f of the molecules is very small compared to $q^{-1} = (q\lambda_f \ll 1)$. For the values of q probed in light scattering this is always the case for dense fluids but may not be valid in dilute gases. For example, a gas consisting of atoms of diameter 3 A° at S T P has $\lambda_f = 2.07 \times 10^{-5}$ cm, giving for a typical $q = 10^{+5}$ cm^{-1}, $q\lambda_f = 2.07$, which clearly does not satisfy the condition required for the application of hydrodynamics. How then can we calculate the light-scattering spectrum of a dilute gas? Nelkin and Yip (1966) provided an answer to this question. They pointed out that the linearized Boltzmann equation.

$$\frac{\partial f(\mathbf{r}, \mathbf{p}, \mathbf{t})}{\partial t} + \frac{\mathbf{p}}{m} \cdot \nabla_r f(\mathbf{r}, \mathbf{p}, \mathbf{t}) = J[f] \tag{14.3.1}$$

should be used to calculate the spectrum[2] (see Uhlenbeck and Ford (1965)).

It can be shown (Van Leeuwen and Yip (1965)) that the density–density correlation function $G(\mathbf{r}'' - \mathbf{r}', t) \equiv \langle \delta\rho(\mathbf{r}', 0)\, \delta\rho(\mathbf{r}, ''t)\rangle$ is given by

$$G(\mathbf{r}, t) = \int d^3r\, f(\mathbf{r}, \mathbf{p}, t) \tag{14.3.2}$$

where the distribution function $f(\mathbf{r}, \mathbf{p}, t)$ is the solution of Eq. (14.3.1) subject to the initial condition

$$f(\mathbf{r}, \mathbf{p}, 0) = \phi_m(\mathbf{p})\, \delta(\mathbf{r}) \tag{14.3.3}$$

where $\phi_m(\mathbf{p})$ is the Maxwell momentum distribution function.

Nelkin and Yip (1966) suggested that light-scattering experiments on dilute gases could be used as a test of the validity of the Boltzmann equation for the description of time-dependent phenomena. Prior to this the equation had been checked by only measuring transport coefficients and sound propagation.

Unfortunately it is difficult to solve Eq. (14.3.1) subject to these boundary conditions for realistic intermolecular potentials. Even for the idealized model of hard spheres this program has not to our knowledge been carried to completion. $S(\mathbf{q}, \omega)$ has been calculated for Maxwell molecules—molecules that interact with a potential which is proportional to r^{-4} (Sugawara et al., 1968). Rather than applying this difficult integral differential equation, Nelkin and Yip used the Krook equation—a well known approximate equation in which the integral operator $J[\cdot \cdot]$ in Eq. (14.3.1) is replaced by a relaxation time term.[3] This equation, like Eq. (14.3.1), gives for $q \to 0$ the hydrodynamic equation and for $q \to \infty$ the free-particle result (see Section 5.6). In all of their calculations these authors use a parameter y which is proportional to $(\lambda_f q)^{-1}$ and is defined as

$$y = \left(\frac{\alpha}{q}\right)\left(\frac{M}{2k_BT}\right)^{1/2} \tag{14.3.4}$$

where α is an effective collision frequency. The value of α is chosen so that the Krook equation gives the correct sound absorption coefficient. Clearly when y is sufficiently large all of these equations give the hydrodynamic results and when y is very small they give the Doppler line shape. For intermediate values of y there is a significant difference between the kinetic and hydrodynamic results. This is shown in Fig. 14.3.1. It is important to note that y can be varied by changing either q or the gas pressure.

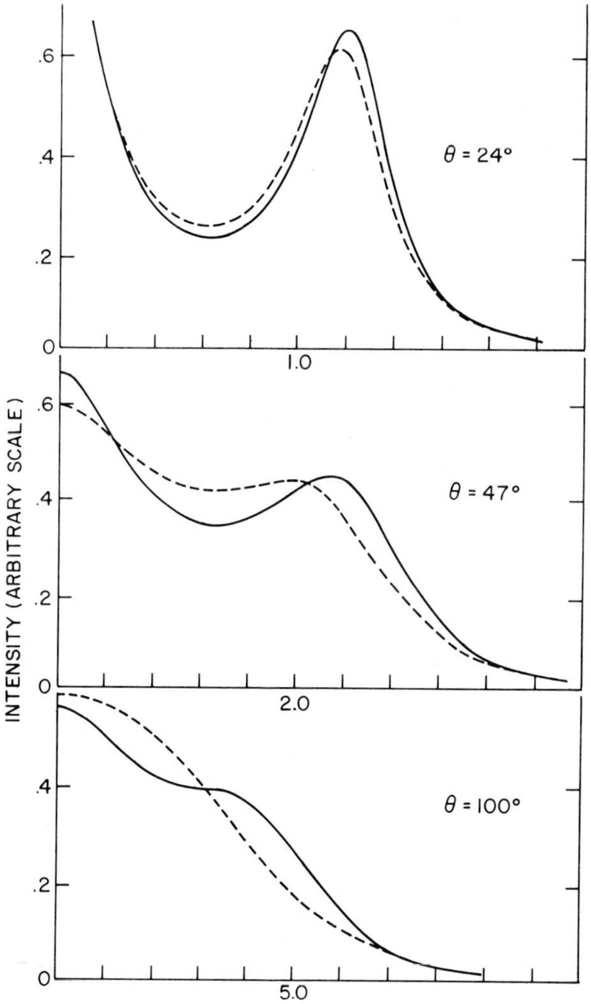

FIG. 14.3.1. Intensity of Rayleigh scattering by argon gas at one atmosphere and 0 °C as a function of frequency change for different scattering angles θ. Solid and dashed curves are calculations based on the Krook model and hydrodynamic equations, respectively. (From Nelkin and Yip, 1966.)

Soon after Nelkin and Yip (1966) suggested the experiment, Greytak and Benedek (1966) performed experiments on xenon (and CO_2) and observed the deviations from the hydrodynamic line shape predicted by Nelkin and Yip. More recently Clark (1970) studied Xe and found the theoretical spectra to be in very good agreement with experiment. The theoretical calculations have now been extended to mixtures of gases by Boley and Yip (1972) and to polyatomic molecules by Desai et al. (1972). Experiments on helium—xenon mixtures by Clark (1970) and Gornall and Wang (1972) confirm the theoretical calculations. These latter calculations reduce to the results of Sections 5.6 and 10.4 respectively in the $y \to \infty$ limit.

NOTES

1. 1 amagat is the density of the gas (in this case Ar) at 1 atm and 273.16 °K, or in other words the density when the gas is at STP.
2. $f(\mathbf{r}, \mathbf{p}, t)d^3 r\, d^3 p$ is the number of atoms in the neighborhood d^3r of the point **r** with momentum in the neighborhood d^3p of **p** at time t. In Eq. (14.3.1) $J[\cdot\ \cdot]$ denotes a linear integral operator.
3. Like

$$-\frac{1}{\tau}[f(\mathbf{r}, \mathbf{p}, t) - f_{eq}(\mathbf{r}, \mathbf{p})]$$

REFERENCES

Alder, B. J., Strauss, H. L., and Weis, J. J., *J. Chem. Phys.* **59,** 1002 (1973).

Alder, B. J., Weis, J. J., and Strauss, H. L., *Phys. Rev.* **A7,** 281 (1973).

Berne, B. J., Bishop, M., and Rahman, A., *J. Chem. Phys.* **58,** 2696 (1973).

Boley, C. D., Desai, R. C., and Tenti, G., *Can. J. Phys.* **50,** 2158 (1972).

Boley, C. D., and Yip, S., *J. Phys. (Paris)* **33,** Cl-43, (1972).

Clark, N. A., Ph. D. Thesis, Massachusetts Institute of Technology (1970).

Fleury, P. A. and Boon, J. P., *Adv. Chem. Phys.* **24,** 1 1(974).

Fleury, P. A., Daniels, W. B., and Worlock, J. M., *Phys. Rev. Lett.,* **27,** 1493 (1971).

Gelbart, W., *Adv. Chem. Phys.* **26,** 1 (1974).

Heller, D. F., Harris, R. A., and Gelbart, W. M., *J. Chem. Phys.* **62,** 1947 (1975).

Gornall, W. S., and Wang, C. S., *J. Phys. (Paris)* **33,** Cl-51 (1972).

Greytak, T. J. and Benedek, G. B., *Phys. Rev. Lett.* **17,** 179 (1966).

Lallemand, P. M., *Phys. Rev. Lett.,* **25,** 1079 (1970).

McTague, J. P. and Birnbaum, G., *Phys. Rev. Lett.* **21,** 661 (1968).

McTague, J. P. and Birnbaum, G., *Phys. Rev.* **A3,** 1376 (1971).

McTague, J. P., Fleury, P. A., and DuPre, D. B., *Phys. Rev.* **188,** 303 (1969).

Nelkin, M. and Yip, S., Phys. Fluids **9,** 380 (1966).

Boley, C. D., Desai, R. C. and Tenti, G., Can. J. Phys **50,** 2158 (1972).

Sugawara, A., Yip, S., and Sirovich, L., *Phys. Fluids,* **11,** 925 (1968).

Uhlenbeck, G. E. and Ford, G. W., *Lectures in Statistical Mechanics,* American Mathematical Society (1963).

Van Leeuwen, J. M. J. and Yip, S., *Phys. Rev.* **A 139,** 1138 (1965).

CHAPTER 15

OTHER PROBES OF MOLECULAR DYNAMICS

15 · 1 INTRODUCTION

Light scattering is one of several methods used to probe the dynamics of thermal fluctuation in fluids. It is the purpose of this concluding chapter to mention some of these techniques and the correlation functions that are measured by them. Table 15.1.1 lists the experimental techniques, the important dynamical quantity involved and the time-correlation function probed by the technique. A few words should be said about some of these techniques.

15 · 2 NEUTRON SCATTERING

The theory of neutron scattering (Marshall and Lovesey, 1971) is very closely related to that of light scattering. Neutrons are scattered by the nuclei of atoms and molecules

TABLE 15.1.1

Some Experimental Probes of Time Correation Functions

Experiment	Dynamical Quantity	Time-Correlation Function
Neutron scattering	$\mathbf{r}_l$, position of lth nucleus in a fluid	$\langle \Sigma \exp(-i\mathbf{q}\cdot\mathbf{r}_l(o)) \exp(i\mathbf{q}\cdot\mathbf{r}m(t) \rangle$ln
Raman scattering; depolarization of fluorescence	$\mathbf{u}$, unit vector along molecular transition dipole	$\langle P_2(\mathbf{u}(o)\cdot\mathbf{u}(t))\rangle$
Infrared absorption	$\mathbf{u}$, unit vector along molecular transition dipole	$\langle \mathbf{u}(o)\cdot\mathbf{u}(t)\rangle$
Dielectric relaxation	$\mathbf{u}$, unit vector along permanent dipole	$\sum_{l,m} \langle \mathbf{u}_l(o)\cdot\mathbf{u}_m(t)\rangle$
Spin-rotation relaxation time	$\mathbf{J}$, angular momentum about molecular center of mass	$\langle \mathbf{J}(o)\cdot\mathbf{J}(t)\rangle$
NMR line shape	M_x,x component of the magnetization of the system	$\langle M_x(o)M_x(t)\rangle$
Self-diffusion coefficient	$\mathbf{V}$, center-of-mass velocity of tagged molecule	$\langle \mathbf{V}(o)\cdot\mathbf{V}(t)\rangle$
Rotational diffusion coefficient	$\boldsymbol{\Omega}$ angular velocity about molecular center of mass	$\langle \Omega_\alpha(o)\Omega_\beta(t)\rangle$

and the frequency shift of the scattered neutron over that of the incident neutron is due to the thermal motion of the scattering nuclei. While light scattering is usually coherent, neutron scattering may be either coherent or incoherent, depending on the nucleus considered. In fact each nucleus has a scattering cross section for coherent scattering and incoherent scattering. In incoherent scattering interactions between the neutron spin and the nucleus randomize the phase of the scattered neutrons, while in coherent scattering this phase is preserved. The incoherent double-differential scattering cross section for scattering into a solid angle element $d\Omega$ and frequency range $d\omega$ is given by

$$\left(\frac{d^2\sigma}{d\Omega d\omega}\right)_{\text{inc}} = N\sigma_{\text{inc}}\frac{k_f}{k_i}S_s(\mathbf{q}, \omega) \tag{15.2.1}$$

where σ_{inc} is the total incoherent cross section for the scattering nucleus and $\hbar\omega$ and $\hbar\mathbf{q}$ are, respectively, the energy and momentum losses of the neutron

$$\hbar\omega = E_i - E_f$$

$$\hbar\mathbf{q} = \mathbf{k}_i - \mathbf{k}_f$$

Note that the definitions of ω and $\mathbf{q}$ are identical with those for light scattering.

The scattering function $S_s(\mathbf{q}, \omega)$ is simply the space–time Fourier transform of the function $G_s(\mathbf{r}, t)$ introduced in Section 5.4 or equivalently the time-Fourier transform of $F_s(\mathbf{q}, t)$ defined in the same section. The expressions for light scattering corresponding to Eq. (15.2.1) are derived in Section 5.4. Note that the light scattering is "incoherent" in this case because of the assumed lack of correlation between the space–time positions of the scatterers.

For coherent neutron scattering, the double-differential scattering cross section is

$$\left(\frac{d^2\sigma}{d\Omega d\omega}\right)_{\text{coh}} = \sigma_{\text{coh}}\frac{k_f}{k_i}S(\mathbf{q}, \omega) \tag{15.2.2}$$

where $S(\mathbf{q}, \omega)$ is the spectral density of the number density fluctuation.

A major difference between light scattering and neutron scattering for atomic fluids lies in the range of q sampled. Light has long wavelengths and hence small values of q while the neutrons used in these experiments have wavelengths of the order of a few angstroms and hence have very high values of q. Thus neutrons will sample much shorter wavelength (and usually faster) fluctuations than light scattering. In many neutron-scattering experiments, for instance, one cannot use hydrodynamics to compute the scattered spectrum (see Chapter 10 and Section 14.3).

For molecular systems the theory of neutron scattering can be developed in analogy to that for isotropic light scattering from macromolecules. For instance, the theory of light scattering from rigid rods described in Chapter 8 may be applied to small molecules containing coherently scattering nuclei arranged along a line (e.g., see Yip, 1974).

15 · 3 RAMAN AND INFRARED BAND SHAPES

In Chapters 7 and 12 it was shown how depolarized Rayleigh scattering can be used to probe molecular rotations in fluids. Measurements of the band shapes of infrared vibrational and Raman lines may also be used to obtain information about the dynamics of molecular rotations. The theory of Raman bandshapes (cf. Section 3.3, Gordon, 1968) is more straightforward than that of the infrared shapes. Let the polarizability tensor of a molecule in the Heisenberg picture of quantum mechanics be given by $\boldsymbol{\alpha}$. Following the procedure of Appendix 7. B, we divide $\boldsymbol{\alpha}$ into its scalar part α and its anisotropic part $\boldsymbol{\beta}$ so that

$$\boldsymbol{\alpha} = \alpha\,\mathbf{I} + \boldsymbol{\beta} \tag{15.3.1}$$

where $\alpha = \frac{1}{3}\,Tr\boldsymbol{\alpha}$, $Tr\boldsymbol{\beta} = 0$, and $\mathbf{I}$ is the unit tensor. Then it is easy to show (Appendix 7. B) that

$$\mathbf{I}_{VV}\,(\omega_f) = \mathbf{I}_{ISO}(\omega_f) + \frac{4}{3}\,\mathbf{I}_{VH}(\omega_f)$$

where

$$\mathbf{I}_{ISO}(\omega_f) = \frac{N}{2\pi}\int_{-\infty}^{+\infty} dt\,\langle\alpha(0)\,\alpha(t)\rangle \exp - i\omega_f t \tag{15.3.2}$$

and

$$I_{VH}(\omega_f) = \frac{N}{2\pi}\int_{-\infty}^{+\infty} dt\,\frac{1}{10}\langle Tr\,\boldsymbol{\beta}(0)\cdot\boldsymbol{\beta}(t)\rangle \exp - i\omega_f t$$

where ω_f is the frequency of the Raman-scattered light. If we expand α and $\boldsymbol{\beta}$ in the vibrational normal coordinates of a molecule and keep only the linear term, we obtain

$$\alpha\mathbf{I} = \alpha_0\mathbf{I} + \sum_i \left[\frac{\partial\alpha\,\mathbf{I}}{\partial Q_i}\right] Q_i + \ldots \tag{15.3.3}$$

and

$$\boldsymbol{\beta} = \boldsymbol{\beta}_0 + \sum \left[\frac{\partial\boldsymbol{\beta}}{\partial Q_i}\right] Q_i + \ldots \tag{15.3.4}$$

where the derivatives are evaluated at the equilibrium separation. The α_0 and $\boldsymbol{\beta}_0$ terms are the "isotropic" and "anisotropic" components of the molecular polarizability tensor evaluated at the equilibrium internuclear separation. They give rise to the Rayleigh scattering that has been discussed in this book. The terms dependent on the normal coordinates give rise to vibrational Raman scattering. Note that $\partial\alpha\mathbf{I}/\partial Q_i$, is independent of molecular orientation while $\partial\boldsymbol{\beta}/\partial Q_i$ depends on molecular orientation. Thus when Eqs. (15.3.3) and (15.3.4) are substituted into Eq. (15.3.2) we see that the I_{VH} Raman spectrum depends on both the vibrations and rotations of a molecule, while I_{VV} contains a term depending solely on vibrational motion in addition to I_{VH}.

A normal mode i is "inactive" or "active" in the Raman spectrum depending on whether $(\partial\alpha\mathbf{I}/\partial Q_i)$ and $(\partial\boldsymbol{\beta}/\partial Q_i)$ are zero. The symmetry of the normal mode may in most cases be used to determine which modes are active and which are "isotropic" (only $(\partial\alpha\mathbf{I}/\partial Q_i)$ different from zero).

In Raman scattering it is usually assumed that the vibration and rotation of a molecule are mutually independent so that the I_{VH} spectrum in Eqs. (15.3.2) factors into a product of independent terms, one depending solely on rotation and the other on vibration. This assumption has not yet been justified by experiment. When this assumption is made it may be seen (cf. Appendix 7. B) that for vibrations of cylindrically symmetric molecules that preserve the molecular symmetry, I_{VH} is proportional to $\langle P_2(\mathbf{u}\,(0)\cdot\mathbf{u}\,(t))\rangle$ where $\mathbf{u}$ is a unit vector along the molecular symmetry axis.

The theory for infrared absorption (Fulton, 1971) proceeds by showing that the dipole–dipole timecorrelation function $\langle \boldsymbol{\mu}(0) \cdot \boldsymbol{\mu}(t) \rangle$ is related to the imaginary part of the dielectric constant $\varepsilon''(\omega)$ at frequency ω by

$$\frac{\langle \boldsymbol{\mu}(0) \cdot \boldsymbol{\mu}(t) \rangle}{\langle |\boldsymbol{\mu}(0)^2| \rangle} \propto \int_{-\infty}^{+\infty} d\omega \left\{ \frac{\exp(-i\omega t)}{1 - \exp\left(\frac{-\hbar\omega}{k_B T}\right)} \right\} \varepsilon''(\omega) \tag{15.3.5}$$

In infrared absorption experiments, one measures the absorption coefficient $\kappa(\omega)$ as a function of ω. Since the complex refractive index is $N = n + i\kappa$, it may be shown that $\varepsilon'' = 2n\kappa$ (Landau and Lifshitz, 1960). Thus in order to relate κ, the measured quantity, to ε'' and then to the dipolar correlation function [Eq. (15.3.5)], one must know how the refractive index $n(\omega)$ changes through the band.

The molecular dipole moment may be expanded in terms of the normal coordinates in a manner similar to that for the polarizability

$$\boldsymbol{\mu} = \sum_i \left[\frac{\partial \boldsymbol{\mu}}{\partial Q_i} \right] Q_i + \dots \tag{15.3.6}$$

where we assume that there is no permanent dipolement ($\boldsymbol{\mu}_0 = 0$) and where the sum is over all normal modes. For a given normal mode, assuming that vibrations and rotations are uncoupled, we obtain

$$\langle \boldsymbol{\mu}(0) \cdot \boldsymbol{\mu}(t) \rangle \cong \left\langle \left(\frac{\partial \boldsymbol{\mu}(0)}{\partial Q} \right)_{\text{rigid}} \cdot \left(\frac{\partial \boldsymbol{\mu}(t)}{\partial Q} \right)_{\text{rigid}} \right\rangle \langle Q(0) \cdot Q(t) \rangle \tag{15.3.7}$$

The $\langle Q(0) \cdot Q(t) \rangle$ term depends only on vibrational motion while the $(\partial \boldsymbol{\mu}/\partial Q)$ terms depends on the rotational motion of the rigid frame.

It is easy to show that if we consider a vibration of a symmetric top molecule which preserves the symmetry of the molecule (Appendix 7. B)

$$\langle \boldsymbol{\mu}(0) \cdot \boldsymbol{\mu}(t) \rangle = \left| \frac{\partial \mu}{\partial Q} \right|^2 \langle \mathbf{u}(0) \cdot \mathbf{u}(t) \rangle \langle Q(0) \cdot Q(t) \rangle$$

Thus infrared absorption measurements give $\langle P_1(\mathbf{u}(0) \cdot \mathbf{u}(t)) \rangle$ while Raman measurements give $\langle P_2(\mathbf{u}(0) \cdot \mathbf{u}(t)) \rangle$.

In the above discussion, it has been implicitly assumed that both Raman and infrared scattering are "incoherent," that is, that the dipole moment or polarizability of one molecule is uncorrelated with the same quantity of another molecule. This assumption reduces in many cases, to saying that normal modes on one molecule are uncorrelated with those on another molecule. It appears from comparisons of different methods for measuring P_1 and P_2 that this assumption is usually good in practice, at least for liquids at low pressures. Note then that the Rayleigh depolarized spectrum (Chapter 12) measures the collective polarizability time-correlation function, while the infrared and Raman techniques usually measure the single molecule dipole moment and single molecule polarizability time-correlation functions, respectively. Thus the methods complement each other. It should be noted that a difficulty with the infrared and Raman methods is in subtracting the contribution of vibrational relaxation from the measurements. Rayleigh scattering does not suffer from this complication.

15 · 4 DIELECTRIC RELAXATION

There has been much controversy in the past several years concerning the relation of the dispersion of the dielectric constant to the molecular dipole-moment correlation function (see Titulaer and Duetch, 1974). Fatuzzo and Mason (1967) have shown that the autocorrelation function of the net dipole moment of a *sphere* imbedded in a medium of the same dielectric constant is related to the frequency-dependent dielectric constant by

$$L\left(-\frac{d\phi}{dt}\right) = \frac{[\varepsilon(\omega) - 1][2\varepsilon(\omega) + 1]}{[\varepsilon_0 - 1][2\varepsilon_0 + 1]\,\varepsilon(\omega)} \tag{15.4.1}$$

where $\varepsilon(\omega)$ and ε_0 are respectively the complex and static dielectric constants; $\phi(t)$ is the normalized dipole moment autocorrelation function

$$\phi(t) \equiv \frac{\langle \sum_{lm} \boldsymbol{\mu}_l(0) \cdot \boldsymbol{\mu}_m(t) \rangle}{\langle \sum_{lm} \boldsymbol{\mu}_l(0) \cdot \boldsymbol{\mu}_m(0) \rangle} \tag{15.4.2}$$

and L denotes a Laplace transform. Thus from Eq. (15.4.1) it may be seen that: (a) dielectric relaxation measures a collective reorientation time and (b) it measures essentially $\langle \sum_{l,m} P_1(\mathbf{u}_l(0) \cdot \mathbf{u}_m(t)) \rangle$ where $\mathbf{u}_l$ is a unit vector pointing along the molecular dipole moment.

15 · 5 OTHER METHODS

The most useful techniques other than light scattering for probing time-correlation functions of fluids are those utilizing magnetic resonance. Magnetic resonance is a rich and complex field adequately described in many books (e.g. Abragam, 1961; Carrington and McLachlan, 1967) and articles. A discussion of magnetic resonance is beyond the scope of this book.

In Chapter 7 an application of C–13 NMR to study rotational motion of molecules is briefly described. When combined with depolarized light scattering, the magnetic resonance results yield values for the components of the rotational diffusion tensors of some symmetric top molecules. In some circumstances, NMR and ESR methods allow measurement of the relaxation times of the molecular angular momentum (e.g., see McClung and Kivelson, 1968).

Fluorescence depolarization is described briefly in Section 7. B (see also Chuang and Eisenthal, 1972). This technique can be used to measure $P_2(\mathbf{u}(0)\cdot\mathbf{u}(t))$ and thus gives information similar to Raman scattering except that the technique is generally confined to longer times.

Various transport coefficients can also be related to time-correlation functions. For instance, as was shown in Section 5.9, the translational self-diffusion coefficient is proportional to the area under the time-correlation function of the velocity of the center of mass of the particle.

$$D = \frac{1}{3}\int_0^\infty \langle \mathbf{V}(0) \cdot \mathbf{V}(t) \rangle \, dt$$

Other zero-frequency transport coefficients (thermal conductivity, viscosity, etc.) may also be expressed as areas under time-correlation functions by use of the methods described in Chapter 11.

The relationship between spectroscopy, transport coefficients, and time-correlation functions is derived using linear response theory. The excellent review by Zwanzig (1965) provides a simple didactic introduction to this subject and contains a broad bibliography of the important papers on this subject. Berne and Forster (1971) describe progress in this and related areas and give a review of various computer experiments used to calculate time-correlation functions. In a recent manuscript Forster (1975) presents an up-to-date treatment of topics relevant to light scattering.

REFERENCES

Abragam, A., *The Principles of Nuclear Magnetism,* Oxford, University Press (1961).

Berne, B. J. and Forster, D., *Ann. Rev. Phys. Chem.* **22,** 563 (1971).

Carrington, A. and McLachlan, A. D., *Introduction to Magnetic Resonance,* Harper and Row, New York (1967).

Chuang, T. J. and Eisenthal, K. B., *J. Chem. Phys.* **57,** 5094 (1972).

Egelstaff, P. A. (ed.), *Thermal Neutron Scattering,* Academic, New York (1965).

Fatuzzo, E. and Mason, P. R., *Proc. Phys. Soc. Lond.* **90,** 741 (1967).

D. Forster, Manuscript in Press, W. Benjamin, Co (1975),

Fulton, R. L., *J. Chem. Phys.* **55,** 1386 (1971).

Gordon, R. G., *Adv. Mag. Res.* **3,** 1 (1968).

Landau, L. D. and Lifshitz, E. M., *Electrodynamics of Continuous Media,* Addison-Wesley, Reading, Mass. (1960).

Marshall, W. and Lovesey, S. W., Theory of Thermal Neutron Scattering, Oxford, (1971).

McClung, R. E. D. and Kivelson, D., *J. Chem. Phys.* **49,** 3380 (1968).

Titulaer, U. M. and Deutch, J. M., *J. Chem. Phys.* **60,** 1502 (1974).

Yip, S. in *Spectroscopy in Chemistry and Biology,* Academic, New York (1974).

Zwanzig, R., *Ann. Rev. Phys. Chem.* **16,** 67 (1965).

INDEX